Methods and Protocols in Food Science

Series Editor
Anderson S. Sant'Ana
University of Campinas
Campinas, Brazil

For further volumes:
http://www.springer.com/series/16556

Methods and Protocols in Food Science series is devoted to the publication of research protocols and methodologies in all fields of food science.

Volumes and chapters will be organized by field and presented in such way that the readers will be able to reproduce the experiments in a step-by-step style. Each protocol will be characterized by a brief introductory section, followed by a short aims section, in which the precise purpose of the protocol will be clarified.

Consumer Research Methods in Food Science

Edited by

Carlos Gómez-Corona

Firmenich & Cie SAS, Neuilly-sur-Seine, France

Heber Rodrigues

Plumpton College, Brighton, UK

Humana Press

Editors
Carlos Gómez-Corona
Firmenich & Cie SAS
Neuilly-sur-Seine, France

Heber Rodrigues
Plumpton College
Brighton, UK

ISSN 2662-950X ISSN 2662-9518 (electronic)
Methods and Protocols in Food Science
ISBN 978-1-0716-3002-0 ISBN 978-1-0716-3000-6 (eBook)
https://doi.org/10.1007/978-1-0716-3000-6

This Humana imprint is published by the registered company Springer Science+Business Media, LLC, part of Springer Nature.
The registered company address is: 1 New York Plaza, New York, NY 10004, U.S.A.

Preface

"*Et voilà*," this book has finally seen the light of day, after a long process of being shaped under the moonlight! I say that because it was mostly during evenings that this book was sculpted and refined, as a piece of art, by Carlos and myself. Certainly, it wasn't an easy task to perform, coupled with both of our jobs and personal lives, but no doubt it was a pleasurable and interesting adventure.

When we were first approached in 2019 to create this book, we both lived in the American continent, Carlos in Mexico and myself in Brazil. And, it was there, in Latin America, we designed for the very first time the structure of this book. We wanted to make something different, which escaped from the traditional models of experimental protocols manuals we find in every library. The first step was to work on the content of the book, in order to prepare something that could reflect our initial aspirations. Thus, we decided to bring to light old consumer studies methodologies and techniques, as well as innovative and ultra-contemporary approaches. A mix of past and present to build the methodological future of the consumer scientific area.

The next step was to build a list of authors. We decided whose papers reflected originality, precision in details, and the highest quality. Authors who would be able to inspire the next generation of scientists of food science. And we are very proud of our contributors. Coming from a wide diversity of backgrounds and nationalities, they helped to shape this book, complimenting each other in this scientific puzzle we've created. We would very much like to thank them all warmly for their perseverance, for keeping their motivation and enthusiasm to a high standard, even in the sad times we traversed these last two years with this violent global pandemic. We definitely could not arrive at this final configuration without your brains. Thank you immensely for entrusting them to us!

For the final step, the final edits and when this book was finalized, Carlos and I, were both now living in Europe, he in Paris, France, and myself in Brighton, England. Carlos was tireless in making this book move forward, and I need to express how grateful I am for sharing this journey with him, which we dreamed of since the beginning of the 2010s when we were PhD students at Burgundy University, in France. Thank you, my colleague and friend, for your enthusiasm and your hard work, and now for this piece "Gómez-Corona & Rodrigues, 2022"! A massive thank you!

Context, in consumer research, is a background effect that impacts notably the way consumers respond to different stimuli. Therefore, I chose to write this preface in Scotland, this magical country in the northern part of Great Britain, and for adding a little context to that, I chose a historical pub called "Doctors," in the heart of the sumptuous Edinburgh, to write these lines. The building that homes "Doctors" was erected to mark the opening of the Old Royal infirmary hospital in the end of the nineteenth century. An old plaque stands out on the doors, telling the history of the building and welcoming passers-by. Next to my table, a couple of tourists from the far way Alaska, a physician and a Chef, were telling me the reasons what made them arrive in this place, and among them, the historical-scientific facts surrounding the building. The historical scientific reason that was aligned with my inspiration to come here, to be inspired for writing this preface. I hope that in the future we could have, somewhere in the world, a place like this one, that celebrates us, Doctors on Consumer Science. We do an invisible work, but a work that is fundamental to help people to eat better and to be conscious of their consumption choices—and how it impacts the way we preserve our environment and our history.

The last sentences of this preface are addressed to our readers, the next generation of consumer scientists. We truly believe that the efforts of every author in the following chapters will guide you to approach your objectives, and respond to your research questions, and/or test your hypothesis, accordingly. Be innovative, creative, and inspired. Science already runs through your veins. We hope this book will be your best inspirational friend on your journey of discoveries and scientific advances.

Neuilly-sur-Seine, France *Carlos Gómez-Corona*
Brighton, UK *Heber Rodrigues*

Preface to the Series

Methods and Protocols in Food Science series is devoted to the publication of research protocols and methodologies in all fields of food science. The series is unique as it includes protocols developed, validated, and used by food and related scientists, as well as theoretical basis are provided for each protocol. Aspects related to improvements in the protocols, adaptations, and further developments in the protocols may also be approached.

Methods and Protocols in Food Science series aims to bring the most recent developments in research protocols in the field as well as very well established methods. As such, the series targets undergraduates, graduates, and researchers in the field of food science and correlated areas. The protocols documented in the series will be highly useful for scientific inquiries in the field of food sciences, presented in such way that the readers will be able to reproduce the experiments in a step-by-step style.

Each protocol will be characterized by a brief introductory section, followed by a short aims section, in which the precise purpose of the protocol is clarified. Then, an in-depth list of materials and reagents required for employing the protocol is presented, followed by a comprehensive and step-by-step procedures on how to perform that experiment. The next section brings the do's and don'ts when carrying out the protocol, followed by the main pitfalls faced and how to troubleshoot them. Finally, template results will be presented and their meaning/conclusions addressed.

The Methods and Protocols in Food Science series will fill an important gap, addressing a common complain of food scientists, regarding the difficulties in repeating experiments detailed in scientific papers. With this, the series has a potential to become a reference material in food science laboratories of research centers and universities throughout the world.

Campinas, Brazil ***Anderson S. Sant'Ana***

Contents

Contributors

RAMÓN ÁLVAREZ-ESTEBAN • *Department of Economics & Statistics, University of León, León, Spain*
MÓNICA BÉCUE-BERTAUT • *Departament Estadística i Inv. Operativa, Universitat Politècnica de Catalunya, Barcelona, Spain*
JORGE HERMAN BEHRENS • *School of Food Engineering – Department of Food Science and Nutrition, University of Campinas, Campinas, Brazil*
ERIC BONETTO • *Aix-Marseille Université, UR ADEF 4671, Marseille, France*
CHRISTINA BOUTSOUKI • *Department of Economics, Aristotle University of Thessaloniki, Thessaloniki, Greece*
JOAN-MIQUEL CANALS • *Department of Biochemistry and Biotechnology, Universitat Rovira i Virgili, Tarragona, Spain*
FABIANA M. CARVALHO • *Department of Food and Nutrition, School of Food Engineering, University of Campinas, Campinas, SP, Brazil*
ISABELLE CAYEUX • *Firmenich SA, Satigny, Switzerland*
DONATO CEREGHETTI • *Firmenich SA, Satigny, Switzerland*
SYLVIE CHOLLET • *Junia, UMRT 1158 BioEcoAgro, Lille, France*
COLOMBA CODACCIONI • *Aix-Marseille Université, UR ADEF 4671, Marseille, France*
SYLVAIN DELPLANQUE • *Swiss Centre for Affective Sciences, University of Geneva, Geneva, Switzerland*
NICOLAS DEPETRIS CHAUVIN • *HES-SO Haute Ecole de Gestion de Genève, Carouge, Switzerland*
JONAS DI VITA • *HES-SO Haute Ecole de Gestion de Genève, Carouge, Switzerland*
ADRIANA GALIÑANES-PLAZA • *Repères, Paris, France*
LUIS BENITO GARCÍA ÁLVAREZ • *University of Oviedo, Cátedra de la Sidra de Asturias, Oviedo, Spain*
AGNÈS GIBOREAU • *Institut Paul Bocuse Research Center, Ecully, France*
IGNACIO GIL-PÉREZ • *Group in Product and Project Management, Escuela de Ingeniería y Arquitectura, Universidad de Zaragoza, Zaragoza, Spain*
CARLOS GÓMEZ-CORONA • *Firmenich & Cie SAS, Neuilly-sur-Seine, France; Human Insights Department, Firmenich SA, Neuilly-sur-Seine, France*
JORGE URÍA GONZÁLEZ • *University of Oviedo, Cátedra de la Sidra de Asturias, Oviedo, Spain*
ANNE HASTED • *Qi Statistics, West Malling, UK*
LEONIDAS HATZITHOMAS • *Department of Business Administration, University of Macedonia, Thessaloniki, Greece*
IVÁN LIDÓN • *Group in Product and Project Management, Escuela de Ingeniería y Arquitectura, Universidad de Zaragoza, Zaragoza, Spain*
GRÉGORY LO MONACO • *Aix-Marseille Université, UR ADEF 4671, Marseille, France*
KOSTOULA MARGARITI • *Department of Economics, Aristotle University of Thessaloniki, Thessaloniki, Greece*
KELSI MATWICK • *Department of Journalism, University of Florida, Gainesville, FL, USA*
KERI MATWICK • *Language and Communication Centre, Nanyang Technological University, Singapore, Singapore*

NICOLAS MINVIELLE • *Audencia Business School, Nantes & Making Tomorrow, Fr-Be, Nantes, France*

SANDRINE MONNERY-PATRIS • *Centre des Sciences du Goût et de l'Alimentation, CNRS, INRAE, Institut Agro, Université de Bourgogne Franche-Comté, Dijon, France*

CHARLES JAMES NEGUS • *Plumpton College, The UK Centre For Excellence on Wine Education, Training and Research, Lewes, UK*

HOA NGUYEN • *Independent Consultant, Digital Data Intelligence, Geneva, Switzerland*

KATHRIN OHLA • *Firmenich SA, Satigny, Switzerland*

TOBIAS OTTERBRING • *School of Business and Law, Department of Management, University of Agder, Kristiansand, Norway*

KAAT PHILIPPE • *Centre des Sciences du Goût et de l'Alimentation, CNRS, INRAE, Institut Agro, Université de Bourgogne Franche-Comté, Dijon, France*

NICOLAS PICHOT • *Centre de Recherche en Psychologie de la Connaissance, du Langage et de l'Emotion – EA 3273, Aix Marseille Université, Aix-en-Provence cedex, France*

CHRISTELLE PORCHEROT • *Corporate Research & Development, Firmenich SA, Geneva, Switzerland*

RUBÉN REBOLLAR • *Group in Product and Project Management, Escuela de Ingeniería y Arquitectura, Universidad de Zaragoza, Zaragoza, Spain*

HEBER RODRIGUES • *The Secret Vine, Chelmsford, United Kingdom; Plumpton College, The UK Centre For Excellence on Wine Education, Training and Research, Lewes, UK*

DANIEL SALAZAR LAMA • *Archéologie des Amériques, UMR 8096– CNRS/Université de Paris 1 – Panthéon Sorbonne, Paris, France*

CHARLES SPENCE • *Crossmodal Research Laboratory, University of Oxford, Oxford, UK*

ANA MARÍA ULLOA • *Department of Anthropology, Universidad de los Andes, Bogota, Colombia*

DOMINIQUE VALENTIN • *Centre des Sciences du Goût et de l'Alimentation, AgroSup, CNRS, INRAE, Université Bourgogne Franche-Comté, Dijon, France*

OLIVIER WATHELET • *Users Matter, Belgium & Making Tomorrow, Fr-Be, Mehaigne, Belgium*

Chapter 1

Consumer Research in Food Science – An Introduction

Carlos Gómez-Corona and Heber Rodrigues

Abstract

This introductory chapter of "Consumer research in food science – an introduction" is a general view of what it is to perform research in this domain. From the idea of food as identity, we build the structure of the book. Considering that food brings more to the body than just nutrients, we explore the idea of food as meaning and identity, and by doing so we invite researchers from multiple disciplines to answer research questions on what, why, and how consumers approach different categories of food or food-related situations. The eyes and mind of food scientists, anthropologists, historians, art historians, psychologists, marketeers, agronomists, economists, linguists, statisticians, designers, and gastronomists are among the different authors to contribute in different chapters of the book, to answer with different methods and objectives our questions around food consumption. In this introduction, we focus on the origins of consumer research as we understand it today, the different objectives of doing research, and guiding ethical practices and principles.

Key words Consumer research, Food science, Anthropology, History, Arts, Psychology, Marketing, Econometrics, Sensometrics, Design, Linguistics, Culinary arts

1 Introduction

Eating and drinking carry as many nutrients to the human body as they carry meaning. Every bite and every sip will nourish our body, as well as nourishing our psychic apparatus. Only a human in the wild, or in extreme conditions, will limit themselves to finding nutrients in food. Others will share emotions through food, without limiting themselves to the functional act of eating. Eating is seen from long time now as a way of expression, where the act of eating is shaped by the norms of a group which not only defines a social/cultural group, but even a nation. Eating chilies or drinking wine is sharing a cultural identity. Eating habanero chili or drinking Chablis wine was learned, through a cultural norm. At an early age, Mexicans learned to transform the strong pungent sensation into a highly pleasant sensation. On a similar way, the French learn to appreciate the minerality of a Chablis wine. Chili and wine can draw

Carlos Gómez-Corona and Heber Rodrigues (eds.), *Consumer Research Methods in Food Science*,
Methods and Protocols in Food Science, https://doi.org/10.1007/978-1-0716-3000-6_1,

a cultural limit, within a cultural group, and it is that identity in food that makes consumer research of food and beverages so unique, intriguing, and different from other types of consumer research (e.g., consumer research in technology, home care products, etc.).

Identity is not a new concept, it is as old as human beings themselves, but it is at this time that the phenomenon of combining identity and food consumption has become so relevant. Food as an identity has been widely explored, and according to Fishler [1], food is central to our sense of identity, and it combines at least two dimensions: A first dimension that goes from the biological (nutritional function) to the cultural (symbolic function); the second links the individual to the collective, from the individual (psychological) to the social. Rozin and Fallon [2] advanced a step further, exploring the concept of incorporation as a basis for defining identity; the incorporation is "the action in which we send a food across the frontier between the world and the self, between outside and inside our body". The old adage "you are what you eat" is biologically true, as the food we absorb will provide both energy and the nutrients that maintain the biochemical composition of the organism. It is equally true that the act of incorporation will trigger two additional effects; the first one is psychological and the second is cultural. In the psychological, the food that crosses the barriers of the skin is assimilated as its own, the skin is the body's envelope, our border, and the limit with others [3]. Our hands can touch the food and decide when that food stops being on our outer body, to ingest it, and become something of our own. This act entails a whole series of psychological effects on the perception and representation of what we want to ingest. Nevertheless, the second variable, which is the cultural one, where we study not only what passes through the barrier of our skin, but also what the group decides to share, and how to create norms linked to the act of eating. Food becomes a binder that brings people together at a family (e.g., "in this house we eat like this"), religious, and national level.

It is in these dimensions of eating that it is possible to explore food from different angles and to go beyond the nourishing aspects of eating. Agronomists can acknowledge their contribution on the creation of the food matrix. Anthropologists can claim the cultural act of eating as a possession of their field. Cognitive and social psychologists can claim the research dealing on the implicated mechanisms of food perception and representation. On the other hand, sensory scientists can claim the space of the description of food, and marketeers will focus on the uses and habits related to food products. Food and identity together give space to this wide range of disciplines looking to study our relationship with food and understand why and how we eat and drink, which is far more complex than we can imagine, and it can therefore not be studied only by one discipline. In this book, we decided to focus on five

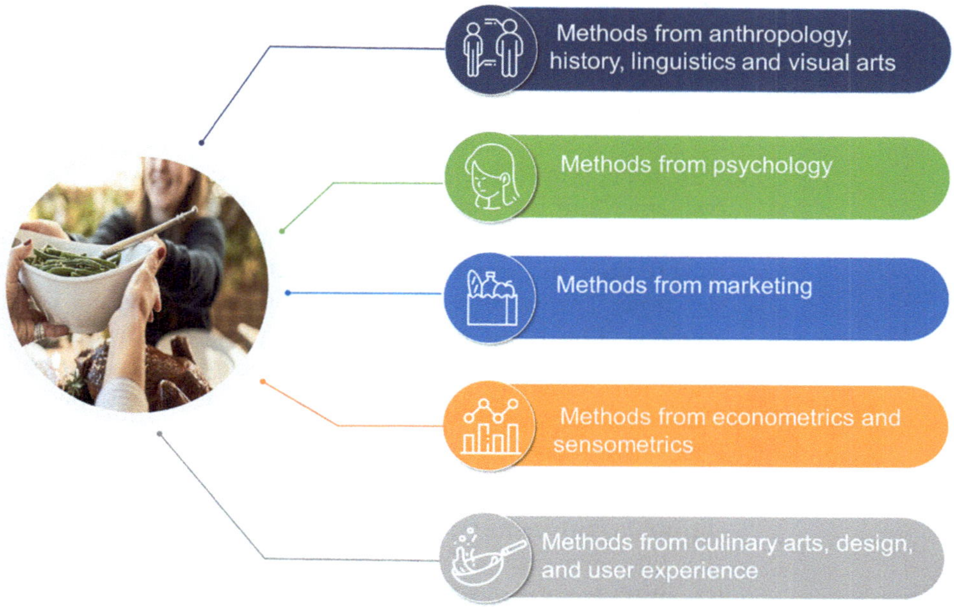

Fig. 1 Disciplines that perform consumer research in food science, divided into five groups of disciplines. (Image from the authors (Microsoft 365 creative and Flaticon: Royalty free))

groups of disciplines that study food consumption (Fig. 1). Each discipline brings their own approach of studying food consumption, and the methods they use are as varied as they are complementary. Some disciplines have been studying food consumption for centuries (e.g., Gastronomy), others for many decades (e.g., Anthropology and Psychology), and other disciplines are newcomers (e.g., Industrial Design). Regardless of time, the applied methods from each discipline will bring us more clarity with which to better understand the act of eating, what it means, and what it is today to buy, cook, and share food with others.

2 The Origins of Consumer Research

To study the consumption-related aspects of cooking, eating, and sharing food, we can approach different disciplines, such as these ones we mentioned previously (Fig. 1). Studying the consumption of food may not be the same as doing consumer research in general. By definition, consumer research in food science, implies the study of food consumption from a market perspective. Whether it is in academia, nonprofit organizations, or industry, consumer research is framed in the act of buying and selling which defines the hypermodern world in which we live [4]. Even in small, traditional, and isolated communities, we can perform consumer research as it is also framed by the act of buying and selling. Although buying food and supplies from a local and traditional street market is not the same as online shopping, both activities are the subject of studies in consumer research in food science. That is another difference between the study of consumer research in food science, versus the study of food in pure anthropological or psychological studies, with food symbolism for example, where we do not study the consumer, but the person, and their relationship with food.

In the case of consumer research, its origins and influences are hard to track to a single source. If we think of consumer research as our home, we must see that every pillar comes from a different architect. Our home is not the masterpiece of one mind, it is the work of many [5]. The Editors of this book think that the origins of consumer research as an independent academic discipline can be tracked back to the 1900s, where the first university courses in these disciplines started. Consumer research probably has its origins in the early courses of Political Economics in German Universities on the nineteenth century, probably during the decade of 1870 [6]. At that time, marketing ideas were embedded in the field of economics. In those early days, marketing looked different and, of course, professors were trained in economics rather than what we call modern marketing today: marketing science, consumer behavior, and marketing strategy. Finally, in 1946 the Survey Research Center (SRC) was established by the University of Michigan with the support of Rensis Likert, Donald Campbell, and other well-known social scientists [7], after that time, the face of consumer research changed completely.

Contrary to what a sensory scientist may think, consumer research does not originate with the invention of the 9-points "hedonic scale method of measuring food preferences" in 1957. What Peryam and Pilgrim [8] were measuring was not preference, they were measuring liking (or acceptance). Although (very) similar, preference and liking are definitely not synonyms. If we go back to the psychological roots of consumer research, we will see that the difference relies on both the semantic nature of the question and the scale (*What do you prefer...?* vs. *How much do you like...?*), as well as the nature of the variable being measured. Scale construction and its psychometric considerations are elements that we borrow both from sociology, psychology, and of course from statistics [9]. The mathematical foundation for scale construction methodology can be tracked to as far back as Francis Galton [10]. Hence, the beauty (and let us say, the complexity) of consumer research is the influence it has from psychology to anthropology, and from economics to statistics. Today, all the scientists working in the consumer research of food should acknowledge our rich and diverse past, as well as foreseeing a multidisciplinary field, just as the contributors of this book do.

3 Consumer Research Today

Proud of our multidisciplined legacy, where should consumer researchers look into the future? What will the challenges and theoretical foundations be in the years to come? We still do not know the challenges we may face in the years ahead, but we can already play with future scenarios (*see* Chapter 23). Today we see a contrasting, unequal, and paradoxical world. Extremely poor on the one hand, and hypermodern, hyperconsumerist on the other. We see a consumer in search of basic needs, more specifically, we see 1.3 billion people in search of basic needs, which represents a human challenge [11], and 3.6% of the population that migrate in a time of international travel restrictions and economic and political constraints [12]. We see people affected by climate change, and 690 million people who are hungry today, as we write this introduction [13]. In a world with one billion people (15% of the world population with disabilities) living with disabilities, there are only 25 published articles on Food Quality and Preference (one of the leading journals in consumer research, from a sensory science approach) with the keyword "disabilities". Why are these human challenges not in our research agenda? Why do these problems not keep us awake at night, wanting to find a solution?

Besides these human and environmental challenges, consumer research can also be influenced by the current social movements. According to Roudinesco [14], the great social movements seem to have changed direction. We ask ourselves more how to change the world to protect it from inequalities, as well as from the effects of climate change. Consumption becomes a tool to show our opinion, our support, or our censorship. Brands quickly suffer from the media effect caused by their consumers on social networks and the boycott has become a mechanism of change for the consumer, who seeks to express his point of view and through consumption seeks to express an identity [15–17]. This constant search for identity, for constant self-affirmation of the self (or hypertrophy of the self in Roudinesco's words) is the distinctive sign of our times, where each of us seeks to be himself, like a king, and not like another [14, 18]. This constant search for identity and showing how we really are, influences all activities and facets of consumption. The consumption of craft beer, for example, is part of this larger social movement where the consumer seeks to generate an identity; the consumers do not drink the product for its functional attributes, they consume it for what it means, and consequently they build an identity, perceived as more authentic and unique, in comparison to mainstream industrial beer consumption [19].

Another important identity milestone can be found back in 1949 when Simone de Beauvoir [20] wrote a sentence in her book *The Second Sex* that would radically change all social and literary research that sought to distinguish gender from the body; "we are not born women: we become one". The current idea of gender is completely different from that of 1949, and gender identity and expression touch elements of consumption [21]. Gender expression ceased to be a dichotomous variable to become a continuum of femininity – masculinity that overlaps with other elements of gender: biological sex, expression, identity, and attraction [22]. Of course, these gender identities are expressed more in certain regions or countries than in others, and what happens in certain Western countries cannot be extrapolated to deeply religious or conservative countries. In countries where there is this new gender identity, conducting studies with consumers that seek to separate gender into men versus women become completely irrelevant, and will have to make room for new ways of categorizing gender, and its effect, on the perception and food selection. We will have to give space to new ways of understanding the effect of gender identity and consumption [23].

To sum up, identity in food is one of the most important elements to highlight in our research agenda to understand food consumption. And this identity, understood in a wider sense in consumer research, is bringing different types of variables to be studied such as food and gender, food and expression, food and health, food and well-being, etc. In the years to come, the research agenda of academia and industry might focus on different aspects than before, such as health and well-being and social responsibility, but will also focus on a new relationship with food, especially on online shopping, online food sharing, virtual reality meals, etc. However, preparing a research agenda for the following years is not an easy task and may be subject to multiple scenarios that we do not yet know and that may occur.

4 Consumer Research Objectives, Ethics, and General Principles

Whatever our objective is in consumer research of food science, and whatever our references are (psychology studies, marketing studies, etc.), we must be sure that we follow basic ethical standards. There is no such thing as common sense (in the social psychology definition as a shared set of beliefs of a specific social group [24]) in consumer research, there are guiding principles and methodological foundations. Briefly, we will explain some of them.

4.1 Consumer Research Objectives

One of the challenges that a researcher faces in the quest to understand something related to food consumption is the use of a different approach, or a new method that has not been used before. The objective of this book is to include a wide range of methods in consumer research but also a minimum of theoretical framework that can help the reader to understand the challenges and questions

Table 1
Research challenges and its key discipline to search for methodological resources

Challenge	Discipline	Chapters
Can I understand food consumption through historical documents?	History	2
How can I use new ethnographic approaches?	Anthropology	3
Consumer research through cookbooks and cooking shows	Linguistics	4
Is it possible to analyze images from consumers?	Visual arts	5
What is a social representation of food and beverages?	Social psychology	6
What is perception and mental representation?	Cognitive psychology	7
How can we measure the emotional response triggered by food?	Experimental psychology	8
What are the implicit association tests?	Experimental psychology	9
Perception of food through multiple senses	Experimental psychology	10
Can I do consumer research with kids?	Experimental psychology	11
What is creativity and how can I measure it?	Social psychology	12
What are the guidelines for analyzing qualitative data?	Social sciences in Marketing	13
How can I use social media research in consumer research?	Social Media research	14
How can we study food and beverages in advertising?	Advertising	15
Going beyond the lab... How to do field studies?	Marketing & Business	16
Applying econometric models in food science	Econometrics	17
How can we understand the ideal set of products for one brand?	Sensometrics	18
Analyzing text responses in consumer research... How?	Lexicometry	19
How to test concept and packaging in the food industry?	Industrial design	20
How to measure desirability?	User experience	21
Measuring not only product but meals	Culinary studies	22
How to study future scenarios in consumer research?	Design & Anthropology	23

of consumer research focused on food (Table 1). It is important to understand that not because we can measure something does it signify that it has a meaning, and not because we have data from consumer research does it mean that we can do any type of statistical analysis. We need to understand the nature of the variable, for example if we measure emotions with consumers triggered by a food stimulus (*see* Chapter 8), we need to understand the psychological implications of its measurement, and not because we can ask consumers for their emotions at different moments while consuming a food product, does it mean that we can make a temporal driver's analysis out of a set of emotional questions. Let us think, how much do our own emotions change while drinking strawberry flavored water? Not much....

The Editors of this book recommend the readers get close to an experienced researcher in a certain field if the method used is far from their own discipline, to avoid common errors. For example, a sensory scientist may try to quantify the responses of a qualitative study instead of analyzing the richness of language, or the principal themes in a conversation (*see* Chapter 13), or a marketeer may underestimate the privacy data in consumer research in social media. We cannot disclose the user's information just because the data are already public in social media (*see* Chapter 14).

4.2 Ethics in Consumer Research

Examples of studies with major ethical issues in research with humans are multiple and well documented, among them, those related to human behavior. The one from Milgram [25] and Zimbardo [26] is perhaps the most famous. In Milgram's obedience experiment, the objective was to understand the phenomenon of obedience to an authority figure. The researcher recruited a group of participants to take part of an apparent memory experiment at Yale University. Two participants were needed for the study. The participant applying the memory test (called here participant), and another participant who was performing the memory task, but we will call him the "confederate", as he was prerecruited by Milgram, disguised as the person doing the memory test. The participant applied the memory test consisting of reading a set of words that the confederate needed to memorize. Participant and confederate were separated by a physical cabin. When the confederate gave an incorrect response, then the participant needed to give him an electric shock that went from low (15 volts), to high (450 volts) depending on the errors being made by the actor. The experiment was filmed and Milgram himself appeared in the videos. The confederate gave multiple incorrect responses, and therefore Milgram asked the participant to give him an electric shock, while the confederate pretended to suffer and shout from the "fake electric shock". On the other side, the participant thought that he was giving real electric shocks to the memory test participant. Error by error, the intensity of the electric shock augmented, and the screams and suffering from the confederate were so real, that participants on multiple occasions asked Milgram if they could stop the experiment because the other person was suffering. The videos show anxious participants giving higher intensities of electric shocks while the test moves forward [27]. Again, and again, Milgram asked the participants to continue giving the electric shocks. Despite the apparent suffering of the confederate, 65% of the participants in the study applied the maximum shock intensity of 450 volts. Milgram's ethical mistakes are multiple: not showing the real nature of the study, lying multiple times to the participants, and playing with fake human suffering to demonstrate a hypothesis of obedience toward authority.

On the other hand, the experiment of Zimbardo, which also focuses on obedience, was so famous that a movie was made recently [28]. The objective of Zimbardo was to understand the role of authority, and to do that, he recruited a set of participants, all students of psychology at the University of Stanford. The most ironic part is that the study took place in the basements of the Faculty of Psychology of the University. In this study, half of the 24 students needed to play the role of fake prisoners, and half of the students played the role of the fake prison guards. The study was also filmed and transmitted in a CCTV to the psychology lab where Zimbardo and his collaborators could see the study live, as it was happening in the basement of the faculty. The students were given complete liberty to play both the guard and the prisoner roles. However, the study did not go as planned and the students playing the role of guards became completely violent with the prisoner students, to the extent that the faculty members and Zimbardo stopped the study. After only 6 days, the experiment was abruptly canceled due to the sadistic behavior of the students playing the role of prison guards. Zimbardo's ethical mistakes are quite similar to the ones of Milgram.

More examples can be explained, but instead of focusing on what a consumer researcher should not do, we might focus more on what he/she should do. A good starting guide can be the guidelines provided by the American Psychological Association (APA) for conducting research with humans [29]. Some general principles for consumer research are stated below, following the recommendations of Cozby [30] for psychology research, and additional recommendations given by the authors of this book. These are the ten key principles:

1. Be responsible for respecting and following ethical standards for research with humans
2. Respect local laws and regulations
3. Conduct research competently and with concern for the dignity and welfare of the participants
4. Avoid stress and psychological harm
5. Use informed consent in all studies (including the use of photographs and videos)
6. Use a debriefing when needed (e.g., in Design Fiction *see* Chapter 23)
7. Protect privacy and confidentiality of the study participants
8. Participation in the study must be voluntary

9. Inform both participants and research institutes about the use and sharing conditions of data from the study
10. Honor the commitments that have been made to research participants and research institutes (see Gabelica et al. [31])

We encourage the readers of this book to read further research guidelines, as some studies may require more attention than others. The readers can also contact their local Ethics Committee at any University.

4.3 Consumer Research General Principles

Consumer research is so diverse, that it is difficult to give a set of general principles that may apply to any type of research. On the one hand, the ethical considerations should always be applied, but on the other hand specific principles could be drawn, and then revised to match the requirements of each method being used, as well as the research objectives. Some general principles applicable in consumer research are given in Table 2.

5 Final Considerations

The objective of this introduction to consumer research was to bring together the general ideas around consumer research. From the origins of consumer research as we understand it now, to the new ideas and needs of consumer research today. No consumer research is possible without the guiding principles of ethics in human behavior research. This general introduction is not intended to be specific, but is a first glance into what it is to do consumer research in food science. The general objectives of doing consumer research are given in this introduction and linked to the questions that can be answered by each one of the chapters in this book.

The methods expressed in this introduction and in this book are just a selection of both the most common methods, and the innovative ones. And our objective is to bring a food science perspective to each of the chapters in the book. Other books may be complementary to this book, and we expect that the methods selected will be relevant for many years to come in the consumer research of food science.

Table 2
General principles in consumer research, in terms of definitions and current practices

	General principle for consumer research
Person studied	
Consumer	Most of the times we study the "consumer" who is the person eating/drinking the food product but may not necessarily buy the product
Shopper	The shopper is the person buying the product but may not consume it. A typical example is a mother & children. A mother will buy, and the children will consume
User	The user is a term that is borrowed from Design [32]. It is the same as a consumer, but it generally implies a use of the product, for example the user of an oven, of a coffee machine, etc.
Object studied	
Product	In many cases, researchers are interested in an object per se, like the perception of beer [33], or wine [34]
Perception & Representation	Concepts, ideas, and attitudes are part of the human representation toward something. We may be interested in studying a representation rather than a product itself like flavor pairing principles [35], the perception of masculine flavors [23], or a high order sensory descriptor such as the words 'elegance' and 'complexity' [36, 37]
Qualitative studies	
Sample	In a qualitative study, the minimum sample size can be as low as 2 (2 focus groups, 2 in-depth interviews), in order to compare results. However, the common practice is to establish a saturation point, the moment in which we do not find any new data [38]. Common practices are to have an average of 4 focus groups, 8–12 ethnographies
Analysis	Analysis is done in terms of emerging themes, text, and meaning (*see* Chapter 13). We should avoid quantification of the results. Qualitative studies are used to explore, to understand, but not to validate a hypothesis
Quantitative studies	
Sample	It is generally accepted that the sample size should be a minimum of 60 consumers. And to split the data (e.g., women vs. men), in each split a minimum of 60 consumers is expected, in this case 120. Read Hough et al. [39], for more details
Analysis	The analysis can be as varied as the scale or methods used. In quantitative studies nonparametric or parametric statistics should be done, and multivariate analysis is recommended when needed. Avoid the use of only descriptive methods
Convenient samples	A convenient sample is defined as a specific population which is not the target population. The most common cases are students and faculty members in academic research, and employees in the case of industry Students are usually biased in their response and could have a higher education compared to a general population. Employees are usually not the target consumer group that the researcher is interested in and are also biased in their response We recommend the use of nonconvenient samples as far as possible

(continued)

Table 2 (continued)

	General principle for consumer research
Special populations	In consumer research, the special populations are considered: low-income consumers, people in other vulnerable conditions, people with physical or mental health problems, and in need of special care (e.g., multiple disabilities, considerably older consumers, etc.)
Children	The study of children should be done by experienced researchers and may need the authorization of an Ethics Committee. In the Industry it is common to interview children starting at 6–8 years, below that age, very specific methods are needed to perform consumer research

References

1. Fischler C (1989) Food, self, and identity. Soc Sci Inf 27(2):275–292
2. Rozin P, Fallon AE (1981) The acquisition of likes and dislikes of for foods. In: Solms J, Hall RL (eds) Criteria of food acceptance. Forster Verlag, Zurich
3. Anzieu D, Séchaud E (1985) Le moi-peau. DUNOD, Paris
4. Lipovetsky G (2006) Le bonheur paradoxal: essai sur la société d'hyperconsommation. Gallimard, Paris
5. Gómez-Corona C, Rodrigues H (2020) Editorial overview: sensory science and consumer perception: from a multi-disciplinary perspective. Curr Opin Food Sci 33:iii–v
6. Parrish JB (1967) Rise of economics age handbook of marketing as an academic discipline: the formative years to 1900. South Econ J 34:1–12
7. Kassarjian HK, Goodstein RC (2009) The emergence of consumer research. In: Maclaran P, Saren M, Stern B, Tadajewski M (eds) The sage handbook of marketing theory. Sage, London
8. Peryam DR, Pilgrim FJ (1957) Hedonic scale method of measuring food preferences. Food Technol 11(Suppl):9–14
9. Dawis RV (2000) Scale construction and psychometric considerations. In: Tinsley HE, Brown SD (eds) Handbook of applied multivariate statistics and mathematical modeling. Academic
10. Galton F (1961|1868) Co-relations and their measurement, chiefly from anthropometric data. In: Jenkins JJ, Paterson DG (eds) Studies in individual differences. Appleton-Century-Crofts, New York, pp 17–26
11. UN News: new UN poverty report reveals 'vast inequalities' between countries (2019)
12. McAuliffe M, Triandafyllidou A (2021) World migration report 2022. International Organization for Migration (IOM), Geneva
13. FAO (2022) Climate change. FAO's support to countries facing climate change. Retrieved from: https://www.fao.org/climate-change/our-work/what-we-do/en/. Apr 2022
14. Roudinesco E (2021) Soi-meme comme un roi. Seuil, Paris
15. Klein JG, Smith NC, John A (2004) Why we boycott: consumer motivations for boycott participation. J Mark 68(3):92–109
16. Kakashekh HM, Murad HR, Ahmad AR, Saud M (2021) Effects of the Facebook boycott campaign on Turkish products and its influence on demand of Kurdish consumers 2020. UKH J Soc Sci 5(1):98–106
17. Salma SY, Aji HM (2022) What drives Muslims to boycott French brands? The moderating role of brand judgement and counterargument. J Islam Market. https://doi.org/10.1108/JIMA-04-2021-0128
18. Ricoeur P (2015) Soi-même comme un autre. Média Diffusion
19. Gómez-Corona C, Escalona-Buendía HB, García M, Chollet S, Valentin D (2016) Craft vs. industrial: habits, attitudes, and motivations towards beer consumption in Mexico. Appetite 96:358–367

20. De Beauvoir S (1949|2021) Le deuxième sexe (Tome 2)-L'expérience vécue. Gallimard Audio
21. Rodrigues H, Goméz-Corona C, Valentin D (2020) Femininities & masculinities: sex, gender, and stereotypes in food studies. Curr Opin Food Sci 33:156–164
22. The Genderbreadperson (2022) Genderbread person v4.0. A teaching tool for breaking the big concept of gender down into bite-sized, digestible pieces. Retrieved from: https://www.genderbread.org/. Apr 2022
23. Gómez-Corona C, Abdi H (2020) Men, masculinity, and flavors: a multidimensional social representation. Oral presentation at Sensometrics 2020, Online conference
24. Juárez J, Arciga S, Mendoza J (2012) Memoria colectiva. Procesos psicosociales. UAM/Miguel Angel Porrua, Mexico
25. Milgram S (1963) Behavioral study of obedience. J Abnorm Soc Psychol 67(4):371
26. Zimbardo PG, Haney C, Banks WC, Jaffe D (1971) The Stanford prison experiment. Zimbardo, Incorporated
27. PsychHub (2015) Milgram experiment 1963 [Video]. YouTube. https://www.youtube.com/watch?v=Kzd6Ew3TraA
28. Emery B (Producer), Alvarez KP (Director) (2015) The Stanford prison experiment [Video file]. Retrieved from https://www.netflix.com
29. American Psychological Association (2016) Revision of ethical standard 3.04 of the "ethical principles of psychologists and code of conduct" (2002, as amended 2010). Am Psychol 71(9):900
30. Cozby PC (2007) Methods in behavioral research. McGraw-Hill
31. Gabelica M, Bojčić R, Puljak L (2022) Many researchers were not compliant with their published data sharing statement: mixed-methods study. J Clin Epidemiol 150:33. https://doi.org/10.1016/j.jclinepi.2022.05.019
32. Gómez-Corona C, Valentin D (2019) An experiential culture: a review on user, product, drinking and eating experiences in consumer research. Food Res Int 115:328–337
33. Gómez-Corona C, Chollet S, Escalona-Buendia HB, Valentin D (2017) Measuring the drinking experience of beer in real context situations. The impact of affects, senses, and cognition. Food Qual Prefer 60:113–122
34. Rodrigues H, Ballester J, Saenz-Navajas MP, Valentin D (2015) Structural approach of social representation: application to the concept of wine minerality in experts and consumers. Food Qual Prefer 46:166–172
35. Arellano-Covarrubias A, Gomez-Corona C, Varela P, Escalona-Buendia HB (2019) Connecting flavors in social media: a cross-cultural study with beer pairing. Food Res Int 115: 303–310
36. Shepherd H, Parr W, Lo Monaco G, Rodrigues H (2023) The meaning of the word elegance as a wine descriptor: effect of expertise and wine type. Food Res Int 164:112399
37. Parr WV, Mouret M, Blackmore S, Pelquest-Hunt T, Urdapilleta I (2011) Representation of complexity in wine: influence of expertise. Food Qual Prefer 22(7):647–660
38. Gomez-Corona C (2019) Estudios cualitativos tradicionales. In: Gomez-Corona C, Lelièvre-Desmas M (eds) Introduccion al Analisis Sensorial y Estudios con Consumidores. Nuevas perspectivas. XOC Estudio, Ciudad de Mexico
39. Hough G, Wakeling I, Mucci A, Chambers EIV, Mendez-Gallardo I, Rangel Alves L (2006) Number of consumers necessary for sensory acceptability tests. Food Qual Prefer 17:522–526

Part I

Methods from Anthropology, History, Linguistics and Visual Arts

Chapter 2

The Historical Development of Food Systems and Heritage

Jorge Uría González and Luis Benito García Álvarez

Abstract

Historians assumed long ago that food is a reality that cannot be examined in isolation, since its study encompasses different dimensions that can be social, ideological, or economic. Therefore, when addressing these issues from the historical field, it is necessary to frame it within a broad context that allows to combine multiple variables, since it is linked to rather diverse political, cultural, and institutional traditions, and especially to everything related to the gastronomic activity.

Eating entails more than the merely nutritional, the performance of a social and heritage act, so much that food is linked both to agricultural activities and the rural universe, as well as to the sociability practices typical of urban areas. Consequently, food is part of identity practices and constitutes a strong ethnic marker. We must also note the rich material culture associated with production and consumption.

Key words Food history, Food systems, Food heritage, Gastronomy, Commensality

1 Introduction

Considering the social, ideological, or economic dimensions within the historical analysis accounts for a radical change experienced in recent research, although interest in this field can be traced back to the first decades of the last century, and even to the scholarship of the eighteenth century. However, it was in Poland where the research developed considerably thanks to the Institute and the Journal of Material Culture History.[1] Established in 1953, the institution has achieved a relevant position in the historiography of food, although its research did not emphasize consumption as much as production, trades and techniques, management and supply problems, or shortages, hunger, and malnutrition, in addition to the economic, social, and political factors that framed it. The proposal of the *Annales* in the 1960s had a greater impact in the Western world, following interests already pointed out by M. Bloch

[1] Among the first approaches, we can note those carried out by the Academy of Material Culture History of the USSR, promoted by Lenin in 1919.

Carlos Gómez-Corona and Heber Rodrigues (eds.), *Consumer Research Methods in Food Science*,
Methods and Protocols in Food Science, https://doi.org/10.1007/978-1-0716-3000-6_2,

and L. Febvre, promoting research on these issues in other countries. The *Annales* historians insisted on the biological and nutritional aspect of food and tried to offer a nutritional explanation of the demographic, technical, and intellectual facts, although relegating such aspects as the psychological.

In the final decades of the twentieth century, the rising research through publications and congresses demonstrated the extraordinary convening power of a subject where various disciplines capable of contributing their vision in this field converged, currently enjoying this academic aspect a wide credibility among the scientific community while firmly establishing itself in the historiographical backdrop.

The first historians who really approached the study of food only analyzed its material and nutritional dimension: caloric balances, individual and collective rations, hunger, surpluses, undernourishment…, consequently considering nutrition as a mere material condition necessary for social development. A later cultural shift induced the appreciation of aspects such as ideas about nutrition or the ability to choose, or the values, symbols, and messages that are transmitted associated with eating. The emergence of these new concerns is undoubtedly related to the influence of anthropology on historical science, showing that it is not enough for a *product* to be edible to be consumed, and highlighting the necessity of a whole series of cultural conditioning factors for such a step. These novel circumstances refer to exclusions and preferences; everything we call *taste*, which is determined not only by environmental and economic conditions, but also by mentality, social rituals, the value of the messages exchanged while food is consumed in company, the ethical and religious values, or the individual and collective psychology that determines all these factors.

Indeed, in the 1960s, judicious criticism was being made on the biological aspects of the historiography of food, standing out Aymard [1] for his reflections on the dangers and difficulties when assessing the strictly nutritional aspect of past diets. In fact, starting a biological assessment of food is still a process with issues difficult to overcome, because the claim of reaching precise quantitative determinations or caloric assessments of past diets is practically unsustainable, since it is impossible to apply current dietary values to the nutrition and diets of other historical contexts.

Obviously, trying to reconstruct a history of taste, its evaluation criteria, preferences, or judgment ways in its measurable features is problematic, since they are factors that we can get to know very little. Conversely, it can be claimed that food constitutes a framework for sociability, communication, and exchange, and that its importance in this sense is beyond measurable scales. As Montanari maintains [2], food is culture when it is produced, when it is prepared, and, above all, when it is consumed, because humans, even when can eat anything, chooses with criteria linked to the economic and nutritional dimension of taste, but also to symbolic values of food.

In this regard, the cultural appetite for food is not strictly related to digestive processes. Nutrition is a physiological activity, eating and drinking are social and cultural phenomena. The "food units" significant from the sociocultural and historical point of view, which are also the ones who condition the consumers choices, seem to be a much subtler evidence than the parameters generally handled by economists or statisticians.

Certainly, except for some luxury products, no food derives its meaning solely from its intrinsic characteristics, but rather depends on the cultural associations that society attributes to them. The choice of food can be explained by technical or economic reasons, taste, or flavor and, often, by beliefs related to its goodness or badness. The status of food is not always a function of its nutritional value; eating and drinking provide important elements of community identity and are social activities that express how individuals project their identities. Provided the symbolic meaning of food, it is possible to identify people according to what they eat and drink, just as they "built themselves" around the table. Likewise, the behavior of people in the presence of food tends to leave the strictly functional dimension to assume a communicative value. The social vocation of men immediately translates into the attribution of a meaning to gestures made while eating.

In short, it could be claimed that the act of feeding transcends the pure need for survival, and it is full of meanings and emotions, constituting a behavior that goes beyond its own physiological purpose. Each social situation has a food expression, whereas food represents one of the preponderant elements of cohesion in a community. Culture creates the food communication system that dictates what is edible and what is not, it creates rules and taboos.[2] This "construct" is influenced by religious, medical, gender or age beliefs.

It seems that the sociocultural constraints related to food are powerful and complex: the uses given to each food, their combinations, the order, composition, number and times of the different meals, everything is codified in a precise way; and this codification is the result of a social and cultural process whose meaning, and reason can be found essentially in history. All along, food has remained linked to status, being a means of asserting oneself in front of others or of acquiring prestige; even in today's consumer society, food and drink are a relevant means of asserting oneself socially, the possibility of choice is highly valued, and the

[2] On the creation of rules and taboos from a materialist perspective is significant [3], and from an opposite one, where religion determines ecology [4].

impossibility is considered as a socially negative quality. Of course, the cuisine reproduces and expresses hierarchical and class relations.[3]

2 Materials in Historical Food Research

In parallel to this shift, there is a change in the nature of the sources for the study of food. Historians handle a considerable mass of data from the past. In part it is extracted from *traditional* sources such as notarial archives, with wills that include *postmortem* inventories of assets; municipal archives that include data on municipal taxes on necessities, accounting, or statistics on food markets; documentation of old noble houses, diets from hospitals or hospices, convents or inns, and there are even large surveys carried out on large national territories at the request, as in the French case, of the State administration. Also are considered the printed data of old recipe books, diets prepared by doctors or philanthropists, or scholarly works that begin to record consumption habits in this regard in an increasingly systematic way at least since the eighteenth century. Moreover, the *cultural turn* imposed in recent decades on historiography has revalued the information contained in aspects previously little valued, such as realist literature or painting, popular prints, or photographic archives. The press, through advertising, its traditional articles, or reports, is also another inexhaustible source of information. And the recent anthropological perspective present in historiography means, without abandoning the statistical rigor of the large series—when there are any—, a comprehensive study of the qualitative information of folkloric or ethnomusicological nature, the rites of passage of birth, adolescence or marriage, or death, and those of religious festivals; almost always solemnized with specific commensality rituals. In short, all the typical available iconographic information, often with a high symbolic value, are reused with good results by the new cultural approaches or those of ethnohistory. The intersection of all these data sources, and their own methodologies, is also possible in an increasingly internationalized space of study, where the specifics of history tend to dissolve more and more in the common space of *food studies*, and its editorial platforms, as the *Food & History* magazine, dependent on the French Institut Européen d'Histoire de l'Alimentation, combine articles in several languages with very different methodological

[3] Bourdieu [5] suggests that the true principle of the differences found in the sphere of consumption is the contrast between the tastes of luxury and the tastes of necessity, differentiating consumption at the level of social stratum: the upper classes deprive themselves to achieve thinness, while the working class retains a certain morality of the good life at the table.

orientations; and where merge the contributions of the active foci of food history in the English speaking world, the no less important contribution of the solid French tradition, or the thriving research of the Spanish and Latin American space (Table 1).

3 Methods in Historical Research

3.1 Methods and Techniques

Regarding the methodology, when dealing with the history of food, the procedure is substantially alike in any other sector of historiography. It is evident at first in the choice of subject and justification, understanding that it responds to the social and scientific demands of the community. Next, the materials are collected, a process that, as has been pointed out, in the case of food, comes from very different types of sources. Logically, the basic steps of the scientific method are observed, such as the elaboration of hypotheses, description and analysis, and validation or verification.

To achieve this, the historian uses both quantitative and qualitative research techniques. Among the first, procedures such as documentary observation can be noted, to understand and contrast the information, using archaeological, philological, oral techniques, etc. As for the latter, they use procedures such as statistics, quantified textual analysis or graphic techniques.

3.2 Building Systems and Food Heritage

One of the most relevant lines in current research is related to the construction of food systems and heritage, which corresponds to a whole vision of the world since food represents a solid anchor for community identities and occupies a preferential place in the field of collective representations, building from religious beliefs to national pride, encouraged through certain culinary preparations that, at the same time, reflect and reaffirm them.

Despite its fragility, intangible heritage is a relevant factor in maintaining cultural diversity in the face of growing globalization. In the case of food capital, this support is crucial to contain the onslaught of homogenizing and standardizing phenomena brought about by food industrialization such as *McDonaldization*, "fusion cuisine" or "miscegenation cuisines". On the other hand, the development of mercantile networks and a technology exclusive to the sector have diluted the seasonal rhythm of food and the traditional self-sufficiency previously typical of the peasant family. Indeed, one of the characteristic principles of a food culture, virtually extinct, attributed a preponderant meaning to the calendar in this regard.

It is a phenomenon in constant growth since the end of the Second World War has meant the transit of diverse ecosystems to hyperspecialized environments integrated into international agri-food production networks, resulting in the disappearance of

Table 1
Sources of information in historical food research

Primary source	Advantages	Disadvantages	Examples[a]
Traditional archival sources	They offer information on the foods produced and consumed in each period and territory, as well as data on the material culture associated with those food systems or the taxation applied to them	Information tends to be more quantitative than qualitative and, in most cases, is restricted to the hegemonic parts of society	Burnett [6] Floud et al. [7]
Recipe books and gastronomic literature	It provides information on the methods and techniques of preparation and elaboration, and on the evolution of the forms of consumption, as well as changes in taste and the reception and adaptation of culinary novelties	In most cases until relatively recent times, these texts were aimed at the most favored social sectors	Pérez Samper [8] Díaz Rodríguez [9]
Press	It offers a huge amount of information about all aspects related to food, commensality, or gastronomy	Handling this volume of documentation involves an arduous task—Especially for subjects that cannot be sufficiently limited	García Álvarez [10] Díez Rodríguez [11]
Literary sources	They are a very vivid testimony of the social reality and daily life of a given period, often providing a vision on the popular sectors nourishing, and is extremely useful when reconstructing the social uses of space	It is a source that must be addressed taking a series of precautions, such as the ideological conditioning of the author or his social origin. The possible filters established due to these limitations must be considered	García Álvarez [12] Hernández Olmos [13]
Visual and pictorial sources	They allow to clearly visualize the food contexts. They can be key to study the material culture and the use of artifacts of the past, their disposition, management, and social dimension	They are the same as in the case of literature. Here, it is also important to know how to discern between the literal and the symbolic message. Moreover, in the case of photography, it is a source only available for the contemporary era	Burke [14] Castro Flórez [15]
Folklore and ethnographic testimonies	They offer a view of the world and popular sectors of society different from the "official" one. They are a significant testimony of the "common sense of the people"	The testimonies are sometimes too typified and tend to accentuate the "picturesque". They can hide messages of nationalist or socially harmonious idealization	Schmidt-Leukel [16] Lévi-Strauss [17]

(continued)

Table 1 (continued)

Primary source	Advantages	Disadvantages	Examples[a]
Oral sources	Testimonies that can be very thorough and provide different points of view on the same reality, which greatly facilitates adopting a comparative approach. They can bring to light messages nonexplicit in other types of sources (trades, gender, social conflicts. . .)	It is a source developed only in recent times and its gathering is obviously limited to the human life cycle	Gidonet Riera [18] Díaz Méndez and Gómez Benito [19]
Demographic and economic studies	They allow to evaluate by means of variables such as height or mortality, the standards of living or food safety	Does not assess the weight of hereditary, environmental, or cultural factors in the diet	Fogel et al. [20] Gómez Mendoza and Pérez Moreda [21]
Historical studies of medicine and nutritionism; medical topographies	They value the historical evolution of some consumption patterns, deficient nourishment or diet types	They are often strongly descriptive, tending to not include qualitative or cultural assessments	Tribin [22] Carbonell et al. [23]
Supply systems of cities and food markets	They assess the food supplies critical to the urban diet, and their supply problems	They do not evaluate other aspects of the diet. They ignore nourishment with products from barter, harvest, or the self-consumption of the small production	Brown [24] Calero Delgado [25]
Local and national food statistics municipal consumption taxes	They can evaluate the foods of greater commercial circulation in urban areas, and therefore of greater significance in the popular diet	They do not value rural food or food quality; as any tax source can suffer from adulterations and concealments; they only include foods of significant commercial circulation	Segura and Segura [26] Conard and Lovett [27]
Statistics, large series, and records of maritime, rail or customs traffic	They allow to analyze the evolution in the prices and the demand of basic consumer products not produced locally. They assess the complementarity of large food production markets. They detect the food products with the highest consumption and nutritional need	They cannot value local food production or self-consumption. They disdain the commercial exchanges of food between bordering and little fiscally monitored areas. They do not necessarily explore the social or cultural causes of consumer trends	Medina [28] Bello and Rivero [29]

(continued)

Table 1
(continued)

Primary source	Advantages	Disadvantages	Examples[a]
Sources from states polices and internal surveillance services	They allow to evaluate the sociability and socializing functions of some places of commodified consumption of food, especially alcoholic beverages	They hardly penetrate the domains of individual behavior and its values, or into the areas of consumption and family sociability	Fernández Labbé [30] Brennan [31].
Manuals of urbanity, books of protocol, and rhetorics of commensality	They allow to reconstruct the rituals of consumption, the ceremonial values of food, and the food staging of class, status, or power	They are meticulous for the upper classes, but little accurate for the lower ones. They may not be significant from the dietitian or nutritionist point of view	Samper [32] Remedi [33]
Inventories of material culture	They allow to examine, from the material culture generated at the table and kitchens, consumption habits, value systems, and patterns of commensality of a certain era	They tend to descriptive and taxonomic cataloging and not always are sensitive to the sociocultural contextualization of material objects, nor to their chronological location	Garric [34] del Pueblu dAsturies [35]

[a]The table highlights articles and monographs where the sources referred to have been used essentially, or prominently; however, there is a large quantity of works that simultaneously use several of them, some representing authentic referents of this study. Such is the case of some that appear in the bibliographic section of L. B. García Álvarez, K. F. Kiple and K. C. Ornelas, S. Mennel, M. Montanari, J. F. Revel or J. Uría. See notes for complete references.

numerous plant and animal species that constituted the dietary basis of many local communities. The traditional relationship of human groups with their territory has deteriorated, and the domestic dimension of food has been seriously weakened by the stalking of an industry that is also supported by a powerful marketing apparatus aimed at creating a new mass gastronomic universe.

These processes translate in the food field into the hardly remediable loss of the diversity of cultural practices associated with cooking and of agricultural and ecological wealth. In response to this significant decrease, the concept of food sovereignty has been born—understood to a large extent as a defense and struggle mechanism—, formulated by the defenders of peasant family farming; and social movements such as *Via Campesina* or Slow Food have been cemented.

It should be noted that cooking is one of the factors that invariably accompanies migrant communities and population movements, resulting in an exceptional way of transmitting recipes, techniques, and materials, which sometimes give rise to the formation of cultural hybridisms that can contain worrying elements in heritage terms if they catch on within dominant cultures.[4] In any case, migrant groups have shown to be particularly reluctant to abandon their culinary practices and the social networks created in the places of destination, palpably realized in the meetings that frequently underpin an emotional remembrance of their communities of origin, keeping alive the food heritage of many communities. In this context, it is worth highlighting the importance of sensory memory, derived from the community construction of a culinary remembrance that shapes the palates.

Regarding the formation of this cultural capital, when identifying and defining a specific cuisine, it is unavoidable to analyze a series of basic factors such as the limited number of foods chosen among those offered by the environment, and whose selection is usually guided by ease of access; the method of preparation; the principles of seasoning; and the adoption of a set of rules regarding the distribution of meals throughout the day, the group or individual way of eating, the reservation of certain foods for ritual, religious or festive occasions, and the observance of some taboos.

Therefore, the cuisine of a certain community can be conceived, as suggested by Contreras and Garcia Arnaiz [36], as "a set of practices, representations, rules, norms, based on classifications, one of whose essential functions is, previously, the resolution of the paradox of the omnivore". Once "cooked", and therefore subject to the established rules, the food has a private identification stamp.

In this frame of reference, certain food items become, apart from merchandise that can be essential for the economy of a region, key symbolic references in the collective identity of their communities. The sets of practices linked to the processes of production, distribution, and consumption, which in many cases have been developed since ancient times, can result in the configuration of food particularisms derived from the relationships and links generated around a product, a territory, and its inhabitants who, through generations, and in many cases subject to the influences of external agents, create specific practices, spaces, and languages that denote their cultural richness.

Furthermore, forms of commensality and sociability are also forged that can be found spaces and occasions usually linked to festive, family, neighborhood events—associated with mutual aid circuits in many cases—or labor with ludic or commodified elements. This set of practices appears as a community identifier that reinforces social ties and, to a large extent, contributes to provide a sense of belonging and ensure the continuity of the very essence of community. All this is part of a generational learning sequence, to

[4] Think, for example, of the role played by the hamburger or pizza in the overly exported system of the *American way of life*.

which it is possible to add the novelties and the transmission through oral or written means, whether in a family, informal or professional environment. Beyond the social and cultural dimension, an economic circuit is also generated between producer, processor, and consumer, through which ancestral techniques and recipes are preserved, regardless of the incorporation of novelties. The heritage is often preserved thanks to the activity of professional associations, voluntary groups, institutions, or public policies aimed at its conservation, such as competitions, championships, or gastronomic or culinary days. In this case, evidence and lines of research have realized reaching the university sphere, and even gave rise to some of their own disciplines; all of this along the campaigns which tend to visualize the culinary practices, some of them within UNESCO. And in the shaping of these realities, there is no doubt that the mechanisms of selection and re-elaboration of meanings play a fundamental role, without dismissing other scenarios of idealization, evocation, or commodification such as, for example, the very fashionable tourism linked to *live experiences*.

It must be noted that, when examining this type of content, the problem entailed by the "traditional" idea of tradition must be assessed. As Burke [37] points out, the apparent innovation can mask the persistence of the tradition and at the same time the problem of the "internal conflict of the traditions" arises, the inevitable contest between the universal rules and the concrete and ever-changing situations. In other words, what is transmitted changes during the transmission, resulting in a new reality.

Hobsbawn and Ranger [38] have established in turn the basic assumption that "traditions" that appear or claim to be old are frequently quite recent and sometimes invented, trying only to connect with a historical past that suits them.[5] In any case, it is evident that all the invented traditions use History as a legitimizer. This circumstance can be easily related to the construction process of national and regional cuisines, and the invention is more evident when a rapid transformation of society occurs, which destroys or weakens the models in which the "old" traditions operated. An extremely interesting aspect of the invention of traditions is the use of old materials for new purposes, since a great deal of such elements accumulates in the past of any community, and always has an elaborate language of symbolic practices and communication.

In summary, the identification of a product perceived as "traditional" with a food system and a certain place, and its historical link with the territory, has made possible to establish in contemporary societies which elements have the character of an *emblem*, not just of a particular cuisine, but also of the community, and therefore,

[5] In the case of food, this may be the remembrance of a lost happy Arcadia, where prevailed hunger for most of society.

stands as one of its heritage assets. Of course, different cultures have developed diverse food systems. Culture acts in this regard, endowing these systems with regularity and specificity, shaping the establishment of the daily order of meals or the norms of urbanity. These conditions make necessary a learning which, in some cases, must be undertaken from childhood,[6] for which an effective transmission within the individual's social networks (family, ethnic group, social class, local community or nation) is essential. It is not surprising then that food, both in its material and symbolic practices, constitutes one of the first social learnings of the human being.

These behaviors form the basis of food culture and, as Fribourg [39] points out, eating serves as a sign exchanged between the participants in the dining occasion, since it constitutes a mark of belonging, as well as of social inclusion and exclusion. That is, the dishes tasted are considered as *own*, which helps to define the group opposite to the *others*. In turn, food culture also serves as a source of knowledge of the other.

Within the framework of the development of the modern nation state, whether due to the invention of its identities or the exaltation of regionalisms and particularisms, there was a "reconstitution of ethnic groups around the table" that helped decisively to the codification of regional cuisines or the proliferation of gastronomic clubs and brotherhoods' cultivators of the peculiar and self-defining.

In many countries, family agricultural production was mostly allocated for self-consumption, and tended to be as diversified as possible in the stage prior to the irruption of the market economy in the countryside. Their economies used to be in many cases mixed—agricultural and livestock. The basic pillars of survival were represented for a long time by the products of the land, and milk and its derivatives, eggs (with an outstanding mercantile aspect) or honey. As for meat, pork was key in innumerable cultures, while poultry was frequently associated with luxury or destined for the sick or newly delivered mothers.

In any case, the delicacies that would end up becoming the most representative of these regional cuisines involved the exceptionality of the festive diet or one associated with rites of passage, phenomena of great importance for the development of the community. The act of commensality developed around banquets was a very effective means of reinforcing ties of kinship or friendship, although, on the other hand, it should be noted that all the festive meals of the year imply quantitative and qualitative progress with respect to the ordinary ones. The overcoming of a subsistence economy, with the consequent improvement of the diet, did

[6] This happens, for example, in situations where it is feasible to expose the people who do not belong to the group or to smear social upstarts.

considerably lower the dietary and ritual significance of these events; but, in any case, in many places constituted an indispensable occasion for the mass consumption of meat until well into the twentieth century. The best of the family production was reserved for feasts, —retired from the ordinary diet—and people ate *what they did never eat*. The food associated with feasts that responded in a not-too-distant past to a logic of replenishment (which explains the constant presence of heavy dishes), in the society of abundance obeys a logic of pleasure. Likewise, traditional foods that previously responded to exceptionality or social positioning, today are coated with meanings associated with identity.

Throughout modernity, and within the framework of the historical construction of food systems, the dissemination of the cuisine of society's upper layers made by gastronomes among the public from different strata of the bourgeoisie and the nascent middle classes contributed in part to keep the elitist codes of the Old Regime, but also assumed a democratizing task. Once the bourgeoisie was consolidated as the new hegemonic class, its cuisine, which drank from different influences both from above and below, did obviously become the model of the new national cuisine.

Authors such as Revel et al. [40] claim that cuisines come from two essential sources, one popular and one cultivated, one manifesting as peasant (or seafaring) cuisine, and the other as courtesan. The one considered *wise* rests on invention and experimentation, imagines things, and takes risks; the one of the people is based on repetition. Hence the importance of the dialogue established between the two modalities, not exempt from elements of tension. With the consolidation and growth of the urban middle classes in the eighteenth and nineteenth centuries, an alliance was established between the two cuisines, giving rise to bourgeois gastronomy that was codified in numerous treaties and that, while maintaining the solidity and aromas of peasant cuisine, added the restlessness of haute "class" cuisine.[7]

In Revel's [40] opinion, there are no national cuisines: there is an international cuisine, necessarily flexible, and regional cuisines. From this approach, the basic gastronomic cell is the region, and in many cases the national cuisines are nothing more than a summation. The *regional* ones, to which historical values are often attributed—that they have never possessed—, would come from the industrialization process. Thus, although it is not an *ex novo* creation because there have always been local particularities—as

[7] We must consider than in the Middle Ages, cuisines hardly differed from each other and, in many aspects, did not have characteristic features. Accordingly, the culinary borders did not separate nations, but estates. The subsidiary layers did not eat what they wanted, but the food they could get, and we can typically speak of monotonous and poor diets. Also, the existing difficulties for the transmission of knowledge are added, since aside the fact that the printing press did not yet exist and literacy was frankly meager, communications were extremely deficient.

Montanari [41] points out—, territoriality as a notion and as a positive fact is a recent invention that nonetheless stands as an absolute reference value in the choice of food.

And once more, history can offer relevant elements of analysis in the constitution of these food heritages. This is the case, for example, of products from America and the importance of fruits such as tomatoes in the formation of Provençal or Italian cuisine; or the relevance of potatoes as a "social stabilizer" as defined by Salaman and Burton [42], since they ensured subsistence with the lowest possible salary. And similar assessments could be made about products such as beans, corn, turkey, or chocolate.

In principle, cultures are hostile toward new culinary influences and are full of exclusionary prejudices. But the existence of agents capable of breaking down cultural barriers and internationalizing food, such as war, colonization, or famine, is systematically confirmed. Elements like the ones above explain why the evolution of food taste and cooking is so slow, since it is a very complex mechanism that requires collective acceptance. Many times, certain foods are incorporated when resilient traditional agrarian and social systems are just outlined—or at least some of their defining elements.

From another perspective, authors such as Fieldhouse [43] consider that a national cuisine refers fundamentally to the foods and ways of preparing them that are considered *normal*, typical, or specific to a given country. And to the same extent it constitutes an aspect of their identity as a group. The community considers it a given and only deviations from the *norm* are recognized as such. For example, it is common that estrangement occurs when contacting with another culinary culture and comparisons are made with that is recognized as one's own.

Be that as it may, it must be noted that the so called "national cuisines" have been developed recently, are an eminent bourgeois creation and associated with their values and ways of life. In this regard, it should be noted that the concept of identity—which is often misused—is frequently associated with cuisines understood as national or regional. In fact, the food culture of a certain territory is conceived, sometimes in a worryingly mechanistic way, as a sign of identity without reflecting too much on what this might mean. The latter often leads to heritage elements being *out of place* with respect to the way of life to which they should be linked, being adopted not so much by the groups that carry their values, but by other agents such as public administrations, tourism, and hospitality companies, associations, and a long etcetera. It is significant that cultural heritage is found in contemporary societies increasingly linked to one of its most defining phenomena: leisure.

The development of its own literature has undoubtedly contributed in turn to the configuration of modern culinary codes, which, essentially since the mid-eighteenth century, have spread throughout all social strata, albeit especially oriented to the new middle sectors that cannot afford their own cooks, but are eager to play a good role in the field of social distinction and appearances, avoiding complex preparations and excessive expenses. The techniques and flavors continued being the same of the aristocracy, but there could be differences in terms of the quality and quantity of the ingredients, with expensive starters or gimmicky decorations tending to disappear.[8]

Moreover, regarding the change of valuation, redefinition, and dignity of some products, it has sometimes been related to the change of contexts and meanings in terms of production and consumption. This is the case, for example, of beer in England, which was considered a fundamental food to undertake any heavy task and its manufacture was essential for the domestic economy,[9] and currently is much more linked to the recreational dimension and social time, even penetrating decisively the circuits of gourmet products; and all this without ceasing to constitute an evident food heritage that even UNESCO has recognized for the Belgian case. Evidence like these are still phenomena linked to the transition from the society of hunger to the society of abundance, leading to consider a sign of high cultural and social quality products not long ago cataloged as rustic and related to the *culture of necessity.*

To the equation, extant between the popular and the elitist, between modernity and tradition and social origin, the element of territorial plurality must be added. From this point of view, national cuisines—which are based on the compilation of the best regional specialities—are rich and diverse. Snobbery has assumed a noteworthy role in establishing culinary changes, understandable in the context of social distinction strategies; by contrast, it is evident that some very stable correspondences between types of food and consumers still survive today.

Likewise, it has been made easier for the largest social strata to taste occasionally foods that traditionally could not have access to. In this regard, the efforts of chefs, gastronomes, and authorities of diverse status have multiplied to "recover" or "invent" traditional regional and national cuisines that sometimes have little to do with the daily meals of the communities they claim to represent; but they do reveal a reaction against the homogenization and relocation experienced by modern food when separated from its context.

Indeed, one of the main problems found when dealing with the new historiographical orientations lies in discerning within food systems what remains, what changes, what adapts to new contexts

[8] Beyond the recipe books and treatises published by renowned chefs, there are texts such as those written by English tavern cooks from the end of the eighteenth century, or domestic cooks generally from the bourgeoisie.

[9] Hence, an increase in taxes could be considered as an incitement to revolution.

and how, how novelties are incorporated and even to what extent they are pertinent or respond to interests of a commercial or political nature.

This is how the vindication of the autochthonous, of certain emblematic preparations and raw materials, has forced even the all-powerful "junk food" chains to incorporate them into their repertoire. Be that as it may, as authors like Jegou [44] point out, the industry has provided a "flow of food without memory" in which the symbolic dimension of food is no longer the result of a slow process but precedes it.

There are numerous cases in which changes in taste are closely linked to the circumstances of the historical context. A well-studied example is the so-called *English beverage revolution* of the seventeenth and eighteenth centuries. This variation took place due both to the boom of new beverages such as coffee, chocolate, rum, or gin, and to the growing preference for fuller-bodied wines (port, sherry), in opposition to the predominant light wines hitherto led by the *French Claret* from Bordeaux. The permutation was influenced mainly by the prohibition of importing French wines between 1678 and 1686, and the high tariff imposed in 1689 because of the Nine Years' War, but also by the greater demand for quality due to the general increase in the standard of living and the thickening of the middle sectors, especially the urban ones.

Such a circumstance, apart from the falsification of numerous Gallic wines to pass as wines from other sources and penetrate the British market, did stimulate the transcendental transformation of the Bordeaux features toward a higher quality, both to be profitable in Anglo-Saxon circuits, and become an essential part of the French food identity.

Another aspect that confers identity, related to an interesting process of historical construction, has to do with the dimension of commensality, which provides food with a communicative value of belonging or exclusion. Food, as Barthes [45] pointed out, takes on a "circumstantial value" that in a *society of abundance* acquires a scope beyond the properly nourishing. But this fact took on a deeper meaning in contexts of need when it was often a vital complement to a brief and monotonous diet of mere survival.

The greatest complication of the ceremonial to behave at the table is directly related to this dimension, and urbanity usually exceeds any other ritual of social life on this point. The manuals on the subject, popularized especially since the nineteenth century, were unanimous on the matter and constitute a valuable source of study along, as for the rest of the aspects related to food and its social magnitude, literature or painting, plausible testimonies when reconstructing contexts and scenarios of commensality.

The question lies, however, in considering the reason for this obsessive attention to the social norm at the table. An obvious answer can be attributed to the mere pleasure of eating, and gastronomy, a recently coined term, turned it into a complex knowledge toward the end of the nineteenth century. But the truth is that in the dining rooms of the socially favored, there were other circumstances, being a privileged observatory and a place to test the distinction and good manners of the diners through an elaborate verbal and body language of long apprenticeship. The practice of commensality was, therefore, strongly regulated, leaving the banquets shielded with their rigid protocol to the penetration of social upstarts and automatically free from the presence, not only of members of lower social groups, but even of those "nouveau riche" newcomers who had not had time to delve into the rigid regulations of bourgeois education and so would be exposed to ridicule. Education at the table thus evidenced the cachet of the individual and his group or class. Eating could be a good example of the many filters opposed to a quick social ascent, functioning as elements of distinction between contiguous social categories whose transfer was more difficult than it might seem at first. In short, albeit obvious, it must be noted that the table was also an outstanding element of differentiation between social classes very far apart each other in the social structure.

Conversely, a whole series of discourses and practices referring to the supply, preparation, and consumption of food have been configured, which show the interest of public authorities in integrating gastronomy and tourism, a phenomenon that did become evident past the halfway point of the twentieth century. Some advertising mechanisms of "traditionalization" and "cultural authentication" of food were developed anticipating the process of its official codification as a heritagization phenomenon. As a result, sometimes it is complex to discern the food heritage from its mercantile aspect, something related to what some authors have called *folk revival*, one of the ways of "recovering" the past by altering some of its meanings.

4 Conclusion

Food behaviors and traditions are by no means stagnant and immutable aspects, and in the history of food a fluid and paradoxical dialogue is established between what is permanent and what is subject to profound transformations, where the culinary systems of a territory could change due to internal developments or external influences.

Each culture, tradition, and identity are a dynamic and unstable product of history, generated by complex exchanges and hybridization processes that enrich them. Ideas such as Certau's [46] are of great interest in this regard, that in each food there is an intersection of invisible stories, and each way of tasting, cooking, and sharing food is not only a passive way of reiterating codes, traditions or messages, but also an active and adaptive way of forming meaning and building it in which uses and reuses, assimilation and mixing of procedures end up forming not only a space for following norms, but also for resistance and creativity. Hence the configuration of living and operating systems.

To undertake studies related to this entire social fact, it is necessary to resort to the methods and techniques of historical research, but always considering the multiplicity of primary sources that must be collected, as well as the number of dimensions of knowledge that intersect and to which it is connected. These fields range from production and consumption to the political and social, without forgetting the enormous cultural implications that entail.

In short, a food culture is the result of a long learning process. The way of eating, the products that are consumed, and the way they are cooked are related to local resources, its climate, its soil, the agricultural practices, and forms of provisioning; at the same time, it is closely related to lore, knowledge, and cultural practices.

References

1. Aymard M (1975) Pour l'histoire de l'alimentation: quelques remarques de méthode. In: Annales. Histoire, Sciences Sociales, vol 30, No. 2–3. Cambridge University Press, pp 431–444
2. Montanari M (2006) La comida como cultura. Trea, Gijón
3. Harris M (1989) Bueno para comer. Alianza, Madrid
4. Douglas M, Simons E (1973) Pureza y peligro: un análisis de los conceptos de contaminación y tabú. Siglo Veintiuno de España, Madrid
5. Bourdieu P (1991) La distinción. Una crítica del gusto. Taurus, Madrid
6. Burnett J (1966) Plenty and want: a social history of diet in England from 1815 to the present day. Routedgle, London
7. Floud R, Wachter K, Gregory A (1990) Height, health, and history: nutritional status in the United Kingdom, 1750–1980 (No. flou90-1). National Bureau of Economic Research
8. Samper MÁP (1997) Los Recetarios de mujeres y para mujeres Sb sobre la conservación y transmisión de los saberes domésticos en la época moderna. Servicio de Publicaciones UCM, Madrid
9. Díaz Rodríguez JM (2014) El género literario gastronómico en el Reino Unido en la primera mitad del siglo XIX. In: Uria J, Garcia Alvarez LB (eds) Comensalidad y alimentación y en la época contemporánea, Historia Contemporánea, p 48
10. García Alvarez LB (2013) Sidra y manzana en Asturias. In: Sociabilidad, producción y consumo (1875–1936). KRK, Oviedo
11. Díez Rodríguez F (1980) Prensa agraria en la España de la Ilustración. In: El seminario de Agricultura y Artes dirigido a los Párrocos (1797–1808). Servicio de Publicaciones Agrarias, Madrid
12. García Alvarez LB (2008) Las representaciones de la sidra. El contexto social de la sidra a través de la literatura y la pintura asturianas (1850–1939). Red de Museos Etnográficos de Asturias, Gijón
13. Hernández Olmos C (1994) Comer y escribir. Una historia de la alimentación a través de la literatura española. Mercasa, Madrid
14. Burke P (2001) The uses of images as historical evidence. Reaktion Books Ltd, London
15. Castro Flórez F (2013) Arte y gastronomía: la cocina de la pintura. Instituto Valenciano de Arte Moderno, Valencia
16. Schmidt-Leukel P (2000) De Religionen und das Essen. Heinrich Hugendubel Verlag, München
17. Levi-Strauss (1981) Mitológicas III. El origen de las maneras a la mesa. Siglo XXI, Ciudad de Mexico
18. Gidonet Riera A (2011) Memoria oral y alimentación. Estrategias de supervivencia

durante la Guerra Civil Española (1936–1939) y la Posguerra (1939–1995). Arxius 24
19. Díaz Méndez C, Gómez Benito C (2008) Alimentación, consumo y salud. Fundación "La Caixa", Barcelona
20. Fogel RW, Engerman SL, Floud R et al (1983) Secular changes in American and British stature and nutrition. J Interdiscip Hist XIV:445
21. Gómez Mendoza A, Pérez Moreda V (1985) Estatura y nivel de vida en la España del primer tercio del siglo XX. Moneda y Crédito, Madrid, p 174
22. Tribin HAS (1993) Epidemias de escorbuto y beriberi en la historia de Colombia. Medicina 15(2):34–36
23. Carbonell E, Cáceres I, Lozano M, Saladié P, Rosell J, Lorenzo C et al (2010) Cultural cannibalism as a paleoeconomic system in the European Lower Pleistocene: the case of level TD6 of Gran Dolina (Sierra de Atapuerca, Burgos, Spain). Curr Anthropol 51(4):539–549
24. Brown J (1986) The English market town. A social and economic history 1750–191. The Crowood Press, Ramsburry
25. Calero Delgado ML (2014) Hygiene and food: the meat supply in Huelva (1855–1870). International University of Andalusia, Sevilla
26. Segura XC, Segura RG (2007) La transición nutricional en la España contemporánea: las variaciones en el consumo de pan, patatas y legumbres (1850–2000). Investigaciones de historia económica 3(7):69–100
27. Conard P, Lovett A (1969) Problèmes de l'évaluation du coût de la vie en Espagne. Mélanges de la Casa de Velázquez 5(1):411–441
28. Medina AMR (2009) Del mar a la taberna: el vino en Bilbao (ss. XV–XVII). Itsas Memoria: Revista de Estudios Marítimos del País Vasco 6: 615–627
29. Bello MDCE, Rivero JG (2015) El bacalao en España (1850–1914): importación y política comercial. Historia Agraria: Revista de Agricultura e Historia Rural 67:43–78
30. Fernández Labbé M (2006) Los usos de la taberna: renta fiscal, combate al alcoholismo y cacicazgo político en Chile. 1870–1930. Historia (Santiago) 39(2):369–429
31. Brennan TE (2014) Public drinking and popular culture in eighteenth-century Paris. Princeton University Press, Princeton
32. Samper MDLÁP (2003) La alimentación en la corte española del siglo XVIII. Cuad Hist Mod 2:153–197
33. Remedi FJ (2017) Modernidad alimentaria y afrancesamiento. Ciudad de Córdoba (Argentina) en el tránsito del siglo XIX al XX. Hist Crít (65):71–92
34. Garric J-M (2014) The table is mise. Lemusée des Arts de la Table. Abbaye de Belleperche, Cordes-Tolosannes
35. del Pueblu dAsturies M (2005) Los asturianos na cocina: la vida doméstica nAsturies 1800–1965. In: Catálogo de la Exposición. MPA, Gijón
36. Contreras J, Garcia Arnaiz M (2005) Alimentación y cultura: perspectivas antropológicas, vol 392. Ariel, Barcelona
37. Burke P (2006) ¿Qué es la historia cultural? Paidós, Barcelona
38. Hobsbawm E, Ranger T (eds) (2012) The invention of tradition. Cambridge University Press, New York
39. Fribourg J (1996) Fêtes et cuisine traditionnelle en Espagne. In: Bataille MC, y Cousin F (Comp.) (eds) Cuisines. Reflets des société. Éditions Sépia-Musée de lHomme, París
40. Revel JF, Gavarrón L, Armiño M (1996) Un festín en palabras: historia literaria de la sensibilidad gastronómica desde la Antigüedad hasta nuestros días
41. Montanari M (1993) Historia, alimentación, historia de la alimentación. In: Sánchez, Montanari M, Fernández E, Gelman J, Dumoulin M et al (eds) Problemas Actuales de la Historia. Universidad de Salamanca, Salamanca, pp 19–28
42. Salaman RN, Burton WG (1985) The history and social influence of the potato. Cambridge University Press, Cambridge
43. Fieldhouse P (2013) Food and nutrition: customs and culture. Springer, New York
44. Jégou F (1991) Nourritures Artificielles: le rôle du design dans l'avenir de l'industrie alimentaire. Ministère de l'Agriculture et de la Forêt, DGAL-Dâlt, Paris
45. Barthes R (1961) Pour une psycho-sociologie de l'alimentation contemporaine. In: Annales. Histoire, Sciences Sociales, vol 16v. Cambridge University Press, pp 977–986
46. De Certeau M (1996) La invención de lo cotidiano: artes de hacer. I, vol 1. Universidad Iberoamericana, México

Further Reading

Afinoguénova E (2019) Una gastrocracia de película: el turismo y la modernización del sistema alimenticio en el cine desarrollista español. In: Sociedades y culturas: IX Congreso de Historia Social. Treinta años de la Asociación de Historia social. Comunicaciones. Oviedo, 7–9 de noviembre de 2019, Asociación de Historia Social, pp 825–835

Benito LEA (2005) Mitologías alimentarias cotidianas. Una relectura de Roland Barthes. Rev Int Sociol 63(40):79–107

Back KW (1977) Food, sex and theory. In: Fitzgerald TIM (ed) Nutrition and anthropology action. Van Gorcum, Amsterdam

Barrau J (1983) Les hommes et leurs aliments: esquisse d'une histoire écologique et ethnologique de l'alimentation humaine. Temps actuels, Paris

Con Redondo JA (2019) Bueno para escanciar, buena para pensar. Elementos para el análisis de las atribuciones culturales y patrimoniales dadas a la sidra en la Asturias contemporánea. In: Castillo S, Uría J (eds) Sociedades y culturas. Treinta años de la Asociación de Historia Social. IX Congreso de Historia Social, Oviedo

Contreras JJ (1993) Antropología de la alimentación. EUDEMA, Madrid

Elías N (2011) El proceso de civilización. Fondo de Cultura Económica, México

Fernandez Armesto F (2004) Historia de la comida. Alimentos, cocina y civilización. Tusquets, Barcelona

Fischler C (1982) Alimentation, cuisine et dietité: le identification des alimentes et l'identité du manger. Reserches et travaux de l'Institute d'Ethnologie 6

Fischler C (1995) El (h)omnívoro. El gusto, la cocina y el cuerpo. Angrama, Barcelona

Fischler C (1995) Gastronomía y gastroanomía. Sabiduría del cuerpo y crisis biológica cultural de la alimentación contemporánea. In: Contreras J (ed) Alimentación y cultura. Necesidades, gustos y costumbres. Publicaciones de la Universidad de Barcelona, Barcelona

Fonte M (1991) Aspectos sociales y simbólicos en el funcionamiento del sistema alimentario. Agric Soc 60:165–185

García Alvarez LB (2013) Sidra y manzana en Asturias. In: Sociabilidad, producción y consumo (1875–1936). KRK, Oviedo

García Alvarez LB (2021) La construcción histórica del patrimonio alimentario. In: VVAA, multiplicidades del patrimonio. Universidad Mayor, Santiago de Chile

García Alvarez LB (2008) Sidra y manzana en Asturias. Sociabilidad, producción y consumo (de la Guerra Civil a nuestros días). Study carried out for the Principality of Asturias as a documentary basis for the candidacy of the cider culture as a World Heritage Site

García Alvarez LB (2005) Beber y saber. Una historia cultural de las bebidas. Alianza Editorial, Madrid

García Álvarez LB (2011) Comensalidad, sociabilidad y rituales de consumo. La "Espicha" en Asturias en el primer tercio del siglo XX. Hist Soc:21–40

García Álvarez LB (2018) La sociabilidad sidrera en Asturias durante el franquismo. Hist Soc (92): 47–62

García Alvarez LB (2014) Solidaridad, sociabilidad y comensalidad en el ciclo festivo asturiano (1850–1936). Hist Contemp 48(I)

González Turmo I (1996) Introducción: modelos ideales y realidad en alimentación. In: González Turmo I, Romero de Solis P (eds) Antropología de la alimentación: nuevos ensayos sobre la dieta mediterránea. Universidad de Sevilla, Sevilla

Goody J (1995) Cocina, cuisine y clase. Gedisa, Barcelona

Hobsbawm E, Ranger T (2002) La invención de la tradición. Crítica, Barcelona

Kiple KF, Ornelas KC (2000) The Cambridge world history of food. Cambridge University Press, Cambridge

Krosmeyer C (2002) El sentido del gusto. Comida, estética y filosofía. Paidós, Barcelona

Lévi-Strauss C (1972) Mitologícas I. Lo crudo y lo cocido. Fondo de Cultura Económica, Ciudad de México

Lévi-Strauss C (1972) Mitológicas II. La miel y las cenizas. Fondo de Cultura Económica, Ciudad de México

Lévi-Strauss C (1972) Mitológicas III. El origen de las maneras a la mesa. Siglo XXI, Ciudad de México

Maldonado Roso J (2014) Cambios de consumo y de gusto de los vinos de Jerez en el Reino Unido y sus consecuencias en la zona de producción entre mediados de los siglos XVIII y XIX. In: Uría J, García'Alvarez LB (eds) Comensalidad y alimentación y en la época contemporánea, Historia Contemporánea, p 48

Mennel S (1985) All manners of food. Eating and taste in England and France from the middle ages to the present. Basil Blackwell Ltd., Oxford

Massimo Montanari (1993) El hambre y la abundancia: historia y cultura de la alimentación en Europa. Crítica, Barcelona

Rotberg RI, Rqbb TK (1975) El hambre en la historia. Siglo XXI, Madrid

Rozin E, Rozin P (1981) Some surprisingly unique characterictics of human food preferences. In: Fenton A, Owen T (eds) Food in perspective. J. Donald Publishers, Edinburgh

Thompson EP (1989) La formación de la clase obrera en Inglaterra. Crítica, Barcelona

Toussaint-Samat M (1991) Historia natural y moral de los alimentos. Alianza, Madrid

Uría J (1994) Ocio, espacios de sociabilidad y estrategias de control social: La taberna en Asturias en el primer tercio del siglo XX. In: Redero M (ed) Sindicalismo y Movimientos Sociales. Siglos XIX y XX. UGT, Madrid

Uría J (2001) La cultura popular y la historiografía española contemporánea: breve historia de un desencuentro. In: Ortiz D, Ruíz M, Su'qrez I (eds) Movimientos sociales y Estado en la España contemporánea. Ediciones de la Universidad de Castilla la Mancha, Cuenca

Uría J (2000) De la fiesta tradicional al tipismo mercantilizado. Asturias a principios del siglo XX. In: Bulletin d'Histoire Contemporaine de l'Espagne, pp 30–31

Uría J (2001) Los lugares de sociabilidad. Espacios, costumbre y conflicto social. In: Castillo S, Fernández R (eds) Historia social y ciencias sociales. Milenio, Lleida

Uría J (2014) Una nueva historia de la alimentación. In: Uría J, García Álvarez LB (eds) Comensalidad y alimentación y en la época contemporánea. Historia Contemporánea, p 48

Uría J, García Álvarez LB (2014) Presentación. Nuevas áreas de investigación y consolidación de la historia de la alimentación. In: Uría J, García Álvarez LB (eds) Comensalidad y alimentación y en la época contemporánea. Historia Contemporánea, p 48

Chapter 3

Sensory and Multisited Ethnographic Methods for Consumer Research

Ana María Ulloa

Abstract

In recent decades, ethnographic methods have gained increased attraction in different fields outside of anthropology. Contemporary approaches to ethnographic research have pushed traditional methods to address new research subjects and purposes in a collaboratory and participatory manner. One such new area is consumer and market research. In this chapter, the reader will be introduced to contemporary ethnographic methods with a special attention to sensory ethnography and multisited ethnography—strands particularly fitting for consumer research. In addition, in order to make the case for the need of ethnography within the food industry, an example of doing multisited and sensory ethnographic research on flavor will be presented. From this example, important insights for consumer research will be drawn as well as practical advice for those new to ethnography and interested in this descriptive and interpretative approach for data collection and analysis.

Key words Qualitative research, Ethnography, Sensory ethnography, Multisited ethnography, Consumer research, Participatory methods

1 Introduction

No longer is ethnography an exclusive matter of anthropological work and reflection [1]. In recent decades, as a method of participatory practice, ethnography has gained increased attraction in diverse fields ranging from the social sciences and the humanities to organizational, marketing, and consumer research. While crossing these and other disciplinary boundaries, ethnography traditionally conceived as the process of writing about other people by observing and participating in their lives has transformed into an experiential and experimental ground open to different tools and approaches.

A classical ethnography can be briefly understood as an interpretive story about a group of people and their way of living. It stands for both a practice (a way of knowing and conducting research) and a product (a text that results from the research).

Carlos Gómez-Corona and Heber Rodrigues (eds.), *Consumer Research Methods in Food Science*, Methods and Protocols in Food Science, https://doi.org/10.1007/978-1-0716-3000-6_3,

Following Bronislaw Malinoswki's *Argonauts of the Western Pacific* (1922) – one of the founding ethnographies of modern anthropology–, the classic ethnographic approach established the need for "extended experience in close proximity among those about whose lives they wished to speak" [2]. This long-term and face-to-face approach to fieldwork common to anthropological practice while still prevalent has adapted and changed accordingly to different circumstances.

Because of its interest in understanding human culture, a complex concept at the core of anthropology, there has never been a fixed or single definition for ethnography, as there are no predetermined rules to conduct ethnographic research. Ethnography has never been a stable entity; it remains open to debate and contestation [3]. Even if it is possible to talk about some general aspects of what makes ethnography a method, as I will discuss here, each ethnographic endeavor is highly dependent on the research circumstances and thus irreproducible [4].

While ethnography in anthropology, sociology, and other akin disciplines has always been marked by diversity, in recent decades what constitutes fieldwork has become even more diverse. Doing fieldwork among "one's own" or what others have called "anthropology at home" [5, 6] has multiplied possible research sites and the objects and subjects of anthropological research. Ethnography is now simultaneously urban, multisited, virtual, visual, and sensory.

Ethnographies of hospitals, senior care centers, cafes, university canteens, public transportation, neighborhood organizations, public streets, and any other urban place where people gather in some sort of group, are increasingly common and relevant for consumer studies [7]. Other researchers, in response to cultural transformations occurring globally, have moved out from single sites and designed multisited ethnographies to examine the circulation of cultural meanings, objects, and identities in different places and scales [8]. Concomitantly, various anthropologists taking the Internet as a fieldsite for analysis started studying mass media and digital technologies, while developing new ethnographic approaches for studying virtual spaces [9, 10]. In consumer research, technology such as website development, cellphones, and video have become common means of accessing and transmitting field data [11]. Finally, scholars with an interest in visual culture and the lived experience of the body in social and cultural practices have taken into consideration new methodological tools to study sensory perception in communication and practice [12]. These developments within and outside of anthropology, among many others, have raised interesting challenges for classical ethnography and have proven useful for new areas such as consumer research.

This chapter will focus on ethnography as practice and method relevant for consumer research. It will start by pointing out what it means to think a research problem ethnographically. Then, it will pay special attention to sensory, multimodal, and multisited ethnography—strands particularly fitting for food-oriented consumer research. Later, in order to make the case for the need of ethnography within the food industry, an example of doing ethnographic research inside/of a flavor and fragrance company will be presented. I will use this example to illuminate key ethnographic considerations as well as practical advice for those new to ethnography and interested in this descriptive and interpretative approach for data collection and analysis.

2 Thinking Ethnographically

Ethnography is a place to ground concepts and assumptions about consumer behavior. It is an opportunity for understanding what people's practices are and the meanings people attach to those practices. Analyzing what people do and what they say they do in any given context is equally important for an ethnographic study. But instead of thinking that the reason behind ethnographic methods is to catch discrepancies between what people say and what they do, it is better to think that the job of the ethnographer is to understand the tensions and divergences that may occur between them. How people present themselves, their likes and dislikes, is an important source of information about meanings people ascribe to things and their actions. Not only does ethnography capture informants' words; but it also strives to uncover what is known as tacit knowledge, referring to largely unarticulated social understandings manifest in routines, skills, postures, gestures, silences, and so forth. These are all cultural matters that are so familiar as to go unnoticed.

Ethnography is an embodied, empirical, and experiential field-based method centered around participant observation [13]. Participant observation is widely considered the main characteristic feature of the ethnographic approach and it involves doing research by watching and listening to other people and partaking in their activities as they go about their daily business. But before going into detail about ethnography as method, it is important to express what thinking ethnographically means.

Ethnography is a way of knowing that uses the self as the instrument of knowing [13, 14]. It is subjective in character and does not attempt to be an objective account of reality [12]. Contrary to qualitative methods, ethnography "embraces the more humanistic, naturalistic, creative, and intuitive ways of acquiring knowledge and making sense of the world" [15]. To know something ethnographically is to perceive something differently and through a partial perspective. As Suderland and Denny explain, "our questions, our presence, our assumptions, our views of the situation provide never-ending filters for the questions we ask, what we observe, and what we conclude" [16]. Therefore, many scholars have described the ethnographic approach as a way of seeing, listening, writing, and being with people [17, 18].

Being ethnography a type of situated knowledge [19], there is no recipe for how to conduct fieldwork or an order for the development of different fieldwork abilities or ethnographic thinking. Each ethnographer or collective can find a way of conducting fieldwork adjusted to the research to be done as well as to his, her, or the group's personal qualities. However, no matter the project or its length, it is indispensable to keep in mind three related aspects for embarking on ethnographic research.

First, go to the field with a clear research question and make sure the ethnographic approach serves your research question. This research question or problem can be either academic or pragmatic in nature, that is, inflected by theoretical concepts or responses to a business problem in trying to understand consumer behavior. This is a necessary first step to know what to pay attention to, what to look for, and what to ask to/of people. For consumer research, usual questions are related to how products are used and perceived as well as to how people experience their homes or a public space in relation to products. These type of questions not only allows us to distinguish what is relevant from what is not but it also guides the work of the ethnographer making visible cultural aspects or meanings that would otherwise remain hidden [4]. When the observation is adequate, it can make visible things that appear obvious or trivial and tend to pass largely unnoticed.

Second, if ethnography is a way of listening to others, this listening is guaranteed if there is a close interaction with people and the ethnographer(s) develop ethical informant relationships. This empathetic relationship is part of what is called ethnographic sensibility. Participants need to be trustfully enrolled in the research and made active participants. This is what active ethnographic engagement entails.

Third, not only is the ethnographer attentive to what people say, but also to how they say it, and under which circumstances. In this way, the purpose of ethnographic research is to focus, by listening, seeing, feeling, and partaking with others, on singularities and small details that contribute to comprehending and contextualizing a larger phenomenon, in this case, related to the experience of the consumer. In practice, this means documenting diligently and copiously. As anthropologist Clifford Geertz argued, if ethnography is "an elaborate venture in 'thick description', –borrowing the notion from Gilbert Ryle—, it is a description that is microscopic" [20]. Small matters can speak of wider contexts. To catch these minuscule details, the more familiarized with the context, people, or situation under study the researcher is, the better. Site visits, observational and participatory encounters, and formal and spontaneous talks, can all guarantee greater detail.

In general, ethnographic thinking is something that is acquired over time in practice and training. The application of an "ethnographic eye" requires openness to exploration, on the ground familiarity, adaptability, and curiosity over the why and how of peoples' doings. With many unknowns to discover, ethnographic thinking can be, as Hasbrouck argues, "an incredibly effective means for making sense of human complexity, developing deep cultural insight, and reshaping perspectives" [21]. Appreciating the value of ethnographic thinking can thus be a first step toward establishing the need for qualitative forms of consumer research.

3 Sensory and Multisited Strategies for Ethnographic Research

By the 1980s and early 1990s, ethnographic methods were increasingly applied to consumer and marketing research [15, 22–24]. They were a departure from in-house consumer studies and allow researchers to approach product usage in daily life situations at home or at the shop. Instead of a single ethnographer doing long-term fieldwork, consumer and marketing research has encouraged teams of researchers doing focus participant observations over short periods to account for discrete practices like doing the laundry, drain cleaning, shopping cooking, eating in a public space, among others. Working collaboratively has proven the advantage of having multiple viewpoints for attempting to acquire an insider's view and analyzing the collected data [16]. Each person may be thinking about or attending to different things about the same situation, practice, or environment.

Out of the different currents of contemporary ethnography, sensory, multimodal, and multisited are of special relevance for consumer research focused on food and beverages. All these iterations have made collaboration gain more relevance and traction for theory, fieldwork, and different applications [25]. Instead of excluding one another, these flexible currents share a common ground, reinforcing each other. Multisited research designs can incorporate a sensory approach to fieldwork, or vice versa, and be simultaneously multimodal in its presentation and dialogical in its analysis. In this section, I will describe some techniques developed within these approaches for data collection and analysis.

3.1 Data Collection

There is not a single way to start collecting data ethnographically. Ethnographic research might begin by selecting a site or sites for participant observation, by carrying out some initial conversations or interviews, by examining an artifact or a product, or simply by way of introspection [16]. Regardless of how inquiry starts, every ethnography needs to ensure a significant amount of participatory learning. For this, it is necessary to establish rapport with informants and negotiate access to daily situations in which the practices in question take place. In some situations, just simply getting permission to enter a fieldsite can take a considerable amount of time and effort. As Raymond Madden argues not only is ethnography a way of writing about people, but it is also a way of being with people [1].

In general, collecting data ethnographically to "make the familiar strange" involves taking copious notes of the context, of what people say and do, as well as of what the ethnographer perceives, feels, or thinks while in the field. Fieldwork can be either sustained over a period or be partial and follow an intermittent pattern. The usual form for this notetaking is the use of a diary that can be complemented with photographs, audio, or video recordings. Direct experiences and sustained observation or an overall immersion are frequently coupled with interviews and informal conversations that inquire how consumers do what they do and make decisions about it. Despite the copiousness that is required in notetaking, it is important to recognize that fieldnote texts are inevitably selective and are not intended to provide a 'complete' record, as in Fig. 1 [26]. The often "messy" form notes that are taken usually reflect what the researcher finds significant to register.

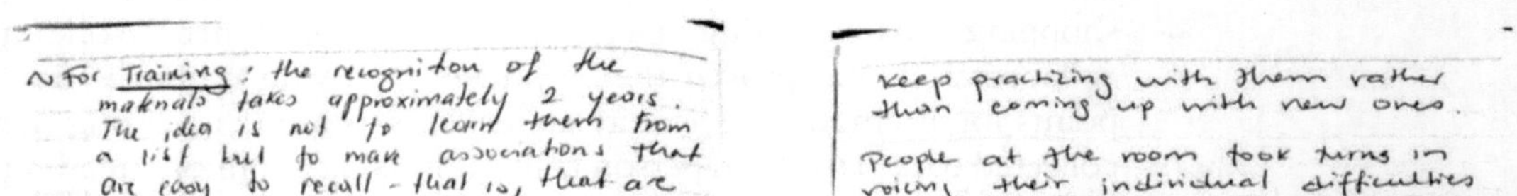

Fig. 1 Fieldnotes written by the author on a sensory panel discussing vanilla flavors for Greek yogurt. (Image: Ana Maria Ulloa)

Sensory ethnography –an approach that "both seek out knowledge about the senses and uses the senses as a way of knowledge" [12]—focuses on techniques for data collection that stress materiality and practice. It is a reaction to the privileged status of vision (and hearing in second place) in modern culture. It has been used in consumer research projects that study food consumption and production, personal care and cleaning practices, and other sensorial experiences to make them more pleasurable and effective. For example, ethnographic research on cleaning practices among Hispanic Americans found the importance of olfactory cues such as the smell of pine as a sign of real cleanliness [15], and research on sensory meanings attached to domestic laundry in English homes found how important are not only the sight and smell of clean laundry but also the feel and texture of items like shirts and towels [27].

These types of consumer studies have proliferated as personal experiences and sensory embodiment have become more relevant for marketing and advertising processes [11]. This shift toward 'experiential marketing' has found fruitful the open-ended character of ethnographic methods, including sensory ethnography. As the editors of this volume make clear, changes in consumption patterns must be understood from a cultural perspective that addresses the shift from a consumer as possessor of goods to a consumer as a collector of experiences [28].

For ethnographers who seek to understand what others are experiencing and how sensations get to be shared, there is more profit, as these examples attest, in learning with others in practice than relying exclusively on interviews about meanings attached to these sensations. If the goal is to grasp knowledge that is largely embodied, it is best to seek ways that can elicit sensorial engagements for both informants and researcher(s), rather than purely discursive accounts. For this, every ethnographer needs to develop creative approaches or research material that can facilitate ways of being close to the nonverbal and tacit [12].

This approach which calls for going further than interviewing and observing, asks of ethnographers to open themselves up and experience through sight, sounds, taste, touch, and smell their surrounding environment [29]. As with seeing, the ethnographer needs to learn how to use all the different senses properly to capture intensely and intimately how her own experience is situated next to other people's sensory experiences. In practical terms, this means registering how different tastes, smells, or sounds have an effect on the ethnographer as well as on research participants.

Among some of the concrete techniques devised for multisensory participation, there are simple things such as eating with, walking with, sensing with, as well as using material objects (for instance, sounds, foods, music, or video) to elicit sensory responses, as in Fig. 2 [12]. Some of these strategies have been combined as in the "walking with video" approach in which the

Fig. 2 Tasting an unknown species of sea urchin at a fisherman's boat and in the kitchen of a high-end restaurant. (Photos taken by the author/ethnographer. Image: Ana Maria Ulloa)

camera is used (either held by the ethnographer or the participant) "as a tool through which to explore informants' experiences of and engagements with the environment" [30].[1] The idea is that through these quotidian activities, like sharing a meal, ethnographers can learn empathetically about other people's experiences and how these experiences contribute to the perception and appropriation of places, new knowledge, products, or brands.

In addition, partaking in activities where it is possible to learn other people's craft and sensory skills is also a useful way for registering the senses at work. As Cristina Grasseni [33] suggests, participant observation can also involve a process of sensory apprenticeship. Some of these activities concerning food include sitting through training sessions [34–36], partaking in vocational classes [37], hanging out in bars and people's homes [31, 38], waiting tables in a restaurant [39], or lessons about the "taste" of Japanese cuisine [40]. Through these practices, ethnographers not only learn about other people's sensory experiences, but also about their identities, values, and beliefs and how something like food is valued.

---

[1] There are other related tools. First, what Pink [31] has termed a "video tour", a research strategy that consists of recording research participants as they show their homes and the things they do in them to a researcher or team of researchers. Second, Adams and Bruce's soundwalk method [32] is a method that rather than video uses sound and differs from the idea of walking with people along their own routes. Here the researchers predesign a route and try to stimulate people's hearing using different sounds along the way, so that later research participants can discuss these new experiences.

Other data collection strategies that can be useful for consumer research come from multisited ethnography. This approach became popular in the 1990s as a response to the deterritorialization of traditional anthropological sites and subjects [25]. As first conceptualized by Marcus [8], multisited ethnography foregrounds different strategies for following connections and relationships across spaces. Instead of selecting a single site for conducting fieldwork, this "mobile ethnography" holds a more constructive approach to fieldwork. This means studying multiple sites to try to understand the connections or associations that bind those sites together [41]. The ethnographic field is, thus, not a place to describe but a "bounded" network made of relations to map or make visible. And as with other ethnographic endeavors, the research site emerges out of the relationships established between researchers and participants.

For tracing connections, anthropologists have relied on different objects or subjects to "follow" as a device to structure fieldwork [8]. Instead of choosing a fieldsite, for instance, street vending cars or supermarkets, the ethnographer can trace where and how people, objects, or ideas travel. Under this approach, the notion of site is highly flexible and not bounded to physical frontiers. For instance, they can include virtual spaces, advertisement and promotional material, discourses, images, or flows of information and resources.

To follow the people means shadowing the movements of a particular group of initial subjects, for example, a family of migrants that arrive at a new community. This shadowing technique has been used in different settings and is, of course, something that participants need to agree upon. It is, in the end, a continuation of the classic anthropological technique of simply following people where and how they go about their business.

To follow the thing means tracing the circulations of objects like commodities, food, gifts, or works of art through different contexts. These different contexts can be equal to chains of production, distribution, and consumption or other more unexpected chains of discard and reuse. In tracing this movement, it is possible to see the shifting status or meanings of things as they "change hands" and belong or are used or produced by one group or the other. As Kopytoff and Appadurai [42, 43], have argued, things, like people, may be said to have biographies or a social life and go through various phases from the moment of their production to their consumption, destruction, or reuse. For this reason, an object that belongs to one person can also acquire different meanings depending on the phase is going through.

Finally, to follow an idea can mean multiple things. In the realm of consumer research, the most useful strategies might be those related to following a metaphor in discourse, a meaningful event in somebody's life or community and the memories associated with it, or how an object/subject figures in different practices. For example, Bettany and Daly's multisited ethnography of consumption attached to companion species/human relations traced how the Afghan hound figures as simultaneously heart hound, savage, artifact, and cyborg [44]. These figures were found initially by going through Web forums, Internet-discussion groups, breed magazines, and journals, and kept reappearing in the fieldwork carried out at breeders' homes and dog shows.

Multisited ethnography has fewer how-to guidelines in the literature. However, some challenges have been identified and should be considered when trying to accommodate this type of ethnography in a short-time consumer research [41]. First, the time consumed in accessing a site can become even more strenuous if there is a continuous need to negotiate access to multiple sites. Ethnographic research, in general, is known for its time and labor intensiveness. Second, given the greater flexibility of the approach and its open interpretation of a field, it can be hard to determine a priori the limits of the fields to follow. Third, and relatedly, sometimes it is not clear which trails to follow and which not, to not "drown in a sea of endless connections" [41]. Regardless of these challenges, multisited ethnography can offer new insights to consumer research that are worth further exploring and adapting.

3.2 Data Analysis

No matter the approach, ethnographers are rarely explicit about their analytic process. However, generally cultural analysis, as Sunderland and Denny argue, "is about going back through one's materials and thinking as hard and as long as possible about what has been said or done and what it all implies" [16]. While techniques for this going back and thinking hard remain somehow implicit, the idea is to work analytically as closely as possible to the data collected and reflections from the field. Typically, ethnographers work through their diary notes as well as through close transcriptions of their interviews. Some rely on coding as a systematic way of going through notes and interviews and produced "grounded" analyses tied to this corpus of original texts [45, 46].

There are different ways of coding qualitative data. Some consider this strategy as a final step after fieldwork is conducted and all information have been collected. This methodological strategy is typically found in research projects that use analytic modes suggested in grounded theory and software programs that support them, such as ATLAS.ti and NVivo [45]. Others have emphasized the importance of what Emerson et al. [46] referred to as "in-process analytic writing", suggesting that initial analytic or interpretative insights should be written actively in the fieldnotes while doing fieldwork. The more explicitly the ethnographer identifies themes, the better able he or she will be to put attention to observations or words that confirm, modify, or reject these interpretations that are a work in progress. Also, this approach can be helpful to identify whether some information or document is missing and should be looked after during fieldwork. In both cases, research moves back and forwards between inductive analysis and deductive analysis.

No matter the place of analysis and its actual mechanics (whether codes, asides, commentaries, or memos are used as annotations and how they relate to larger themes), ethnographers should use social and cultural theory "to improve understandings, to solve problems, to build more complex stories and to generate new questions" [1]. The process of analysis involves "moving reflexively between data exploration, interpretation, and conceptualization" [28], and for this, some familiarity with conceptual tools that address consumer research is needed.

Next to these general and more traditional ways of analyzing data, in the last decades, we have witnessed the proliferation of innovative ways to communicate findings as a result of multimodal research. Today, videos, presentations, performances, hyperlinks, recipes, and even email exchanges are all both materials to analyze and forms to represent ethnographic understandings. Photographs and video have become central in the work of consumer research [16], while other media have been less explored. For instance, asking participants to keep diaries (in form of video, audio, photo, or notes) have constituted a good strategy to lengthen the ethnographic inquiry given the usual time limitations in consumer research. However, the challenge remains in better integrating multimedia projects so as to provide robust cultural analysis on consumer practices [12].

4 Example: A Multisited and Sensory Ethnography of Flavor Production

There is a growing recognition of the need for practicing anthropologists for conducting consumer research inside food and beverage industries. Getting closer to consumers, listening to their voices, and learning from their experiences have constituted a powerful strategy for understanding consumer behavior and preferences. Valentin and Gomez-Corona offer a detailed example of conducting a short-term sensory ethnography in a couple of restaurants, shops, markets, and street food cars to explore a national cuisine such as Peru's in terms of flavors, pairings, ingredients, and habits of consumption [28]. However, ethnography is not only a valuable method for this type of consumer research.

Consumer research does not exist in a vacuum. It is part of a cultural context in which companies and corporations and the business practices supported by them define their scope, applications, and potential. Inside the industry, qualitative research typically needs to be fast, precise, and yield short-term results. In conducting this mode of research, anthropologists working for different types of companies have found themselves to be at the interface between organizations and consumers [47]. This middle zone has turned life inside an organization as relevant for consumer understanding as life outside it. Ethnographic methods can thus also offer a "mirror-effect" that reflects organizational practices, values, and expectations in relation to the consumer they seek to comprehend. The same methods can be used to study producers at one end or consumers at the other.

To offer an example that combines a sensory with a multisited approach to ethnography, I will describe in what follows my own research inside different types of organizations that are in the business of understanding and producing flavor. Although this was long-term research, taking up over 12 months of fieldwork, the strategies used can illustrate well what it takes to conduct a multisited ethnography of flavor as sensory experience. The rationale behind the study was to understand how expert practices and organizations turn flavor into an object of knowledge, commerce, and craft.

Soon, in the process of thinking about the project and noting the salience of flavor for different practices, I decided to follow flavor as an "object" in scientific, industrial, and gastronomic settings. Access is always a problem in business or organization research, and this was no exception. I started the project with a strong interest in accessing first the site that concentrates the most comprehensive knowledge about flavor: the flavor and fragrance industry—a multibillionaire global business dedicated to making aromas for inclusion in the most quotidian and mass-produced objects ranging from cleaning products, sodas, bread, or yogurts, to your most expensive and cherished bottle of perfume. Inside the industry, it would be possible to see in situ the negotiations around flavor and the expertise necessary to make it tangible, translatable, and objective. However, it was difficult to negotiate direct access and I had to restructure my fieldwork plan.

The industry does not stand alone in the making of flavors. It responds to a market driven by large food companies that are engaged in the constant struggle to keep on top of new trends, it maintains a close relationship with basic taste and smell science, and it reaches all types of food businesses, including high-end restaurants, for inspiration. Elucidating different facets of flavor as an object of knowledge and technology was a first impulse for this project and hence its multisited nature. Thus, I looked at the connections between science, industry, and gastronomy and noticed that their shared interests in the empirical world of cooking, eating, and drinking have pushed for the cross-fertilization of people, methods, instruments, and approaches.

It was by tracing online some of the relationships that the flavor and fragrance industry sustain with other organizations that I arrived at the site that would become my point of entry to the professional world of tastes, smells, and flavors: the Monell Chemical Senses Center. Monell is one of the most productive and well-known institutes worldwide that is devoted exclusively to the sciences of taste, smell, and chemosensory irritation. The center began in 1968 as a joint venture involving the academy, industry, and government in a time when such collaborations were looked down upon. But from then on, Monell's infrastructure and collaborations with industry have only grown, turning it into a hybrid place where sensory science (an industrial endeavor by principle) and basic science meet. Through Monell, I was in a better position to reach and continue my research in the flavor and fragrance industry.

At Monell, I did not exclusively follow the work done at any one division but was rather in contact with people from different laboratories. I spent my time in between experiments, talks, conferences, workshops, sponsor meetings, outreach events, lab meetings, interviews, and informal talks. I did not have an opportunity to follow the execution of a single project, as they span the course of several months, sometimes years, but I made sure I could sit and serve as a volunteer in as many taste and smell experiments as possible.

Contrary to my involvement with Monell—structured, self-contained, and fluid according to the center's tempo and the generosity of its members—my fieldwork in the flavor industry was more difficult to secure. I ended up visiting many different flavor companies, large and small, in the US and Spain, at different times and under different conditions. After several access negotiations, I was approved to pay daily visits to a middle-size flavor company in Philadelphia. However, if I were allowed on their premises it would have to be under nondisclosure conditions that assured that no intellectual property of the company would be revealed in the research.

There, I observed and participated in the activities of different departments: flavor creation, culinary development, sensory testing, and consumer research. Over the course of 2 months, I paid weekly visits to this Philadelphia company and got to volunteer in a major event that they hold every 2 years, a showcase of innovative products and flavors organized by the company for its clients. These experiences were an excellent opportunity to familiarize myself with the industry and the conjoined work behind the production of natural and artificial flavorings. But the relative ease with which I was able to pay such visits would later prove exceptional when I tried the same approach with larger flavor and fragrance companies.

After numerous failed attempts at contacting large flavor and fragrance companies, I finally got people working for one of these companies interested in the project and willing to grant me access to their work premises in Spain. There, I followed the flavor creation team for 3 months, spending full days at the company twice a week. On top of my learning from the daily activities (meetings, tastings, formulations, calls, mentoring sessions, etc.), I also contributed to the development of a small project where an anthropologist like myself could be of help.

Finally, my third main site was the research space of El Celler de Can Roca, a Spanish high-end restaurant. The restaurant opened in Girona, the hometown of the owners, the Roca brothers, in 1986. It has held three Michelin stars since 2009 and has been listed in The World's 50 Best Restaurants by the British magazine *Restaurant* since 2006, occupying first place in 2013 and 2015. I was able to secure an internship at the restaurant and participate in the life of La Masia—the house in front of the restaurant they have devoted to research and development. As with the flavor and fragrance company, I went to the restaurant twice a week for a period of 3 months. Actually, it was by moving back and forth between the company and El Celler that the esthetic aspect of flavor creation gained prominence in my research, as it was in these two places that I witnessed close interactions between practitioners and the objects of their craft: observing kitchen work, going out to collect foodstuffs, sharing meals, conducting experiments with the restaurant's staff, applying powder flavors to snacks, smelling and tasting alongside flavorists, and talking repeatedly with them, as well as with cooks, waitresses, and sommeliers, as shown in Fig. 3.

Fig. 3 Kitchen staff during service. (Photo taken by the author. Image: Ana Maria Ulloa)

At the restaurant, my focus was on the things that were being done at La Masia, where different people develop projects associated with the restaurant such as cookbooks, classes, or fermentation and distillation processes. As in the industry, I developed a small project to ensure contact with the restaurant's personnel. This was a strategy that was born in the field as to compensate for not having access to the restaurant's kitchen. With the restaurant's project manager, we developed an experiment to evaluate and describe the sensory abilities of the restaurant's team [48]. For this, I was inspired by both a sensory ethnographic approach and some of the knowledge I have learned at Monell, the research center on taste and smell.

The boundaries between science, industry, and gastronomy are ever more flexible and it is precisely through understanding more about their relationships, as I started unraveling them through fieldwork, that one can capture the multidimensionality of flavor, as an object of expertise, in practice. What were initial contacts became important interlocutors, and through our sustained conversations I was able to provide a cultural analysis about the role of science and technology in shaping flavor as an object of knowledge, craft, and commerce.

Taking a multisited approach to the study of flavor production guaranteed tracing unexpected connections by placing the scientific next to connoisseurship, standardization next to craftsmanship, and the industrial next to the esthetic. By closely studying the sensory skills of scientists, flavorists, technicians, chefs, and sommeliers, it became clear that the food industry also requires of an esthetic disposition toward food as a medium of communication and cultural meaning.

5 Advice for New Ethnographers

Having exemplified the conduct of a multisited fieldwork and sensory-oriented ethnography, I will then sum up some of the basic take away messages of this chapter and turn to some useful advice for people interested in carrying an ethnographic approach to corporate consumer research.

Remember there is no recipe or strict rules to follow.

Understand and reinforce the difference between ethnographic research and its commitment to context and other qualitative methods widespread in consumer research (surveys, semi/structured interviews, focus groups, etc.).

Be prepared for anxieties raised and rejections incurred in conducting ethnographies.

Do not try to determine the ethnographic situation you find yourself in. Contrary to laboratory methods, ethnographers do not want to control what happens in their field situation. An ethnography is not replicable in a laboratory sense.

Familiarize yourself with the place you are carrying research at, and get as early as possible a first-hand feel for how life or a particular practice (shopping, eating, cooking) is for others.

Try to capture multiple voices. While the figure of informants is of use to ethnography, it is almost always desirable to talk to different people in different situations.

Register dutifully and carefully what you observe as you participate in the field. Even the simplest of things or that thing you think is too obvious for you to forget. Ethnographic work requires discipline, attentiveness, and systematicity in keeping a diary.

Make sure your notes speak of a wider context. Do not register data from interviews or other material as divorced from contexts in which they are produced or used [3].

Don't wait until the end to make sense of your data. Try to actively and constantly make meaning from it. Carefully register these annotations and interpretations next to your fieldnotes.

Let the field surprise you. Don't come with too many preconceptions or use ethnography as a hypothesis-driven enterprise.

If conducting multisited research, *remember* to trace connections between places and actors from the beginning. Open-ended questions lead to unexpected conclusions.

Regardless of whether you follow a sensory ethnography approach or not, *keep in mind* that "fieldwork is necessarily an embodied activity" [49].

Consider that ethnographic research and thinking are not linear processes. Instead, these are processes that move in loops, can have stops and starts, and require constant fueling.

Account for the ethical issues that are raised by specific cultural contexts and the culturally and personally specific moralities of research participants. Guarantee discretion and protection of personal and sensitive information.

Don't let the pace compromise your reflexivity. If you need to conduct short-time research, make sure you reflect on your position an your relationship with research participants. Promote, as much as possible, reciprocity in your research.

References

1. Madden R (2017) Being ethnographic, 2nd edn. Sage, London
2. Boellstorff T (2008) Coming of age in second life: an anthropologist explores the virtually human. Princeton University Press, Princeton, p 4
3. Atkinson P, Coffey A, Delamont S, Lofland J, Lofland L (eds) (2001) Handbook of ethnography. Sage, London
4. Restrepo E (2016) Etnografía: alcances, técnicas y éticas. Envión editores, Bogotá
5. Jackson A (ed) (1987) Anthropology at home: ASA monographs 25. Tavistock, London
6. Madden R (1999) Home town anthropology. Aust J Anthropol 10(3):259–270
7. Ocejo R (ed) (2013) Ethnography and the city: readings on doing urban fieldwork. Routledge, New York
8. Marcus G (1995) Ethnography in/of the world system: the emergence of multi-sited ethnography. Annu Rev Anthropol 24(1): 95–117

9. Ginsburg F, Abu-Lughod L, Larkin B (eds) (2002) Media worlds: anthropology on new terrain. University of California Press, Berkeley
10. Miller D, Slater D (2001) The internet: an ethnographic approach. Berg, Oxford
11. de Waal MT (2009) Understanding the rise of consumer ethnography: branding techno-methodologies in the new economy. Am Anthropol 111(2):201–210
12. Pink S (2009) Doing sensory ethnography. Sage, London
13. McGranahan C (2018) Ethnography. In: Callan H (ed) The international encyclopedia of anthropology. Wiley
14. Ortner S (2006) Anthropology and social theory: culture, power, and the acting subject. Duke University Press, Durham
15. Mariampolski H (2006) Ethnography for marketers: a guide to consumer immersion. Sage, London, p 7
16. Sunderland P, Denny R (2007) Doing anthropology on consumer research. Left Coast Press, Walnut Creek
17. Cardoso de Oliveira R (2004) El trabajo del antropólogo: mirar, escuchar, escribir. Avá Revista de Antropología 5:1
18. Walcott D (2008) Ethnography: a way of seeing, 2nd edn. Altamira Press, Lanham
19. Haraway D (1988) Situated knowledges: the science question in feminism and the privilege of partial perspective. Fem Stud 14(3): 575–599
20. Geertz C (1973) The interpretation of cultures. Basic Books, New York, p 6, 21
21. Hasbrouck J (2018) Ethnographic thinking: from method to mindset. Routledge Press, New York
22. Belk R (2014) The art of using ethnography. Int J Mark Res 56(4):551
23. Sherry J (1995) Marketing and consumer behavior: an anthropological sourcebook. Sage, London
24. Belk R, Fischer E, Kozinets RV (2012) Qualitative consumer and marketing research. Sage, London
25. Boyer D, Marcus GE (eds) (2021) Collaborative anthropology today: a collection of exceptions. Cornell University Press, Ithaca
26. Atkison P (1992) Understanding ethnographic texts. Sage, London, p 17
27. Pink S (2005) Dirty laundry: everyday practice, sensory engagement and the constitution of identity. Soc Anthropol 13(3):275–290
28. Valentin D, Gomez-Corona C (2018) Using ethnography in consumer research. In: Methods in consumer research, vol 1. Woodhead Publishing
29. Stoller P (1994) Embodying colonial memories. Am Anthropol 96:634–648
30. Pink S (2007) Walking with video. Vis Stud 22(3):240–252, p 245
31. Pink S (2004) Home truths: gender, domestic objects and everyday life. Berg, Oxford
32. Adams M, Bruce N (2008) Soundwalking as methodology for understanding soundscapes. Proc Inst Acoust 30(2):552–558
33. Grasseni C (2004) Video and ethnographic knowledge. In: Pink S, Kürti L, Alfonso A (eds) Working images: visual research and representation in ethnography. Routledge, London
34. Smith V (1990) Managing in the corporate interest: control and resistance in an American Bank. University of California Press, Berkeley
35. Pierce J (1995) Gender trials: emotional lives in contemporary law firms. University of California Press, Berkeley
36. Butler E (2018) Tasting off-flavors: food science, sensory knowledge and the consumer sensorium. Sens Soc 13(1):75–88
37. Fine G (1996) Justifying work: occupational rhetorics as resources in restaurant kitchens. Adm Sci Q 41(1):90–115
38. Wells M (1996) Strawberry fields: politics, class, and work in California agriculture. Cornell University Press, Ithaca
39. Paule G (1991) Dishing it out: power and resistance among waitresses in a New Jersey restaurant. Temple University Press, Philadelphia
40. de St. Maurice G (2018) Making the multidimensional taste of Japanese cuisine public. In: Counihan C, Højlund S (eds) Making taste public: ethnographies of food and the senses. Mloomsbury Academic, London
41. Van Duijn S (2020) Everywhere and nowhere at once: the challenges of following in multisited ethnography. J Organ Ethnogr 9(3): 281–294
42. Appadurai A (ed) (1986) The social life of things: commodities in cultural perspective. Cambridge University Press, Cambridge
43. Kopytoff I (1986) The cultural biography of things: commodification as a process. In: Appadurai (ed) The social life of things, vol 5, Cambridge, pp 64–94
44. Bettany S, Daly R (2008) Figuring companion-species consumption: a multi-site ethnography of the post-canine Afghan hound. J Bus Res 61(5):408–418

45. Glaser B, Strauss A (1967) The discovery of grounded theory: strategies for qualitative research. Aldine, Chigaco
46. Emerson R, Fretz R, Shaw L (1995) Writing ethnographic fieldnotes. University of Chicago Press, Chicago
47. Hirsch E, Gellner D (2001) Introduction: ethnography of organizations and organizations of ethnography. In: Gellner DN, Hirsch E (eds) Inside organizations: anthropologists at work. Berg, Oxford
48. Ulloa A, Roca J, Vilaseca H (2017) From sensory capacities to sensible skills: experimenting with El Celler de Can Roca. Gastronomica 17(2):26–38
49. Coffey A (1999) The ethnographic self: fieldwork and the representation of identity. Sage, London, p 59

Chapter 4

Consumer Research Through Cookbooks and Cooking Shows: Linguistics, Cultural Studies, and Media Studies

Keri Matwick and Kelsi Matwick

Abstract

This chapter provides tools on how to conduct consumer research through cookbooks and cooking shows from a linguistics perspective informed by cultural and media studies. Three methodological approaches: multimodal discourse analysis, narrative studies, and critical discourse analysis are described and related to consumer research. Cookbooks and television cooking shows influence consumers' lifestyle habits and food choices from—what and who should cook, why, when, and how—to what comprises high/low cuisine, everyday/luxury, local/global, and traditional/modern. Understanding the sources and reflecting on these values give insight to consumer research. To succeed in the market, a product has to function well, taste good, or promise to fulfill an aspiration. Consumer research seeks a deep understanding of the needs and desires of consumers, an understanding that is guided by a humanities approach as outlined in this chapter.

Key words Food discourse, Consumer research, Cookbooks, Cooking shows, Qualitative research methods, Food media, Cultural studies, Food and language

1 Introduction

Global appetite for cookbooks and cooking shows accelerates with the coronavirus pandemic stoking print cookbook sales and production of one of television's fastest growing subgenres. Consumers hunger for food and food products that satisfy physiological, emotional, and psychological fulfillment. As the pandemic roars, restaurants close and quarantine periods lengthen, forcing people to cook at home. Many turn to baking and cooking as a way to pass the time. Cookbook sales on baking and bread, two traditionally emotion-laden foods, swell in the United States, growing 145% in 2020 [1]. Individuals make food choices every day based on convenience, food flavor, personal preference, health, experiences, ethics, and background. These decisions include what and how much to eat, with whom, and when.

Carlos Gómez-Corona and Heber Rodrigues (eds.), *Consumer Research Methods in Food Science*,
Methods and Protocols in Food Science, https://doi.org/10.1007/978-1-0716-3000-6_4,

Cookbooks and cooking shows are among the food-centered discourses that have been investigated across the disciplines. For example, fields from linguistic anthropology [2, 3], communication [4, 5], and linguistics [6, 7] have examined how food and language are a social and cultural practice. Despite the continued consumption of cookbooks and food on television, there remains a sparse body of consumer research on how these food texts relate to food products and services and consumer responses. An in-depth understanding about customers can help organizations create and promote products based on customer research, thus producing increased purchases.

Cookbooks and television cooking shows offer insight to consumer research, with food and language serving as a way to share, interpret, and influence food choices. Throughout history, cookbooks guide readers in procurement, preparation, and serving of food. Documenting and transmitting knowledge about food, from recipes and cooking methods to lifestyle, are also through television cooking shows. With the global interest in food, food on television is a mainstay, with programs devoted exclusively to food and cooking shows, nationally and internationally, reaching millions of viewers, such as *Food Network* and *Cooking Channel* in the U.S., *Lifestyle Food* in Australia, and *BBC Food* in the UK. More broadly, food proliferates across media, from video on-demand services of Netflix, online video sharing with YouTube, and social media, such as TikTok and Instagram. Educating and entertaining about food and lifestyle, cookbooks, and television cooking shows are sources of authority around food and consequently play a determining role in individuals' perception about food.

Examining best-selling categories are ways for consumer researchers to identify what is trendy or who is the most influential at the time. For example, Fig. 1 on the best-selling books in the category of "cooking" in the U.S. (2017) illustrates latest cooking equipment, i.e., pressure cooker, Instant Pot, Air Fryer, and diets like Keto are among the top food trends. Food blogger and celebrity chef Ree Drummond (*The Pioneer Woman*) and lifestyle megastar Oprah Winfrey frequently rank as top cookbook authors, as they do in 2017.

Besides cookbooks, food social media accounts are also excellent sources of consumer information on food and beverage. Instagram, a photo and video sharing app, indicates food as the most tagged and shared category. Figure 2 shows the most-followed food influencers on Instagram worldwide as of June 2020. British celebrity chefs Gordon Ramsay and Jordon Oliver dominate the top two spots followed by Americans Ree Drummond and Martha Stewart and Canadian baker Yolanda Gampp.

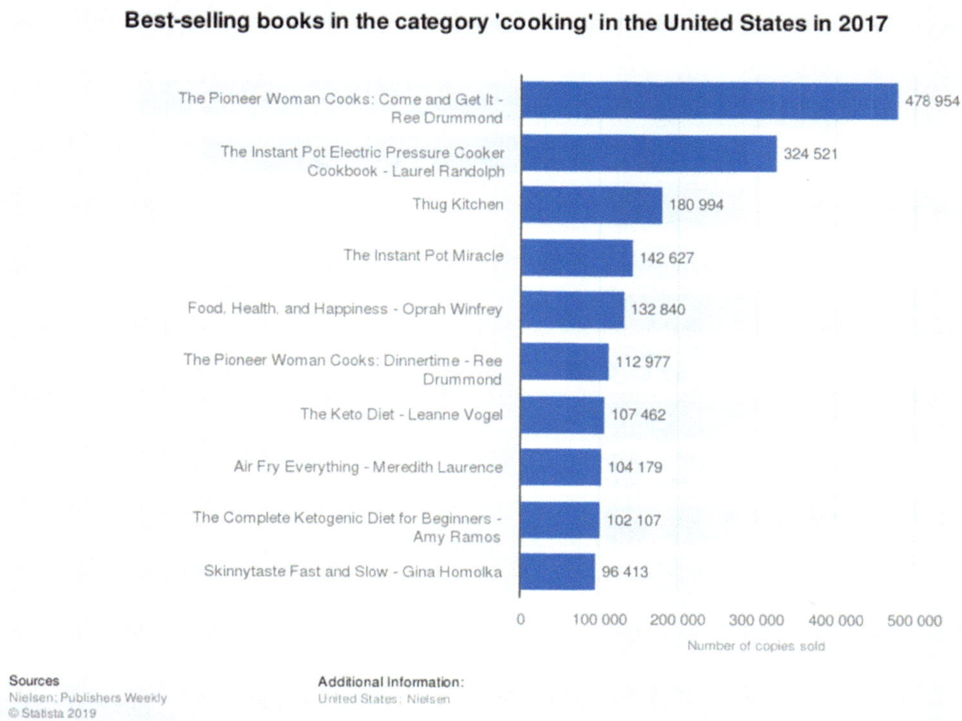

Fig. 1 Best-selling books in the "cooking" category in the United States in 2017

With these statistics, researchers can better understand the numerical differences across food communication channels. For example, the younger generation is a higher user of newer social media platforms (i.e., Instagram, TikTok) while the older generation favors Facebook. Depending on the target audience, researchers can accordingly select the social media platform most used and examine the language, visuals, and values displayed.

Studying the relationship between food and language increases understanding in consumer research of consumer values. Our consumption of food, from satiation for the body to nourishment for the mind, occurs and revolves around food. Given the complexity of food as a discursive phenomenon, the study of food and discourse requires attention to cultural influences that favor certain ways of eating and of negotiation in both connection and power, and the multiple interrelations among food, discourse, and social meaning.

Understanding how and why the production and reception of cookbooks and cooking shows is appealing to consumer research, because food is a social and semiotic fact and language plays a key role in its construction, definition, circulation, and consumption. Moreover, the act of eating and drinking is central for human

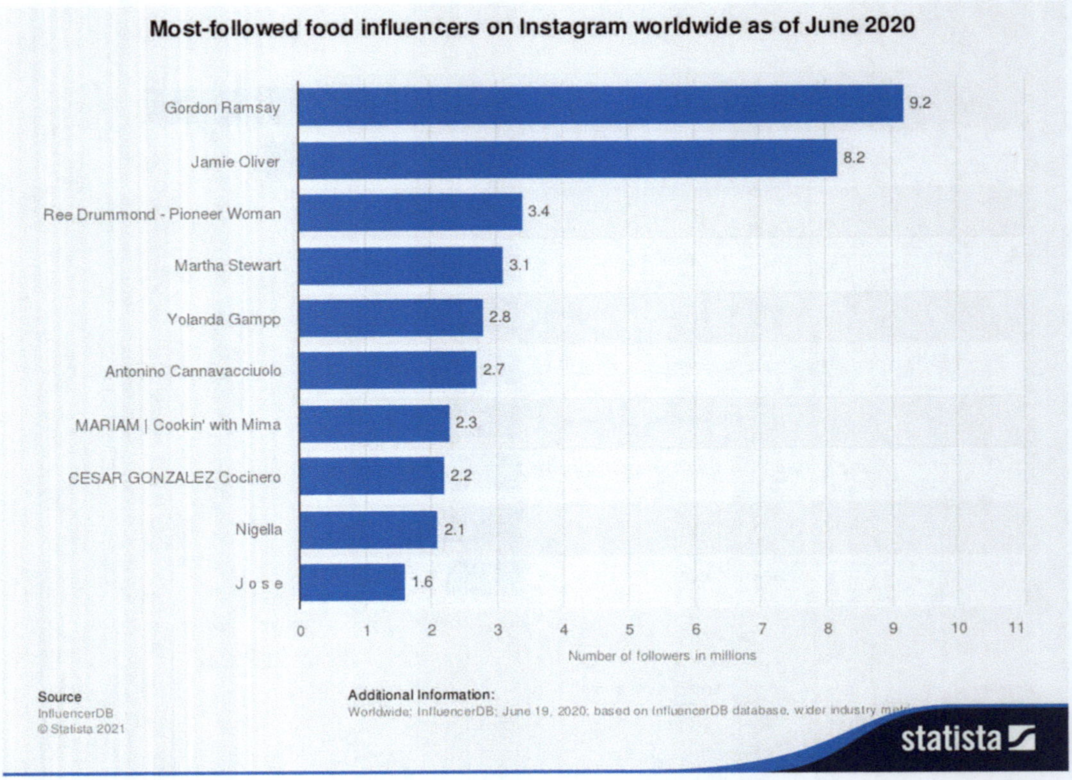

Fig. 2 Most-followed food influencers on Instagram, worldwide, June 2020

behavior, whether for physiological needs, pleasure, expression, or community. These acts, seemingly simple, are the result of a complex phenomenon of choice which is influenced by a set of factors of nature (e.g., social, psychological, biological, and more) that can be studied independently or taking into consideration their interactions. It seems that discussion about the relationship among food, language, and consumerism is still neglected in our daily lives, including empirical work involving food, language, and consumer perceptions. Thus, the aim of the present review is to provide an overview of the advances of food studies and an indication of the direction of future research for consumer studies.

2 Laying the Foundation: Food and Taste

Language and food intersect and co-occur and are dynamically co-constitutive. Banal events such as cooking dinner, watching cooking shows, shopping for ingredients, and listening to food radio involve and are shaped by language. How food and language are choreographed communicate values, identity, social relations, and more. Eating organic mixed green salad conveys values of fresh and nutritious foods but also status and higher-economic standing. Brillat-Savarin [8], one of the earliest food writers, claimed, "Tell me what you eat, I will tell you what you are." Linguistics and social scientists have shown that identity is constructed through language. Food has both cultural social and cultural functions [9] and its own unique language use.

Both food and language are used to signify identity, who people are, how they see themselves, and how they wish to portray themselves to others. These attributes are not necessarily innate but can be earned. In the field of sociology, the social assets of a person are known as capital, a concept largely made famous by Pierre Bourdieu [10]. In his seminal work *Distinction: A Social Critique of the Judgement of Taste*, Bourdieu describes three kinds of capital: economic capital (what you have; e.g., financial wealth), cultural capital (what you know; e.g., education, skills, knowledge), and social capital (who you know; e.g., membership in formal/informal groups; relationships with friends, coworkers). All the capitals are linked together:

> The volume of the social capital possessed by a given agent thus depends on the size of the network of connections he can effectively mobilize and on the volume of the capital (economic, cultural or symbolic) possessed in his own right by each of those to whom he is connected [10].

Bourdieu's definition suggests that one's total capital can be quantified in terms of the volume of its combined capitals. Key to Bourdieu's argument is the importance of everyday cultural forms like music, art, and food in creating and upholding one's social status. In accessing these interconnected forms of capital, individuals can acquire social esteem and power.

Bourdieu's cultural capital in particular has inspired numerous studies of cultural analyses, including ones applied to food. Assuming that values are not fixed, Naccarato and LeBesco [11, p. 4] use the term "culinary capital" to "understand how and why certain foods and food-related practices connote, and by extension, confer status and power on those who know about and enjoy them." Celebrity chefs for instance are influential in ascribing value to certain foods and food items. "The Delia Effect" was coined in the 1970s when British celebrity chef Delia Smith recommended a lemon zester and shortly after the show, there was a shortage of lemon zesters. However, product placement on food programs does not necessarily transfer successfully across international programming. As Bonner [12, p. 42] notes, UK celebrity chef Jamie Oliver's commercial relationship with Sainsbury does not draw extra attention to Australian viewers who can neither buy the produce much less go to the supermarket. Locally made programming is more successful in the products embedded on the show because they are more readable by viewers and accessible to purchase.

Like celebrity chefs, food writers also have influence over food trends. Studying three gourmet food magazines, Baumann and Johnston [13, p. 62] argue that the articles do much more than offer didactic lessons on how to make dinner; "rather than instructing readers about what to do, these implicit instructions teach them how to be. By suggestion, these preferences and dispositions are held by a person sophisticated enough to enjoy gourmet food." The lavish prose and sumptuous photos prescribe the quest for getting food from the source (e.g., truffles from Alba) or dining at a Michelin-starred restaurant (e.g., Le Bernardin). Expanding their idea of a gourmet food discourse, Baumann and Johnston [13] write that foodies negotiate two ideologies: democracy and distinction. Authenticity is valued at the one end while exoticism on the other end; mediating this is the desire for meritocracy but both authenticity and exoticism require cultural and/or economic capital. Cookbooks and cooking shows are vehicles of this capital.

Consumers can acquire more capital by reading cookbooks and food magazines and watching cooking shows. The act of decoding or the ability to "see" the knowledge in these discourses indicates capital. Informed, one makes food choices, creating boundaries between those that have and do not have. This "culinary differentiation" [14, p. 2] is linked to consumer motivation and purchase behavior. Food and television are also consumed, as Sarah Murray [15, p. 188] notes, adding that "Identity and consumption are subsequently linked to class-based notions of taste hierarchies, cultural distinction, and good consumer citizenship through lifestyle-based food television." Through their choice of ingredients, equipment, kitchen decor, menu, and occasion of the meal, celebrity chefs as "cultural intermediaries" provide models on how to acquire culinary capital.

Research on cookbooks and cooking shows revolves around an objective to identify the social, cultural, and political implications of food. Researchers explore how and why food is deployed discursively in written and oral form, and more specifically, how food is linked to class, race, and gender in cookbooks and on television [16–18]. The way food is prepared and who is in the kitchen for example usually reflect a society's gender roles. The home kitchen traditionally is run by females, while the professional restaurant kitchen is dominated by males, resulting in a dichotomy of private-domestic-feminine-cook and public-professional-masculine-chef. Yet, cases such as Jamie Oliver, whose images in his cookbooks are highly feminized [16], challenge the rigidity of these gendered divisions. As manuals for cooking, cookbooks and cooking shows present guides for eating and living. They are not simply a collection of recipes or for entertainment but are a transformative device in telling readers or viewers what is the preferred lifestyle.

Discussion of capital and distinction leads to a broader body of work on class, consumption, and citizenship. Culinary education from cookbooks and cooking shows consists of how to purchase, cook, and consume as a foodie. Literature makes a connection between food and lifestyle television, suggesting that food television contributes to the "trends in the way we understand ourselves as citizen and consumer" [19, p. 14]. Food television within the lifestyle television frame focuses on the makeover, encouraging individual change to align with society in a capitalistic economy. "Adopting particular eating habits, recipes, and food-related lifestyle becomes the key to social mobility, stability, and civic empowerment" [20, p. 17].

Furthering research on food and consumer culture, De Solier [5] explores how foodies negotiate the morality of making a self through consumption. Identity formation is linked to fine dining and shopping—two consumption practices that are central to foodie lifestyles, which, as [5] argues, have different moral registers within each practice. According to [5, p. 24], on the one hand, foodie consumption in fine dining restaurants places higher value on moral ascetism, which restricts consumption and spending money; on the other hand, foodies are governed by a "morality of quality" and are willing to spend more money, perceiving expensive food not only as "right." How moralities of consumption differ in public spaces versus private spaces and from a range of moral aspects of cognitive decision-making would be fruitful [5, 21]. More discussion on the moral dimensions of consumption, including the study of ethical consumption, fair trade, and sustainability, can offer a helpful approach for better understanding the underlying nature of consumption practices.

3 Consumer Research Methods: Food, Language, and Media

Various consumer research methods can approach food, language, and media through cooking shows and cookbooks. This section focuses on *multimodal discourse analysis, narrative studies, and critical discourse analysis.*

The following Table 1 Linguistic Approaches to Consumer Research Methods on Food provides an overview of three possible research methods, their key points, advantages, disadvantages, important theoretical pieces, and sample applications. These methods are then discussed in detail.

3.1 Multimodal Discourse Analysis

Cookbooks and cooking shows relay information multimodally, or conveying meaning in more than one form. Cookbooks rely on written text and images, while cooking shows also use audio and gestures. Research on multiple modes is motivated to determine how and in what way food is deployed discursively and formally in

Table 1
Linguistic approaches to consumer research methods on food

Research method	Key points	Advantages	Disadvantages	Theoretical works	Example Sample use in food texts
Multimodal discourse analysis	Considers multiple modes of communication as part of the meaning-making; modes consist of visual, audio, spatial, gestural, textual	Applicable to traditional text, i.e., cookbooks, and new forms of food text, i.e., Instagram, Facebook	Analysis tends to be tabular, which is laborious and tedious; tends toward generalizations, especially of limited samples	Kress and Van Leeuwen [22]; Kress [23]	Studies the color, text, image, of a food advertisement to see how they are put in a "syntax" to produce a message
Narrative studies	Studies the story or a series of related events and/or experiences	Applicable across texts; interdisciplinary; pieces together segmented parts to reveal a deeper understanding of a situation	Setting boundaries to the beginning/end of a story may be difficult; personal stories as recounted by individuals may be unreliable	Barthes [24]; Labov [25]; De Fina et al. [26]	Examines the story in the introduction of a cookbook or recipe and what identity is constructed
Critical discourse analysis	Takes a critical approach to discourse produced within specific social contexts; aims to promote social change	Interdisciplinary, flexible approach; aims to uncover power and structural dynamics	Researcher may be predetermined to find a particular outcome	Fairclough and Wodak [27]; Fairclough [28]	Makes transparent the way texts attempt to construct official knowledge, such as how food texts can promote certain values, e.g., vegetarianism, plant-based foods, keto diets

print and on the screen. Multimodal discourse analysis is typically qualitative and can be used by consumer research to examine how language is used across various formats, genres, and spaces to make meaning.

Media has changed the way in which text is presented, offering an enhanced multimodal text with sound, images, and body language. Kress and Van Leeuwen [22, p. 177] define multimodal texts as "any text whose meanings are realized through more than one semiotic mode." The study of multimodal text, or how meaning is communicated in more than one way, has been largely situated within the sociocultural context. Mode is defined by Kress [29, p. 5] as, "culturally and socially produced resources for representation." The mode is analyzed in its form and in context with other modes. For example, written text may appear on the cooking show screen with a list of ingredients and culinary tips that accompany the spoken discourse of the show host. Cookbooks like Ree Drummond's *The Pioneer Woman* series offer step-by-step photos along with the instructions.

Multimodal discourse analysis can be used to show how discourses in cookbooks and cooking shows are structured to convey meaning. The connection between the mode's form and its meaning, specific to its social environment, is called "social semiotics" [30]. Five semiotic systems constitute design elements in the meaning-making process: (1) language, both oral and written, such as words, grammar, paragraphs, (2) visual, both still and moving images, and includes aspects such as viewpoint, color, page layouts, screen formats, (3) auditory, both music and sound effects, such as rhythm and volume, (4) gestural, both facial expressions and body language, including speed of movement and stillness, and (5) spatial, such as the layout organization of objects and space, including the direction and position. Studying more than one semiotic system recognizes its interrelationship with the other semiotic systems. With this knowledge of the various formats, consumer research can refine its audience and product targeting.

Importantly, it is the idea of visual consumption, rather than the actual consumption, that drives peoples' consumption practices. Close-up, tantalizing images of food, or "food porn," is increasingly normalized across food texts, from cookbooks to cooking shows to social media platforms. When food porn or "gastro porn" was first used by Signe Rousseau in 1971, it described eroticized food imagery depicting desires of fantasy and the unattainable [31]. Food porn can be described as a site of excess, an overly sexualized visual aesthetic. Suggesting that food porn can be more than the erotic, Cheri Ketchum [32, p. 217] argues that the Food Network cooking channel "offers the possibility of pleasure through creating the fantasy of an intimate connection to viewers and the promise of satisfaction through consumption." Visual imagery in cooking shows and cookbooks can evoke voyeurism, but it can also motivate a consumer to take action. In a panel held by McBride [33, p. 46], Alan Madison, television producer of food shows, claims that "food porn in itself cannot sate desire; it must inspire to action." In this respect, looking at highly stylized food should prompt one to go into the kitchen and cook or buy the product.

The evoking of desires through sensual food imagery has become the norm, so consumer research needs to take this in to account when designing a campaign or launching a product. Design and styling in food photography can be used to build evocative spaces and compelling narratives of the food or product. Elaborate plating, manipulating, and embellishing food through glazing and dressing, and post-production tweaks are among the artistic strategies that have been historically used and still continue in contemporary food commercial images [34]. Across four stylistic frameworks: innovation, entertainment, mastery, and authenticity, Taylor and Keating [34, p. 308] identify the use of "creative disruption," a term used to describe art techniques that intentionally challenge the expectations of an audience, and how it effectively creates a sense of intimacy and draws the gaze of the viewer. Originally attributed only to food porn, visual tropes of indulgence, desire, fantasy, and sensuality are prevalent across popular food imagery.

Multimodal discourse analysis also considers the materiality of the world, allowing for concrete studies of how we move around spaces [35]. For consumer research, questions can be made about the design, materials, and aesthetics of cookbooks and the space created on the set of cooking shows.

- For cooking shows, what effect does lighting/music/decor have in creating a specific mood on set? How can variation of close, middle, and far shots be indicators of power? How do the various camera angles position the celebrity chef and the viewer at home?
- For cookbooks, what materials are used in the printing? i.e., matte, silk, or gloss finish, uncoated or coated, smooth or vellum finish? What font is used? Size? Are there photos per recipe? Is it digital? How does this affect the intended use of the cookbook?

These questions help answer how these choices reflect the personality of the chef and his/her culinary style. For instance, serif, script, and handwritten typefaces give an elegant look, appealing to customers wanting luxury. Bubbly, wobbly fonts are playful and childish. *Food Network* celebrity chef Ina Garten, beloved for her no-nonsense, dependable recipes, always uses the same lowercase font (Modern No. 216) for her cookbook titles [36]. One of her latest *Modern Comfort Food* [2020] provided the answer during the pandemic as many turned to cooking at home as a coping mechanism. Studying the brand of another *Food Network* star Guy Fieri, Contois [37, p. 147] describes the symbolism behind his use of tattoos: "While once nonconformist symbols of rebellion, Fieri's tattoos—on his body, printed in his cookbook, and branded into his cookware—become ambivalent representations of both working-class resistance and corporate capitalism, of self-expression and mass production." Fieri's brand embraces values of freedom and abundance as well as nostalgia for America, reflecting the recent rise of populist sentiment in the United States. Understanding how celebrity chefs and their products influence consumers are important aspects for consumer research.

What mode best engages the consumer and how the product is intended to be "read" are to be considered when designing a multimodal approach to consumer research study.

3.2 Narrative Studies

Studies of narrative, or the structure of the story, can be applied to cookbooks and cooking shows [38]. We use narrative structures for a wide variety of purposes: to recall events, order and organize the past, explain human action, understand ourselves, create expectations about the future, and provide scripts that tell us how to act with others [39]. Narratives are a set of expectations, or norms, which set the standard as ordinary and as appropriate. At the same time, narrative gives the possibility of deviations from these norms. Cookbooks and cooking shows provide models or "social scripts" that create expectations in our own memories and ways to behave [39]. When understanding narrative ideas, consumer researchers can recognize narrative structure and thus see the norms of the consumers' world recreated on the page and screen.

Within narrative studies, functional approaches identify the components that make up a story. Four elements or parameters: character, action, prop, and setting are generally the basic analytical used for narrative studies, such as [40–42]. The elements are applicable for functional analysis of media texts, such as cooking shows. The character is the cooking show host appearing in the narrative of the cooking show season. The action involves cooking and interacting with guests on the show or with viewers via the camera. Props include cooking ware and ingredients. Setting is the location or where the plot occurs, such as at the Hamptons in New York. Food shopping and procuring ingredients from the garden may also be considered action in the narrative while establishing the setting.

Cooking shows have varied settings, from the home kitchen to the supermarket to the community. Setting is the determining feature of a type of cooking show, food travelogues, which are programs in which a chef travels around a country and promote its food. Shows such as Jamie Oliver's *Jamie's Great Britain* act as gastrodiplomatic texts, though foods selected to represent the nation are not unproblematic [43]. Shows such as *Nigella Bites* featuring Nigella Lawson are set in the home kitchen yet transport viewers to "that fantasyland where the impossible is made possible—a land in which it is possible to not only *have your cake and eat it*—but a land in which we are drawn into a particular version of feminine identity, a branded subject..." [44, p. 486]. The *Nigella* brand includes cookware, from red measuring cups to cookie cutters to an all-purpose pot, conveying a "swanky" and functional kitchen [ibid], a land of taste, and distinction [10]. In other cooking shows, ones in which food preparation occurs, the setting may be predominantly used as a way of characterization; for example, a kitchen stadium is used in competition cooking shows like *Iron Chef*, *Iron Chef America*, and *Beat Bobby Flay*; a cattle ranch sets the background for the instructional cooking show *The Pioneer Woman*, appealing to the romantic wild west.

Stories must have characters to play the role within the narrative. As the heroes, or the person on a quest to solve the problem [40], cookbook authors and cooking show hosts try to solve the problem of their audience, such as what to eat [45]. Offering recipes is a way to resolve this tension. Advocating certain labeled and certified food products is another. Examining the media campaigns of British celebrity chefs Jamie Oliver and Hugh Fearnley-Whittingstall, Phillipov and Gale [46] note the increased sales of their heavily endorsed use of "free range" eggs. However, Phillipov and Gale [46] argue that the celebrity chefs' endorsement is too simplistic and overlooks the complexity of the certification systems. Further, pressuring consumers to make "right" choice of food products places celebrity chefs as influencing consumers' ethical values and animal welfare expectations.

Moreover, in telling stories, cookbook authors and cooking show hosts illuminate how they perceive the world. As De Fina and Georgakopoulou [47, p. 276] note, "Narrative is a privileged communication mode for making sense of the self." What is important to someone comes up in stories. Some identity aspects are made more salient in the story than others, resulting in *storied selves* [48]. These literate identifications consist of the language, cultural models, and relationships enacted. In speaking the same language, sharing similar values, beliefs, and background knowledge, individuals develop a sense of communal identity and become a "folk group," a concept central to contemporary folkloristic approaches to people who have shared experiences and shared traditions [49]. Shared traditions and stories define the group's identity, with food often acting as a medium through which groups construct and define themselves. Consumer research studies can examine the stories being told and how this might influence the audience.

3.3 Critical Discourse Analysis

Critical discourse analysis [CDA] offers a qualitative approach to critically describe, interpret, and explain cookbooks and cooking shows. Developed by Fairclough [50], CDA is a critical theory of language that sees the use of language as a form of social and cultural practice. The approach is problem-oriented, dealing mainly with ideologies that reproduce or challenge existing power structures. For consumer research, CDA is a good tool to focus on the strategies of manipulation or legitimation that are being used to influence food and beverage consumers.

CDA has been applied to the study of cookbooks. Examining Slovene celebrity chef cookbooks, Tominc [51] uses Van Leeuwen's [52] legitimation strategies to understand how authority is constructed through discursive strategies. Focusing on the introductions, Tominc [51] illustrates how the celebrity chefs are both members and authorities of the rising new middle class. The use of the first-person plural "we" and going by a first-name basis create an illusion of intimacy and closeness or "synthetic personalization" [28] between the audience and the celebrity chefs. There is an appeal to solidarity with fellow Slovenians. Other discursive strategies include intertextual references to world nutritionists and food writers that align the celebrity chefs with expert authority. Authority is also made through reference to tradition, sliding the celebrity chefs as tastemakers of Slovene national cuisine. Also studying celebrity chef cookbooks through CDA, Kelsi Matwick [53] identified gendered discourses about the roles of women in the kitchen. Traditional discourses of caring for others through food alternate with discourses of cooking as self-fulfillment and achievability. Language used to build trust with readers are discourses of expertise, politeness strategies, and confessions [54].

Inspired by media studies, CDA has been used to examine how language in the food media constructs authenticity. We are reminded by media scholar Dyer [55, 56] that stars or celebrity chefs in this case are produced and commodified. According to Dyer [55], our fascination in stars is paradoxical. We look for the individual, searching for the behind-the-scenes clues of what the star is really like behind the glitz and glam. As Dyer [57, p. 139] famously suggested, celebrity culture is a quest for the authentic and the "true" personality of the performer. For Rachael Ray, legendary *Food Network* celebrity chef, her "girl-next-door" persona embodies the everydayness, representing the ordinary woman in her cooking shows [58] and cookbooks [59]. How the viewer sees the star-as-performer versus the star-as-person across multiple platforms could be an area of further investigation for consumer research.

The celebrity chef performance on television as ordinary and authentic is constructed in part by language. Through storytelling, celebrity chefs create synthetic personalization, constructing themselves as equals to viewers yet experts [60]. While preparing ham salad, for instance, Rachael Ray tells a story about the recipe origin. Her addition of jalapeños, an otherwise normal ingredient, is made extraordinary in the context of the story. The recipe telling and storytelling take on more authority as being on television offers legitimation. With the use of direct address and inclusion of ordinary individuals, the television has become a source of trusted information and advice around everyday concerns such as food choices.

Besides cookbooks and cooking shows, CDA can be applied to the study of other food texts. Examining British celebrity chef websites, Cesiri [61] analyzed the introductory paragraphs preceding each recipe. The CDA was categorized according to four approaches found in Ancient Greek philosophy: Platonism, Asceticism, Hedonism, and Epicureanism. The latter, described as "a balanced fulfilment of desires" and "tasty nourishment" was the most predominant [61, p.5]. Besides websites, CDA is also well-suited to analyze magazines and advertising that feature food. Fuller, Briggs, and Dillon-Sumner [62] examined the food discourses in women's and men's health magazines and discovered that the discourses about food represent and reproduce stereotypes about ideal men and women and their gendered foodways. Women are portrayed as having the goal of improving appearance, while for men it is more about improving performance. The connection between food and body is an area that consumer research can examine further.

The increased volume of information available on the celebrity chefs' activities, including food shopping, can influence the viewer, their consumer behavior, and even their speech. As Dyer [56] notes, a star's image is intertextual, extending to public appearances, interviews, biographies, and what people say and write about her/him, and how the star is used in advertisements. The star can become part of everyday speech, such as the use of their names on a first-name basis, e.g., Martha (Stewart). Moreover, the celebrity chef's own expressions and way of talking may influence everyday language. For instance, Martha Stewart's trademark phrase—"it's a good thing"—names a segment of her show besides being evaluative [63]. Fans may mimic Martha in their own evaluation of food as a "good thing." Rachael Ray's neologisms is another example of how celebrity chefs influence language. Her shortened form of extra virgin olive oil—EVOO—made it to the list of new words on the Oxford American College Dictionary [64] and later the Oxford English Dictionary in 2018 [65]. While EVOO may have been coined previously, Rachael's public influence accelerated the adoption of the term.

4 Future Consumer Research on Cookbooks and Cooking Shows and Beyond

Future consumer research on cookbooks and cooking shows is paramount, as the literature has shown how ordinary people are influenced by celebrity chefs. In educating viewers on particular tastes and lifestyle choices as well as what foods are considered to be ethical, celebrity chefs seek to "responsibilize" consumers on how their choices relate to larger food system problems [66, 67].

Future research angles for consumer research may be:

- Audience-centered studies: Conduct focus groups, interviews, and surveys with audiences of cookbooks and cooking shows to assess their impressions and values.
- Cross-cultural comparisons: Examine cross-cultural, cross-linguistic differences of cookbooks and cooking shows.
- Millennials, Gen Zs: Identify the values of the rising generation (i.e., innovation, gutsy, competitive, unique) and test these out on new products.

- Social media: Examine how food is talked about or written about in food media (e.g., on Instagram, a social networking platform that encourages the posting of photos, food is one of the most posted categories).
- Race, ethnicity: Examine how cooking shows and cookbooks represent diversity; a discussion of the potential of these mediums can act as a form of critical intervention.
- Food values: Carry out studies of other food dichotomies that may be morally based (good/bad), skill-based (expert/novice), gendered (female/male), or nutritionally evaluated (healthy/junk).
- Artificial Intelligence: Address how non-human technology will affect the way food is made and consumed. For example, questions may be: How will robots affect the way we perceive and relate to food? Will food taste the same if we know it is prepared by a robot versus a human?

Consumer research may want to also consider other sources of food media and social media for food and beverage audience research. As food is aesthetically pleasing, food photos are among the most popular content on social media platforms such as Instagram, Facebook, YouTube, and TikTok.

Here are some statistics to help consumer researchers begin their research of top food influencers across different media platforms.

For Instagram:

- Food is the most photographed subject.
- There are 484 million posts tagged #food on Instagram (as of 2022).
- The most-followed food influencers on Instagram are UK celebrity chefs Gordon Ramsay (13.6 m) and Jamie Oliver (9 m), followed by American blogger Ree Drummond (4.1 m) and domestic lifestyle entrepreneur Martha Stewart (3.9 m) (as of 2022).
- The most popular cuisine on Instagram is Japanese with 18.6 million tags (#japanesefood), likely because of its aesthetically forward approach to food (as of 2022).

For YouTube:

- Mega-channel cooking and food shows are BuzzFeed's Tasty (20.9 m), First We Feast (11.2 m), Travel Thirsty (7.54 m), and Bon Appetit (6 m) (as of 2022).

- Individual food bloggers with the most followers are Gordon Ramsay (19 m), Rosanna Pansino (13.3 m), Andrew Reyes' Babish Culinary Universe (9.66 m), and Jamie Oliver (5.67 m) (as of 2022).
- One of the most famous food and travel vloggers in the world is Mark Wiens (8.8 m) (as of 2022).

For Magazines:

- As of 2017, the most popular cooking magazines in the United States are *Relish* (8.86 m), *Women's Day* [3.28 m], *Taste of Home* [2.53 m], *Martha Stewart* [2.09 m], and *Food Network Magazine* [1.8 m] [68].
- To reach a high-end clientele, consider the American *Food & Wine* magazine. In 2021, the audience reach includes print readers (IPSOS Spring 2021, 10 m), digital uvs (11.4 m), social media followers (15.4 m), circulation (937 k), MRI audience, average age: 48, average HHI: $119,915 K, male: 36%, female: 64%, IPSOS audience: average age: 43, HHI: $373,009, male: 60%, female: 40%], F&W.Com audience: average age: 51, average HHI: $111,043, male: 29%, female: 71% [69].

5 Notes

Cookbooks and cooking shows are conveyers of communication created for and by a particular culture. Multimodal discourse analysis, narrative studies, and critical discourse analysis are apt for qualitative studies that aim to capture the nuances of culture as understood by consumers. Case studies and interviews, which enable informants to express their lived experiences and articulate their subjective interpretations [70], may be used in conjunction with the textual analysis of cookbooks and cooking shows. As researcher biases are inherent to qualitative research, one must be wary of confirmation bias when designing the study. Time constraints are also to be considered so that the study remains feasible.

With enthusiasm, we welcome consumer researchers to join the study of cookbooks and cooking shows and beyond to other genres of food texts. The table is abundant and ripe for rich discussion and intellectual pleasure.

References

1. McLynn K (2020) Bread cookbooks rise in the time of COVID-19, the NPD group says, NPD, October 20. https://www.npd.com/news/press-releases/2020/bread-cookbooks-rise-in-the-time-of-covid-19-the-npd-group-says/
2. Cavanaugh JR, Riley KC, Jaffe A, Jourdan C, Karrebaek M, Paugh A (2014) What words bring to the table: the linguistic anthropological toolkit as applied to the study of food. J Linguist Anthropol 24(1):84–97

3. Riley KC, Paugh AL (2019) Food and language: discourses and foodways across cultures. Routledge, New York
4. Albala K (2012) Routledge international handbook of food studies. Routledge, London
5. De Solier I (2013) Food and the self: consumption, production and material culture. Bloomsbury, London
6. Gerhardt C, Frobenius M, Ley S (2013) Culinary linguistics: the chef's special. John Benjamins, Amsterdam
7. Jurafsky D (2015) The language of food: a linguist reads the menu. W.W. Norton & Company, New York
8. Brillat-Savarin J-A (1925, [original 1825]) The physiology of taste or meditations on gastronomy. Doubleday
9. Counihan C, Van Esterik P (2013) Food and culture: a reader, 3rd edn. Routledge, New York
10. Bourdieu P (1986) The forms of capital. In: Handbook of theory and research for the sociology of education. Greenwood, Westport, pp 241–258
11. Naccarato P, Lebesco K (2012) Culinary capital. Bloomsbury, London
12. Bonner F (2005) Whose lifestyle is it anyway? In: Bell D, Hollows J (eds) Ordinary lifestyles: popular media, consumption and taste. Open University Press, Glasgow, pp 35–46
13. Baumann S, Johnston J (2009) How to be a foodie. Contexts 8(1):62–64
14. Belasco W (2002) Food matters: perspectives on an emerging field. In: Belasco W, Scranton P (eds) Food nations: selling taste in consumer societies. Taylor & Francis, Ch 1
15. Murray S (2013, Ch 17) Food and television. In: Albala K (ed) Routledge international handbook of food studies. Routledge, London, pp 187–197
16. Brownlie D, Hewer P (2007) Prime beef cuts: culinary images for thinking 'men'. Consum Mark Cult 10(3):229–250
17. Hollows J (2003) Feeling like a domestic goddess: postfeminism and cooking. Eur J Cult Stud 6:179–202
18. Matwick K, Matwick K (2018) Women's language in female celebrity chef cookbooks. Celebr Stud 9(1). https://doi.org/10.1080/19392397.2017.1325761
19. Ouellette L, Hay J (2008) Better living through reality TV: television and post-welfare citizenship. Blackwell, Malden
20. McCarthy A (2010) The citizen machine: governing by television in 1950s America. The New Press, New York
21. Caruana R (2010) Morality and consumption: towards a multidisciplinary perspective. J Mark Manag 23(3–4):207–225
22. Kress G, Van Leeuwen T (2006) Reading images: the grammar of visual design, 2nd edn. Routledge, London
23. Kress G (2010) Multimodality: a social semiotic approach to contemporary communication. Routledge, London
24. Barthes R (1975) Introduction to the structural analysis of narrative. New Lit Hist 6(2): 237–272
25. Labov W (1972) The transformation of experience in narrative syntax. In: Language in the inner city: studies in the Black English Vernacular. University of Pennsylvania Press, Philadelphia, pp 354–396
26. De Fina A, Schiffrin D, Bamberg M (2006) Discourse and identity. Cambridge University Press, Cambridge
27. Fairclough N, Wodak R (1997) Critical discourse analysis. In: van Dijk T (ed) Discourse studies: a multidisciplinary introduction, vol 2. Sage, London, pp 258–284
28. Fairclough N (2001) Language and power, 2nd edn. Longman
29. Kress G (2001) Multimodal discourse: the modes and media of contemporary communication. Arnold Publishers, London
30. Adami E, Kress G (2014) Introduction: multimodality, meaning making, and the issue of 'text'. Text Talk 34(3):431–237
31. Rousseau S (2014) Food "porn" in media. In: Thompson P, Kaplan D (eds) Encyclopedia of food and agricultural ethics. Springer, Netherlands, pp 748–754
32. Ketchum C (2005) The essence of cooking shows: how the food network constructs consumer fantasies. J Commun Inq 29:217–234
33. McBride AE (2010) Food porn. Gastronomica 10(1):38–46
34. Taylor N, Keating K (2018) Contemporary food imagery: food porn and other visual trends. Commun Res Pract 4(3):307–323
35. Kress G (2011) Multimodal discourse analysis. In: Gee JP, Handford M (eds) The Routledge handbook of discourse analysis. Routledge, pp 35–48
36. Maynard M (2007) The barefoot impresario. The New York Times, Mar 18. https://www.nytimes.com/2007/03/18/business/yourmoney/18cook.html
37. Contois E (2018) Welcome to Flavortown: guy Fieri's populist American food culture. Am Stud 57(3):143–160. https://muse.jhu.edu/article/713427

38. Matwick K, Matwick K (2019) Food discourse of celebrity chefs of food network. Palgrave Macmillan, Cham
39. Balkin J (1998) Cultural software: a theory of ideology. Yale University Press, New Haven
40. Propp V (1968, originally published 1928) Morphology of the folk tale, 2nd edn. University of Texas Press, Austin
41. Barthes R (1977) Image-music-text. Fontana, London
42. Chatman S (1978) Story and discourse: narrative structure in fiction and film. Cornell University Press, New York
43. Buscemi F (2014) Jamie Oliver and the gastrodiplomacy of simulacra. PD Magazine, 46–50
44. Hewer P, Brownlie D (2009) Culinary culture, gastrobrands, and identity myths: 'Nigella', an iconic brand in the baking. Adv Consum Res 36:481–487
45. Matwick K, Matwick K (2019) Humor and performing gender on TV cooking shows. Humor 32(1):125–146. https://doi.org/10.1515/humor-2017-0093
46. Phillipov M, Gale F (2020) Celebrity chefs, consumption politics and labelling: exploring contradictions. J Consum Cult 20(4):408–418
47. De Fina A, Georgakopoulou A (2008) Introduction: narrative analysis in the shift from texts to practices. Text Talk 28(3):275–281
48. Georgakopoulou A (2006) The other side of the story: towards a narrative analysis of narratives-in-interaction. Discourse Stud 8(2):235–257. https://doi.org/10.1177/1461445606061795
49. Long L (2015) The food and folklore reader. Bloomsbury, London
50. Fairclough N (1995) Critical discourse analysis. Longman
51. Tominc A (2014) Legitimising amateur celebrity chefs' advice and the discursive transformation of the Slovene national culinary identity. Eur J Cult Stud 17(3):316–337
52. Van Leeuwen T (2008) Discourse and practice: new tools for critical discourse analysis. Oxford University Press, Oxford
53. Matwick K (2007) Language and gender in female celebrity chef cookbooks: cooking to show care for the family and for the self. Crit Discourse Stud 14(5):532–547
54. Matwick K, Matwick K (2017) Women's language in female celebrity chef cookbooks. Celebr Stud 9(1):53–68. https://doi.org/10.1080/19392397.2017.1325761
55. Dyer R (1979) Stars. British Film Institute
56. Dyer R (1986) Heavenly bodies: film stars and society. St. Martin's Press
57. Dyer R (1991) A star is born and the construction of authenticity. In: Gledhill C (ed) Stardom: industry of desire. Routledge, London, ch. 11
58. Nathanson E (2009) As easy as pie: cooking shows, domestic efficiency, and postfeminist temporality. Telev New Media 10:311–330
59. Matwick K (2016) A critical discourse analysis of food and gender in cookbooks by female celebrity chefs. Doctoral dissertation, University of Florida
60. Matwick K, Matwick K (2014) Storytelling and synthetic personalization in television cooking shows. J Pragmat 41:151–159
61. Cesiri D (2019) Philosophical tenets in the construction of culinary discourse: the case of British celebrity chefs' websites. Poetics 74:1–11. https://doi.org/10.1016/j.poetic.2019.04.005
62. Fuller J, Briggs J, Dillon-Sumner L (2013) Men eat for muscle, women eat for weight loss. In: Gerhardt C, Frobenius M, Ley S (eds) Culinary linguistics: the chef's special. John Benjamins, Amsterdam/Philadelphia, pp 261–279
63. Davies CE (2003) Chapter 11: Language and American 'good taste': Martha Stewart as mass-media role model. In: Aitchison J, Lewis D (eds) New media language. Routledge, London, pp 146–156
64. Chowhound (2006) The EVOOing of America. Chowhound Dec 5, https://www.chowhound.com/food-news/1479/the-evooing-of-america/
65. Oxford English Dictionary (2018) New words list January 2018, Oxford English Dictionary. https://public.oed.com/updates/new-words-list-january-2018/
66. Bell D, Hollows J, Jones S (2017) Campaigning culinary documentaries and the responsibilization of food crises. Geoforum 84:179–287
67. Hollows J, Jones S (2010) 'At least he's doing something': moral entrepreneurship and individual responsibility in Jamie's Ministry of Food. Eur J Cult Stud 13(3):307–322
68. Watson A (2020) Most popular cooking magazines in the U.S. 2017, by circulation. Statista 2021, Nov 20. https://www.statista.com/statistics/714387/most-popular-cooking-magazines-by-circulation/
69. Food & Wine Media Kit (2021). https://www.meredith.com/Food-and-Wine-Media-Kit.pdf
70. Graebner ME, Martin JA, Roundy PT (2012) Qualitative data: cooking without a recipe. Strateg Organ 10(3):276–284

Chapter 5

The Analysis of Images in Consumer Research

Carlos Gómez-Corona and Daniel Salazar Lama

Abstract

The world in which we live is full of visual stimulation. Everywhere we go and every task we do implies seeing or using an image. Moreover, the rise of social media in the past few years has bombarded us with millions of images where every person seems to have the desire to communicate and share images. However, the academic publications that study social media rely more on text than images. Text has received a lot of attention and different automatic techniques have been used to study social media content. To give a different view of social media content analysis, this chapter focuses on the analysis of images from social media. And as the academic publications dealing with images in consumer research are scarce, the authors have decided to search for the information in the domain of visual arts.

The first section of the chapter is an introduction to what images are, how images have become hegemonic in our society, and the ability of images to communicate. A specific point is made about how images are a reconstruction (in the eyes of the creator of the image), and how an image contains meaning for the eyes of those who look at it. The second section of the chapter focuses on some methodology of semiotics, semantics, and specifically of rhetoric. A comparison is made between verbal and visual rhetoric to exemplify how visual rhetoric functions in social media images. And finally, the third section deals with a draft step-by-step analysis of social media images, and while thought as a method, it is just a first proposal. In the fourth section, the methodology is illustrated with some case studies from Instagram images.

Key words Visual arts, Rhetoric, Social media, Semiotic analysis, Semantic analysis

1 Introduction

Why is a picture worth a thousand words? Is this old adage still valid, and what is the value of these images? Even in the field of social media research, there are authors that mention that an Instagram image is worth a thousand Twitter words [1, 2]. But let's start with the essentials... An image is a visual representation loaded with meaning, while text is a statement. Images can be full of symbolic meaning, while text tends to be more straightforward (most of the time). And images can give information that text cannot. Let's take a simple example: the famous photograph of 1950 "Kiss by the Hotel de Ville." The title of the image simply tells us that a kiss takes place in front of the Hotel de Ville in Paris, but the image can

Carlos Gómez-Corona and Heber Rodrigues (eds.), *Consumer Research Methods in Food Science*,
Methods and Protocols in Food Science, https://doi.org/10.1007/978-1-0716-3000-6_5,

Fig. 1 "Baiser de l'Hôtel de Ville" or "Kiss by the Hotel de Ville" by Robert Doisneau, 1950. (Image: Picture from postcard, Carlos Gómez-Corona)

reveal much more. The photograph is taken from a cafe (we can see a person's head from behind), and the image shows us an already hectic Paris of 1950, the design of the cars and the clothes of the people can give us information about the temporality of the picture. This kiss, perfectly centered, seems to be spontaneous at the Hotel de Ville, people passing without stopping in the afternoon (because of the brightness of the photograph), and the time of year (Fig. 1).

This famous picture that appears to have been taken at the perfect moment, and at the perfect place on Rue de Rivoli, is a staged image where the photographer asked two actors to pose for him. The photographer was working on a project to photograph the romantic side of Paris after World War II. And just as this image can be "prepared" or "modified," it is loaded with meaning, and today it is an icon of Paris [3]. Just as photographers can prepare or modify an image, visual artists also have the power to own and modify images, or visual representations.

Over the centuries, visual representations were carried out by artists, and with their creations we saw the world, different objects, and people immortalized in large format paintings or engravings. For centuries, we were limited to seeing the world through the eyes of the artist, and those eyes could modify and enhance certain elements of the visual representation. An artist could change the light of a landscape and add or remove certain elements to better tell or to enhance a story in what we see. The objects represented could become highly complex images, sometimes difficult to understand for the layman's eye. To understand those images, art history developed iconography as a branch of art dedicated to the study of images [4].

Iconography became the discipline dedicated to the identification, description, and interpretation of the content of images. The study of images, at least in the domain of art history, is expected to bring to light useful information about the meaning of the image. Different levels of analysis can be applied to the images, and different aspects of the images can be analyzed. Simple changes in the composition of the image can bring different elements to the interpretation. For example, different sizes within a painting, different postures, and the way that the persons in the image are dressed can often be assumed to mirror different status, identities, cultural belongings, or actions. But what does studying these images from art have to do with consumer research? Well... We have many things to learn from iconography, as the mother discipline that studies images. We can learn from their different methods and the theory behind the image analysis. But we must apply them carefully to consumer research, and we need to establish the framework and define the images we are going to use in consumer research. Perhaps the closest analysis of images in consumer research that has been published is the symbolic meaning of advertisement, and within advertisement, one of the most famous examples is Panzani advertisement from the 1950s (Fig. 2).

Roland Barthes published in his famous book on the rhetoric of images [5] an analysis of an apparently simple advertisement for Italian pasta (Fig. 2 left). Packets of pasta, a box, a sachet, tomatoes, onions, peppers, a mushroom, all coming out of a half-opened fabric bag, in yellow and green hues on a red background. The image immediately delivers a first message, a linguistic one, in the lower corner of the image (pasta, sauce, parmesan, a l'Italienne deluxe). Roland goes further and mentions that the message itself can still be decomposed, because the Panzani advertisement not only delivers the name of the brand, but also, by its assonance, an additional meaning which is "Italianism." The nest containing the ingredients seems to talk about a "return" from the market with fresh ingredients, iconic ingredients used in Italian gastronomy, and the mix of the ingredients with the red tomato, the pepper and the white onion seems to represent the Italian identity. This image rooted in certain touristic Italian stereotypes according to Roland Barthes, can be easily found nowadays in current Italian restaurants in Rome (Fig. 2 right. Image taken from first author outside a restaurant during a visit to Rome in 2018), where a table at the entrance of the Roman restaurant gives the same image of Italianism, identity, freshness, etc. as the Panzani advertising gave it 71 years earlier.

Fig. 2 Left: Panzani pasta advertisement in 1950. Right: Table outside an Italian restaurant in Rome, August 2018. (Images: Carlos Gómez-Corona)

But what about the images used in advertising nowadays? In the hypermodern and strongly aesthetic society in which we live [6], a single image is enough to trigger a sensory (non-visual) stimulus in the observer. This is the case of the fine fragrance advertising, where the images are full of meaning (Fig. 3). Whoever sees the advertisement of a fine fragrance in the street cannot smell the fragrance but can imagine it through the visual stimulus. Through the different visual stimuli, the consumer obtains different sensory aspects of the perfume, the freshness of the environment, the elegance of the person who wears it, her sensual and daring look that have a very specific sensory characteristic. The person in the advertising image represents the entirety of the fragrance, which the observer can reach through a (not so simple) visual stimulus. Even during an evening walk, the advertisement of the fine fragrance is available, and with the perfect illumination, the advertisement becomes even more attractive to the human eye (Fig. 3—left).

Fig. 3 Left—Miss Dior fragrance advertisement. Right: Dior Sport Homme advertisement. Both images are an illustration based on the original image. (Images: Carlos Gómez-Corona, October 2022)

While looking at the fine fragrance advertisement, we can imagine that Miss Dior is not a heavy woody or amber fragrance, but rather a more elegant flowery perfume, maybe with dominant rose notes. When the first author sees the image, he thinks of a fresh perfume, opening with citrus like lemon or mandarin, covered by a flower bouquet. The sensual look of the person in the image might imply a sensual accord of rosewood that make this a fresh, modern, and sensual fragrance. On the other hand, in a similar advertisement, but during a daytime walk (Fig. 3—right), we can see a men's fragrance advert. A young man, not looking at the camera, represents the fragrance of Dior Sport. By looking at the black and white image, we can see a sporty and chic man, touching his neck. This should also be a fresh fragrance as this is a sportsman, but with the leather jacket it seems that the fragrance might have some spicy elements in it. Perhaps the fragrance of this man has a ginger note as the heart of a fresh woody and spicy scent. Black and white should also bring some classical men's citrus notes like an ageless bergamot scent or a Sicilian lemon, a fragrance that is fresh and spicy, rather than fresh and elegant.

That is what two images from a fine fragrance advertisement can transmit through visual codes that are already learned in our culture and that we put into motion unconsciously. With a semantic and semiotic approach, we can find how these strategies of meaning work. However, in this chapter we will focus not on advertising, art, or photography, but on the images produced by consumers, and how we can analyze them. More specifically, we will focus on digital images that we can find on social media, as it may be currently the biggest repository of images that a consumer researcher may have access to. Additionally, social media images may or may not have another purpose and meaning, and we should develop special methodologies to analyze images shared in social media that can help consumer researchers to better understand food and beverage consumption. This chapter is a first attempt to propose certain guidelines to analyze images from social media, and the first author borrows different aspects of the art iconology methodologies published by the second author, outside the consumer research field [7, 8]. Both authors hope that more studies will follow and propose more specific guidelines and examples of how to analyze the images produced by consumers in social media.

1.1 The Hegemony of Images

Besides their meaning and what they symbolize (e.g., the romantic kiss in Paris), images are simply all around us; we live in a visually centered society where we are forced to always see and think of images. As René Magritte said, "to think of an image means to see an image" [9], and the same might apply to food and drink. To think of a special dish, implies to see it. Going further, when we think of an event (even the act of cooking for example), we also see an image; our memory does not film, memory photographs. Under this assumption from a literary inspiration [10], our memory is full of thousands of mental photographs, of static images that show certain characteristics of the events we lived. Our memory is highly visual, and with the birth of photography, it just became more evident, and images so massively available. And while photography might seem to have been around us for centuries, it does not have such a long history.

"View from the Window at Le Gras" is said to be the world's earliest surviving photograph (Fig. 4). It was taken in 1826 in a French town by Nicéphore Niépce. Niépce himself called it heliography, or "light writing" [11]. Almost 200 years later, we doubt that Nicéphore could have ever imagined what photography represents in our current society. In the beginning, photography needed a considerable amount of time to be prepared (8 hours in the case of View from the Window at Le Gras) and highly specialized equipment. Over the years photography started to get closer to the consumer, and in 1888, Kodak Company sold the first commercial camera. With the slogan "you press the button, we do the rest," George Eastman put the first automatic camera into the hands of a world of consumers [12]. At that time, a person would take pictures and send the camera back to the factory for the film to be developed and printed.

With the first commercial camera on the market in 1888, the conception of images changed completely. The image (photography in this case) in the hand of the consumer became a clear visual representation that manifests the exact appearance of the object or of a person. After the commercial camera revolution, photographs

Fig. 4 View from the Window at Le Gras, by Nicéphore Niépce, 1826. (Image: CreativeCommons—juandesant, CC BY 2.0. (Royalty free))

started to have a place in the life of millions of people and lived on the walls, shelves, and in family albums. Usually, this photography was from professional photographers as it was not until the late 1950s that the technology of automatic cameras was available for mainstream mass consumption. At this moment, the omnipresence of cameras persuaded people that life consisted of interesting and photograph-worthy events and moments [13], life and experiences became a "Kodak moment," but above all, it became a tangible image of what the moment represented.

As we stated before, images always have an underlining meaning, and this photograph taken from the consumer, this "kodak moment" is not an "innocent" image; we consciously or unconsciously load it with meaning so it can better express what we want to be communicated through the image: a happy family, a tender puppy, or a delicious Sunday meal. Little by little, photographs alter and broaden our notions of what is worth looking at, and what we have a right to observe [13]. Photographs started to teach us visual codes and let us collect the world while we were collecting photographs. It would not be wrong to speak of a compulsion to photograph everything and transform the experience itself into a way of seeing. The need to confirm reality and expand the experience through photographs is an aesthetic consumerism to which today everyone is addicted. Fundamentally, having an experience becomes identical to photographing it. Nowadays, the ease of photographing with a smartphone and the possibility of spreading images instantly through social media constitute the hyperbole of that need to collect experiences (and images), merged through social media, in an image-experience: two inseparable elements in the contemporary consumer. Figure 5 shows an example of a photographed experience shared in Instagram with the hashtag #NotSharedNotLived. In this example of a trip to a Caribbean Island in 2019, the post is as much of an experience as it is a proof of being there.

1.2 Images as a Communication (Social) Media

After the digital image of the 1990s, and the networked image (digital images on the Web) of the 2000s, we now have the social media image as the dominant cultural visual form of images [14]. According to Prada [15], today's world is anxiously intended to reduce all information to a headline or a tweet, but most importantly, it is desired that everything adopt a visual configuration, to be able to place everything at a glance. Now that the camera lens of a smartphone points at the face of the one who used to look through the viewfinder, that person wants to form a central part of the scene to be photographed and that is how we witness the selfie phenomenon nowadays.

Never before has a process of self-design of our own exteriority been so demanded of us [15]. The selfie, understood as that photograph of "me with someone" or "me somewhere," has become an authentic social phenomenon, a mandatory and recurring practice for all smartphone owners around the globe. The rise of smartphones, as well as Internet-based chat applications, has changed both our forms of written communication and our expectations about public behavior. Smartphones allow individuals to be linked to two places at one time, raising new questions about acceptable behavior in each place, and the relationship between individuals' physical and virtual presence [16].

Selfies have become a compulsive and stereotyped activity, an expression of narcissism. Confessional uses of the Internet seem to fit this definition; they would be renewed manifestations of the old autobiographical genres. The self that speaks and is tirelessly displayed on the web is usually triple: it is at the same time author, narrator, and character [17]. Sharing these types of images should be understood as a practice of self-branding, of personal promotion, almost always aimed at offering the best vision of oneself, trying to show one has been in wonderful places or with fantastic friends, being happy, feeling relaxed, being positive, and seductive [15]. Our desire to be permanently under the gaze of others encourages us to develop "micro" forms of pseudo-events, situations created to be able to take photographs and share them, small

Fig. 5 Image based on an Instagram account of an Argentinian young man. The image shows how sharing this moment on Instagram constitutes a validation of being at a specific place in time (a Caribbean Island in 2019): #NotSharedNotLived. Note: The original image from social media was replaced by a similar one from the personal image-stock of the author. (Image: Carlos Gómez-Corona (Cancun 2022))

synthetic novelties that may be the reason or occasion of visual records for their sharing the network, which serve us for our continuity on stage [15].

These selfie phenomena are of most importance when doing consumer research and may be one of the main differences when studying images in art history versus images from consumer research. In social media, the consumer researcher should see

Fig. 6 Image rebuild by the authors, based on an Instagram account of a Brazilian influencer living in the UK. The image could be wrongly analyzed without the help of the text "I will confess – I only took that Aperol that ('another tagged person') was drinking, to take a picture... 😂 😂 😂 I hate Aperol Spritz 🤢 I was drinking my basic beer 🍺." (Image: authors)

these images as part of a cultural phenomenon and not just as the image standing alone. When doing consumer research through the analysis of images in social media, the selfies will be a recurrent theme that we will need to analyze in a proper way. Let's take an example, Fig. 6 shows the Instagram image of a young man in a restaurant. Our key words for the search could have been "Aperol Spritz" because we are interested in the uses and attitudes of this beverage. Just by looking at the image, we see a young man with a discreet smile in the restaurant taking his Aperol Spritz. And again, this is not an innocent image, in the sense of a "quick picture," it is an image planned with the right light, an incredible background, the gesture, the look in the eyes. However, there was something missing, an external element, and the young man grabs someone's Aperol Spritz to take the picture. Maybe all the elements were not carefully analyzed at the moment of the image, and it could have been a quick decision, but that decision was full of meaning and cultural codes, learned over the years. In this case, our young man chooses a set of elements that will have a specific connotation in the social media image.

1.3 Images as Reconstruction and Meaning in Social Media

In 1986, Roland Barthes [18] mentioned that journalistic photography is a message, where the receiving public is the public that reads the newspaper. And the channel of transmission, the diary itself, or more precisely, a complex of concurrent messages, the center of which is the photograph and the contours of which are represented by the title, the caption, the page layout, and more abstractly, but no less importantly, the very name of the newspaper (because that name constitutes knowledge that can notably deviate the reading of the message itself. And without a doubt, we can take the same idea of Barthes and apply it to social media. Let's take Instagram as an example again. The image is a message, and the receiving public are the social media followers of a given account. The channel of transmission is Instagram, where the image is usually followed by an explanatory text, which unlike in their journalistic counterpart, the Instagram image caption might not always be clear. The social media images are usually followed with hashtags (which facilitates their search), and comprehensive or incomprehensive captions and emojis.

In social media, the images are almost inseparable from the text, but sometimes one same text can contain multiple different meanings. And that is what Roland Barthes talks about as the denotation and connotation of the image [18].

In simple words, the denotative part of the image is what it represents in its broader sense. For example, the image of a taco denotes a traditional food from Mexico, which is usually a maize (corn) tortilla wrapping a specific filling [19]. Denotation gives the information of what the image is, in an objective way and without giving space for additional interpretations. In other words, a taco is just a taco. Figure 7 represents a special type of taco called "taco de canasta." The text from the image also suggests the same direct information: Tacos de canasta, #tacosdecanasta, #acomer (#toeat), etc.

Figure 8 was searched for in social media under the same key words: tacos de canasta. The image found is from a social media influencer and gastronomist called "Lady Tacos de Canasta." This person rose in social media fame when dressed as a woman in an LGBT march in Mexico City in 2016 [20]. Today, Lady Tacos de Canasta is not only an icon of Mexican Gastronomy and street food,

Fig. 7 Image from an Instagram account in Mexico in 2022. The image shows a type of tacos (tacos de canasta), with the caption "tacos de canasta". Note: The original image from social media was replaced by a similar one from the personal image-stock of the author. (Image: Carlos Gómez-Corona)

Fig. 8 Illustration based on the original image from an Instagram account in Mexico in 2019. The image shows a well-known Mexican personality called Lady Tacos de Canasta (@ladytacosdecasta). This person rose to sudden social media fame when selling tacos dressed as a woman in an LGBT march in Mexico City. Lady Tacos de Canasta has become an icon of tradition, gastronomy, Mexican pride, etc. (Image: Carlos Gómez-Corona)

but also of national pride. Marven, or Lady Tacos de Canasta is a *Muxe,*[1] a living image of a thousand-year-old social and culinary tradition. His photo, also found with the keywords "tacos de canasta" not only denotes a dish, but the connotation is of culinary and gender identity, where some tacos de canasta launch a subjective message where the person who sees the image must interpret it to really understand the symbolic content of the image. The connotation of this image can have multiple interpretations, but the most common will be national identity, social struggle, and gender equality [21]. These denotative and connotative elements are the key for image analysis in iconography and are explained in the following sections of the chapter.

2 Two Methodological Foundations

The modern methodological foundations of the analysis of images in art history are often attributed to Erwin Panofsky. As mentioned before, iconography is the branch of art history that studies the meaning of images. In a famous essay, published in 1939 [22], Erwin Panofsky pointed out that the iconography description of the images can be studied in three basic levels of analysis: "pre-iconographic," "iconographic in a narrow sense," and the "iconographic or iconological in a deeper sense" [23].

On the one hand, the pre-iconographic level for Panofsky is the general visual descriptions of the forms, the objects, and the expressions found in what we are looking at, which are carriers of certain natural meaning to the image. This is the denotative level of imaging approximation according to other authors. On the other hand, the iconographic description in a narrower sense concerned the conventional meanings assigned to themes or concepts that are present in the images and that can be, for example, the stories and allegories to which they refer. And finally, the iconological description was the most controversial of the three, being subjectively based on the philosophy of symbolic forms. Panosfky wrote that iconology attempts to ascertain those underlying principles which reveal the basic attitude of a nation, a period, a class, a religious or philosophical persuasion—unconsciously qualified by one personality and condensed into one work. Panofsky's schema, famous as it is, has of course begun to show its age and has been subjected to much criticism [4].

[1] *Muxe* is a Zapotec term that refers to a gender that defines a man that assumes female roles in society. In the Zapotec tradition in the Southwest of Mexico, the last son of the family is raised as a woman. The grown-up man has a woman's expression, but not necessarily a woman's identity. Their sexual attraction is also personally dependent, and their biological sex is of a man.

For art historians, there are at least two main limitations to the use of Panofsky's approach to analyzing images [24]. The first one is that at no point did Panofsky create a methodology; he published a set of ideas to separate the analysis of the images into three steps. In the Introduction of his publication, the three steps are just briefly explained [23]; however, they do not represent any specific methodology. And the second one is that the approach of Panofsky was conceived for the study of religious images from the Renaissance in Europe. Therefore, Panofsky's approach to studying images should be restricted to these case studies. The same will apply to the analysis of images in consumer research, where there is no formal approach, guidelines, or recommendations. In the case of consumer research, we will need to borrow different approaches and methodologies, as we have done in the past from methodologies used in Psychology (*see* the book Part II: Methods from Psychology). In this chapter, the authors' initial recommendation is to focus on visual rhetoric. Before that, let's move on to a small theoretical point for semiotics and semantics, and why they are important in image analysis.

2.1 Semiotic and Semantic Approximation of Images' Analysis

According to Umberto Eco [25], semiotics studies all cultural phenomena as communication processes. Regarding the image, semiotics addresses the internal structures that produce meanings and the internal functioning that governs the meaning of any type of visual construction. Visual semantics, on the other hand, is in charge of the meaning. The meaning of an image is the entire universe of connotations that is supported by the denoted (or denotative) level, which is the literal meaning of an object within an image, that is, its most obvious meaning [26]. For example, in Fig. 5, every user of social media would recognize the pure meaning of the objects represented: palm trees. Now, outside the context of this image, the palm trees could convey an additional meaning (beyond the denotative, making it clear that the denotative in Fig. 5 is "palm trees"). An additional meaning of palm trees can be the realization of a desire to travel, to know the Caribbean beach, where palm trees are a symbol of beautiful beaches or of a tropical environment. This is the semantic level, while the semiotic will look for the means by which palm tree means vacation in the tropics and goes beyond its literal meaning.

Semiotics and semantics go hand in hand, that is, the first deals with the structure, while the second deals with the sense or the meaning transmitted. Our intention in this chapter is not to offer a rigid method to analyze images in social media, but rather to offer some guidelines on how to search for the meaning that images convey. To this end, we believe it is useful to outline a rhetorical approach toward the image.

2.2 Visual Media Rhetoric

One of the main tools used to analyze images is visual rhetoric [27]. Rhetoric has long been studied when applied to language, and it is a discipline that studies the rules and principles of speaking and writing, the way we say something that carries an additional or different meaning. In the case of images, visual rhetoric applies the same principles as in language rhetoric, to explain how meaning is introduced in visual images. For example, nobody who sees an image of a knife and a fork while driving along a road would think that they sell cutlery nearby, but rather that they sell food (a visual metonymy). In the same way, when one sees a drawing of a glass on a cardboard box, one would not think that the box strictly contains glasses, but that the content of the box is fragile, like a glass (a visual metaphor) [28]. In the next section, we will define some visual rhetorical strategies, including metaphor and metonymy.

2.2.1 Metaphor

Metaphors are probably one of the most common rhetorical figures in verbal and image content. In a simple definition, we can say that in a verbal metaphor, one term is replaced by another. The term that "replaces" the original content has a relationship content of a partial similarity. The features of the metaphor are substitution and similarity. For example, a popular singer can sing "shine bright like a diamond" (although we as scientists know that diamonds do not shine, but reflect light), or an ordinary person can say that something is "clear as water." In either case, the characteristics or functionality of one element is replaced by another: diamonds shine, water is clear. In a similar construction, a visual metaphor is a substitution of the image of an object with that of another by virtue of a partial similarity, for example: the world is melting (like an ice cream). Fig. 9 shows a melting world as a metaphor for a melting ice cream.

Fig. 9 A melting world, visual metaphor illustration. (Image: Hand-drawing by the authors based on a social media post)

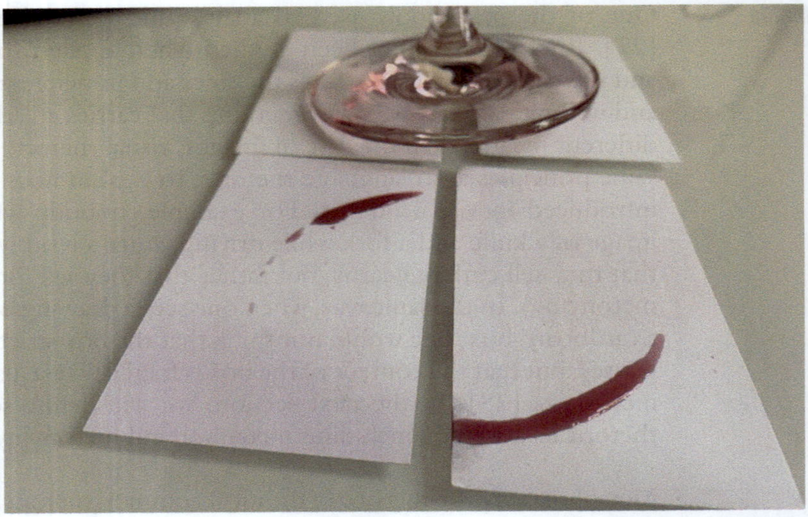

Fig. 10 Design and process of a sommelier presentation card using wine stains to evoke the relationship with wine. (Image: Authors)

2.2.2 Metonymy

To use metonymy in a verbal construction, a term of the sentence is replaced by another, but the final meaning of the phrase is given in terms of a linguistic proximity. For example, "raise our glasses" (cheers), "vamos por unas copas" (lets go for some glasses | Spanish), "boire un verre" (drink a glass | French). In three different languages, the glasses in which someone drinks are replaced by the action of drinking. Similarly, in a visual metonymy, an image is replaced by a detail of another, because both are part of the same original shape (or very close proximity). The detail of the image using metonymy can show the big picture by the mere fact of having been chosen. (Fig. 10) shows a design of a proposed business card of a sommelier using wine stains as a metonymy of wine expertise.

2.2.3 Hyperbole

The verbal hyperbole is a rhetorical figure consisting of the exaggeration of a description of reality by means of expressions that can amplify its scope, by excess and by default in some cases. When speaking, the hyperbole also involves emphasis. For example, in verbal hyperbole we could say that someone is "drowning in a glass of water" or a "sea of tears." In a visual hyperbole, the image presents an exaggerated aspect of reality in either a quantitative or qualitative way. In the advertising industry, the hyperbole is quite common as it tends to emphasize the functionality of a given product. Some car advertisements can be very clear at using hyperbolic visual rhetoric to show that a car is faster, safer, more technological, or more elegant. In social media, the hyperbolic use of images is very common, especially in memes where certain elements of the images are completely exaggerated (Fig. 11).

Fig. 11 A hyperbolic image in social media. A giant cup of coffee (#cafegigante) is used in a visual rhetoric of all the coffee apparently needed for a year with a lot of work. Note: The original image from social media was replaced by a similar one from the personal image-stock of the author. (Image: Carlos Gómez-Corona)

Fig. 12 Visual onomatopoeia. Left: BAM. Recreates the sound of someone slapping another person's face. Right: Keep calm and OINK OINK. Recreates the sound of a pig. (Images: Recreated by the authors)

2.2.4 Onomatopoeia

This usually refers to words that have a sound very close to the object they represent. For example, a common French word to say a sofa bed is "clic clac" [29], or the Italian word for cow is "mucca," from a clear onomatopoeic origin. Less formal words are also formed with onomatopoeic origin like "hahaha" or "jajajaa" (laughing in Spanish), "glu glu" (drinking in Mexican Spanish), or "toc toc" (for knocking at a door, also in Mexican Spanish). In the case of visual onomatopoeia, the images are usually accompanied with an informal onomatopoeic word such as glu glu or toc toc. It evokes the impression of a sound without corresponding to any lexical meaning (Fig. 12).

3 Important Aspects in the Analysis of Social Media Images

The past two Subheadings (2.1 and 2.2) are intended to bring a theoretical framework to the readers of this chapter. Although the review is not exhaustive, they are some initial elements to have more theoretical elements when analyzing the images of social media. It is important to take into consideration that the image of social media cannot be treated as professional photography or as an artistic means of aesthetic expression since it does not fulfill that function but rather of registration or exhibition of something or someone. The theoretical background of history of art extrapolated to the analysis of social media images should be done carefully, as the production of the images does not follow the same principles and objectives. We are immersed in a predominantly visual medium and each of the images we see on a daily basis is loaded with meaning and conveys meaning. In social media, many of these images lack expert knowledge in the way the meanings are generated. It is simply an unconscious knowledge that we reproduce following cultural patterns and codes.

Going further, in the past examples, especially in the Visual Media Rhetoric section, where most of the examples used come from social media, we can begin to see that the images in social media are not an isolated structure, as it can be an image from advertising, or from art. Social media images communicate with another structure, which is the text associated with it in the social media platform (title of the image in Instagram, or comments in Twitter). Additional elements of the image, which are not part of the initial message or objective but still communicate, are the social media interactions: likes, comments, and sharing (sharing image or retweeting).

The social media image is therefore very close to the image–text interaction that Roland Barthes [19] defines in journalistic photography. For Barthes, in journalistic photography the totality of information is supported by two different structures (one of which is linguistic). These two structures are concurrent, but since their units are heterogeneous, they cannot mix; in one case (the text) the substance of the message is made up of words; in the other (photography), by lines, planes, tints. This binomial effect of image and text is a key element in social media research that should not be set aside in the analysis. And the weight of one element can be different from post to post, or from one social media platform to another. For example, we can say that in Twitter, the image tends to be the supporting content, while for Instagram, the text tends to support the image.

3.1 Images' Source and Selection

The source of social media images can come from multiple platforms such as Instagram, Twitter, Facebook, Pinterest, etc. (for further social media platforms *see* Chapter 14). Some platforms are more common in Western countries, while China is a world in itself for social media platforms. For extraction methods and tools also refer to Chapter 14; however, in simple terms, the extraction can be done by means of specific social media monitoring tools, which are expensive paid licenses (e.g., Synthesio, Falcon, Digimind, etc.), and open specialized sources such as twitte**R**, (R Project software) and manual search engines. The last two options will imply a considerable amount of time in manual cleaning and selection of final data.

Once the images have been identified using a specific social media monitoring tool, the next step is to define the number of images needed for a specific analysis. This specific point of the number of images needed for a social media image analysis has never been published, and therefore, the authors suggest two options. The first one, from a qualitative approach, is to analyze the images until the researcher achieves a saturation point of information (no additional information is captured when adding one or more individuals to the dataset). This saturation cut-off point is inspired by qualitative interviews or ethnographies [30, 31]. The second option consists of attaching traditional quantitative techniques, where a standard of 120–150 consumers is generally seen as the minimum acceptable [32]. However, this apparently standard number for quantitative methods is not always respected for social media, where sometimes thousands of users are expected to be analyzed. But going back to the objective of the chapter, thousands of images cannot be manually analyzed, and previous studies suggest that some hundreds of images are enough for image analysis in social media [2].

3.2 Social Acceptance and Sharing

In the world of art, some images, more specifically some paintings, cross the barriers of the cultured world into the mainstream one. In other words, images valued by the critics will not necessarily be valued by the mainstream consumer, and sometimes images which are not valued at all by critics can have a huge social value for the same mainstream consumer. Let's take an example. Figure 13 shows two images well-known by art critics. The first one (left) is Spring by Botticelli at the Uffizi Gallery in Florence, a complex painting to analyze and understand [33]. It is also a spotlight for massive social media pictures and instant sharing (#primaverabotticelli with 1481 publications in Instagram, or just #Botticelli with 245,896 publications in March 2022). On the other hand, the image on the right from Gustave Doré exhibited in Strasbourg (Les Martyrs Chrétiens) is one of the masterpieces on the use of light and color, but goes unnoticed by social media users. These two paintings receive completely different social acceptance and sharing, at least in social media.

Fig. 13 Left shows Spring from Botticelli, and the right-hand image shows Les Martyrs Chrétiens, from Gustave Doré. Two masterpieces for art critics, with different social media valorization. (Images: Carlos Gómez-Corona)

Just like in art, the same happens with social media images. Beautiful, complex, and interesting images from a user can go unnoticed, but an influencer in a swimsuit can cause a stir, be shared by millions of people, and unleash a series of comments throughout the world. Several studies have focused on the sharing effect of social media (*see* Cheung et al. [34]). On a smaller scale, images of "earthly" consumers shared on social media are also socially valued. The most commented images, with most likes become, to a certain extent, more "valuable" to the user uploading the image, but not necessarily to the consumer researcher, as the number of likes, shares, and comments is always very relative. Toubia and Stephen [35] found that most non-commercial users go through two phases, where intrinsic utility from posting is greater than image-related utility when they have fewer followers, but image-related utility becomes greater than intrinsic utility as they amass more followers. However, the question as to whether social acceptance and interaction in social media analysis should be analyzed is a question that has not yet been answered, and should be the subject of future research on social media analysis.

4 A Case Study of the Analysis of Images in Social Media

The following cases illustrate the analysis of social media previously presented at the Pangborn Symposium 2019 [31], to illustrate the previous theoretical background exposed in this chapter. The set of images analyzed were from Instagram accounts in both Mexico and the United Kingdom. The images were extracted and separated into whisky and mezcal (Mexican drink distilled from agave). To illustrate this chapter, only two images were selected containing the hashtag #whisky.

Fig. 14 The following image was posted in a social media channel. Image contained the hashtag #whisky. Note: The original image from social media was replaced by a similar one from the personal image-stock of the author. (Image: Carlos Gómez-Corona)

4.1 Case Study 1—Whisky in Mexico

In the following image (Fig. 14), the most important element is the bottle of whisky in the center, with a brighter color than all the other elements, so it immediately attracts the eye. It is the point from which the eye begins to see and understand an apparently simple and innocent image. The whisky bottle is only understandable if it is seen in relation to the objects around it. For that we have carried out the analysis following the steps detailed below.

Composition

Image composition refers to the way the objects in the image are arranged in the plane. In the case of our image, two important points of the composition can be distinguished: the first plane, where the credit cards and the wallet are, and the second plane with a preponderance in the center, which is where the bottle of whisky (William Lowson's) is located. At the level of visual hierarchies, these two elements predominate in the entire composition of the image, and we could expect that the greatest semantic weight falls on them. In a third and fourth plane (displaced from the center of the composition) are the complementary elements (sausages, sauces, snacks, and bottles of carbonated water).

Image Rhetoric

Thanks to the hashtag #whisky, it is understood that the meaning behind credit cards is the purchasing power of the people who own them. The card and the wallet, in this sense, work as a metonym for purchasing power. This idea is corroborated by the text of the Instagram post (#creditcards, #banks, #money); however, the image is explicit enough to convey that message, as long as we understand that the credit card here means something more than its denotative value (which is a credit card). In both the image and the text, whisky becomes the symbol of social prestige through the purchasing power of the individual who shares this image. At no time is there any mention of other elements of whisky other than the symbolic value of prestige and money, for example, there is no mention of the taste of whisky, its origin, or its preference over other drinks. On the contrary, an ideal scenario is built in the image to show a social status, and if the composition of the image or its elements are not enough, they are reaffirmed through the aforementioned hashtags. The only additional elements shown in the text refer to the #wife #friends #drunkness social moment. The conclusion of the analysis in a nutshell would say "I have enough money to buy whisky and have a social drunk night out."

4.2 Case Study 2—Whisky in United Kingdom

The following image is an act of socialization around whisky, just like the previous image (Fig. 14). In this sense, the image is more obvious than the previous one; however, they have a similar (but not identical) message created through the very different use of visual resources. The center of the image is again the whisky, but due to a perspective composition, toward a vanishing point, the visual/semantic weight is shared by both the whisky and the people around the table. That is to say, unlike in the previous image, the leading role is not given to whisky and the purchasing power that you have to have to buy it, but to what whisky leads to, to socialization, to coexistence through the knowledge of whisky as a special drink. The message of social status that is preponderant in the previous image, in this one is relegated to a third or fourth semantic plane, and the text of the image reinforces the sense of knowledge of the drink #class #taste #tryout.

In this apparently more direct image, the company's text expands the interpretative character, since only thanks to the text do we know that it is a tasting or a class, and that all the people gathered have the sensory discovery of whisky as their main interest, contrary to the preceding image, where the sensory characteristics of the product were irrelevant.

5 Conclusion and Further Thoughts

Let's imagine what an image involving whisky, friends, palm trees, and the Hotel de Ville of Paris could symbolize (and no, we're not just referring to #parisplage). What could the whole series of combinations of these elements be and what would their integral sense be at the end of the analysis? What meaning would each of the represented elements take on and what rhetorical strategies could be used in these images? In our case, in the last two images analyzed using similar analytical strategies, we have realized that two images (Figs. 14 and 15) that share some hashtags, and in which the protagonist of the image is whisky, are capable of transmitting two different messages about whisky and what it symbolizes in the eyes of the person who produces the image and shares it on social networks.

To sum up, this chapter is a first attempt to bring the tools that visual arts and art history have used for the analysis of the content and meaning of images closer to consumer research. The main theoretical background is shared by semiotic, semantic, and visual rhetoric, and this information will bring the consumer researcher closer to the analysis of the images. However, the consumer

Fig. 15 A whisky guided tasting in Paris. Note: The original image from social media was replaced by a similar one from the personal image-stock of the author. (Image: Carlos Gómez-Corona)

researcher is not an expert in this field, and much more is yet to be explored. In the case of this chapter, the first author is a sensory and consumer researcher, while the second one is a theorist working on the study of images in art. The authors believe that the combination of both professional backgrounds is suited to put forward this first proposal of image analysis.

Future development in the analysis of social media is needed in consumer research to better understand the meaning of images and how they can be studied in order to identify and answer specific research questions. The theoretical formation is needed in consumer research, as well as new digital tools that will enable a more automatic analysis of the images, while making an extraction for a specific study. Computer-based pre-analysis can already be done today, but enabling just the identification of elements (e.g., image with faces, image with specific objects, etc.) However, the detailed analysis of the composition of the image and the real meaning behind it still relies heavily on the human eye to give the final interpretation.

Acknowledgments

The authors wish to thank Gafa Films and Araceli Arellano-Covarrubias for their support in images recreation. And Pauline Lebrun, Emma Berreby, Lyne Devanlay, Alessandra Avanzi, and Elke Dewolfs for images shooting.

References

1. Pittman M, Reich B (2016) Social media and loneliness: why an Instagram picture may be worth more than a thousand twitter words. Comput Hum Behav 62:155–167
2. Arellano-Covarrubias A, Escalona-Buendia H, Gómez-Corona C, Varela P (2022) Pairing beer and food in social media: is it an image worth more than a thousand words? Int J Gastron Food Sci 27:100483
3. Monde L (2005) Enchères passionnées pour le "Baiser de l'Hôtel de Ville". L'un des très rares tirages originaux du "Baiser de l'Hôtel de Ville" (1950) de Robert Doisneau a été l'objet d'enchères passionnées, lundi 25 avril, à Paris. Retrieved from: https://www.lemonde.fr/culture/article/2005/04/26/encheres-passionnees-pour-le-baiser-de-l-hotel-de-ville_642966_3246.html. Feb 2022
4. Taylor P (ed) (2008) Iconography without texts, vol 13. Warburg Institute
5. Barthes R (1964) Rhétorique de l'image. Communications 4(1):40–51
6. Liopovetsky G, Serroy J (2013) L'esthétisation du monde. Gallimard, Paris
7. Salazar Lama D, Garcia Barrios A (2019) Imágenes mayas: iconografía, discurso y contexto. Introducción al Dossier. Revista Española de Antropología Americana 49:127–129
8. Salazar Lama D, Valencia Rivera R (2017) The written adornment: the many relations of text and image in classic Maya visual culture. Vis Lang 51(2)
9. Magritte R (1955) Les mots et les images. Espace Nord, Bruxelles
10. Kundera M (1990) La inmortalidad. TUSQUETS, Barcelona
11. Frizot M (2005) Nicéphore Niépce, inventeur: un Prométhée rétrospectif. Communications 78(1):99–111
12. Eastman Kodak Company (2021) History. Retrieved from: https://www.kodak.com/en/company/page/history. Nov 2021

13. Sontag S (1973) Sobre la Fotografía. DEBOLSILLO, Ciudad de México
14. Hochman N (2014) The social media image. Big Data Soc 1(2)
15. Prada JM (2018) El ver y las imágenes en el tiempo del internet. AKAL/Estudios Visuales, Madrid
16. Stillerman J (2015) The sociology of consumption. A global approach. POLITY, Cambridge
17. Sibila P (2017) La intimidad como espectáculo. Fondo de Cultura Económica, Ciudad de México
18. Barthes R (1986) El mensaje fotográfico. In: Lo obvio y lo obtuso. Imágenes, gestos, voces. Ediciones Paidós Ibérica, S.A., Barcelona, pp 11–27
19. Gómez-Corona C, Casotti L, Pino FJ, Deliza R (2022) Consumer acceptance of Latin American foods and beverages: a case study of Mexico, Brazil, and Argentina. In: Costa AI, Monteiro MJP (eds) Sensory evaluation and consumer acceptance of new food products: principles and applications. Royal Society of Chemistry, Cambridge
20. Canana Films (Producers) (2019) Las Crónicas del Taco (TV Series). Netflix. https://www.netflix.com/watch/81040719?trackId=13752289
21. Gómez-Corona C (2021) Sensory and consumer research for good: Our contribution to social responsibility. Plenary presentation at 14th pangborn sensory science symposium, Online, 9–12 August, 2021. https://doi.org/10.13140/RG.2.2.20522.95687
22. Panofsky E (1939) Introduction. In: Studies in iconology: humanistic themes in art of the renaissance. Routledge, New York
23. Panofsky E (1955) The history of art as a humanistic discipline, vol 11. Meaning in the Visual Arts
24. Salazar Lama D (2014) Aj k'an witz. Montañas, antepasados y escenas de resurrección en el friso de Balamkú. Campeche Master dissertation, Universidad Nacional Autónoma de Mexico
25. Eco U (2011) La estructura ausente. Introducción a la semiotica. DEBOLSILLO, Mexico
26. Lizarzo Arias D (2009) Introducción. In: Lizarzo Arias D (ed) Semantica de las imágenes. Figuracion, Fantasía e Iconocidad. Siglo XXI, Mexico
27. Mariani M (2019) What images really tell us. Visual rhetoric in art, graphic design, and advertisement. HOAKI, China
28. Carrere A, Saborit J (2000) Retorica de la pintura. Catedra, Madrid
29. Darty (2022). www.darty.com search: "CLIC-CLAC". Retrieved from: https://www.darty.com/nav/achat/meuble/clic_clac_bz/clic_clac/index.html. Mar 2022
30. Gomez-Corona C (2019a) Analysis of image in social media. What can we learn from semiotics and arts? Workshop at 13th pangborn sensory science symposium, Edinburgh, United Kingdom, 28 July–1 August 2019
31. Gómez-Corona C (2019b) Estudios Cualitativos Tradicionales. In: Gómez-Corona C, Lelievre-Desmas M (eds) Introducción al análisis sensorial y estudios con consumidores: Nuevas perspectivas. Mexico City, XOC ESTUDIO
32. Hough G, Wakeling I, Mucci A, Chambers E IV, Gallardo IM, Alves LR (2006) Number of consumers necessary for sensory acceptability tests. Food Qual Prefer 17(6):522–526
33. Galeria degli Uffizi (2022) La Primavera di Botticelli. Retrieved from: https://www.visituffizi.org/it/opere/primavera-di-botticelli/. Mar 2022
34. Cheung M, She J, Wang N (2017) Characterizing user connections in social media through user-shared images. IEEE Trans Big Data 4(4): 447–458
35. Toubia O, Stephen AT (2013) Intrinsic vs. image-related utility in social media: why do people contribute content to twitter? Mark Sci 32(3):368–392

Part II

Methods from Psychology

Chapter 6

A Structural Approach to Social Representations of Food and Beverages: Some Methods to Scientifically Study Lay Thinking

Grégory Lo Monaco, Colomba Codaccioni, and Eric Bonetto

Abstract

This chapter addresses the question of food and beverage studies through the theoretical framework of social representations and more particularly that of the structural approach of representations. We will present some elements relating to the study of food and beverages according to this approach and then present the conceptual framework of social representations and that of the structural approach. We will show how this is relevant for understanding the socially constructed meanings about these important objects of everyday life. Then, a large part of this chapter will be devoted to methodological aspects to grasp the content, structure, and organization of social representations. This part will present the ways of collecting cognitions, the methods to grasp their structuring by virtue of the theoretical aspects related to the structural approach, and we will show how it is possible to highlight their organization. Finally, a part will be reserved for the unspoken of social thought, for socially unspeakable aspects related to questions about normative pressure. In this context, it will be a question of presenting the techniques that have been used in order to be able to unmask socially proscribed representational content and to show all the relevance of such viewpoint in the context of food and beverage studies.

Key words Social representations, Structural approach, Food and beverage studies, Methodology, Mute zone, (Un)masking strategies

1 Introduction

This chapter aims to address methodological issues of food and beverage studies from a social representation perspective. Whether in our way of evaluating our own consumption or that of others, whether it is a question of building meaning on objects related to food in general, a set of social constructions are at work. Indeed, when it comes to considering the meaning that individuals and groups develop about the products they consume and their consumption behaviors, the question of social representations is raised. Any study focusing on the meanings constructed in and by the

Carlos Gómez-Corona and Heber Rodrigues (eds.), *Consumer Research Methods in Food Science*,
Methods and Protocols in Food Science, https://doi.org/10.1007/978-1-0716-3000-6_6,

social system in fact involves social interactions regulated by a set of individual, interpersonal, intergroup, and ideological factors. This is why the question of the group will appear as fundamental as social representations are both guides for behaviors and judgments, but they constitute social and cognitive systems which are able to reflect an identity.

The social representations perspective will be approached from the structural model. First, a theoretical introduction will point out the stakes of such an apprehension of the phenomena considered in this chapter. Then, a methodological section will be proposed in order to indicate ways of collecting the content of social representations, the identification of their structure, and the highlighting of the social regulations at play in the social constructions differentiated from the meaning given to the objects under study.

If the literature concerning social representations of foods and beverages is growing up [1–11], no contribution was made on the methods available and, as we shall see, there are many ways for studying these aspects. Indeed, whether it is a question of collecting the content, the structure, the social regulations of the construction of these contents, and their sociocognitive organization, methodological options are available. Far from being untied from a theoretical set, they are based on conceptual considerations and are in this sense a guarantee of formulations of hypotheses.

By relying on the theoretical framework of social representations, we aim to question common sense as a backdrop to the sociocognitive processes that are at work in the context of judgment, decision-making, and behavioral issues. Actually, social representations are considered as guides for action both in terms of their prediction and in terms of their understanding. Initially proposed by Moscovici [12] about the study of the diffusion and reappropriation of a scientific theory (i.e., psychoanalysis) by common sense, the theory of social representations has since been the subject of important theoretical and methodological developments. First of all, from a theoretical point of view, since the first formulation of the theory, several approaches have been proposed. The reader can refer to Rateau, Moliner, Guimelli, and Abric [13] for a synthesis of these approaches. They refer in turn to the sociogenetic model, the structural model, and the socio-dynamic model. To these first three models are added dialogical and discursive approaches [14].

2 A Structural Approach to Social Representations

Abric proposed an approach by formulating the hypothesis of social representation as a set of structured and organized sociocognitive elements [15, 16]. According to the author, some elements of representation play a more important role than others with regard to the attribution of meanings and the organization of the social representation field. Abric thus proposes to distinguish, within social representations, the central elements of the peripheral ones. The former plays an essential role in terms of assigning meaning and organizing the representational field as a whole, while the latter fulfills a role of interface between the representation and the concrete reality. Apart from the points we have just pointed out, the central and peripheral elements differ in many respects. Central elements are consensual, strongly shared, very stable, resistant to change, and very insensitive to variations in immediate context. Peripheral elements, for their part, allow individual modulations thus allowing individualized worldviews while maintaining a consensual coherence relating to the sharing of central elements. They are also more fluid, less resistant to change. They are therefore defined as context-dependent, thus giving the representation a capacity to adapt to external constraints. Beyond these characteristics, Abric also assigns functions specific to the central core and the peripheral system [15, 16]. According to him, the elements composing the central core would perform two functions. These are the functions that generate and organize meaning. Because of these functions, the central elements are indispensable with regard to the recognition of the object as the object of representation. Without them, or when they are called into question, it is impossible for individuals and groups sharing this representation to recognize the object as the object of representation [17, 18]. In view of these different elements, the "structural theory" proposes a model capable of reflecting the places of consensus, but also the individual variations caused by exposure to situational and daily constraints that individuals and groups must face [19]. Moreover, despite the necessarily abstract nature of this conceptual modeling, this theory reflects a constant concern to be confronted with the daily and concrete reality of individuals and groups.

3 Methodological Pathways Related to the Structural Approach

Since the first work carried out on the theory of social representations, some contributions have focused more on presenting the different methodologies used in the framework of this theory [13, 20–29]. We have seen that we owe the idea of a structure composed of a central core and a peripheral system of social representation to Abric [15]. The challenge for researchers working in this field has been to be able to create methodologies in order to reveal the structure and the internal organization of the representational field. Thus, several methods and techniques have been developed. Among these methodological options, some make us able to work on the content, others on the structure, and others both on the content and the structure or on the organization of the content (organization regulated by a set of variables referring to social anchoring; [30, 31]).

3.1 The Collection of the Content

Verbal association tasks are one of the main methods of collecting the content of social representations. This point is illustrated by the number of works carried out on various objects of social representation whose contents have been revealed by verbal associations. They consist in asking individuals to produce the words or phrases that come to mind concerning an object of representation. According to Flament and Rouquette [25], in practice, the number varies between 3 and 5. Two techniques use this type of task: the free associations and the hierarchical evocations techniques [20, 25, 27, 32]. These methods are based on the hypothesis of the salience of the central core of the representation [33]. Also, in the context of free associations, the analysis carried out on the data collected is part of the prototypical analysis [34]. It makes it possible to highlight the salience of the elements of the representation by crossing two independent indices: the frequency of appearance of an element and its rank of appearance [33, 34]. Thus, apart from asking the subject to produce *n* words to the inductor (i.e., the object of representation), no other constraint is then posed, which is not the case of the method of hierarchical evocations (for syntheses on this topic, [25, 35–38]. Indeed, within the framework of the hierarchical evocation technique, participants have the opportunity to reconsider their production and to rank the elements associated according to an importance criterion. Participants are of course not obliged to change the initial order produced.

Regarding the criteria in the context of the data analysis, the first (i.e., the frequency) is considered as a quantitative and collective criterion; the second (i.e., the rank of appearance in the context of free associations vs. the rank of importance in the context of hierarchical evocations) is considered as a qualitative criterion that is expressed by the participant through the order of appearance of the evocations (i.e., free associations) or the ranking task in terms of importance (i.e., hierarchical evocations).

By distinguishing strong and weak values in these two indices, one can thus construct a double-entry table. The box comprising the most frequent elements and those that were mentioned in the first productions of the associative chain (low rank of appearance or strong importance) will therefore be the most salient; it is then necessary to speak of candidates for the centrality diagnosis. The other boxes in the table include elements with a high frequency and a high or low appearance rank, items with a low frequency of occurrence and a low or high average appearance rank, and finally, elements with a low frequency and a high or low average appearance rank.

In a study of the social representations of wine, Lo Monaco and Guimelli [39] used the hierarchical evocations technique in order to formulate a set of hypotheses relating to the structural status of the elements composing this representational field. One hundred and thirty-four participants were interviewed and 530 evocations were collected. The results of the crossing of frequency and importance are presented in Table 1 below. The results reflect the possibility of formulating the hypothesis of a central core composed of six elements. These six elements are as follows: "alcohol," "party," "qualities," "pleasure," "alcoholism," and "accidents." Therefore, these results show an opposition between two representational universes: positive and negative. In addition, the authors asked participants following the order of importance to evaluate each of the verbal associations using Likert scales ranging from −3 (very negative) to +3 (very positive) in order to assess the attitudinal orientation of the elements and the representational field.

It appears that two of the terms likely to be part of the central core are evaluated negatively; these are the elements "alcoholism" ($M = -2.80$) and "accidents" ($M = -1.86$). On the other hand, three other terms are evaluated more positively, these are the elements "party" ($M = 1.53$), "pleasure" ($M = 1.73$), and "qualities" ($M = 1.92$), and finally only one term has an average approaching 0, it is the element "alcohol" ($M = -0.95$).

Table 1
Prototypical analysis related to the study of the social representations of wine (Ref. [39], p 40)

		Average importance	
		High	**Low**
Frequency	High	Alcohol Party Alcoholism Qualities Pleasure Accidents	Red Vine Grape Meal Culture Bottle Vineyard
	Low	Drunkenness Drink Tasting Moderation drunk religion Cellar	White Bordeaux Harvest Glass Danger Cheese Rosé wine Vintage Abuse Headache To drink Drunk Winemaker Vintage Restaurant Heat

As we have just mentioned, the hierarchical evocation technique provides access only to structural hypotheses. It is therefore impossible, as Lo Monaco et al. [36] point out, to conclude on the structural status of representational elements on the basis of the results obtained by means of this technique. It is therefore essential to use methods able to revealing the structure according to the dichotomy established by Abric [15]. To do this, several techniques have been developed. We will focus on two of them, and the reader interested in more details can refer to the literature on the topic in which recent syntheses have been proposed [36, 40]. We will therefore now address the Attribute-Challenge Technique (ACT) and Test of Context Independence (TCI) as methodological tools allowing the systematic identification of the representational structure.

3.2 Locating the Structure of Social Representations

3.2.1 The Attribute-Challenge Technique (ACT)

Developed by Moliner [17, 18], the ACT is the technique that allowed for the first time the systematic identification of the central versus peripheral status of representational elements. The ACT is based on the symbolic property of the central elements (i.e., assignment of meaning to the object). These elements, as we have seen, are indispensable (non-negotiable; [41]) to define the object. Without them, it is impossible for the individual or the group to recognize the object, it loses its meaning. This technique therefore makes it possible to test the non-negotiable nature of representational elements. It is based on a double-negative principle and makes it possible to know whether the absence of a link between the representational element and the object of representation (i.e., first negation) is unacceptable (i.e., second negation) or acceptable (i.e., absence of second negation). If the absence of the link between the element and the object of representation is unacceptable to the majority of respondents, it means that this element participates in a non-negotiable way in the definition of the object and is therefore central. (In your opinion, can a drink that does not refer to pleasure be called "wine"?)

3.2.2 The Test of Context Independence (TCI)

Faced with the various constraints associated with the various structural diagnosis techniques, Lo Monaco, Lheureux and Halimi-Falkowicz [42] proposed a technique that is based on another property of the central elements: context-insensitivity [32]. Considering therefore that a central element is trans-situational, it should be characteristic of the object of representation whatever the situation. Thus, it is possible to ask participants if in their opinion, "wine" is a drink that always refers, in any case, to pleasure ("In your opinion, is 'wine' a drink that always refers, in any case, to pleasure?"). Faced with this question, if participants answer in the affirmative, then the element can be considered central. Through their presentation of this technique, Lo Monaco et al. [42] were able to demonstrate that it made it possible to collect the same results as a challenge test, while representing a lower perceived cognitive cost for participants.

3.3 Identification of the Organization of the Representational Field

The characterization questionnaire is directly inspired by the Q. Sort technique, a Qualitative Sorting (Sort) Questionnaire (Q.) that was originally proposed by the American statistician Stephenson [43], as a new instrument for collecting information more finely than questionnaires or scales of attitudes, subjective assessments, and personal reflections. From the point of view of the procedure, as it is illustrated in Fig. 1 below, participants are led to make choices in blocks from a list of items. For example, for a list with 20 items, they must identify the 4 items that best match their object design and assign them the score of "+2." Then, from the remaining items, they must select the 4 items that are furthest from their design and assign them the score of "–2." Then, again, they must identify the 4 items that still correspond quite well to their conception of the object (score of +1). Then, among the remaining items, they must select the 4 items that match quite badly and assign them the score of "–1." There are then 4 items that are denoted "0." Each item is thus weighted on a scale with five modalities. It should be noted, however, that it is also possible to proceed in three blocks instead of five. The characterization questionnaire presents several possibilities for processing the data it provides. In this sense, it is interesting in that it not only makes it possible to obtain an average for each of the items tested as part of the procedure but, above all, it makes us able to carry out other analyses in order to reveal the internal organization of the representational field. By way of illustration, we note Guimelli's work on

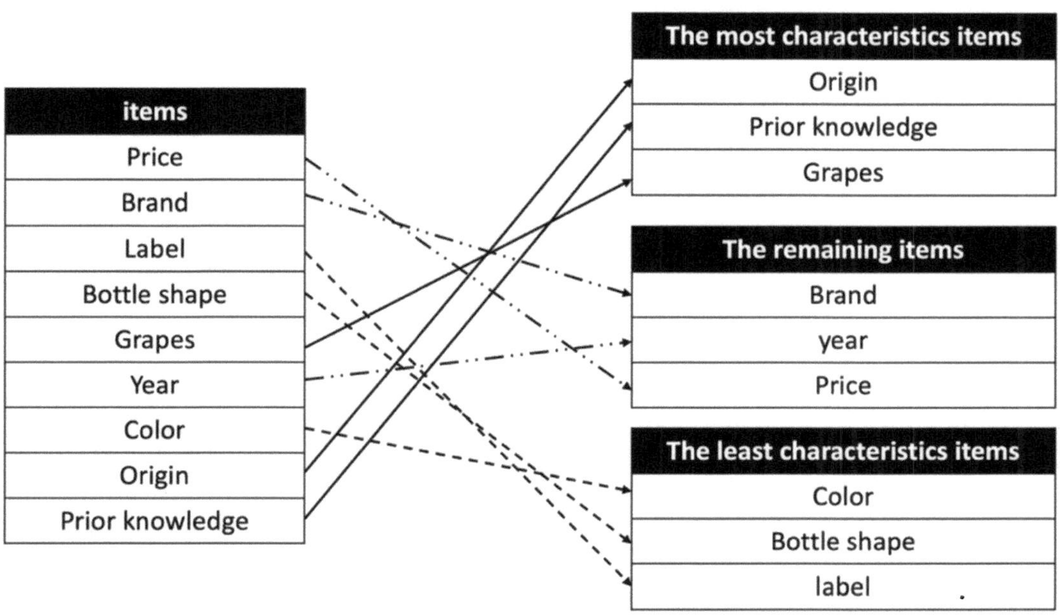

Fig. 1 Illustration of the Q. Sort technique principle related to quality cues when purchasing wine [47]. (The figure is inspired by Lo Monaco et al. [36])

the social representation of hunting [44, 45]. The author used the similarity analysis [46] to see what relationships the items could maintain in his study carried out among a group of Languedoc (i.e., south of France region) hunters. In addition, this technique makes it possible to observe the oppositions within the field and the regulations at play in the context of the sharing of certain representations in connection with the objects studied. In this context, it is possible to mobilize the correspondence factor analysis by considering as categorical dependent variables the response frequencies on each of the three response modalities and by examining the correspondences between the frequencies relating to these response modalities and the independent variables mobilized in the context of the research [47]. Let us add that this work can also be carried out in an experimental perspective by studying the correspondences between the frequencies relating to the modalities of the items and the different experimental conditions of a research, the correspondences factor analysis can be used for this purpose [48].

The Q. Sort technique was therefore used in the field of wine marketing by Lo Monaco et al. [47]. The authors were able to demonstrate that when it came to relying on certain quality cues in order to buy a good wine, this technique was able to reveal a hierarchy of these cues. In addition, by proposing a coding of these data from the Q.Sort in order to carry out a correspondence factor analysis, the authors were also able to highlight the social regulations of these choices in terms of quality cues thus showing effects of structural homology between the positions occupied (i.e., gender and socio-economic status) in the social field and the choices in terms of quality cues. Note that these results have recently been confirmed by Lo Monaco, Tavani and Codaccioni [49] by mobilizing other data processing and in particular the Guttman effect in principal component analysis [50–54].

If concerns about the identification of representational content have mobilized the research activity of researchers for many years since the initial formulation of the social representation theory by Moscovici [12], several works have no longer focused on the contents verbalized spontaneously by the participants, but on the unspoken.

4 The Unspoken Common Sense: Theoretical and Methodological Stakes Concerning the Study of Lay Thinking

Social representations are developed and constructed through social interaction which differentiates them from a Durkheimian perspective considering that common sense was established with regard to social pressure. Consequently, and according to Guimelli [45], we must consider what it is expressed, and what it is not expressed in the framework of these interactions. Indeed, certain areas of the representational field would not be verbalized spontaneously by the participants under standard survey conditions. According to the author, certain aspects that are prey to normative issues are likely to be masked by participants when questioned. It is a whole field of research that has developed from the first empirical results published by Guimelli and Deschamps [55]. The issue is both theoretical and methodological. Theoretical because it is important to theoretically situate this area qualified as "mute zone" by Guimelli and Deschamps in the theory of social representations [55]. The challenge is also methodological in that it was important for the researchers to be able to find means to reveal the socio-representational contents not expressed by participants in certain situational and contextual issues.

A set of studies has been conducted on alcohol and its consumption both to try to show the normative but also identity issues concerning this lively question of research with societal scope. Indeed, in the context of a first work Lo Monaco et al. [56] questioned a population of students about their social representation of alcohol. Based on an abundant literature in the field of prevention in France [57–59], the authors were interested in the dynamics of the expression of students concerning this object. In order to try to highlight the normative and identity issues taking part in the expression of these participants in relation to this object, Lo Monaco et al. [56] carried out an experiment mobilizing two between-subject factors. The first factor referred to the instruction of expression where some of the students had to answer in their own name, in a classical way and the other part of these students, in all comparable points, were invited to express themselves as students would do in general. This factor (or independent variable) refers to the substitution technique proposed by Guimelli and Deschamps [55], a technique inspired by the work carried out by Jellison and Green [60] concerning the identification paradigm. The second factor was to vary the presentation of the experimenter. In a direct connection with the work done on the experimenter effect [61, 62], it was a question of manipulating the status of the experimenter (in this case, the interviewer who proposed the questionnaire in the context of this study). According to the experimental condition, the interviewer presented himself either as a master's student in sociology who carried out a final work on the relationship maintained by young people to alcohol, or as a health worker who was a member of the Marseilles alcohol abuse center (a fictitious structure created for the purpose of the experimentation).

Participants were invited to associate the first four words that came to mind when they were told the word "alcohol." They were also instructed to rank these words in order of importance [20, 36, 37]. The authors considered the words that were both the most important and the most frequent and compared the results in the four experimental conditions obtained by crossing the two independent variables as follows: 2 (Instruction: normal vs. substitution) × 2 (Presentation of the interviewer: student in Master in Sociology vs. Health Worker). The results show that the only condition where students link alcohol to poor health is when they answer on their own and in front of a health worker. This result shows the interest of making these variations of instructions when it comes to questioning people about objects that are able to take on normative and identity issues.

In the field of food studies, these instructions are particularly interesting to mobilize when it comes to work on food objects in connection with innovation or in connection with new practices. This is the case in particular in our opinion of new foods or organic. Ultimately, as soon as the object is new and/or refers to what innovation breaks with usual practices or comes to what people consider to be natural, it is possible to formulate the hypothesis of these masked cognitions. However, leaving them in the shadows and not revealing them would be a mistake because, it is very likely that the masked, more besides that the verbalized elements are in order in the orientation and triggering of the action. Experimental evidence remains to be provided despite encouraging results obtained by Hidalgo [63] or Guimelli et al. [64].

5 Conclusion

The work carried out in the field of study of social representations therefore offers great prospects for trying to scientifically investigate objects related to foods and beverage studies. Whether from a perspective of health prevention or brand management in the field of marketing and in particular marketing segmentation on the basis of beliefs considered as social psychological variables, it seems to us that the methodological tools offered by this field of study are important levers to operate. Indeed, some works have shown that social representations were strong predictors of willingness to try GMOs and organic foods [2, 3]. This is also the case of the results obtained in relation with new foods and new functional drinks [65, 66]. These examples reveal how it is important to be clear concerning methodologies that are able to be used to work on social representations both for studying their content and structure. According to us, as theoretical and methodological ones are unrelated, working on social representations reclaims to be scientifically cautious concerning the use of methodological tools. This was the objective of this chapter to make the reader aware of the stakes, the constraints, and the possibilities when we have to connect food and beverage studies with the social representation theory.

Finally, this field of study is not limited to considering that what people say is the most important thing to know in order to understand their behaviors. A vast field of work is available to researchers who would like to explore more precisely the role played by the masked cognitions of lay thinking, those masked ideas or beliefs about which one can assume a preponderant role when it comes to trying to understand the consumers' preferences, choices, and behaviors.

References

1. Araujo MV, Lo Monaco G, Bruch KL (2021) Social mobility and the social representation of sparkling wine in Brazil and France. Wine Econ Policy 10:89–100
2. Bäckström A, Pirttilä-Backman AM, Tuorila H (2004) Willingness to try new foods as predicted by social representations and attitude and trait scales. Appetite 43:75–83
3. Bartels J, Reinders MJ (2010) Social identification, social representations, and consumer innovativeness in an organic food context: a cross-national comparison. Food Qual Prefer 21:347–352
4. Chen MF (2018) Social representations of genetically modified foods and public willingness to consume such foods in Taiwan. J Sci Food Agric 98:5428–5434
5. Onwezen MC, Bartels J (2013) Development and cross-cultural validation of a shortened social representations scale of new foods. Food Qual Prefer 28:226–234
6. Lo Monaco G, Bonetto E (2019) Social representations and culture in food studies. Food Res Int 115:474–479
7. Lo Monaco G, Bonetto E, Codaccioni C, Araujo MV, Piermattéo A (2020) Alcohol 'use' and 'abuse': when culture, social context and identity matter. Curr Opin Food Sci 33:9–13
8. Melendrez-Ruiz J, Arvisenet G, Laugel V, Chambaron S, Monnery-Patris S (2020) Do French consumers have the same social representations of pulses as food industry professionals? Foods 9:147
9. Patinho I, Saldaña E, Selani MM, Teixeira ACB, Menegali BS, Merlo TC et al (2021) Original burger (traditional) or burger with mushroom addition? A social representation approach to novel foods. Food Res Int 147: 110551
10. Rodrigues H, Ballester J, Saenz-Navajas MP, Valentin D (2015) Structural approach of social representation: application to the concept of wine minerality in experts and consumers. Food Qual Prefer 46:166–172
11. Rodrigues H, Cielo DP, Goméz-Corona C, Silveira AAS, Marchesan TA, Galmarini MV, Richards NSPS (2017) Eating flowers? Exploring attitudes and consumers' representation of edible flowers. Food Res Int 100:227–234
12. Moscovici S (2008) Psychoanalysis: its image and its public. Polity Press, Cambridge
13. Rateau P, Moliner P, Guimelli C, Abric JC (2011) Social representation theory. In: Van Lange P, Kruglanski A, Higgins T (eds) Handbook of theories of social psychology, vol 2. Sage, London, pp 478–498
14. Lo Monaco G, Delouvée S, Rateau P (2016) Les représentations sociales. Théories, méthodes, applications [Social representations. Theories, methods, applications]. De boeck
15. Abric J-C (1976) Jeux, conflits et représentations sociales [Games, conflicts and social representations]. Thèse de Doctorat d'État de l'Université de Provence, Aix-en-Provence
16. Abric J-C (1987) Coopération, compétition et représentations sociales [Cooperation, competition and social representations]. Del Val
17. Moliner P (1988) La représentation sociale comme grille de lecture [Social representation as a reading grid], Aix-en-Provence, Thèse de doctorat de l'Université de Provence
18. Moliner P (1989) Validation expérimentale de l'hypothèse du noyau central des représentations sociales [Experimental validation of the hypothesis of the central core of social representations]. Bull Psychol 41:759–762
19. Rateau P, Lo Monaco G (2016) La théorie structurale ou l'horlogerie des nuages [Structural theory or cloud watchmaking]. In: Lo Monaco G, Delouvée S, Rateau P (eds) Les représentations sociales. Théories, méthodes et applications [Social representations. Theories, methods, applications]. De Boeck, Louvain-La-Neuve, pp 113–129
20. Abric J-C (2003) Méthodes d'études des représentations sociales [Methods of study of social representations]. Érès

21. Breakwell GM, Canter DV (1993) Empirical approaches to social representations. Oxford University Press
22. Doise W (2005) Les représentations sociales [Social representations]. In: Dubois N (ed) Psychologie sociale de la cognition [Social psychology of cognition]. Dunod, Paris, pp 153–207
23. Doise W, Clémence A, Lorenzi-Cioldi F (1993) The quantitative analysis of social representations. Harvester Wheatsheaf, New York
24. Doise W, Palmonari A (1986) L'étude des représentations sociales [The study of social representations]. Delachaux et Niestlé
25. Flament C, Rouquette M-L (2003) Anatomie des idées ordinaires. Comment étudier les représentations sociales [Anatomy of lay ideas. How to study social representations]. Armand Colin, Paris
26. Jodelet D (1989) Folies et représentations sociales [Madness and social representations]. Presses Universitaires de France
27. Moliner P, Rateau P, Cohen-Scali V (2002) Les représentations sociales. Pratique des études de terrain [Social representations. Practice of field studies]. Presses Universitaires de Rennes, Rennes
28. Moscovici S, Buschini F (2003) Les méthodes des sciences humaines [Methods of the humanities]. Presses universitaires de France, Paris
29. Rouquette M-L, Rateau P (1998) Introduction à l'étude des représentations sociales [Introduction to the study of social representations]. Presses Universitaires de Grenoble, Grenoble
30. Doise W (1990) Les représentations sociales [Social representations]. In: Richard J-F, Ghiglione R, Bonnet C (eds) Traité de psychologie cognitive [Handbook of cognitive psychology], vol III. Dunod, Paris, pp 113–174
31. Doise W, Clémence A, Lorenzi-Cioldi F (1993) Analyse de Données et Représentations Sociales [Data analysis and social representations]. PUG
32. Abric J-C (1994) Les représentations sociales: aspects théoriques [social representations: theoretical aspects]. In: Abric J-C (ed) Pratiques sociales et représentations [Social practices and representations]. Presses Universitaires de France, Paris, pp 11–35
33. Vergès P (1994) Approche du noyau central: Propriétés quantitatives et structurales [The approach of the central core: quantitative and structural properties]. In: Guimelli C (ed) Structures et transformations des représentations sociales [Structures and transformations of social representations]. Delachaux et Niestlé, Paris, pp 233–253
34. Vergès P (1992) L'évocation de l'argent. Une méthode pour la définition du noyau central d'une représentation [The evocation of money. A method for defining the central core of a representation]. Bull Psychol 45:203–209
35. Dany L, Urdapilleta I, Lo Monaco G (2015) Free associations and social representations: some reflections on rank-frequency and importance-frequency methods. Qual Quant 49:489–507
36. Lo Monaco G, Piermattéo A, Rateau P, Tavani JL (2017) Methods for studying the structure of social representations: a critical review and agenda for future research. J Theory Soc Behav 47:306–331
37. Moliner P, Lo Monaco G (2017) Méthodes d'association verbale pour les sciences humaines et sociales [Methods of verbal association for the humanities and social sciences]. Presses Universitaires de Grenoble, Grenoble
38. Moliner P, Lo Monaco G (2019) Métodos de asociación verbal para las ciencias humanas y sociales. Fundamentos conceptuales y aspectos prácticos. biblioteca de educación Herramientas universitarias, Mexico
39. Lo Monaco G, Guimelli C (2008) Représentations sociales, pratique de consommation et niveau de connaissance: le cas du vin [Social representations, consumption practice and level of knowledge: the case of wine]. Les Cahiers Internationaux de Psychologie Sociale 78: 35–50
40. Lo Monaco G, Rateau P (2016) Les méthodes d'étude de l'approche structurale des représentations sociales [Methods of studying the structural approach to social representations]. In: Lo Monaco G, Delouvée S, Rateau P (eds) Les représentations sociales [Social representations]. de Boeck, Bruxelles, pp 135–144
41. Moscovici S (1993) Introductory address to the International Conference on Social Representations, Ravello, 1992. Pap Soc Represent 2:160–170
42. Lo Monaco G, Lheureux F, Halimi-Falkowicz S (2008) Le test d'indépendance au contexte (TIC): une nouvelle technique d'étude de la structure représentationnelle [The test of context independence (TCI): a new technique for studying representational structure]. Swiss J Psychol 67:119–123
43. Stephenson W (1935) Correlating persons instead of tests. J Pers 4(1):17–24
44. Guimelli C (1988) Agression idéologique, pratiques nouvelles et transformation progressive

d'une représentation sociale: la représentation de la chasse et de la nature chez des chasseurs languedociens [Ideological aggression, new practices and progressive transformation of a social representation: the representation of hunting and nature among Languedoc hunters]. Aix-en-Provence: Thèse de doctorat de l'Université de Provence
45. Guim Guimelli C (1998) Chasse et Nature en Languedoc: étude de la dynamique d'une représentation sociale [Hunting and Nature in Languedoc: study of the dynamics of social representation]. L'Harmattan, Paris
46. Flament C (1981) L'analyse de similitude: une technique pour la recherche sur les représentations sociales [The similarity analysis: a technique for the research on social representations]. Les Cahiers de psychologie cognitive 1:375–395
47. Lo Monaco G, Piermattéo A, Guimelli C, Abric J-C (2012) Questionnaire of characterization and correspondence factor analysis: a methodological contribution in the field of social representations. Span J Psychol 15: 1233–1243
48. Piermattéo A, Lo Monaco G, Moreau L, Girandola F, Tavani JL (2014) Context variations and pluri-methodological issues concerning the expression of a social representation: the example of the Gypsy Community. Span J Psychol 17:1–12
49. Lo Monaco G, Tavani JL, Codaccioni C (2020) Social factors and preference for quality cues: towards a social construction of wine quality. Food Res Int 134:109270
50. Flament C, Guimelli C, Abric JC (2006) Effets de masquage dans l'expression d'une représentation sociale [Masking effects in the expression of a social representation]. Les Cahiers Internationaux de Psychologie Sociale 69:15–31
51. Flament C, Milland L (2003) Un effet Guttman dans l'analyse des représentations sociales [A Guttman effect in the analysis of social representations]. In: Abric J-C (ed) Méthodes d'étude des représentations sociales [Methods of study of social representations]. Erès, Ramonville Saint-Agne, pp 201–220
52. Flament C, Milland L (2005) Un effet Guttman en ACP [A Guttman effect in PCA]. Mathématiques et Sciences Humaines 71:25–49
53. Gaymard S (2006) The representation of old people: comparison between the professionals and students. Int Rev Soc Psychol 19:69–92
54. Lo Monaco G, Guimelli C (2011) Hegemonic and polemical beliefs: culture and consumption in the social representation of wine. Span J Psychol 14:232–245
55. Guimelli C, Deschamps J-C (2000) Effets de contexte sur la production d'associations verbales. Le cas de la représentation sociale des Gitans [Context effects on the production of verbal associations. The case of the social representation of gypsies]. Les Cahiers Internationaux de Psychologie Sociale 47–48:44–54
56. Lo Monaco G, Lheureux F, Chianèse L, Codaccioni C, Halimi-Falkowicz S, Cano P (2009) Contexte d'expression, statut social des intervenants de santé et production d'un discours normatif: le cas de l'alcool et des jeunes [Context of expression, social status of health workers and production of a normative discourse: the case of alcohol and young people]. Prat Psychol 15:367–386
57. Ancel P, Gaussot L (1998) Alcool et alcoolisme: pratiques et représentations [Alcohol and alcoholism: practices and representations]. Editions L'Harmattan, Paris
58. Freyssinet-Dominjon J, Wagner AC (2003) L'alcool en fête: manières de boire de la nouvelle jeunesse étudiante [Alcohol in parties: drinking habits of the new student youth]. L'Harmattan, Paris
59. Gaussot L (2004) Modération et sobriété: études sur les usages sociaux de l'alcool [Moderation and sobriety: studies on social habits towards alcohol]. L'Harmattan, Paris
60. Jellison JM, Green J (1981) A self-presentation approach to the fundamental attribution error: the norm of internality. J Pers Soc Psychol 40: 643
61. Gilbert DT, Hixon JG (1991) The trouble of thinking: activation and application of stereotypic beliefs. J Pers Soc Psychol 60:509–517
62. Schuman H, Converse JM (1971) The effects of black and white interviewers on black responses in 1968. Public Opin Q 35:44–68
63. Hidalgo M (2012) Représentations sociales et contextes: études autour de l'expression et des comportements en lien avec les éléments masqués [Social representations and contexts: studies around expression and behaviors related to hidden elements]. Thèse de Doctorat de l'Université d'Aix-Marseille, Aix-en-Provence
64. Guimelli C, Hidalgo M, Piermattéo A, Lo Monaco G, Abric J-C (2012) Zone muette des représentations sociales et comportements [Silent zone of social representations and behaviors]. In: 11e Conférence Internationale sur les Représentations Sociales. Portugal, Evorà

65. Huotilainen A, Pirttilä-Backman AM, Tuorila H (2006) How innovativeness relates to social representation of new foods and to the willingness to try and use such foods. Food Qual Prefer 17:353–361

66. Huotilainen A, Seppälä T, Pirttilä-Backman AM, Tuorila H (2006) Derived attributes as mediators between categorization and acceptance of a new functional drink. Food Qual Prefer 17:328–336

Chapter 7

Perception and Representation: Sorting Task and Projective Mapping

Sylvie Chollet and Dominique Valentin

Abstract

People form mental concepts of categories of objects, which permit them to respond appropriately to new objects they encounter by making inferences from properties of known categories. Among those concepts, food concepts are central to human food perception and acceptation, but how these concepts are represented in memory is unclear and remains a vast field of research to explore. Traditionally, it was though that categories and hence concepts were well defined with clear definitions that specifies what is in and out of the category. However, it has not been possible to find definitions for many familiar categories especially in the food domain where food can be classified in many different ways: food to eat at breakfast, processed food, sweet food, food for kids, protein-containing food, plant food. Thus, it is necessary to use empirical approaches to understand food mental representations. In this chapter, we will focus on how to access consumers' mental representations. We first discuss the notions of categorization, categories, and concepts that are closely related to mental representations. Then, two methods used in sensory evaluation and consumer science to access mental representations are presented: sorting task and projective mapping. These methods are illustrated with applications taken from the literature.

Key words Representations, Categorization, Sorting task, Projective mapping

1 Introduction

Understanding consumers' thoughts and behaviors is an issue that has plagued sensory evaluation researchers for years. Different methods have emerged in the past years to address this issue. Some of these methods come from the fields of anthropology and sociology (*see* Part I) and others from marketing and consumer sciences (*see* Part III) or social and cognitive psychology (Part II). This chapter focus on two methods originally used in psychology to access consumers' mental representations: sorting task [1] and projective mapping [2, 3]. The rationale behind these two tasks is that identifying how consumers organized their knowledge is an open window into their behavior. Their main advantage is that, because they do not rely primarily on language and consciousness,

Carlos Gómez-Corona and Heber Rodrigues (eds.), *Consumer Research Methods in Food Science*,
Methods and Protocols in Food Science, https://doi.org/10.1007/978-1-0716-3000-6_7,

they are less prone to social desirability effects than more explicit methods such as questionnaire or focus groups. We first define the notion of representation, categorization, categories, and concepts which are closely related to mental representations. Then, each method is described and illustrated with examples taken from the literature.

2 Some Key Notions

2.1 What Is a Representation?

A representation is any notation or sign or set of symbols which re-presents something or stands for something in the absence of that thing. The objects of the representation could belong to the external world (i.e., things) or to the internal world (i.e., ideas).

Representations can be of two types: conceptual representations like thoughts, beliefs, or wishes (language-like "propositional" representations) or sensory motor representations like visual, auditory, or any other sensory representations (pain, hunger, etc.).

2.2 What Is a Category?

A category is a set of diverse objects or elements considered similar from a certain point of view. The term category refers to a set of things that exist objectively in the world. Depending on the point of view, an object may or may not belong to a category. For instance, a category could be formed by elements with the same name (clothes: trousers, skirt, pullover, etc.), perceptually close (sweet food: candy, honey, banana, etc.), involving the same action (situations that make you cry: bereavement, onion, wound, etc.), having a common function (sharp objects with elements: knife, scissors, chainsaw, etc.), or included in the same scene (at a movie: ticket, popcorn, screen, etc.).

2.3 What Is a Concept?

A concept is the mental representation of a category. The term concept refers to an abstract notion. For example, the category dog would include large dogs, small dogs, German Shepherds, Yorkies, and long-haired dogs. The concept dog would refer to an animal, which has four legs and a tail, which barks, which has the reputation of being man's best friend, which is a common pet, etc.

2.4 What Is a Categorization Process?

Categorization is a cognitive activity that consists in grouping elements considered similar. These elements can be objects, people, events, ideas, emotions, tastes, smells, etc. Categorization is one of the most studied psychological processes because it has long been seen as a basis for many other cognitive processes. It is involved in object recognition and identification, comprehension, reasoning, and problem-solving.

The benefits of categorization are numerous. First, categorization allows *less costly cognitive processing* of the information we are confronted with on a daily basis by reducing the constant need to learn. Without a concept, each entity would be perceived as unique. We would quickly be overwhelmed by the diversity of things we interact with and we would be unable to *remember* more than a few seconds each of these things. Moreover, we would have to refer to each individual entity with a separate name, so that our lexicon would be considerable and communication as we know it would be impossible. Categorization helps us to *organize our knowledge* or to decide what is an appropriate action in response to a stimulus. For example, if I categorize the dog in front of me as an attack dog, I probably will not try to pet it, but rather move in the opposite direction. Categorization also allows us to *infer characteristics* of an object from the properties of the category to which it belongs. For example, if the animal I see in front of me has feathers, I will categorize it as a bird and I will infer that it might fly away if I get too close to it. The objects have visible properties such as color or shape and non-visible properties such as the name of the object, the knowledge that one has of the object (for example its age, to whom it belongs), its function (what it is used for, how it is used), and the personality (for people). Categorizing allows one to infer properties that cannot be perceived.

The goal of categorization is to provide maximum information with the least cognitive effort. As a consequence, it leads to a simplified version of the world. This simplification concerns both the inductive aspects of categorization, i.e., the recognition of an object as belonging to a category, and its deductive aspects, i.e., the attribution to an object of the characteristics of the category. A major effect of categorization is both an accentuation of the differences between the elements belonging to different categories (contrast phenomenon) and an accentuation of the similarities between elements belonging to the same category (assimilation phenomenon) [4, 5].

3 Sorting Techniques

Originally the sorting task was created in psychology in the 1930s to understand representations (6, 7 for an extensive review) with the main objective of revealing the structure of the representations and interpreting the underlying dimensions. In the following, we present the free sorting task and two of its variants, the directed sorting task, and the hierarchical sorting task.

3.1 Free Sorting Task

3.1.1 General Principle

The sorting task is based on a simultaneous presentation of a stimulus set to the assessors. The stimuli are presented in a different order for each assessor, randomly or according to a Williams Latin square. Assessors have to first look at, smell and/or taste (according to the study objectives) all the stimuli, then sort them into groups mutually exclusive based on their similarity. Assessors choose their own grouping criteria. The sorting task can be stopped at this step or can be followed by a criteria identification step in which assessors are asked to indicate which criteria they used to sort the stimuli. To go further in the understanding of the criteria used by the participants, it is also possible to conduct at the end of the task a retrospective report. This latter could be in the form of elicitation interviews where experimenter ask participants to comment their decisions on particular moments during the sorting task either from memory or with the help of videos of the sorting task. This phase allows to extract the strategies the most frequently used by participants (for technical details, *see* Refs. [1, 6–9]).

3.1.2 Illustration

Sorting tasks have been used in many studies aiming at understanding consumers representations. It began in the field of odor representation with Harry Lawless in 1989 who studied the olfactory space for 18 chemicals or essential oils. He found that this space was represented in three dimensions; "woody," "citrus," and "partially woody and citrus" [10]. Subsequently, other authors applied the same approach to other sets of odors [11–14] and later to many other areas such as food [15–17], beverages [8, 18, 19], fabrics [20], or flowers [21] for instance.

A nice illustration of how sorting task can be used to explore mental representations was published recently by Gómez-Corona and collaborators [22]. The objective of their study was to evaluate the effect of gender and type of consumption on Mexican beer consumers' representations. Four groups of participants, men drinking craft beers, men drinking industrial beers, women drinking craft beers, women drinking industrial beers were asked to visually sort a set of 30 commercial beer pictures depicting 15 industrial and 15 craft beers. The assessors were given the following instructions:

Step 1

> *"Here you have a set of 30 images with 30 different beers. Please look at them and sort them. You can sort them anyway you like and make as much groups as you want."*

Step 2

> *"Here you have some post-its. Please write down the description of each group you have made and stick them near each group."*

The sorting data of each participant were encoded in individual beer × beer distance matrices where the rows and the columns are beers. A value of 0 between a row and a column indicates that the participant put both beers together, and a value of 1 indicates that the beers were not put in the same group. The final matrices were then submitted to DISTATIS analyses [23]. The words used to describe the categories of the sorting task are projected as supplementary points on the DISTATIS RV map (e.g., they do not contribute to the DISTATIS dimensions).

Results are presented Fig. 1. Maps, obtained for the two men's groups, show an organization of the beers that go from the more industrial beers (Sol, Heineken, Modelo), to the darker industrial beers (Victoria, Bohemia, Leon) followed by the specialty industrial beers (Noche Buena, Guinness), and finally the craft beers at the opposite on the map. The main difference between the two maps is

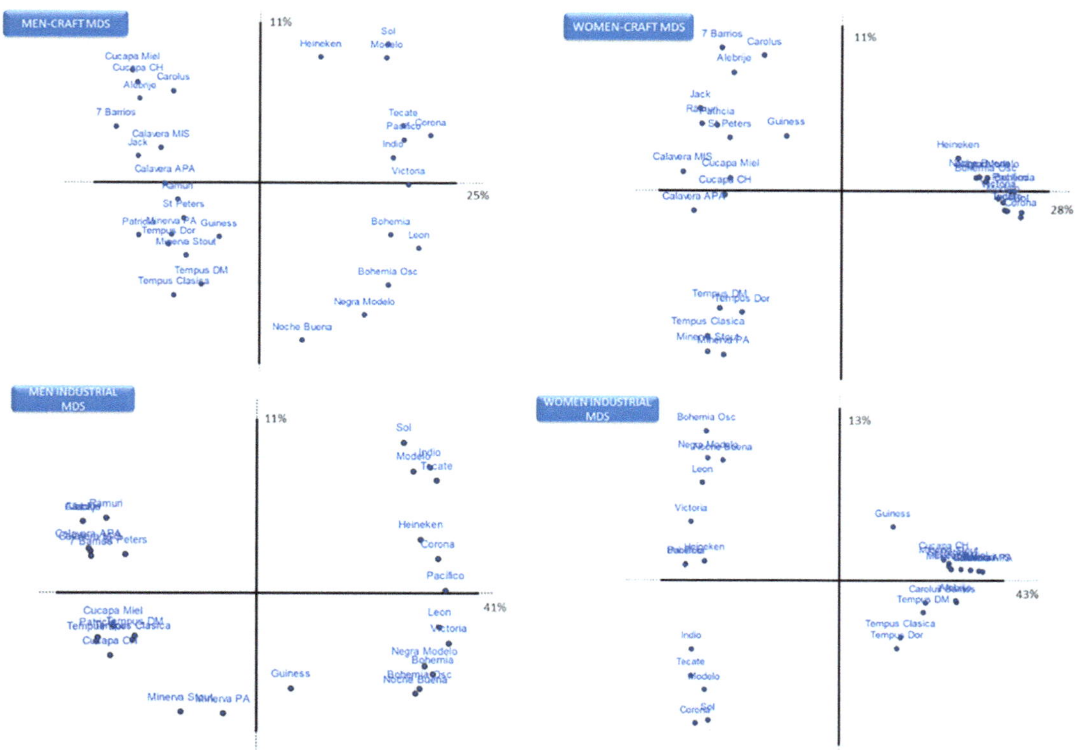

Fig. 1 DISTATIS compromise map of (**a**) craft men, (**b**) craft women, (**c**) industrial men, (**d**) industrial women. (Adapted from Gómez-Corona and collaborators [22])

that while for industrial participants the beers appear to be separated in two major groups (industrial versus craft), they form a gradient for craft participants.

The difference between craft and industrial women is even clearer as their maps show opposite trends. On the craft women map, while industrial beers are closely grouped, showing no distinction, the craft beers are separated along the second dimension in two or more groups. At the opposite on the industrial women map, industrial beers are separated along the second dimension into blond industrial beers (Sol, Corona, Tecate), and dark industrial beers (Bohemia Oscura, Noche Buena), while craft beers are closely grouped showing no distinction among them.

This study reveals an interaction between gender and product consumption. For men, a beer continuum going from the industrial to the craft beers is observed. Women, on the other hand, differentiate the beers they consume (e.g., craft beer) and group all other beers together (e.g., industrial beers). This gender effect suggests that men and women build their mental representation differently. Men's representations seem to be mediated by consumption habits and previous knowledge (cognitive dimension), whereas women's representations seem to be mediated by their attitudes toward the products (affective dimension). This interaction effect between gender and type of consumption highlights the complex relationship that consumers have toward beers.

3.2 Directed Sorting Task

3.2.1 General Principle

The directed sorting task consists in providing information on the number and/or nature of the groups to be formed. It is generally used to validate specific hypotheses on the structure of mental representations.

3.2.2 Illustration

In the last 10 years, this technique has been used to understand the effect of expertise on wine and beer representations [1, 24] as well as to evaluate whether factors such as origin plays a role in the sensory representation of wine [25]. For example, in an often-cited study, Ballester and collaborators in 2009 used a directed sorting task to explore the representation of the aromas of white, red, and rosé wines among participants from three panels: consumers, trained panelists, and wine experts [24]. Participants carried out a ternary sorting task in which they were asked to sort 18 samples (six red, six white, and six rosé) in three groups: red, white, and rosé by smelling them only. The instructions were as follows:

> *"You have in front of you 18 wines. We ask you to smell each wine in the order in which the wines are presented. Then, you have to sort the different samples, only by smelling them, in three groups: red, white, and rosé wines. Please call the facilitator when you have finished."*

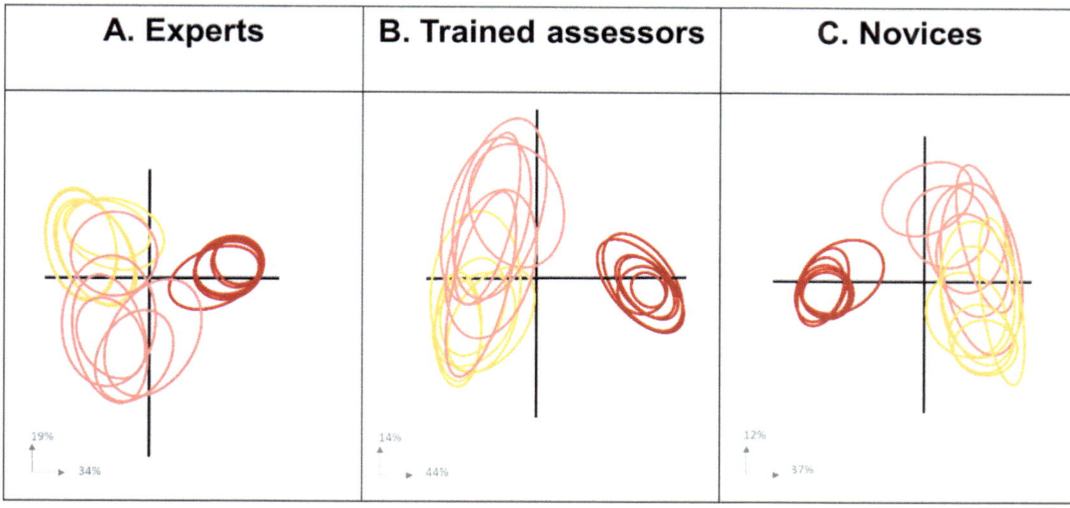

Fig. 2 DISTATIS compromise map for (**a**) experts' ternary sorting task, (**b**) trained assessors' ternary sorting task, (**c**) novices' ternary sorting task. (Adapted from Ballester and collaborators [24])

The results were analyzed using DISTATIS. The compromise maps (Fig. 2) show a very good separation between white and red wines for all panels. This suggests that white and red wine color categories have very salient odor features, that are well represented in memory and thus enable the recognition of the type of wines whatever the level of expertise considered. In contrast, rosé wines are more scattered in the space and tend to be confounded with white wines. None of the participants seemed to have developed a clear representation of the aroma of rosé wines.

3.3 Hierarchical Sorting Task

3.3.1 General Principle

Another variant proposed by Rao & Katz in 1971 is called hierarchical sorting [26]. In this variant, assessors start by a free sorting task; then, they are asked to successively merge the two most similar groups until a single group is formed (hierarchical ascending sorting, [27]), or conversely, to separate each group into finer groups until no separation was possible (descending hierarchical sorting, [28] or both [8, 29].

3.3.2 Illustration

Hierarchical sorting provides information about the structure of representations as illustrated by Honoré-Chédozeau and collaborators in 2017 [8]. This study explores Beaujolais wine knowledge representation using 70 wine labels. It was designed to compare the representation of Beaujolais wine experts to the official PDO Beaujolais wine taxonomy (commonly found in the wine literature). The specificity of this study lies in the fact that it involves an ascendant and descendant hierarchical sorting task followed by a semi-directed interview. During the interview, participants were shown the hierarchy of groups they have made and asked to explain how they sorted the wine labels at each step of this hierarchy. The end goal of this interview was to access the strategies and criteria used by the participants and thus better understand their representation of Beaujolais wines.

The instructions received by the participants are presented below. Depending on the number of groups obtained during the first step (here >4), participants were asked to make ascending or descending sorting tasks.

Free sorting task

> *"You have in front of you a set of wine label pictures. Please group the wine labels that, in your opinion, belong to the same wine category. You can make as many groups as you like, with a minimum of two groups."*

Ascendant sorting task

> *"You have just made x number of wine labels. Now, would it be possible for you to gather together some of your groups?"*

Descendant sorting task

> *"You have just made x number of wine groups. Do you think it would be possible now to redivide some of your groups?"*

Semi-directed interview task

> 1. *"Can you explain to me how you made different groups for each sorting?"* 2. After the assessors stopped speaking, the experimenter asked them *"why"* and *"how"* questions.

The data from the free hierarchical sorting task were encoded in a rectangular matrix where the rows are wine labels and the columns hierarchy levels. Results were analyzed using a hierarchical multiple factor analysis (HMFA, [30]) followed by a hierarchical ascending cluster analysis (HAC). This analysis provides a representation of the consensus among participants as well as a hierarchical ascending clustering of wine labels obtained from collective hierarchies. Figures 3 and 4 show the results obtained for a panel of 30 experts from the Beaujolais vineyard. The consensus map (Fig. 3) shows that the expert panel is divided into two subgroups. To better understand the representation of these two subgroups, separate HMFA were conducted for each subgroup. For each

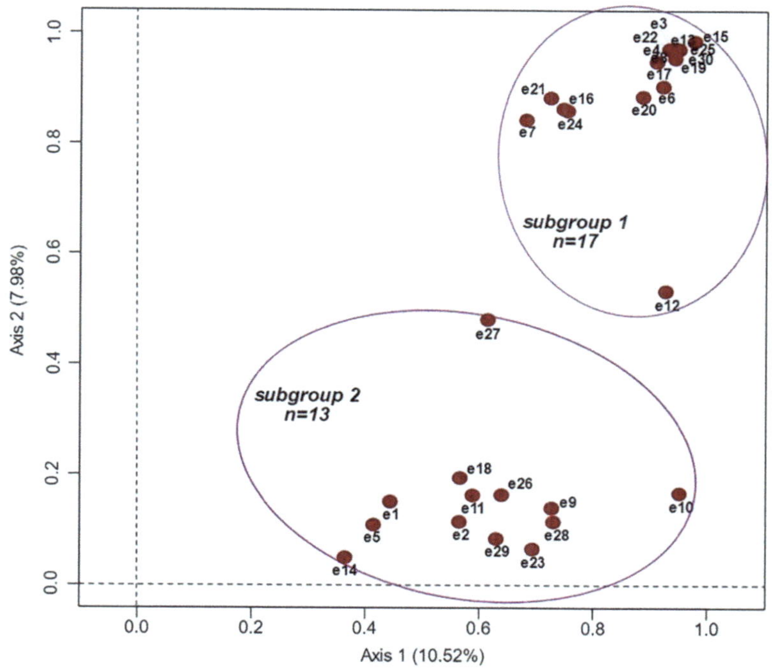

Fig. 3 HMFA graphical representations of assessors (Axes 1 & 2). The dotted ellipses correspond to the two subgroups of experts. (Adapted from Honoré-Chédozeau and collaborators [8])

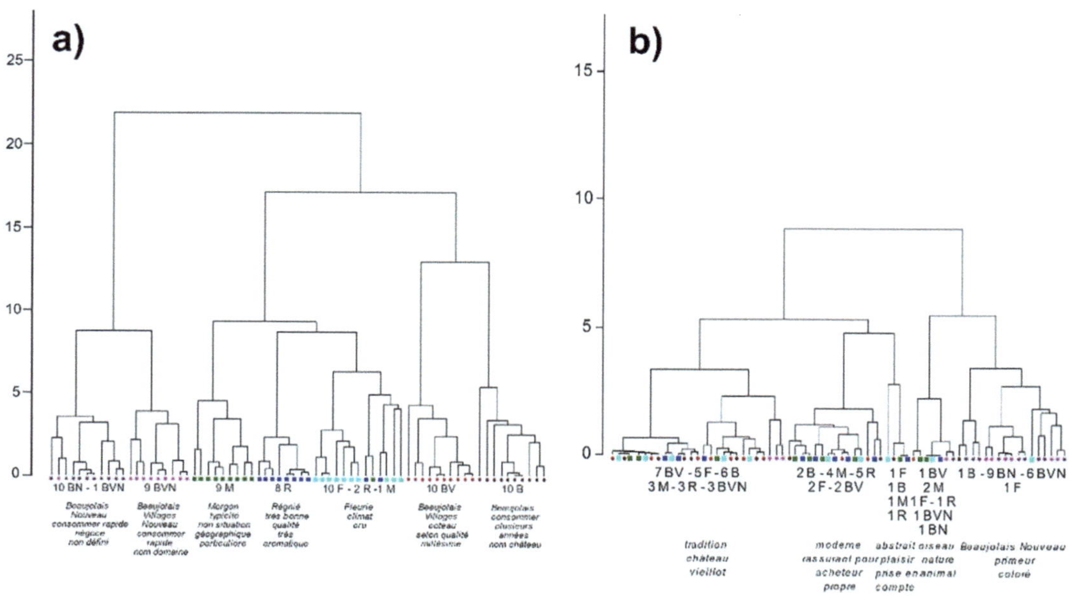

Fig. 4 Hierarchical ascending cluster of wine labels from collective hierarchies for the two subgroups of experts, showing the most characteristic terms for each cluster (**a**): 17 experts; (**b**): 13 experts. B Beaujolais, BN Beaujolais Nouveau, BV Beaujolais Villages, BVN Beaujolais Villages Nouveau, M Morgon, F Fleurie, R Régnié. (Adapted from Honoré-Chédozeau and collaborators [8])

HMFA, an HAC was performed on the first HMFA dimensions explaining 85% of variance (Fig. 4). For the first subgroup of 17 experts, the seven clusters obtained correspond exactly to the PDO Beaujolais wine taxonomy and the first term that best characterizes each cluster corresponds to the name of the wine PDO. However, for the second subgroup of 13 experts, the five clusters do not correspond to this taxonomy. Only the first term for one cluster is related to the name of a PDO (Beaujolais Nouveau), while the terms for the other clusters are mostly related to the visual appearance of the wine labels. These results explain the separation of the expert panel into two subgroups. It suggests two different categorization strategies within the panel: a categorization based on wine taxonomy, and another based on the visual appearance of the wine labels. This difference in strategy has also been reinforced by interview analysis: The subgroup of 17 experts used a top-down strategy based on the PDO taxonomy to sort the wine labels. The second subgroup deliberately avoided the expected taxonomy, preferring ad hoc categorizations, using a saliency-based strategy directed by more personal goals. This alternative categorization scheme suggests that expert wine representations are flexible enough to allow adaptation to different types of situations.

4 Projective Mapping Techniques

The idea of projective mapping was first mentioned by Dun-Rankin [31] as "placement" to describe a technique in which evaluators are asked to evaluate the similarities of a set of stimuli by giving their relative positions on a plane. It was then reintroduced independently in the mid-1990s as "Projective Mapping"by Risvik [2, 3] or "Spatial Arrangement Procedure" (SAP) by Goldstone [32] or Napping® by Pagès [33, 34]. In what follows, we present the projective mapping and two of its variants: the partial projective mapping and the digit-tracking projective mapping.

4.1 Projective Mapping

4.1.1 General Principle

All the stimuli are presented simultaneously and are arranged on a table, randomly or according to a Latin square with a different order for each assessor. Each assessor receives a large sheet of paper, large enough to provide space for the assessor to separate the stimuli easily. Assessors are asked to look at, smell and/or taste (depending on the objectives of the study) all the stimuli, and then position them on the sheet of paper or on a computer screen according to similarities or differences between the stimuli. Assessors are informed that two stimuli should be placed close together if they are perceived as similar and far from each other if they are perceived as different. No instruction on how the samples should be positioned in this space are mentioned and each assessor chooses his or her own criteria. As in the sorting task, to help understanding the positioning criteria, in a second step, assessors may be asked to explain which criteria they used to perform the task. This task can also be followed by an elicitation interview to better understand the strategies used by the participants during the task.

4.1.2 Illustration

Although not as frequently used as sorting tasks, projective mapping is well suited to study consumers' mental representation. For example, Torri and collaborators in 2013 applied it to wine aroma perception by naïve and experienced wine assessors [35], Mitterer-Daltoé and collaborators in 2017 to the representation of fishes by children [36], and Aschemann-Witzel and collaborators in 2019 to the perception of food ingredient [37].

As an illustration, we present an interesting study by Heatherly and collaborators investigating whether the aromas of chardonnay wine are associated with certain colors and shapes [38]. Fifty participants carried out a projective mapping of five chardonnay wine samples spiked with wine aromas (buttery, citrus, floral, smoky, and vegetable). The samples were presented simultaneously with randomized 3-digit codes in covered wine glasses. Participants had to smell the headspace of the wines and place them on a 91 × 91 cm sheet of white paper based on perceived similarities and differences. After the wine placement, participants received a set of 3D objects of different shapes or colors and had to position them on the sheet of paper among the wines, based on how well the objects matched the odor of the wine samples. They received the following instructions:

1. *"You have in front of you five wines and a large sheet of paper. Please smell the headspace of the wines and distribute them on the paper in such a way that samples with similar odors are located near one another and different samples are placed far apart. You are free to evaluate the samples according to any criteria that you choose, and you will not need to specify your criteria. Feel free to use as much of the paper as is necessary to express the differences you may perceive. When you are finished, please mark the location of each sample with the corresponding number. You are free to take all the time that you need, and, if necessary, you can smell the crook of your elbow between each sample."*
2. After wine placement, assessors received one set of 3D objects (colors or shapes). *"You have now to distribute the 3D objects among the wines based on how well the objects matched the odors of wine."*

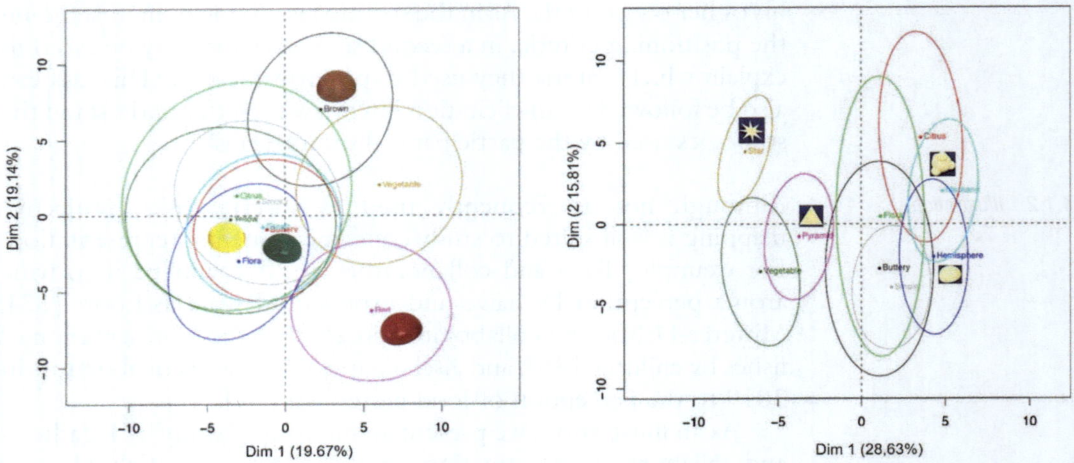

Fig. 5 Projective mapping map for (**a**) color and (**b**) shapes, and odor spiked wines. (Adapted from Heatherly and collaborators [39])

X and Y coordinates were collected for all colors, shapes, and odors across all subjects. Data were analyzed using multiple factor analysis. Figure 5 shows the correspondence between wine odors and color/shapes. For color, yellow and green colors overlap heavily with all chardonnay odors except vegetable, indicating most variants of chardonnay odor were associated with these two colors. For shape, round shapes (hemisphere and methane molecule) heavily overlapped with citrus, floral, buttery, and smoky spiked wines, indicating most of the chardonnay odor profiles were associated with rounder shapes. Sharp shapes (star and pyramid) were moderately associated with the vegetable spiked wine.

4.2 Partial Projective Mapping

4.2.1 General Principle

Projective mapping can be performed on a specific aspect of a stimulus. In this case, we talk about partial projective mapping. For example, Smith and McSweeney in 2019 used a partial projective mapping focused on flavor to evaluate the effect of color (red light vs. white light) on participants' ability to categorize wines based on grape varieties [40]. In that case, participants are asked to position the stimuli on a sheet of white paper, or a computer screen based on a single sensory modality (sight, smell, or texture for instance).

4.2.2 Illustration

Another possibility is to perform in parallel a global and separate partial analysis to better understand the contribution of each sense in consumers' representations. We used this approach in an unpublished study to evaluate the role of vision and olfaction in consumer's representation of curry powder. In this study, 18 assessors carried out successively three projective mapping with 13 curry powders in three conditions: smell condition, vision condition,

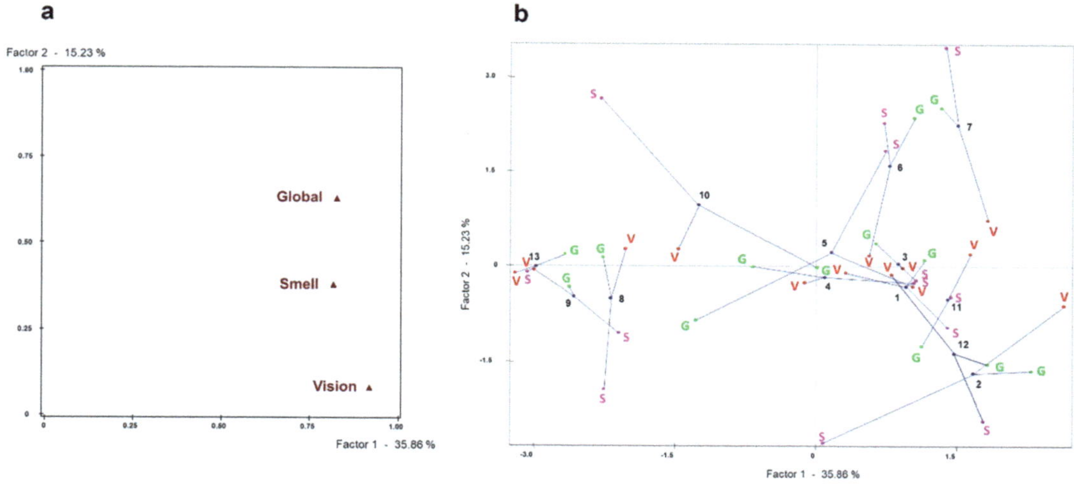

Fig. 6 MFA maps: (**a**) the three conditions: vision, smell, and global conditions and (**b**) curry powders in the three conditions: the blue points represent the barycenter of the three conditions, the green points the global condition, the pink points the smell condition, and the red points the vision condition

and global condition (smell and vision). In the vision condition, the curry powers were presented in lidded petri dishes and in the smell condition in glass jars covered with aluminum foil.

Data were analyzed with a multiple factorial analysis (MFA) considering the three conditions as three active blocks. Results show that the global condition is closer to the smell condition than to the vision condition, indicating that smell had more weight in the global assessment than vision (Fig. 6a). The first MFA dimension opposes the dark curry powders to the yellow curry powders (correlated to instrumental color measurements). In Fig. 6b, some curry powders are very close together (e.g., 9, 11, 13) showing that the condition has no influence whereas other curry powders are completely scattered (e.g., 2, 10, 5) meaning that vision and smell conditions are not in line and create discrepancies.

4.3 Digit-Tracking: Overtime Projective Mapping

The digit-tracking projective mapping is a method developed by Lê and collaborators in 2017 [41]. This tool seems very promising to study the strategy of assessors during a projective mapping and thus to access their representations. Nevertheless, further work is needed to validate this approach. Concretely, this tool aims at studying the evolution over time of stimuli in a product space issued from a projective mapping. The idea is to record chronologically all the steps that lead to an assessor's final configuration. The data are collected on a tactile device via a collaborative environment called Holos. In Holos, all the stimuli are presented randomly on the very left side of the tablet. They are displayed as icons that can

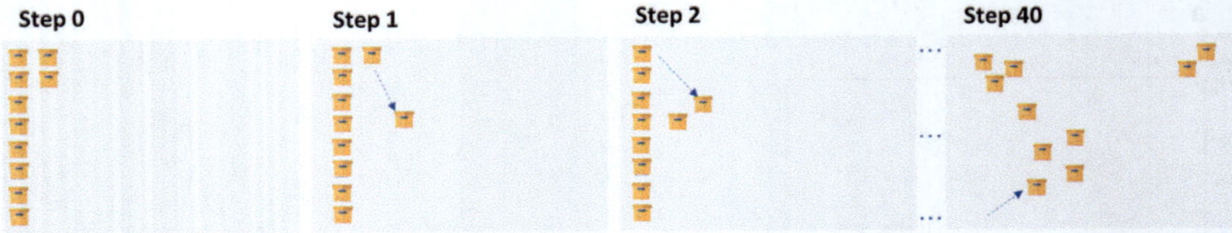

Fig. 7 Representation of the different steps taken by an assessor when performing a projective mapping task on Holos. (Adapted from Lê and collaborators [41])

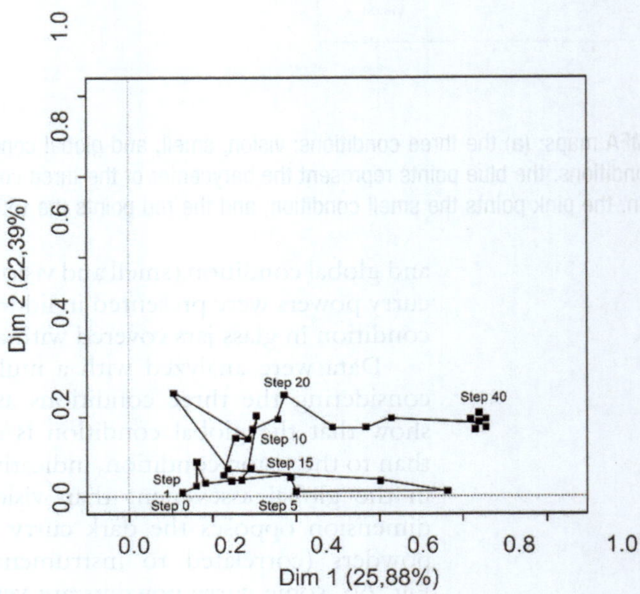

Fig. 8 Example of chronologically steps taken by an assessor during the projective mapping task provided by MFA. (Adapted from Lê and collaborators [41])

be moved (Fig. 7). As the assessor is moving the stimuli on the tablet, their coordinates are recorded. By recording all the coordinates over time, it is possible to reconstitute the trajectories of all the stimuli and analyze the evolution of the sensory dimensions used by an assessor (Fig. 8). Such a tool could provide information about the strategies used by assessors during a projective mapping task: for instance, are their decisions on stimuli positioning instantaneous or do they evolve during the task.

5 FAQs on Sorting Task and Projective Mapping

From a practical point of view, sorting task and projective mapping present many similitudes. In this section, we will answer some practical questions related to both methods.

5.1 Who Can Perform a Sorting Task and Projective Mapping?

A priori everybody! These two methods are very relevant when we want to compare the representations of different groups of assessors. The groups could differ in gender [22] or expertise (e.g., beers [1, 9, 42, 43]; wine [44, 45]; fabrics [20]). For instance, it was showed that whereas novices tend to categorize wines according to basic sensory features, experts tend to use rather higher-level types of categorizations (e.g., grape variety [44]). Likewise, experts were shown to be more precise in their sorting than novices [9, 20]. The only limitation for projective mapping is the capacity to handling spatial information: avoid inviting participants with difficulties to handle spatial representation.

5.2 How Many Assessors Do We Need for Sorting Task and Projective Mapping?

Very few studies have investigated this issue. When you use beers as stimuli, around 20 assessors seem to be enough: Indeed, some studies carried out with beers indicated that stable results could be reached with 20 untrained assessors [1, 43]. However, study with other stimuli suggested that a large number of assessors is required [7]. Blancher and collaborators propose a methodology with a cross-validation procedure to evaluate the number of required assessors and, specifically, suggest using bootstrapping techniques to draw large numbers of samples of different sizes from the original set and compute the average R_V coefficient (which they called R_{Vb}) to determine the number of assessors necessary to obtain stable results. They consider that an average R_{Vb} at least equal to 0.95 is a good indicator of stability [46].

5.3 How Many Stimuli to Use in Sorting Task and Projective Mapping?

Not too much but enough! If you use too few stimuli, the assessors will not be able to make groups or comparisons between stimuli. But on the contrary, if you use too many products, assessors may face sensory fatigue and/or memory problems during the task [9]. Obviously, the optimal number of stimuli to use is stimuli dependent because you will not have the same issues according to the type of stimuli. For instance, for auditory and tactile stimuli, only memory problem could appear whereas for olfactory and gustatory stimuli, in addition to memory problems, problems of sensory fatigue can occur (especially with remanent or alcoholic products). However, if the task is carried out only with visual stimuli, the number of stimuli can be considerably increased as the memory load and fatigue issues are alleviated. Another criterion to take into account is the resemblance between stimuli in the set: the more similar the products are, the more difficult the task is, and memory and fatigue issues are more likely to happen.

6 Pros and Cons of Sorting Task and Projective Mapping

Intuitively sorting task and projective napping are somewhat similar, so they have the same pro and cons. If these two methods have been compared regarding their performance in terms of product description, no comparative study has been done for representations. Both approaches can be used to explore mental representations and the choice of one or the other depends on the objectives of the study and its technical constraints. For instance, if you want to study the link between sensory modalities, the projective mapping by its positioning of the stimuli on a map (and not the formation of groups of stimuli) will be more suitable than the sorting task. On the contrary, if you are interested in the structure of the representation, hierarchical sorting task will be your best option. Moreover, as the sorting task forces the assessors to categorize the products in an all or none fashion and is more constrained than the projective mapping, it probably leads to a decrease in data variability. Finally, sorting tasks may be more suitable to study categorization processes whereas projective mapping seems to be more appropriated to explore information networks. Finally, from a practical point of view, sorting task seems to be easier for assessors than projective mapping.

7 Conclusion

This chapter illustrates how sorting task and projective mapping can be used to access mental representations. These two methods present many advantages. As numerous variations of these two methods have been developed, it allows to meet different objectives and therefore can be adapted to many aspects of mental representations. Moreover, both methods can be performed by various types of assessors such as experts, consumers, adults as well as children. These tasks are, in general, analyzed by statistical techniques providing maps that describe the similarity structure of the stimuli and the similarity structure of the assessors. Thus, these methods allow to compare the representations of different groups of assessors and to study inter- and intra-group variability. In addition, as no language is required in these tasks, it could be easily used to study representations in cross-cultural studies. Finally, these methods can be carried out on face to face, but are also particularly well adapted to online system, which can be quite useful for cultural studies or in the times of pandemic where sensory experimentations were strongly constrained.

References

1. Chollet S, Lelièvre M, Abdi H, Valentin D (2011) Sort and beer: everything you wanted to know about the sorting task but did not dare to ask. Food Qual Prefer 22(6):507–520
2. Risvik E, McEwan JA, Colwill JS, Rogers R, Lyon DH (1994) Projective mapping: a tool for sensory analysis and consumer research. Food Qual Prefer 5(4):263–269
3. Risvik E, McEwan JA, Rødbotten M (1997) Evaluation of sensory profiling and projective mapping data. Food Qual Prefer 8(1):63–71
4. Reed SK (2011) Cognition. De Boeck, Bruxelles, 476 p
5. Sternberg RJ (2007) Manuel de ppsychologie cognitive. Du laboratoire à la vie quotidienne. De Boeck, Bruxelles, 662 p
6. Blanchard SJ, Banerji I (2016) Evidence-based recommendations for designing free-sorting experiments. Behav Res 48(4):1318–1336
7. Faye P, Brémaud D, Durand Daubin M, Courcoux P, Giboreau A, Nicod H (2004) Perceptive free sorting and verbalization tasks with naive subjects: an alternative to descriptive mappings. Food Qual Prefer 15(7):781–791
8. Honoré-Chedozeau C, Lelièvre-Desmas M, Ballester J, Chollet S, Valentin D (2017) Knowledge representation among assessors through free hierarchical sorting and a semi-directed interview: exploring Beaujolais wines. Food Qual Prefer 57:17–31
9. Patris B, Gufoni V, Chollet S, Valentin D (2007) Impact of training on strategies to realize a beer sorting task: behavioral and verbal assessments. In: Ho Chi Minh: University-Ho chi Minh City Publishing House. [cité 22 juin 2021]. Disponible sur: https://hal.archives-ouvertes.fr/hal-00149116
10. Lawless HT (1989) Exploration of fragrance categories and ambiguous odors using multidimensional scaling and cluster analysis. Chem Senses 14(3):349–360
11. Chrea C, Valentin D, Sulmont-Rossé C, Nguyen DH, Abdi H (2005) Semantic, typicality and odor representation: a cross-cultural study. Chem Senses 30(1):37–49
12. Lawless H, Glatter S (1990) Consistency of multidimensional scaling models derived from odor sorting. J Sens Stud 5(4):217–230
13. MacRae AW, Howgate P, Geelhoed EN (1990) Assessing the similarity of odours by sorting and by triadic comparison. Chem Senses 15: 661–699
14. MacRae AW, Rawcliffe T, Howgate P, Geelhoed EN (1992) Patterns of odour similarity among carbonyls and their mixtures. Chem Senses 17(2):119–125
15. Cliceri D, Dinnella C, Depezay L, Morizet D, Giboreau A, Appleton KM et al (2017) Exploring salient dimensions in a free sorting task: a cross-country study within the elderly population. Food Qual Prefer 60:19–30
16. Guerrero L, Claret A, Verbeke W, Vanhonacker F, Enderli G, Sulmont-Rossé C et al (2012) Cross-cultural conceptualization of the words Traditional and Innovation in a food context by means of sorting task and hedonic evaluation. Food Qual Prefer 25(1): 69–78
17. Lawless HT, Sheng N, Knoops SSCP (1995) Multidimensional scaling of sorting data applied to cheese perception. Food Qual Prefer 6(2):91–98
18. Campo E, Do BV, Ferreira V, Valentin D (2008) Aroma properties of young Spanish monovarietal white wines: a study using sorting task, list of terms and frequency of citation. Aust J Grape Wine Res 14(2):104–115
19. Lelièvre M, Chollet S, Abdi H, Valentin D (2009) Beer-trained and untrained assessors rely more on vision than on taste when they categorize beers. Chemosens Percept 2(3): 143–153
20. Soufflet I, Calonnier M, Dacremont C (2004) A comparison between industrial experts' and novices' haptic perceptual organization: a tool to identify descriptors of the handle of fabrics. Food Qual Prefer 15(7):689–699
21. Garbez M, Galopin G, Sigogne M, Favre P, Demotes-Mainard S, Symoneaux R (2015) Assessing the visual aspect of rotating virtual rose bushes by a labeled sorting task. Food Qual Prefer 40:287–295
22. Gómez-Corona C, Valentin D, Escalona-Buendía HB, Chollet S (2017) The role of gender and product consumption in the mental representation of industrial and craft beers: an exploratory study with Mexican consumers. Food Qual Prefer 60:31–39
23. Abdi H, Valentin D, Chollet S, Chrea C (2007) Analyzing assessors and products in sorting tasks: DISTATIS, theory and applications. Food Qual Prefer 18(4):627–640
24. Ballester J, Abdi H, Langlois J, Peyron D, Valentin D (2009) The odor of colors: can wine experts and novices distinguish the odors of white, red, and rosé wines? Chemosens Percept 2(4):203
25. Parr WV, Valentin D, Green JA, Dacremont C (2010) Evaluation of French and New Zealand Sauvignon wines by experienced French wine assessors. Food Qual Prefer 21(1):56–64

26. Rao VR, Katz R (1971) Alternative multidimensional scaling methods for large stimulus sets. J Mark Res 8(4):488–494
27. APM C (1999) Sorting data: collection and analysis. Sage, Thousand Oaks
28. Clark H (1968) On the use and meaning of propositions. J Verbal Learn Verbal Behav 7: 421–431
29. Kirkland J, Bimler D, Drawneek A, McKim M, Schölmerich A (2000) A quantum leap in the analyses and interpretation of attachment sort items. In: Patterns of secure base behavior: Q-sort perspectives on attachment and caregiving in infancy and childhood. Lawrence Erlbaum, Hillsdale
30. Cadoret M, Lê S, Pagès J (2011) Statistical analysis of hierarchical sorting data. J Sens Stud 26(2):96–105
31. Dun-Rankin P (1983) Scaling methods. L. Erlbaum, Hillsdale
32. Goldstone R (1994) An efficient method for obtaining similarity data. Behav Res Methods Instrum Comput 26(4):381–386
33. Pagès J (2003) Recueil direct de distances sensorielles: application à l'évaluation de dix vins blancs du Val-de-Loire. Sci Aliment 23(5–6): 679–688
34. Pagès J (2005) Collection and analysis of perceived product inter-distances using multiple factor analysis: application to the study of 10 white wines from the Loire Valley. Food Qual Prefer 16(7):642–649
35. Torri L, Dinnella C, Recchia A, Naes T, Tuorila H, Monteleone E (2013) Projective Mapping for interpreting wine aroma differences as perceived by naïve and experienced assessors. Food Qual Prefer 29(1):6–15
36. Mitterer Daltoé ML, Breda LS, Belusso AC, Nogueira BA, Rodrigues DP, Fiszman S et al (2017) Projective mapping with food stickers: a good tool for better understanding perception of fish in children of different ages. Food Qual Prefer 57:87–96
37. Aschemann-Witzel J, Varela P, Peschel AO (2019) Consumers' categorization of food ingredients: do consumers perceive them as 'clean label' producers expect? An exploration with projective mapping. Food Qual Prefer 71: 117–128
38. Heatherly M, Dein M, Munafo JP, Luckett CR (2019) Crossmodal correspondence between color, shapes, and wine odors. Food Qual Prefer 71:395–405
39. de Boer J, Aiking H (2019) Strategies towards healthy and sustainable protein consumption: a transition framework at the levels of diets, dishes, and dish ingredients. Food Qual Prefer 73:171–181
40. Smith AM, McSweeney MB (2019) Partial projective mapping and ultra-flash profile with and without red light: a case study with white wine. J Sens Stud 34(5):e12528
41. Lê TM, Brard M, Lê S (2017) Holos: a collaborative environment for similarity-based holistic approaches. Behav Res 49(5):1597–1604
42. Chollet S, Valentin D (2001) Impact of training on beer flavor perception and description: are trained and untrained subjects really different? J Sens Stud 16(6):601–618
43. Lelièvre M, Chollet S, Abdi H, Valentin D (2008) What is the validity of the sorting task for describing beers? A study using trained and untrained assessors. Food Qual Prefer 19(8): 697–703
44. Ballester J, Patris B, Symoneaux R, Valentin D (2008) Conceptual vs. perceptual wine spaces: does expertise matter? Food Qual Prefer 19(3): 267–276
45. Solomon GEA (1997) Conceptual change and wine expertise. J Learn Sci 6(1):41–60
46. Blancher G, Clavier B, Egoroff C, Duineveld K, Parcon J (2012) A method to investigate the stability of a sorting map. Food Qual Prefer 23(1):36–43

Chapter 8

Measure of the Verbal Emotional Responses Triggered by Food Products

Christelle Porcherot, Sylvain Delplanque, and Carlos Gómez-Corona

Abstract

Emotions and food products are intimately related. Individuals' emotions modulate eating behaviors, and negative emotions may induce food disorders. On the other hand, the consumption of food or drink products may trigger different emotional states in individuals. Thus, the relationship between food or drink products and emotion is dynamic and complex. In this chapter, after describing the strong link between food products and emotions and the role of associative learning, we propose a definition of emotion and feelings. The chapter mainly focuses on the measure of verbal emotional responses triggered by food products. This measure refers to feelings, or the conscious verbalization of the emotional response. Although the unconscious emotional response related to physiological and behavioral responses should not be underestimated and is described, the chapter provides a broader review on a methodology developed to measure food-elicited feelings in consideration of cross-cultural differences. A series of examples is presented to illustrate the use of this methodology in sensory and consumer test settings for food product development. We also propose that investigations should be undertaken into the effects of changing the product label, packaging, and color on consumer emotional expectations and on the overall emotional response when the product is experienced. Finally, examples of research conducted in ecological settings and virtual environments are provided.

Key words Emotions, Feelings, Foods, Expectation, Culture

1 Food Product Experiences and Emotions

Emotions and eating behaviors are intimately related and their relationship has been addressed in the literature in two ways: (1) how emotions modulate eating behaviors and (2) the effect of consumed foods or drinks on the emotional state. Eating refers to the consumption of food by an individual, and eating behavior can be measured by food choice and the amount of food consumed [1–4]. The ingestive component of emotional eating can be measured with self-reports of emotional states [5]. An alternative is individuals' subjective estimates of amounts consumed in usual situations, or individuals' subjective desire to eat foods by means of a questionnaire [6, 7].

Carlos Gómez-Corona and Heber Rodrigues (eds.), *Consumer Research Methods in Food Science*,
Methods and Protocols in Food Science, https://doi.org/10.1007/978-1-0716-3000-6_8,

Most studies that investigated how emotions modulate eating behaviors examined the link between the subjective motivation to eat and emotional states [8–10]. Respondents were shown to have higher levels of hunger when they felt angry and joyful than when they felt fearful and sad [11]. This refers to emotional eating, which occurs in response to stress or negative emotions as an attempt to regulate these undesired states [12–15]. The relation between emotion and eating behaviors was therefore studied in the context of eating disorders and obesity mechanisms (*see* Ganley [16] for a literature review on emotion and eating in obesity). More than having higher levels of hunger, individuals had the tendency to eat more to reduce felt negative emotions such as "anger/dominance" and "tension/fear" [17]. In addition, comfort foods that are high in fat and/or sugar were generally chosen to cope with negative emotional states [8, 18], which led to weight increase.

Edwards et al. [19] examined, in a student cafeteria, both how emotions affect dietary choice and meal acceptability and how choices affect emotions. Results showed that positive emotions induced higher meal acceptability, but had no effect on food choice, and the meal eaten had few associations with the subsequent emotion. Nonetheless, eating contributes to emotions and is more often associated with a pleasant experience than with an unpleasant experience [9, 20, 21]. Desmet and Schifferstein [20] found that satisfaction, pleasure, enjoyment, and desire were experienced most often, whereas sadness, anger, and jealousy were experienced least often when participants tasted sweet bakery snacks, savory snacks, and pasta meals. Gibson [9] also showed that eating a meal is generally associated with a reduction in arousal and irritability and an increase in calmness and positive affect. This effect of consumed foods on the emotional state can be explained by the learning and memory processes associated with eating and the manner in which they interact with satiety processes [22]. The emotional state is shaped by what the brain and body have learned during life experiences in terms of experiencing foods and their associated emotional and physiological benefits, as well as in terms of post-ingestive effects. We also know that emotions are a critical source of influence on the learning process [23]. Consequently, unusually small or unhealthy meals may have a negative impact on the affective state in consequence of learning that small meals induce hunger and unhealthy meals induce guilty feelings.

The link between food products and emotions is therefore complex and shaped by individual experiences in certain contexts, which is key in the product experience. The "madeleine" is a typical example described by Proust ("*Du côté de chez Swann,*" published in 1913); the taste of the shell-shaped cake dipped in tea unintentionally caused him to be overwhelmed by emotion and had him reminiscing about a time in his childhood when he was given the same cake. As individuals all have their own experiences, memories, personality traits, and culture, the same freshly cooked cake will not trigger the same emotion in two different people. Durian is a familiar fruit from Asia with a distinctive strong smell of onion; it triggers less disgust in people of Asian cultures than it does in people of US and European cultures [24]. Emotional learning in certain contexts is at the core of these individual and cultural differences because it uniquely shapes an individual's affective response to environmental stimuli on the basis of their experience.

Because of its importance, research on emotional responses to food products has gained much interest [25–30]. Different investigators have shown that emotional responses to foods provide information that goes beyond overall acceptance [26, 31–38] and that even significantly improves food choice prediction [39–41]. The food industry is today paying more attention to the emotional responses triggered by food products, and it is expected that food products may provide a superior-quality consumer experience with a unique emotional experience. This chapter provides examples of methodologies applied to the measure of emotional responses elicited by food products and examples related to food development. The emotional experience can be amplified and maximized when the product experience is in harmony with brand positioning and communication. Examples of alignment with emotional responses triggered by expectation and information provided to consumers (e.g., brand, usage of the product, claim, packaging, color) are also provided. Finally, examples of research conducted in ecological settings and virtual environments are provided in consideration of the diversity of emotional responses triggered by a meal in different contexts.

2 Different Theories on Emotions

The affective phenomena mentioned in food-related studies that influence our psychological state and behavior are varied. It is therefore important to define as well as possible the affective phenomena covered in this chapter: emotion, feelings, and mood.

Although researchers do not agree on the definition of emotion, they agree on the fact that it is a brief affective phenomenon (unlike mood) that is triggered by an event (external or internal to the individual). On this basis, several main families of theories of emotion that have been used in the context of food can be distinguished. The basic (or discrete) emotion theories postulate the existence of a small number of so-called basic emotions characterized by emotion-specific response patterns [42–44]. In this case, depending on the version of the theory adopted by the investigator, the number of proposed emotional terms may vary from 6 (anger, disgust, fear, sadness, surprise, happiness [45, 46]) to 22 (shame, jealousy, fear, anger, sadness, pride, hope, relief, boredom, contempt, admiration, disgust, desire, disappointment, love, dissatisfaction, amusement, stimulation, satisfaction, unpleasant surprise, enjoyment, pleasant surprise [20]). On the other hand, dimensional theories reduce emotions to positions in a two-dimensional valence by arousal space or a three-dimensional space that includes potency (e.g., Refs. [47–49]). In this case, the affective terms are chosen to characterize the underlying two or three dimensions (e.g., arousing, pleasant, powerful: Refs. [50–58]).

Another set of theoretical models, the appraisal models, are much less used but worth mentioning (*see* Moors et al. [59] for an overview). In one version of these models, Scherer [60] defined an "emotion" as a series of synchronized, interconnected changes in an individual's thoughts and body state (unconscious and conscious) in response to a stimulus event (real or imaginary), which prepares the individual by affecting how he or she feels, behaves, and appears. These changes occur in light of what a situation means to the person from past experiences, social context, goals, and desires. Emotions are specific, can be intense, and are triggered by a particular stimulus event.

Another point on which the researchers agree is that several components of the organism are mobilized during an emotional episode: an autonomic response (physiological response, such as activation of the sympathetic nervous system), an action tendency (e.g., approach or avoidance), an expression (facial or vocal expressions, e.g., smiling), and a subjective feeling (e.g., feeling happy) [61]. In practice, it is difficult, if not impossible, to investigate all these components, together with specific measures, to assert that a product or a situation has elicited a true emotion.

According to this componential approach, feelings represent only one of the components of the emotional response: its subjective part. The explicit emotional response can be measured via verbalization or questionnaires. It is generally admitted that to be fully considered as defining true emotion, this measure of feelings needs to be confirmed with measures of implicit or unconscious processes at a cognitive, behavioral, or physiological level (for a discussion of this topic, *see* Refs. [62, 63]). Many questionnaires with different lists of feeling terms were proposed and are further developed in Subheading 4 of this chapter. These lists were adapted to the stimulus of interest (for example, the Geneva Emotion Music Scale [GEMS] [63]; the Geneva Emotion Odor Scale [GEOS] [64]).

Finally, mood is a more diffuse and more lasting affective state. Moods differ from emotions in that they are less specific and often less intense, and they are less likely to be triggered by a particular stimulus. Moods are generally either positive or negative. They are associated with an internal and subjective state, but they often can be inferred from posture and other observable behaviors of individuals and can be measured, for instance, with the Visual Analog Mood Scales [63].

3 Measurement Methods

The scientific community's continually increasing interest in the study of the link between food experiences and emotions has led to the development of different methods to capture consumer emotions elicited by foods [39, 65]. These different methods, either implicit or explicit, were adapted to the measure of the different components of the emotional response (i.e., physiological activation, motor expression, action tendency, cognitive processing, and feelings). Explicit methods are either verbal or visual self-reported measurements that ask participants to report their feelings while seeing, smelling, or consuming food products. Implicit methods are indirect and non-self-reported, and they capture aspects of emotional responses that do not need introspection by the participants [29, 67–69]. These implicit measures encompass psychophysiological, expressive, and some behavioral task measures. They are generally characterized as "unconscious, unintentional, uncontrollable, effortless and fast" [68]. A summary of the different methods is provided in Table 1.

PrEmo, Emotion Measurement Instrument; HRV, heart rate variability; GSR, galvanic skin response; EEG, electroencephalography; fMRI, functional magnetic resonance imaging; EMG, electromyography.

Before reviewing the most commonly used implicit and explicit measures, we feel it is important to warn the potential user of the risk of over-interpreting these measures. An emotion involves the activation of several components of the organism. These components are also mobilized by other psychological or physiological states or by processes that are not emotional. Using measures of only one component to infer emotions could lead to a false reverse inference (i.e., attributing specific variations of components to emotions when they are the result of other psychological or physical processes). As discussed at length by Delplanque and Sander [70], looking for converging evidence among the five components of emotion decreases this risk.

Table 1
Different methods for measuring emotional responses elicited by food products

Method	Explicit	Implicit
Examples	Verbal self-report (questionnaires), visual self-report (PrEmo, emojis, mood boards)	Psychophysiology (HRV, GSR), brain response (EEG, fMRI), expressive measure (EMG), behavioral task (implicit tests)
Principle	Conscious processes, possible control of the answer	Unconscious processes, automatic answers, no control, less biased by social desirability

3.1 Verbal Self-Report

Explicit verbal self-reports consist in the verbalization, or the oral expression, of what individuals feel upon seeing, smelling, or consuming a food product by using specific questionnaires with a list of feeling terms (non-exhaustive examples: Refs. [64, 32, 35]) or by using a free elicitation exercise [71]. This measure has the advantage of being quick in its implementation and data analysis [72]. Many proposed terms derived from basic emotion and dimensional theories appeared not to be relevant to the food domain, which underlines the importance of a term's selection procedure [32]. Chrea et al. [64] also questioned the suitability of these theories to account for the specificity of rich, differentiated emotional feelings experienced in response to odorous substances, and developed a new and specific scale: the GEOS.

Other researchers have developed their own "emotional lexicons," which are mostly feeling lexicons, and have reinforced the idea that the emotional vocabulary highly depends on the products of interest. Many of these lexicons were developed by presenting words to consumers and asking them to select or score the most relevant words to express their feelings as elicited by a particular product. Other emotion lexicons were self-generated by consumers by using qualitative methods. Gmuer et al. [73] systematically compared existing food- and odor-related emotional lexicons. Specific lexicons were created and used for food-specific domains (e.g., EsSense Profile® [32], EmoSemio [35], EmoSensory® Wheel [34]), as well as for more specific products such as blackcurrant squashes [33], chocolate [35, 74], hazelnut spreads [75], photographs of fruit salads [76], cheese [34], wine [77], beverages [78], beer [79, 80], coffee [81, 82], meat [83], and odors and fragrances [20, 58, 84–86; GEOS – 64, UniEOS – 87, ScentMove® – 25]. The comparison between these different emotion lexicons revealed only a small overlap, which could be the result of product specificities in the emotional response, or different approaches to creating the lexicons, or the fact that lexicons were created in different cultures.

Ferdenzi et al. [24] followed a unique cross-cultural approach which underlined differences in language between cultures for expressing the same feeling. A study presented at EuroSense in 2012 [88] therefore showed more appropriateness and discrimination when using culture-specific ScentMove® scales. In sum, this study highlighted the importance of using culture-specific scales, these scales being different in some emotional dimensions (i.e., "melancholy" and "arousal" in the Chinese culture; "pleasure" in the Swiss culture) and in feeling words (i.e., "beaming" and "feeling of plenitude" in the Chinese culture instead of "well-being" and "happiness" in the Swiss culture). However, most of the other studies generated a lexicon in one country and then translated it into other languages when expanding to other countries. Cross-cultural approaches appeared to be a challenge due to translation problems [89, 90]. Indeed, translation may result in a different meaning in another culture or a lack of understanding (e.g., "wild" is commonly used in the United States, but cannot be related to any feelings in European countries).

Moreover, explicit verbal self-report measures may suffer from cognitive bias, which may affect the validity of the emotional assessment [91, 92]. It should not be forgotten that respondents may respond more according to social desirability and their self-representation [93, 94]. This bias may occur in studies conducted on unhealthy or comfort foods, and despite their guilty feeling, participants may not express it. We found that social biases were less impactful when participants expressed feelings elicited by fragrances, which is an easy task to perform.

3.2 Non-verbal Self-Report

Explicit non-verbal self-reported methods mostly use images instead of emotional lexicons to depict different emotions. These methods have the advantage of being easy to use and translation not being necessary [95]. In addition, they may benefit from less involvement of cognitive interpretation. For this reason, several tools have been developed. PrEmo is a self-report approach that relies on the use of a set of cartoon animations that express 12 different emotions, as shown in Fig. 1 (boredom, desire, disgust, dissatisfaction, fascination, fear, hope, joy, pride, sadness, satisfaction, and shame), through facial and body expressions [96–98]. This method has recently been applied to food products such as breakfast drinks [35] and gingerbread and chocolates [99], as well as to odors [100]. However, this method may suffer from a lack of understanding of the meaning of the cartoon and evoked emotions.

Fig. 1 PrEmo2 cartoons. (Image: www.premotool.com)

Fig. 2 Example of pictures provided to consumers in the form of an emotion board. (Royalty free images from pixabay.com)

Fig. 3 Example of emojis used in emotion studies

Emotion boards known as Sens'n Feel® [101] or mood portraits™ [84] use visuals, or a series of photographs, that evoke an emotional response. Respondents select the pictures that correspond to the emotion evoked by the stimulus, i.e., a fragrance (Fig. 2).

More recently, other researchers have used emojis, pictorial representations of emotions commonly used in social media and electronic communications [102–105]. Emojis take the form of facial expressions (or simple symbols) that represent various emotions, as shown in Fig. 3. This method is applicable to cross-cultural studies because facial expressions may be presumed to be common across languages and cultures. Jaeger and Ares [106] compiled the meanings associated with different emojis from internet resources and found good agreement between 15 of them and specific emotions and fair agreement for 10 others. This approach may be particularly interesting for research conducted with children, who may not have the appropriate vocabulary to express their feelings [107, 108], in addition to cross-cultural studies [109]. Although emojis are widely used as a means to communicate feelings, their coding and use in social media could still be improved with machine learning [110]. The question of whether emojis should replace emotion words in questionnaire-based food-related consumer research has been addressed by Jaeger et al. [111]. Regarding enhanced ecological validity, familiarity, and cross-culturally shared meanings, these authors encourage the use of more emoji questionnaires and compare the results with emotion word questionnaires. Still, emotion words may offer more alternatives and precision than emojis do, which offer only a smiling face for happiness, whereas hundreds of synonyms exist [109].

Although explicit measurements are quick and user-friendly, they can still be cognitively biased because each word, cartoon, emoji, or board needs to be rated by the participants from "I do not feel this" to "I do feel this strongly" [92, 95, 112, 113]. This means that the participants need the introspective capacity to identify and verbalize the perceived feelings that they experienced [103, 114, 115]. Therefore, more implicit measures were envisaged.

3.3 Psychophysiological Measures

Many psychophysiological measures are available, such as electrophysiological recordings of galvanic skin response (GSR), electroencephalography (EEG), pupil dilation, heart rate, and functional magnetic resonance imaging (fMRI) [69, 116, 117]. Heart rate variability is generally associated with attention and physical or mental activity, and GSR can be associated with arousal [118, 119]. EEG and fMRI measure brain activity through different techniques, fMRI detects changes in blood flow and is used to investigate where and how information is processed in the brain, and EEG measures changes in electrical activity across the scalp and provides information about how and when information is processed. Physiological measures have the advantage of not demanding any particular attention or specific language skills by the participant to fill in a questionnaire during or after consumption of a product. However, these measures are generally not sensitive enough to distinguish between different types of positive emotions and may be relatively poor at differentiating subtle or low-level emotional responses, which are commonly found in product emotion research [120–123]. Moreover, the emotion specificity for physiological response measurements is debated; Fernández et al. [124] recommend using these measures as markers of arousal and/or valence, but not for any other specific emotion, which makes their interpretation difficult.

These methods also showed difficulties when used for product development [31, 67, 92, 100, 110]. The implicit measures conducted in the laboratory have limited added value for understanding responses toward food experiences, which involve other factors such as the physical, social, and cultural contexts in which foods are consumed in real life. These measures could benefit from being conducted outside traditional laboratories and in real-life contexts [125].

3.4 Implicit Expressive Measures

Implicit expressive measures consist in the capture of facial expressions from facial muscle movements associated with emotions [100, 126–128]. Capture can be automatic with specific software (e.g., FaceReader, nViso, Affidex) or manually coded by trained coders [129]. Electromyography is also capable of recording the movements of two main facial muscles, the corrugator muscle associated with negative emotions and the zygomatic muscle associated with positive emotions [130]. However, in the case of food consumption, the coding of facial expressions is disturbed by facial movements when participants chew and swallow solid and liquid foods. Moreover, these measures seemed to demonstrate more significant results for food products with a low acceptance level, such as fish odors [100], bitter flavors [131], and disliked foods [92], because the body reacts more strongly to negative than to positive stimulation [132].

Another way to measure implicit expression is to study human speeches, which express emotional meaning through semantics and through certain attributes of the voice, such as pitch or loudness. Banse and Scherer [133] examined relations between 14 induced emotions and 29 acoustic parameters. Pitch or frequency, energy/amplitude or loudness, temporal/duration, and quality (stressed/breathy voice) seemed to be relevant paralinguistic features when expressing emotions [134, 135]. Lausen and Hammerschmidt [136] conducted a metanalysis of 1038 emotional expressions according to a set of 13 prosodic acoustic parameters and confirmed that emotions and confidence could be predicted in a reliable manner.

3.5 Implicit Behavioral Performance

It is possible to gather information on emotional processing triggered by products by collecting performance measures and using specific tasks. These measures can be implicit when variations in performance (e.g., speed, accuracy, efficiency) are observed independently of the task asked of the participant. The measures emerged in the field of social psychology with the intent to address issues of social desirability bias and the participants' inability to access unconscious cognitive processes [137–139]. To illustrate our point, let us use an example. Introduced in 1998 by Greenwald and colleagues, the Implicit Association Task (IAT) is a popular task used to measure the degree of association between concepts (emotional or not) and products. The IAT measures reaction times in a computer-based speeded discrimination task between two targets (e.g., "flowers," "insects") and two attributes (e.g., "pleasant," "unpleasant"). Faster reactions are assumed to imply affective congruent relationships [140], so that participants are likely to respond faster and more accurately for stronger associations or compatible blocks (e.g., "flowers" and "pleasant") than for incompatible blocks (e.g., "insects" and "pleasant"). Although the IAT has been largely used in the visual modality, its application in other sensory modalities is still under-investigated. An exception comes from Demattè et al. [141], who used IAT to investigate cross-modal associations between odors and colors. Lemercier-Talbot et al. [142] recently tested the ability of the IAT procedure to explore automatic associations between odors and relaxing/energizing feelings for simple compounds and fine fragrances. Experiments revealed that participants responded faster when a mint odor was paired with energizing feelings and when a vanilla odor was paired with relaxing feelings. These results not only support the existence of automatic associations between odors and feelings, but also highlight the ability of the IAT procedure to capture them. Other implicit methods are based on reaction times and combine both the choice of the participants (the explicit answer) and the time they take to respond (implicit answer) [118].

3.6 Which Methods to Choose?

The choice between the use of implicit or explicit measures depends on the objective of the research, type of food products, budget, and time needed for the project. Explicit methods have been shown to be generally easier and quicker to implement and therefore more cost-effective than implicit methods [143]. On the other hand, implicit methods are not biased by the conscious control of the consumers [68, 92]. As mentioned earlier, one cannot predict how people feel just by observing their neurophysiological responses [70]. The data are not self-explanatory and, most of the time, a single physiological response is associated with more than one mental process. Various studies question the relationship between physiological markers and particular emotions and suggest that physiological measures cannot predict emotions [144, 145]. Trying to predict which cognitive process or emotional state is occurring on the basis of an observed physiological response or brain activity is reasoning backward, called reverse inference, and should not be practiced [146]. Various brain activities could be involved with many different mental processes, and specific and appropriate experimental designs are needed to draw valid conclusions. Combining both implicit and explicit methods is therefore generally recommended in order to consider the complexity of the emotional response and be more predictive of consumer choices [66, 143, 147]. Physiological measures should complement explicit measures, rather than being used as a substitute for them [148]. It is not true that physiological measures are more accurate than other methods. In addition, explicit measures cover many emotions, whereas implicit measures generally cover only a small number of emotions [149]. If the research question can be solved by verbalization, this simplest solution should be preferred. The method chosen in this chapter is based on explicit verbal self-report measures and aims at measuring and comparing the subjective feeling component or verbal emotional response to food odors or food products.

4 Construction of a Verbal Scale to Measure Feelings Evoked by Food and Non-food Odors

Although many researchers worked with different scales that were based on lists of consumer-defined terms, the selection of emotional terms is not clearly described [33, 37, 149, 150]. In this section, we describe the full procedure followed by Chrea et al. [64], which we believe constitutes a rigorous approach based on three steps applied in different cultures. Their objective was to investigate the nature and organization of the semantic space associated with the affective subjective experience or feeling in response to odors from perfumery and food products. They started from a list of 480 terms extracted from the literature on emotions and on olfaction, including the terms derived from the dimensional and the basic emotion models. They narrowed the number of terms from 480 to 73 by asking participants to rate the terms for their relevance in describing an affective state induced by an odor on a continuous scale ("In your opinion, is this term relevant for describing an emotional state you have already experienced when smelling an odor in the past?"). They then further reduced the number of terms from 73 to 36 by presenting 56 odors and asking respondents if "the terms were relevant to describe their emotional state while smelling the presented odor." Notably, the reduction and organization of terms was based on an empirical, data-driven approach (using exploratory and confirmatory factorial analyses) according to the respondents' behavior when presented with the odor alone, without imposing any particular theoretical organization on the data (i.e., basic emotion or two- or three-dimensional emotion theories) or any specific context. The 56 everyday odors were a balanced mix of familiar–nonfamiliar, pleasant–unpleasant, and edible–nonedible, with a large variety of odor qualities (food, perfumery, body, or nature related). These studies suggest that the structure of underlying feelings of odors is highly differentiated and differs from the taxonomy used to refer to habitually experienced emotions in everyday life. For example, guilt, shame, and sadness, which are frequent emotions in everyday life [151], did not appear to be relevant to describe affective states elicited by everyday odors (*see* also Desmet [152] for fine fragrances). More precisely, the resulting emotional model, GEOS, contains 36 representative terms grouped into six dimensions by factorial analyses. The dimensions are illustrated by seven terms related to "Sensuality" (Desire, Romantic, Sensual, In love, Excited, Admiration, and Sexy), five terms related to "Relaxation" (Relaxed, Soothed, Reassured, Light, and Serene), six terms related to "Pleasant feeling" (Pleasant, Well-being, Pleasantly surprised, Feeling awe, Attracted, and Happiness), seven terms related to "Refreshment" (Revitalized, Energetic, Refreshed, Stimulated, Invigorated, Shivering, and Clean), three terms related to "Sensory pleasure" (Nostalgic, Mouth-watering, and Amusement), and eight terms related to "Unpleasant feeling" (Dirty, Unpleasant, Disgusted, Unpleasantly surprised, Dissatisfaction, Sickening, Irritated, and Angry).

As odor-affective values are mainly attributed through associative learning processes, the existence of additional affective facets across different cultures and languages was postulated. Different studies were conducted in six countries: the French-speaking part of Switzerland [64], the United Kingdom and Singapore [153], the United States, China, and Brazil [87]. Of importance is that the odorants presented to build the scales included a common set to assess cultural diversity in eliciting feelings from identical odorants, as well as culture-specific odorants so as not to miss those feelings born of culture-specific odors (e.g., pandan and tea odor for China, banana and coconut for Brazil). The three-step study process, as used for the Swiss culture, was systematically reproduced in five other cultures, and resulted in the development of culture-specific emotion and odor scales. The 480 terms were kept in English for the studies conducted in the United Kingdom, the United States (two cities, one in Arkansas and one in California), and Singapore; translated into Chinese for China; and translated into Portuguese for Brazil by applying a back-translation process. Although some of the affective terms in the emotional categories were culturally shared (i.e., happy, energetic, refreshed, sensual, romantic, relaxed, disgusted, irritated), several culture-specific affective terms also appeared (i.e., for the Happiness/Delight emotional category: "attracted, feeling awe" in Switzerland; "in a good mood" in the United Kingdom; "likeable, satisfaction, smiling, delighted, bliss, fantastic" in the United States; "feeling of plenitude, euphoric, jovial, titillated, beaming" in China; "joyful" in Brazil).

The resulting Emotion Odor Scales, or ScentMove® model (Table 2 and Fig. 4), contains 25 emotional terms grouped into 9 emotional categories. It is available in four languages: English, French, Chinese, and Portuguese (Table 3). The model encompasses the selection of all emotional dimensions from each culture and the selection of the most used terms by all cultures. A minority of the terms were more culture-specific and were also considered in the global model.

Table 2
ScentMove® model contains 25 emotional terms grouped into 9 emotional categories

Unpleasant feelings/Disgust	Disgusted	Irritated	Unpleasantly surprised
Happiness/well-being	Happy	Pleasantly surprised	Well-being
Sensuality/desire	Sensual	Romantic	Desire
Energy/refreshed	Energetic	Refreshed	Revitalized
Soothing/peacefulness	Soothed	Comforted	Relaxed
Hunger/thirst	Mouth-watering	Thirsty	Famished
Interest/amusement	Interested	Amusement	Impressed
Nostalgia/melancholy	Nostalgic	Melancholic	Sad
Spirituality	Spiritual feeling		

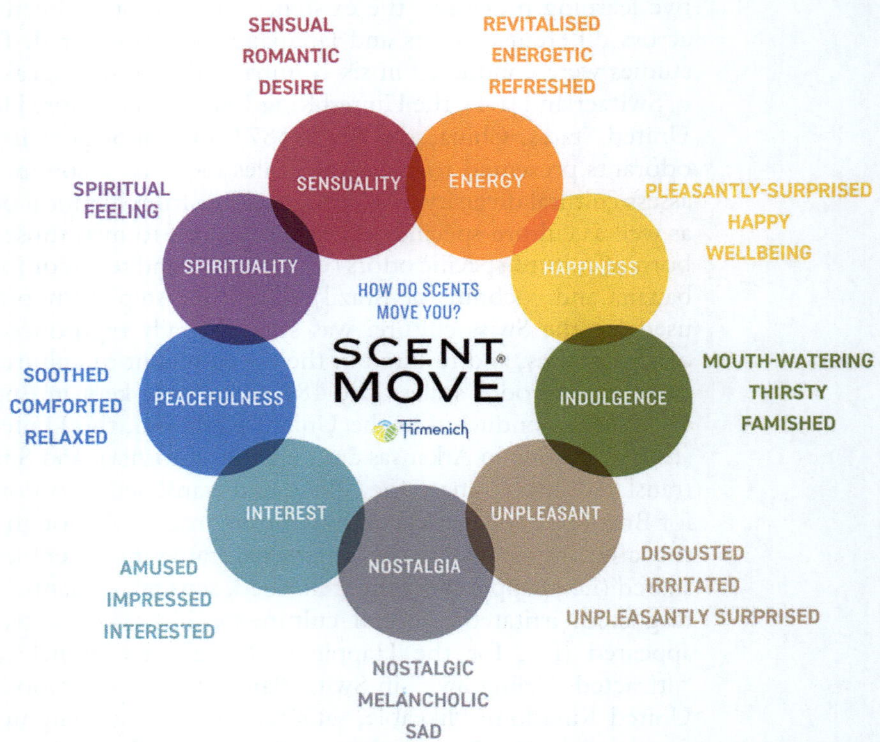

Fig. 4 The ScentMove® model contains 25 emotional terms grouped into 9 emotional categories

Table 3
ScentMove® terms available in French, Chinese, Portuguese, and Spanish

Dimension	English	French	Chinese	Portuguese
Unpleasant feelings/ disgust	Disgusted	Dégoûté	厌恶的	Enojado
Unpleasant feelings/ disgust	Irritated	Irrité	恼怒的	Irritado
Unpleasant feelings/ disgust	Unpleasantly surprised	Désagréablement surpris	不愉快的意外惊喜	Desagradavelmente surpreso
Happiness/well-being	Happy	Heureux	幸福的	Feliz
Happiness/well-being	Pleasantly surprised	Agréablement surpris	惊喜的	Agradavelmente surpreso
Happiness/well-being	Well-being	Bien-être	安宁	Bem-estar
Sensuality/desire	Desire	Désir	渴望	Desejo
Sensuality/desire	Romantic	Romantique	浪漫的	Romântico
Sensuality/desire	Sensual	Sensuel	肉欲的	Sensual
Energy/refreshed	Refreshed	Rafraîchi	恢复精神的	Refrescado
Energy/refreshed	Energetic	Énergique	精力充沛的	Energético
Energy/refreshed	Revitalized	Revitalisé	恢复生机的	Revitalizado
Soothing/ peacefulness	Relaxed	Relaxé	得到安宁的	Relaxado
Soothing/ peacefulness	Comforted	Réconforté	宽慰的	Confortado
Soothing/ peacefulness	Soothed	Apaisé	受安慰的	Sossegado
Hunger/thirst	Mouth-watering	Salivant	令人垂涎欲滴的	Com água na boca
Hunger/thirst	Thirsty	Assoiffé	口渴的	Sedento
Hunger/thirst	Famished	Affamé	极饥饿的	Faminto
Interest/amusement	Amusement	Amusement	娱乐	Diversão
Interest/amusement	Interested	Intéressé	有趣的	Interessante
Interest/amusement	Impressed	Impressionné	印象深刻的	Impressionado
Nostalgia/ melancholy	Sad	Triste	伤心的	Triste
Nostalgia/ melancholy	Melancholic	Mélancolique	忧郁的	Melancólico
Nostalgia/ melancholy	Nostalgic	Nostalgique	怀旧的	Nostálgico
Spirituality	Spiritual feeling	Sentiment spirituel	精神感觉	Sentimento espiritual

This questionnaire has been shown to be relevant for measuring the verbal emotional response triggered by fragranced products [25, 154] and the emotional response triggered by flavored products [155, 156]. However, because the ScentMove® model was developed solely on the basis of odors, it may be limited by the lack of terms related to specific flavored product categories and for specific contexts of use, as underlined by comparison with other questionnaires by Gmuer et al. [73], shown in Table 4. This table highlights that other terms could be relevant in the context of food products, especially "guilty," which was not relevant for odors (a feeling that could be related to the ingestion of treats and comfort foods such as chocolate, desserts, etc.). The "disappointment" feeling may be especially relevant when the food product is given in a certain context and with information, both of which set up an expectation that might not be fulfilled when consuming the

Table 4
Comparison of domain- and product-specific emotion and feeling lexicons in the sensory science literature by Gmuer et al. [73]. These comparisons were conducted on the basis of the English translations provided by the authors when the lists were originally published in a language other than English

Absent minded g, Active a, Admiration/Admiring d,f,g, Adoring f, Adventurous a,g,k, Affectionate a,g, Aggressive a,g,h,k, Alone g, **Amusement**/Amused/Amusing b,c,d,f,h, Anger/Angry c,d,f,i, Annoyed i, Anxious g, Approval i, Arrogant g,k, Astonishment b, At ease i, Attentive i, Attracted c,d,e, Belittled g, Bland h, Bliss c, Boredom/Bored a,f,i, Calm a, Caring g, Cautious i, Charmed/Charming f,g, Cheerfulness c, Clean d,e, **Comforted**/Comforting e,f,i,k, Confident g,k, Confused g,i, Contempt c, Content b, Crabby g, Critical g, Curious h,i, Daring a, Delight b,c, Depressed f, **Desire**/Desirable d,e,f,h,i, Despairing g, Dirty d,e,f, Disappointment/Disappointed b,c,i, Disapproving g, Discontented c,g,i, Disgust/**Disgusted**/ Disgusting a,b,c,d,e,f,g,h,i, Disinterested lethargy g, Displeasure i, Dissatisfaction d, Doubt b, Dreamy e, Drowsy e, Dull g, Eager a, Easygoing k, Elegant h, Embarrassment b, **Energetic** a,c,d,e,f,g,k, Enthusiastic a,c, Envy c, Euphoric h, Excited c,d,g, Exhausted g, **Famished** e, Fascinated f, Fear c, Feeling awe d, Feminine k, Free a,c, Friendly a, Frustration b, Fun k, Furious g, Glad a, Good a,i, Good-natured a, Guilt/Guilty a,b,i, Happiness/**Happy** a,c,d,f,g,h,i, Heart-stricken g, Hesitation b, Horrible f, Horrified g, Impatience b, In a good mood e, In love d,e,f, Indifference b, Infatuation c, Inferior g, Inhibited g, Interest/**Interested**/Interesting a,b,f,h,i, Invigorated d, Irate g, **Irritated** d,f, Jealous g, Joy/Joyful a,c,h, Keen h, Lassitude b, Light d, Light-hearted g, Lively g, Loving a, Lustful e, Luxurious k, Marvel c, Masculine k, Meditative e, Merry a, Mild a, Nauseous e, Neglected g, Nervous g, Nostalgia/**Nostalgic** a,b,c,d,e,g, Not refreshed i, Ordinary k, Overjoyed g, Overwhelming h, Passionate g,h, Passive g, Peaceable/Peaceful a,e,h, Pleasant a,c,d,f,h, Pleasure/Pleased a,b,c,i, Polite a, Powerful k, Pretentious k, Pride b, Protected e, Purposeful g, Quiet a, Reassured g, **Refreshed** c,d,e,f,i, Regret/Regretful b,g,i, Reinsured d, Rejoicing b, Rejuvenated e, **Relaxed** d,e,f,g, Relief c, Religious feeling f, Reminiscence i, Repelled e, Resentment i, Respectful g, **Revitalized** d,e,f, **Romantic** d,e,f, Sadness/**Sad** c,f,g, **Salivating** d,e, Satisfaction/Satisfied a,b,c,i, Scared g, Skeptical i, Secure a, **Sensual** d,e,f,k, Sentimental e, Serene b,c,d, Serious k, Sexually aroused f, Sexy d,e,f, Shivering d, Shocked i, Shy g, Sick/Sickening/Sickly d,e,f,i, Silly g, Sluggish g, Sociable g,k, **Soothed** d,e, Sophisticated k, **Spiritual feeling** f, Steady a, Stimulated d,e, Strange g, Stressed f, Subdued g, Superior g, **(Un-/Pleasant) Surprise/(Un-/Pleasantly) Surprised** c,d,e,f,g,i, Suspicious g, Tacky k, Tame a, Tender a,c, Terrific g, **Thirsty** e, Thrilled b, To feel intimacy e, To like/Liking b,c, Touched g, Traditional k, Trembling c, Troubled c, Trust i, Trustworthy k, Uncomfortable e,f,i, Uncomplicated k, Understanding a, Uneasiness b, Unhappy i, Unpleasant d,e,f, Vigilant b, Warm a,c,i,k, **Well-being** c,d,f, Whole a, Wild a, Willful g, Worried a,i, Youthful k

Domain-specific Product-specific

[a]King and Meiselman [32], [b]Churchill and Behan [84], [c]Pionnier Pineau et al. [157], [d]Chrea et al. [64]– GEOS, [e]Ferdenzi et al. [87] – LEOS, [f]Ferdenzi et al. [153] – SEOS (from ScentMove® in bold character), [g]Thomson et al. [36], [h]Ferrarini et al. [77], [i]Ng et al. [33], [k]Thomson et al. [36]

food product. Because this list is broader than the ScentMove® list, it is recommended that investigators start with this broader list and have consumers select the terms that are the most relevant to them in the context of the studied food product.

We could also ask whether 25 is a good number of terms. The number of emotional terms used within the emotional lexicons varied greatly, ranging from only 5 to 50. Nestrud et al. [158] highlighted that the rating data show significantly higher mean ratings when the list of emotions is shorter. They decided to limit the total number of terms (sensory and emotional) to about 30 to limit potential fatigue and consequently increase the quality of the collected data. However, the number should be adapted to the objectives of the research and the number of products to be tested. A longer list of items might be compensated with a limited number of products under study. In addition, longer questionnaires could capture more subtle variations in experienced emotional consumer heterogeneity.

Finally, use of a randomized presentation order of the lexicon terms is recommended within and across participants to minimize potential order effects [27, 159]. This approach helps avoid the less frequent selection of terms that are positioned at the bottom part of the questionnaire because they are less visually processed [160].

5 Methodology Applied to the Measure of Verbal Emotional Responses

This section explains different methodologies and provides recommendations related to the implementation of emotion scales in questionnaires when conducting consumer tests with food products.

5.1 Respondents

The general recommendation is to follow traditional requirements for conducting consumer tests (i.e., a minimum of 60–100 consumers or more) with different target groups in terms of age, gender, and frequency of consumption of the product in order to allow segmentation. In this case, the sample size should also exceed 100 in each segment. Spinelli et al. [161] demonstrated that segmentation in emotion research provides interesting results, reporting different feeling profiles for different groups, which was not present for the liking scores. We can only recommend that researchers apply the standards for sample size justification. These standards are based not only on the knowledge already accumulated in the field, but also on objective statistical criteria related to the power of the statistical tests used, the significance level chosen, and the expected effect size [162].

5.2 Location

It is important to take into consideration that emotions are context dependent, and the location of the evaluation can influence the results of an emotional measurement. Like any consumer test, these tests can be conducted in the laboratory, as a central location test (CLT), as a home use test (HUT), in real-life settings (bar, restaurant, office, etc.), and in virtual reality environments. This choice of environment is not trivial, differences in measurements being observed between these environments. For example, product differences were overestimated in laboratory settings compared with those in restaurant settings [156, 162]. Danner et al. [112] found that the emotional responses for wines were more positive in a restaurant setting than in HUT and CLT settings. Another study by Schouteten et al. [163] also found differences between a laboratory and a home context for yogurts. However, more studies are in favor of more realistic settings because emotional responses to consumer products obtained in CLT settings do not correspond to any typical eating situation and could be considered artificial. For this reason, several investigators recommended conducting studies outside laboratories and favored more natural settings [164–166]. Some studies also simulated a more real-life environment by testing, for example, in a kitchen [168], a lounge setting [81], or a simulated restaurant or cafeteria setting [26, 39] or by providing a written scenario [72] to provide context and be closer to real-life situations.

The HUT appears to be suitable for emotional product elicitation [e.g., Refs. 112, 163, 169, 170]. Consumers receive only one food product in their home and evaluate it over consecutive days. The questionnaire procedure remains the same, but the feeling response is captured at different times of the product experience, i.e., before, during, and after food consumption. Schifferstein et al. [171] explored consumers' feelings at different stages of the consumption sequence, from shopping to preparation to consumption. They found that different feelings were elicited during the various stages of user-product interactions. Similarly, Giboreau and Meiselman [172] compared emotions before and after the meal and found differences, with a general increase in positive emotions after the meal.

Finally, it is important to consider the appropriateness of the food in relation to the time and eating environment [173]. Different foods might be appropriate in one environment, but not in another; thus, the appropriateness of eating occasion measures should be more considered in research on emotions [166, 174]. Piqueras-Fiszman and Jaeger [37, 38, 175] conducted a series of experiments that showed how the imagined context and appropriateness of different eating situations influenced emotional responses, the reason being that emotional responses could be more related to the eating situations than to the food itself [95].

5.3 Food Samples

There is, of course, a huge variety of food products, a fact that is reflected in the studies that investigate food-related emotions. Despite the wide range of food products used (e.g., snacks, fruits, drinks, dairy, meat, and food odors), most studies were conducted on products with a high level of acceptance. Knowledge in the field of food-induced emotions would benefit greatly from the results of more studies that use a wider variety of food-related product categories. Usually, different food products from the same category are presented in the same session in the case of a CLT setting. The maximum number of samples is defined as a function of the research objective, the nature of the food product, the test procedure, and the test location. In a CLT condition, King et al. [27] showed that the number of significant differences in emotions increased with the number of products; however, the intensity of the rated emotions tended to decrease with an increase in the number of tested samples. The appropriate number of samples in each test depends on the product category, the product, and the signal-to-noise ratio of the measure (e.g., 4 products for savory snack food [27], 6–10 for beverages [176, 177]). The number of samples also depends on the test procedure and how much time is available for the test [178]. The length of the experiment is critical and should be optimized so that sensory fatigue, habituation, and sensory satiety are minimized. Finally, the number of tests should be optimized with the test location so that it corresponds to the natural condition and frequency of consumption.

Different panel sessions or different days could be envisaged so that more products could be tasted by the participants. It is also generally recommended to present one or two products twice to verify the reproducibility of the measure. Moreover, following a balanced presentation order is recommended regarding the first sample position, which tends to result in higher positive emotional scores [72, 179]. More than presenting the samples in blind conditions, one can envisage evaluating an unbranded or branded food product when the name, picture of the product, or its packaging are included. This approach requires the samples to be presented in different conditions, a constraint that has an impact on the total number of samples that can be compared.

5.4 Evaluation Procedure

Measuring self-reported emotional responses to food products is not very different from measuring their sensory and hedonic characteristics. Different approaches have been implemented beside the linear scales: Check All That Apply (CATA) and Rate All That Apply (RATA), yes/no procedure, or Temporal Dominance of Emotions (TDE) for measuring the temporality of the emotional response. The size of the difference between the products or the research objectives may help in the choice of the evaluation procedure. For example, linear scales may better capture more subtle differences between products, and TDE may be more appropriate to capture the temporality of the emotional response over a meal experience or to evaluate lasting food products.

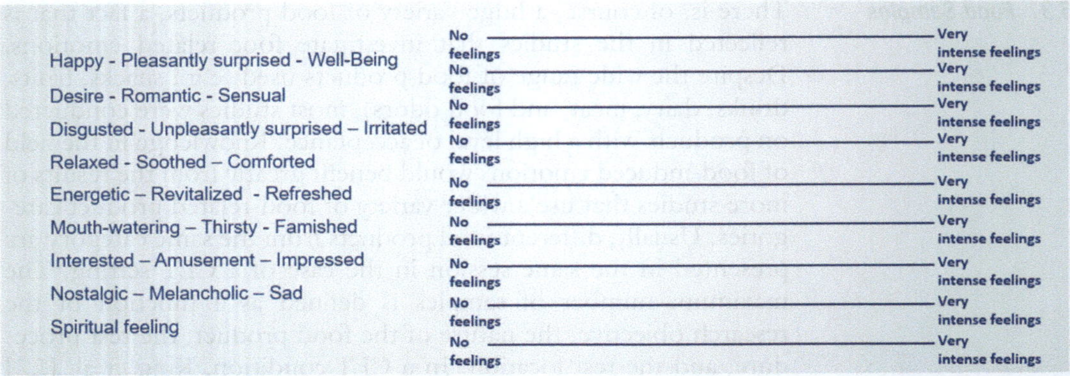

Fig. 5 Example of linear scales of emotions used in a questionnaire

5.4.1 Linear Scales

Respondents are requested to indicate the relevance of each proposed emotional term on linear scales in order to describe their feelings while eating the food product (from left = "no feelings" to right = "very intense feelings"), as described in Porcherot et al. [25]. One scale can be provided for each feeling term. The questionnaires can be further simplified by reducing the number of scales to be rated. In the ScentMove® questionnaire, only one scale was kept for each feeling dimension of the full Emotion and Odor Scale, and each scale was illustrated by the three terms depicting the emotional dimension (Fig. 5). Participants are requested to rate the intensity of their feelings for each series of terms, considering which term(s) is most relevant to them. Spinelli et al. [35] recognized that this approach could help the respondent to better understand the task by clarifying the emotional dimension. In addition, a study conducted on 12 shampoo fragrances showed that the original scale and the simplified scales yielded comparable results [25]. These different psychometrically validated emotion scales have proven to be practical tools and are routinely used in our industry.

5.4.2 CATA Method

Another strategy to reduce the length of the experiments is to consider the CATA method [27, 33, 58, 180]. CATA also has the advantage of requiring lower cognitive involvement and eliciting more spontaneous answers [101, 159]. Consumers are requested to select, or check, as many terms as needed to qualify their feelings and respond to the following question: *"Please check all terms which correspond to your feelings whilst tasting this sample,"* or *"Please check all terms which correspond to your feelings whilst thinking about this flavour"* (in the case of an online survey). We compared the CATA method with a 5-point scale rating method for the comparison of five shampoo fragrances evaluated in six different tests [181]. Multidimensional factor analyses and RV coefficients indicated similar fragrance descriptions among the six tests and different emotional responses among the five shampoo fragrances. In addition, results

Option 1. List

Amusement	
Comforted	
Desire	
Disgusted	
Energetic	
Famished	
Happy	
Impressed	
Interested	
Irritated	
Melancholic	
Mouth-watering	
Nostalgic	
Pleasantly surprised	
Refreshed	
Relaxed	
Revitalized	
Romantic	
Sad	
Sensual	
Soothed	
Spiritual feeling	
Thirsty	
Unpleasantly surprised	
Well-Being	

Option 2. Circle

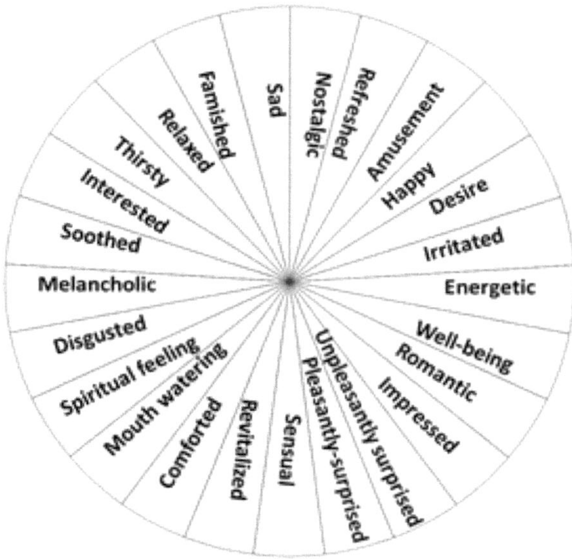

Fig. 6 Two possible presentations of emotional terms given to consumers. A table is most commonly presented if randomization is possible in the software used in the study. For paper-based questionnaires, the use of the circle avoids any carry-over effect linked to the order of presentation of the emotions

showed that the CATA method could provide more discrimination. Better discrimination compared with that on a rating scale was confirmed by King et al. [27]. With such an approach, the feeling terms could be provided in a list (option 1) or in a circle (option 2) (Fig. 6).

When comparing both rating RATA and CATA, Ng et al. [33] suggested that RATA might be a good compromise. RATA questions were used for short lists of up to 16 terms [182] and longer lists of up to 38 terms [183]. RATA can provide a broader picture with more possible statistical analyses than CATA, which provides only binary data.

5.4.3 Yes/No Approach

Another approach is the use of a "yes" or a "no" response, associated with a forced-choice procedure in which respondents indicate "yes" or "no" regarding whether the listed emotion words correspond to their feelings [37, 106, 169, 170, 184–186]. This procedure requires more cognitive effort and reduces satisficing of response behaviors [187–189].

5.4.4 Temporal Dominance of Emotion

More recently, the temporal evolution of the emotional response to products has been implemented by using the TDE approach [185, 186, 190, 191]. This approach is a form of time-intensity measurement [192, 193] adapted to emotions and consists in a temporal record of changes in emotional responses over time. It has been shown to be relevant and appropriate to the study of emotional responses associated with consumption of foods. Future research into new methods of emotion data capture by using mobile electronic devices is important and will become a growing area.

6 Examples of Use

Research on emotions provides relevant information that can be used in food product development, both at the stage of optimization and of innovation. Its relevance for a better understanding of product experience is demonstrated by the growing literature on the topic and by several applications for different product categories. All these studies can capture elicited emotional benefits from the food products themselves, but also with the consideration of the information provided by the product, the context in which it is consumed, and the full meal experience.

6.1 Product Development Purpose

Feeling measures can be used as a relevant source of information in product development projects and can have many practical applications in the food industry throughout the product development cycle [28]. They can be used to detect the impact of small ingredient changes on emotional responses [32]. When feeling responses are collected in parallel with the hedonic response, it is possible to better understand which emotions drive positive hedonic responses and which emotions drive negative responses. Moreover, numerous studies have shown the usefulness of measuring feelings for products in addition to hedonic measures because the products were differentiated on emotional responses but did not differ in liking (Fig. 7) [25, 27, 28, 58, 194–196]. This information can also help in product formulation, which can be adapted to achieve the expected feeling in harmony with product positioning while maintaining the same hedonic level.

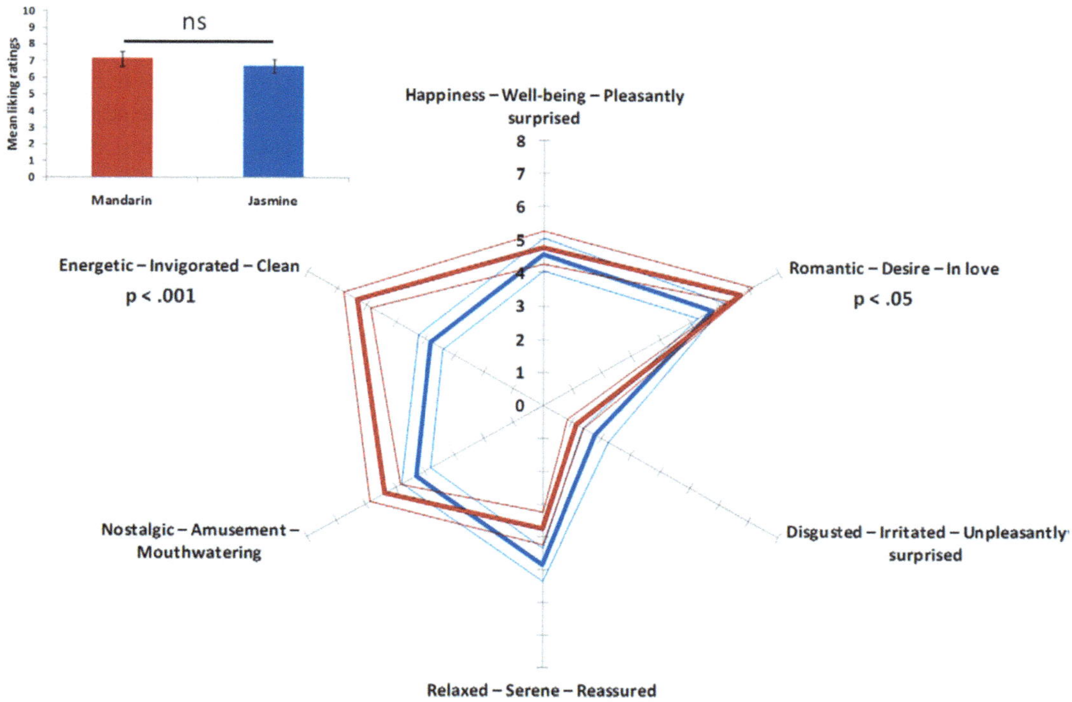

Fig. 7 Mean values and standard deviation for hedonic and emotional responses for two odors (mandarin in red and jasmine in blue)

The addition of feeling terms to questionnaires has been successfully used to measure emotional responses for several different stimuli with different sensory profiles. Thomson et al. [36] worked on the link between feelings and chocolate description and found that cocoa (sensory characteristic) is associated with powerful and energetic (conceptualizations); bitter with confident, adventurous, and masculine; smoky/burnt with arrogant, serious, traditional, and pretentious; vanilla, brown, and initial bite with sensual, fun, and luxurious; and creamy and sweet with fun, comforting, and easygoing. Gomez-Corona et al. [197] also studied the link between the sensory attributes of 10 strawberry yogurts with the emotional response of 150 Mexican consumers. The study underlined that the yogurts with more fleshy, jammy, and sweet flavor attributes elicited happy feelings, and yogurts with milky notes elicited more comfort feelings. Although these results were limited to blind testing and non-contextual evaluations, it was possible to identify the sensory attributes that triggered the emotional response of flavors when using the ScentMove® tool by using multiple factor analysis. In sum, data collected on emotions can be used in combination with acceptability and sensory characteristics data to provide a more complete understanding of product performance and accordingly allow better business decisions to be made. The sensory profile of a product can be modified to maximize specific emotions, such as is performed for drivers of liking.

The link between feelings with product categories, concepts, colors, and ingredients can also be made by means of surveys, in which consumers are asked to match different emotions with flavor tonalities based on their own experience. These associations can be made by consumers on the basis of their experience and expectations without smelling or consuming any product. For example, Parente [198] found that functional hygiene products (toothpaste and deodorant) were associated with well-being, efficacy, simplicity, essential, trust, care, routine, and peaceful, whereas face treatment products (shampoo and cream rinse) were associated with care, essential, well-being, and pleasant. In this study, they emphasized that emotional expectations depend on the product categories. In an unpublished study, our team performed a global survey that aimed at uncovering specific links between feelings and ingredients in several counties around the world. This study underlined that caramel and chocolate were related to playful, pampering, and indulging feelings, whereas mangosteen and pomegranate were related to exciting and thrilling feelings. Another survey was administered in Brazil (unpublished data) for well-known savory ingredients such as spices (garlic and onions), cheeses (mozzarella and parmesan), meat (chicken), and vegetables (tomato and carrot). Results underlined that happy was the most selected emotion among Brazilian consumers in relation to picanha and some cheeses such as parmesan, mozzarella, and requeijão, as well as fresh tomato and oregano spice. Well-being was more related to olive oil at the first point, followed by chicken breast, tomato, onion, and garlic. From the ingredient perspective, this research has shown that olive oil mainly evokes well-being, happy, and revitalized feelings. It was therefore possible to connect savory flavors and emotions, and this study allowed our company to suggest the best savory ingredients according to expectation and marketing story. This information can proactively be used by flavorists when they are asked to develop appropriate flavors in accordance with requested emotional benefits for specific countries.

6.2 Alignment with Emotional Responses Triggered by Expectation

Overall evaluation and choice are not only a result of sensory input from smell, texture, taste, or flavor, which are integrated into so-called perceived quality or bottom-up processing, but also a result of top-down modulation from the external characteristics such as claim, flavor name, package, color, advertising, or price. Therefore, emotional responses may be influenced by both the emotional expectation and the product sensory characteristics (for a review, *see* Ref. [199]). For this reason, consideration of the full experience that comes from the expectation and the product experience is recommended in order to optimize a food product consonant with the brand image or vice versa [171, 200, 201]. As a result, products that appropriately fulfill consumers' emotional expectations will have a higher probability of success in the market

[202] because they avoid disappointment or frustration in the case of a mismatch between expectations and product characteristics [203].

The first studies conducted in this field investigated how and to what extent information could affect the discrepancy between expected and experienced liking and compared their results with Anderson's assimilation and contrast models [204]. More precisely, information related to health and nutrition properties [205–207] or to the nature of the product, product origin, or production [208–210] was reported to have an influence on product liking. In most cases, the influence of assimilation effects was observed, i.e., when the reported liking for a product was presented with information changes in the direction of expectation in comparison with the reported liking in the blind condition, illustrating a positive effect of the information [206, 211]. In fewer cases, a contrast effect was observed when the reported liking for a product was presented with information that moved away from expectation, demonstrating an increased discrepancy between product experience and expectation from a negative effect of the information [211].

Researchers next investigated how best to develop products that are aligned with consumer emotional expectations. To address this question, researchers measure the verbal emotional responses triggered by food products while participants experience them, as well as the verbal emotional responses from the expectation provided by the information. For example, Schifferstein et al. [171] noted that packaging design characteristics are critical, as they can suggest a certain identity that may enhance or interfere with the identification and evaluation of the content. Colors may also better suit some fragrances than others and thus be more congruent with them [57]. Olfactory perceptions were shown to be affected by the visual system, even when the question focuses on perfume perception [154, 212]. In addition, it was shown that congruent associations were more pleasant than incongruent associations [213].

Porcherot et al. [155] examined the influence of food odorant names on self-reported feelings. The flavor name drives strong expectations when it is mentioned on the product packaging because consumers expect to recognize the named flavor [214]. When buying flavored ice cream, consumers select vanilla because of their preference for this flavor and their experience with different vanilla-flavored products, which, in turn, reveals expectations of a particular vanilla taste. In the experiment, consumers reported their feelings while smelling an odorant, which was presented blind or in association with its name (i.e., a mint odor with a "mint" name). In a third condition, consumers reported how they would expect to feel if they smelled a given odorant, with the name of the odorant being presented to them (i.e., "mint"). Results indicated that the information about the odorant name influences the reported feelings, as previously observed by several authors (*see* the review by Deliza and MacFie [204] regarding traditional liking assessments). Experienced feeling intensities were lower than those of expected feelings for some odorants. The quality of the odor seemed to be of crucial importance in that case. The information about the name was sometimes sufficient to decrease the discrepancy between the experience and the expectation. A similar approach was used for food names and tasted foods by Cardello et al. [31] to compare how one expects to feel to how one actually feels. Highly emotion-laden food items (e.g., chocolate) elicited a more intense emotional response to the name than to the tasted product, suggesting that food names may elicit memories of an idealized emotional experience with the food (*see* also Ref. [36]). This observation was confirmed by Spinelli et al. [75] with hazelnut and cocoa spreads, which were presented blind and with their packaging.

We can therefore conclude that consideration of the emotional response for both the blind condition and the informed condition can provide deeper insights into a product's emotional reaction, which helps companies to design products that meet consumer expectations. The challenge is therefore to create food products that are not only pleasant and perform well, but that also reinforce the reported feeling benefits, in harmony with the marketing mix (claim, advertising message, positioning, brand, etc.). However, the distinction between bottom-up processing and top-down modulation is a challenge. Thomson et al. [36] mentioned that "when someone tells us that a product makes them feel 'happy', 'passionate', etc. it is more likely that they are reflecting what the product is communicating to them (emotional conceptualizations) rather than doing to them (emotional consequences)." Ng et al. [196] explained that the intrinsic sensory characteristics have a stronger association with emotions, whereas extrinsic product characteristics may have a stronger association with abstract/functional conceptualizations.

6.3 Consideration of the Emotional Responses Triggered by the Meal Experience

Increasing attention is being paid to ecological validity when studying emotional responses, with the consideration that emotional responses may vary in different stages of product experience and depending on the context in which the product is consumed [167]. Schifferstein et al. [171] showed that the emotional response changed with experience of a dehydrated vegetable food product, from choosing the product on a supermarket shelf, to opening the package, to cooking and eating the food. At the stage of buying, the product was associated with a combination of negative emotions (contempt, sadness) and fascination, which transformed into positive emotions at opening (satisfaction, joy), cooking (attraction), and eating (admiration), but returned to negative emotions when a new buying decision was to be made (aversion, sadness). The emotional response is dynamic and depends on sensory characteristics, which may evolve from meal consumption to digestion [215].

Edwards et al. [19] addressed the extent to which the consumption situation influences the elicited emotion responses. An appetitive context or an overall negative feeling of the day can change the reported perception of the emotion. A negative feeling can weaken a positive emotion or strengthen a negative one.

Many studies have been conducted in laboratory settings, where emotions were measured for different presented stimuli and compared within the same session [20, 27, 36, 194]. However, it is now recognized that conducting emotion research in natural settings is important [19]. Porcherot et al. [156] conducted a study to measure self-reported affective feelings for three Kir variants (i.e., French cocktails with apricot, grapefruit, or raspberry syrups added to white wine) in an ecological setting of a restaurant. One Kir variant was provided as an aperitif before the dinner. Results showed that, despite no differences in the participants' liking of the fruit variants, the raspberry Kir induced significant changes in mood states after the participants consumed the aperitif, which significantly reduced their tense and anxious states during the dinner. This study demonstrated that the measure of feelings can be conducted in an ecological environment, where complete meals are eaten with friends and relatives.

6.4 Measure of Emotional Responses in Virtual Environments

In addition to methods to increase the ecological validity of consumer studies, virtual reality (VR) and various other immersive techniques appear to be good alternatives for taking into consideration the influence of context on the enjoyment and experience of food products. The first question is whether studies in virtual environments are close to those conducted in real settings. Sinesio et al. [216] compared three immersive techniques that reproduced the situation of "drinking beer in a pub": (1) immersive room with a projection of the situation on flat walls with videos at 180° and with a pub setup, (2) VR headset with the projection of the situation from a 360° video, and (3) VR headset with the projection of the situation from three-dimensional modeling and 360° photos (Fig. 8). They compared these three immersive techniques to a traditional CLT testing environment and a test conducted in a real situation of having a beer in a pub. Emotional profile, liking, intention to retaste the beer samples, and consumers' perceived engagement were assessed and compared in these five test conditions with a total of 513 Italian male adults. Less product discrimination for hedonic questions and for emotions in assessments was obtained in the real pub and in the immersive approaches compared with that in the CLT environment. In addition, all immersive technologies tested improved consumers' engagement, and the

Fig. 8 Immersive testing environments. (**a**) Virtual reality using a 360-degree video. (**b**, **d**) Virtual reality using 3D modeling combined with 360-degree photos. (**c**) Immersive room combining video-audio projection from the pub with consistent furniture

frequency of emotions showing a positive relationship with liking was higher for in-context evaluations. Results from the studies conducted in virtual environments were therefore closer to the study conducted in a real pub, especially for the immersive room and VR three-dimensional modelling, with less product discrimination in all these conditions. Moreover, repeatability, measured as the proximity of hedonic scores provided by individual participants to the duplicate products, was better in the real pub and in the immersive room.

Gomez-Corona and Porcherot (unpublished study) also conducted a study to compare the emotional response of three flavored biscuits (chocolate, caramel, and hazelnut) in three test conditions, 360 immersion, CLT, and natural settings, as shown Fig. 9. Significant differences were found among the three test conditions, and the product groups were closer for the 360° immersion and the natural settings. The results of these different studies therefore suggest that immersive techniques are relevant approaches to study the effects of context on product liking and emotional response by maintaining control of test conditions compared with natural settings.

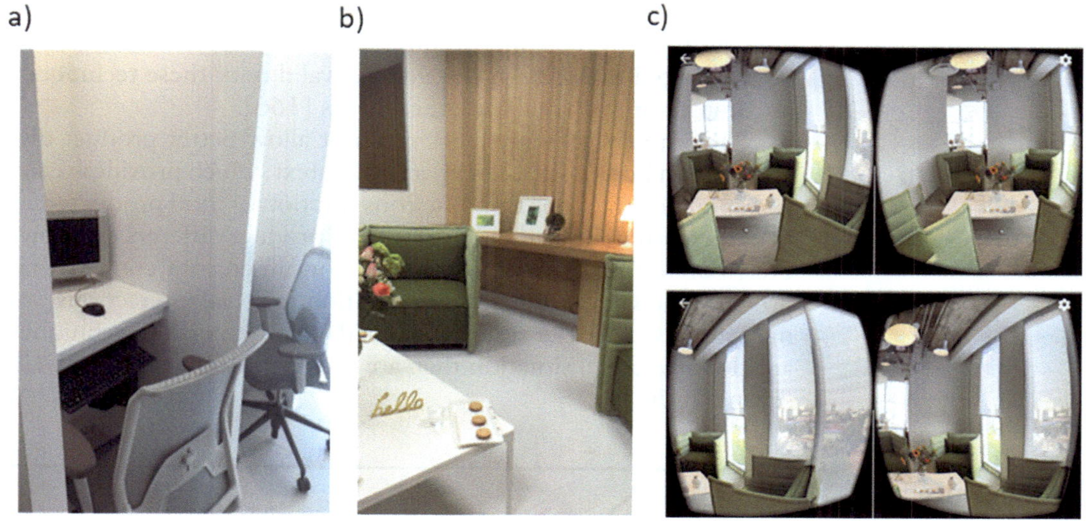

Fig. 9 Pictures of the sensory laboratory (**a**), living room used for the real-life condition (**b**), and 360° image from the Street View app (**c**)

Fig. 10 The two 360° environments: the pool and the park

The second question was whether immersive techniques could improve understanding of the effect of context on emotional responses elicited by food consumption. A second unpublished experiment was conducted by Gomez-Corona and Porcherot for this purpose. Chocolate biscuits were consumed in two virtual contexts, a park and a pool, evoked by 360° videos projected and seen in headsets (Fig. 10). Results showed that the main feelings elicited by the chocolate biscuits for both brands in the pool and park environments were soothed, happy, relaxed, well-being, thirsty, and amusement and that the two contexts had few effects on product liking and emotional responses.

The immersive approaches might therefore underestimate the contextual effect in comparison with what might happen in a real situation due to lack of similarity with real life or consumer personal experiences. In addition, one could question whether these contexts, virtual or real, are realistic enough and relevant to consumers.

Immersive technologies simulating real-life environments can be effective tools in consumer testing. However, more studies need to be conducted to establish the external validity of these techniques in comparison with real-life experiments [166].

The immersive approaches may not allow better product discrimination because of the impact of context, which provides visual and sound distractions from the consumption of food products and may therefore reduce the perception of product differences. However, because products are generally consumed in a context that provides the same distractions in addition to social interactions, we may suppose that the results obtained with immersive techniques are closer to a real situation than are the results obtained in a CLT setting in the lab.

7 Conclusion

This chapter provides many examples of emotional responses that change with food product experiences. It is therefore critical to assess emotional responses triggered by food consumption with different methods, in consideration of consumer expectations and the context in which the product is generally consumed. Many companies have acknowledged the importance of using emotion measurement to help them with the development and optimization of food products and in the understanding of consumer liking and choice. The purpose of this chapter was to provide a view of the measure of emotional responses within a complex and dynamic context related to food consumption. In considering various explicit and implicit measures, we have chosen to present the results obtained from verbalization methods through questionnaires. We indeed recommend this approach, as it has been proven relevant in providing interesting findings and is suitable to address key research questions about expectation and context, or to capture emotional responses in the natural context of consumption.

References

1. Kissileff HR, Klingsberg G, Van Itallie TB (1980) Universal eating monitor for continuous recording of solid or liquid consumption in man. Am J Physiol 238:R14–R22
2. Bellisle F, Le Magnen J (1980) The analysis of human feeding patterns: the Edogram. Appetite 1:141–150
3. de Castro JM, Bellisle F, Feunekes GI, Dalix AM, De Graaf C (1997) Culture and meal patterns: a comparison of the food intake of free-living American, Dutch, and French students. Nutr Res 17(5):807–829
4. Higgs S, Ruddock H (2020) Social influences on eating. In: Meiselman HL (ed) Handbook of eating and drinking; interdisciplinary perspectives. Springer, Cham
5. Macht M, Haupt C, Salewsky A (2004) Emotions and eating in everyday life: application of the experience-sampling method, ecology of food and nutrition. Appetite 43(4):11–21
6. Geliebter A, Aversa A (2003) Emotional eating in overweight, normal weight, and underweight individuals. Eat Behav 3(4):341–347

7. van Strien T, Rookus MA, Bergers GPA, Frijters JET, Defares PB (1986) Life events, emotional eating and change in body mass index. Int J Obes 10:29–35
8. Canetti L, Bachar E, Berry EM (2002) Food and emotion. Behav Process 60:157–164
9. Gibson EL (2006) Emotional influences on food choice: sensory, physiological and psychological pathways. Physiol Behav 89:53–61
10. Macht M (2008) How emotions affect eating: a five-way model. Appetite 50:1–11
11. Macht M (1999) Characteristics of eating in anger, fear, sadness and joy. Appetite 33:129–139
12. Arnow B, Kenardy J, Agras WS (1995) The emotional eating scale: the development of a measure to assess coping with negative affect by eating. Int J Eat Disord 18(1):79–90
13. Heatherton TF, Baumeister RF (1991) Binge eating as escape from self-awareness. Psychol Bull 110(1):86
14. Van Strien T (2018) Causes of emotional eating and matched treatment of obesity. Curr Diab Rep 18(6):1–8
15. Pool E, Delplanque S, Coppin G, Sander D (2015) Is comfort food really comforting? Mechanisms underlying stress-induced eating. Food Res Int 76:207–215
16. Ganley RM (1989) Emotion and eating in obesity: a review of the literature. Int J Eat Disord 8:343–361
17. Macht M, Simons G (2000) Emotions and eating in everyday life. Appetite 35(1):65–71
18. Fong M, Li A, Hill AJ, Cunich M, Skilton MR, Madigan CD, Caterson ID (2019) Mood and appetite: their relationship with discretionary and total daily energy intake. Physiol Behav 207:122–131
19. Edwards JSA, Hartwell HJ, Brown L (2013) The relationship between emotions, food consumption and meal acceptability when eating out of the home. Food Qual Prefer 30:22–32
20. Desmet PMA, Schifferstein HNJ (2008) Sources of positive and negative emotions in food experience. Appetite 50:290–301
21. Macht M, Dettmer D (2006) Everyday mood and emotions after eating a chocolate bar or an apple. Appetite 46:332–336
22. Higgs S, Robinson E, Lee M (2012) Learning and memory processes and their role in eating: implications for limiting food intake in overeaters. Curr Obes Rep 1:91–98
23. Pool ER, Sander D (2021) Emotional learning: measuring how affective values are acquired and updated. In: Meiselman H (ed) Emotion measurement, 2nd edn. Woodhead Publishing, Oxford
24. Ferdenzi C, Roberts SC, Schirmer A, Delplanque S, Cekic S, Porcherot C et al (2013) Variability of affective responses to odors: culture, gender, and olfactory knowledge. Chem Senses 38:175–186
25. Porcherot C, Delplanque S, Raviot-Derrien S, Le Calvé B, Chrea C, Gaudreau N et al (2010) How do you feel when you smell this? Optimization of a verbal measurement of odor-elicited emotions. Food Qual Prefer 21:938–947
26. Gutjar S, de Graaf C, Kooijman V, de Wijk RA, Nys A, ter Horst GJ et al (2015) The role of emotions in food choice and liking. Food Res Int 76:216–223
27. King SC, Meiselman HL, Carr BT (2013) Measuring emotions associated with foods: important elements of questionnaire and test design. Food Qual Prefer 28:8–16
28. Meiselman HL (2015) A review of the current state of emotion research in product development. Food Res Int 76(2):192–199
29. Mojet J, Dürrschmid K, Danner L, Jochl M, Heinio R-L, Holthuysen N et al (2015) Are implicit emotion measurements evoked by food unrelated to liking? Food Res Int 76(2):224–232
30. Walsh AM, Duncan SE, Bell MA, O'Keefe SF, Gallagher DL (2017) Integrating implicit and explicit emotional assessment of food quality and safety concerns. Food Qual Prefer 56: 212–224
31. Cardello AV, Meiselman HL, Schutz HG, Craig C, Given Z, Lesher LL et al (2012) Measuring emotional responses to foods and food names using questionnaires. Food Qual Prefer 24:243–250
32. King S, Meiselman HL (2010) Development of a method to measure consumer emotions associated with foods. Food Qual Prefer 21: 168–177
33. Ng M, Chaya C, Hort J (2013) Beyond liking: comparing the measurement of emotional response using EsSense Profile and consumer defined check-all-that-apply methodologies. Food Qual Prefer 28(1): 193–205
34. Schouteten JJ, De Steur H, De Pelsmaeker S, Lagast S, De Bourdeaudhuij I, Gellynck X (2015) Impact of health labels on flavor perception and emotional profiling: a consumer study on cheese. Nutrients 7:10251–10268
35. Spinelli S, Masi C, Dinnella C, Zoboli GP, Monteleone E (2014) How does it make

you feel? A new approach to measuring emotions in food product experience. Food Qual Prefer 37:109–122

36. Thomson DMH, Crocker C, Marketo CG (2010) Linking sensory characteristics to emotions: an example using dark chocolate. Food Qual Prefer 21:1117–1125
37. Piqueras-Fiszman B, Jaeger SR (2014) The impact of evoked consumption contexts and appropriateness on emotion responses. Food Qual Prefer 32:277–288
38. Piqueras-Fiszman B, Jaeger SR (2014) Emotion responses under evoked consumption contexts: a focus on the consumers' frequency of product consumption and the stability of responses. Food Qual Prefer 35:24–41
39. Dalenberg JR, Gutjar S, Ter Horst GJ, de Graaf K, Renken RJ et al (2014) Evoked emotions predict food choice. PLoS One 9: e115388
40. Gutjar S, Dalenberg JR, de Graaf C, de Wijk RA, Palascha A, Renken RJ et al (2015) What reported food-evoked emotions may add: a model to predict consumer food choice. Food Qual Prefer 45:140–148
41. Schouteten JJ, Verwaeren J, Lagast S, Gellynck X, De Steur H (2018) Emoji as a tool for measuring children's emotions when tasting food. Food Qual Prefer 68:322–331
42. Ekman P (1984) Expression and the nature of emotion. In: Scherer KR, Ekman P (eds) Approaches to emotion. Erlbaum, Hillsdale, pp 319–344
43. Izard CE (1993) Four systems for emotion activation: cognitive and noncognitive processes. Psychol Rev 100(1):68
44. Tomkins SS (1984) Affect theory. In: Approaches to emotion, vol 163. Erl-baum, Hillsdale, pp 163–195
45. Alaoui-Ismaili O, Robin O, Rada H, Dittmar A, Vernet-Maury E (1997) Basic emotions evoked by odorants: comparison between autonomic responses and self-evaluation. Physiol Behav 62:713–720
46. Vernet-Maury E, Alaoui-Ismaïli O, Dittmar A, Delhomme G, Chanel J (1999) Basic emotions induced by odorants: a new approach based on autonomic pattern results. J Auton Nerv Syst 75(2–3):176–183
47. Lang PJ, Greenwald MK, Bradley MM, Hamm AO (1993) Looking at pictures: affective, facial, visceral, and behavioral reactions. Psychophysiology 30(3):261–273
48. Russell JA (1980) A circumplex model of affect. J Pers Soc Psychol 39:1161–1178
49. Wundt VW (1905) Grundzüge der physiologischen Psychologie. Principles of physiological psychology. Nature 71:529–530
50. Bensafi M, Rouby C, Farget V, Bertrand B, Vigouroux M, Holley A (2002) Autonomic nervous system responses to odours: the role of pleasantness and arousal. Chem Senses 27: 703–709
51. Bensafi M, Rouby C, Farget V, Bertrand B, Vigouroux M, Holley A (2002) Influence of affective and cognitive judgments on autonomic parameters during inhalation of pleasant and unpleasant odors in humans. Neurosci Lett 319:162–166
52. Chebat JC, Michon R (2003) Impact of ambient odors on mall shoppers' emotions, cognition, and spending: a test of competitive causal theories. J Bus Res 56:529–539
53. Herz RS, Schankler C, Beland S (2004) Olfaction, emotion, and associative learning: effects on motivated behavior. Motiv Emot 28:363–383
54. Heuberger E, Hongratanaworakit T, Böhm C, Weber R, Buchbauer G (2001) Effects of chiral fragrances on human autonomic nervous system parameters and self-evaluation. Chem Senses 26(3):281–292
55. Jönsson FU, Olsson H, Olsson MJ (2005) Odor emotionality affects the confidence in odor naming. Chem Senses 30(1):29–35
56. Pössel P, Ahrens S, Hautzinger M (2005) Influence of cosmetics on emotional, autonomous, endocrinological, and immune reactions. Int J Cosmet Sci 27(6):343–349
57. Schifferstein HN, Tanudjaja I (2004) Visualising fragrances through colours: the mediating role of emotions. Perception 33:1249–1266
58. Warrenburg S (2005) Effects of fragrance on emotions: moods and physiology. Chem Senses 30:i248–i249
59. Moors A, Ellsworth PC, Scherer KR, Frijda NH (2013) Appraisal theories of emotion: state of the art and future development. Emot Rev 5(2):119–124
60. Scherer KR (2009) Feelings (psychological perspectives). In: Sander D, Scherer KR (eds) Oxford companion to emotion and the affective sciences. Oxford University Press, Oxford, pp 183–184
61. Sander D, Grandjean D, Scherer KR (2018) An appraisal-driven componential approach to the emotional brain. Emot Rev 10:219–231

62. Delplanque S, Chrea C, Grandjean D, Ferdenzi C, Cayeux I, Porcherot C et al (2012) How to map the affective semantic space of scents. Cognit Emot 26:885–898
63. Zentner M, Grandjean D, Scherer KR (2008) Emotions evoked by the sound of music: characterization, classification, and measurement. Emotion 8:494–521
64. Chrea C, Grandjean D, Delplanque S, Cayeux I, Le Calvé B, Aymard L et al (2009) Mapping the semantic space for the subjective experience of emotional responses to odors. Chem Senses 34(1):49–62
65. House ED, Arruda JE, Andrasik F, Grazzi L (2012) The reliability and validity of the visual analog mood scales in non-English-speaking pain patients. Pain Pract 8:626–632
66. Lagast S, Gellynck X, Schouteten JJ, De Herdt V, De Steur H (2017) Consumers' emotions elicited by food: a systematic review of explicit and implicit methods. Trends Food Sci Technol 69(Part A):172–189
67. Danner L, Sidorkina L, Joechl M, Duerrschmid K (2014) Make a face! Implicit and explicit measurement of facial expressions elicited by orange juices using face reading technology. Food Qual Prefer 32:167–172
68. De Houwer J, Moors A (2007) How to define and examine the implicitness of implicit measures. In: Wittenbrink B, Schwarz N (eds) Implicit measures of attitudes: procedures and controversies. The Guilford Press, New York, pp 179–194
69. Delplanque S, Grandjean D, Chrea C, Coppin G, Aymard L, Cayeux I et al (2009) Sequential unfolding of novelty and pleasantness appraisals of odors: evidence from facial electromyography and autonomic reactions. Emotion 9(3):316
70. Delplanque S, Sander D (2021) A fascinating but risky case of reverse inference: from measures to emotions! Food Qual Prefer 92: 104183
71. Bernard HR (2011) Research methods in anthropology: qualitative and quantitative approaches. Rowman Altamira
72. Dorado R, Perez-Hugalde C, Picard A, Chaya C (2016) Influence of first position effect on emotional response. Food Qual Prefer 49: 189–196
73. Gmuer A, Guth JN, Runte M, Siegrist M (2015) From emotion to language: application of a systematic, linguistic-based approach to design a food-associated emotion lexicon. Food Qual Prefer 40:77–86
74. Gunaratne TM, Viejo CG, Fuentes S, Torrico DD, Gunaratne NM, Ashman H, Dunshea FR (2019) Development of emotion lexicons to describe chocolate using the Check-All-That-Apply (CATA) methodology across Asian and Western groups. Food Res Int 115:526–534
75. Spinelli S, Masi C, Zoboli GP, Prescott J, Monteleone E (2015) Emotional responses to branded and unbranded foods. Food Qual Prefer 42:1–11
76. Manzocco L, Rumignani A, Lagazio C (2013) Emotional response to fruit salads with different visual quality. Food Qual Prefer 28:17–22
77. Ferrarini R, Carbognin C, Casarotti EM, Nicolis E, Nencini A, Meneghini AM (2010) The emotional response to wine consumption. Food Qual Prefer 21:720–725
78. Pineau EP, Rytz A, Hudry GJ, Maier A, Alexander E (2010) Do different flavors generate different emotions? A multidisciplinary approach to measure the emotional response related to beverage consumption. In: Poster presented at the 4th European conference on sensory and consumer research, Vitoria-Gasteiz, Spain
79. Chaya C, Eaton C, Hewson L, Vazquez RF, Fernandez-Ruiz V, Smart KA et al (2015) Developing a reduced consumer-led lexicon to measure emotional response to beer. Food Qual Prefer 45:100–112
80. Mora M, Giussani B, Pagliarini E, Chaya C (2019) Improvement of an emotional lexicon for the evaluation of beers. Food Qual Prefer 71:158–162
81. Bhumiratana N, Adhikari K, Chambers E (2014) The development of an emotion lexicon for the coffee drinking experience. Food Res Int 61:83–92
82. Kanjanakorn A, Lee J (2017) Examining emotions and comparing the EsSense Profile® and the coffee drinking experience in coffee drinkers in the natural environment. Food Qual Prefer 56:69–79
83. Rousset S, Deiss V, Juillard E, Schlich P, Droit-Volet S (2005) Emotions generated by meat and other food products in women. Br J Nutr 94:609–619
84. Churchill A, Behan J (2010) Comparison of methods used to study consumer emotions associated with fragrance. Food Qual Prefer 21:1108–1113
85. Rétiveau AN, Chambers IVE, Milliken GA (2004) Common and specific effects of fine fragrances on the mood of women. J Sens Stud 19:373–394
86. Berezina TN (2014) Research of the psychological interconnection between the basic

odors and the emotional state of a human being. Asian Soc Sci 10:159–167

87. Ferdenzi C, Delplanque S, Barbosa P, Court K, Guinard JX, Guo T et al (2013) Affective semantic space of scents: towards a universal scale to measure self-reported odor-related feelings. Food Qual Prefer 30:128–138

88. Porcherot C, Delplanque S, Gaudreau N, DeLaura C, Cayeux I (2012) Cultural specificities in verbal measures of emotions triggered by smells. In Poster presented at the 5th European conference on sensory and consumer research, Bern, Switzerland

89. van Zyl H, Meiselman HL (2015) The roles of culture and language in designing emotion lists: comparing the same language in different English and Spanish speaking countries. Food Qual Prefer 41:201–213

90. van Zyl H, Meiselman HL (2016) An update on the roles of culture and language in designing emotion lists: English, Spanish and Portuguese. Food Qual Prefer 51:72–76

91. Danner L, Haindl S, Joechl M, Duerrschmid K (2014) Facial expressions and autonomous nervous system responses elicited by tasting different juices. Food Res Int 64:81–90

92. de Wijk RA, Kooijman V, Verhoeven RHG, Holthuysen NTE, de Graaf C (2012) Autonomic nervous system responses on and facial expressions to the sight, smell, and taste of liked and disliked foods. Food Qual Prefer 26:196–203

93. Chai J, Ge Y, Liu Y, Li W, Zhou L, Yao L, et al (2014) Application of frontal EEG asymmetry to user experience research. In: International conference on engineering psychology and cognitive ergonomics. Springer, pp 234–243

94. Schwarz N, Oyserman D (2001) Asking questions about behavior: cognition, communication, and questionnaire construction. Am J Eval 22:127–160

95. Köster EP, Mojet J (2015) From mood to food and from food to mood: a psychological perspective on the measurement of food-related emotions in consumer research. Food Res Int 76(Pt. 2):180–191

96. Laurans GFG, Desmet PMA (2012) Introducing PREMO2: new directions for the non-verbal measurement of emotion in design. In: Out of control: proceedings of the 8th international conference on design and emotion, London, UK, 11–14 Sept 2012

97. Desmet PMA (2003) Measuring emotion: development and application of an instrument to measure emotional responses to products. In: Blythe MA, Monk AF, Overbeeke K, Wright PC (eds) Funology: from usability to enjoyment. The Dordrecht, pp 111–123

98. Desmet PM, Hekkert P, Jacobs JJ (2000) When a car makes you smile: development and application of an instrument to measure product emotions. NA-Adv Consum Res 27

99. den Uijl LC, Jager G, Zandstra EH, de Graaf C, Kremer S (2016) Self-reported food-evoked emotions of younger adults, older normosmic adults, and older hyposmic adults as measured using the PrEmo2 tool and the Affect Grid. Food Qual Prefer 51: 109–117

100. He W, Boesveldt S, de Graaf C, de Wijk RA (2016) The relation between continuous and discrete emotional responses to food odors with facial expressions and non-verbal reports. Food Qual Prefer 48:130–137

101. Danilo S, Dreyfuss L, Bremard D (2011) Could a non-verbal emotional method be cross-cultural? In: 9th pangborn sensory science symposium, Toronto, Canada

102. Vidal L, Ares G, Jaeger SR (2016) Use of emoticon and emoji in tweets for food-related emotional expression. Food Qual Prefer 49: 119–128

103. Jaeger SR, Lee SM, Kim K-O, Chheang SL, Jin D, Ares G (2017) Measurement of product emotions using emoji surveys: case studies with tasted foods and beverages. Food Qual Prefer 62:46–59

104. Jaeger SR, Lee P-Y, Ares G (2018) Product involvement and consumer food- elicited emotional associations: insights from emoji questionnaires. Food Res Int 106:999–1011

105. Jaeger SR, Vidal L, Ares G (2020) Should emoji replace emotion words in questionnaire-based food-related consumer research? Food Qual Prefer 92:104121

106. Jaeger SR, Ares G (2017) Dominant meanings of facial emoji: insights from Chinese consumers and comparison with meanings from internet resources. Food Qual Prefer 62:275–283

107. Gallo KE, Swaney-Stueve M, Chambers DH (2017) A focus group approach to understanding food-related emotions with children using words and emojis. J Sens Stud 32(3): e12264

108. Sick J, Spinelli S, Dinnella C, Monteleone E (2020) Children's selection of emojis to express food-elicited emotions in varied eating contexts. Food Qual Prefer 85:103953

109. Schouteten JJ, Meiselman HL (2021) The potential of using emoji instead of words when measuring emotions associated with food. Food Qual Prefer 92:104182
110. Thomson DM, Coates T (2020) Are unconscious emotions important in product assessment? How can we access them? Food Qual Prefer 92:104123
111. Jaeger SR, Vidal L, Ares G (2021) Should emoji replace emotion words in questionnaire-based food-related consumer research? Food Qual Prefer 92:104121
112. Danner L, Ristic R, Johnson TE, Meiselman HL, Hoek AC, Jeffery DW et al (2016) Context and wine quality effects on consumers' mood, emotions, liking and willingness to pay for Australian Shiraz wines. Food Res Int 89: 254–265
113. De Wijk RA, He W, Mensink MG, Verhoeven RH, De Graaf C (2014) ANS responses and facial expressions differentiate between the taste of commercial breakfast drinks. PLoS One 9(4):e93823
114. Köster EP (2003) The psychology of food choice: some often encountered fallacies. Food Qual Prefer 14:359–373
115. Nisbett RE, Wilson TD (1977) Telling more than we can know: verbal reports on mental processes. Psychol Rev 84:231
116. Kreibig SD (2010) Autonomic nervous system activity in emotion: a review. Biol Psychol 84:394–421
117. Norman GJ, Necka E, Berntson GG (2016) The psychophysiology of emotions. In: Emotion measurement. Woodhead Publishing, pp 83–98
118. Porcherot C, Raviot-Derrien S, Beague MP, Henneberg S, Niedziela M, Ambroze K, McEwan JA (2022) Effect of context on fine fragrance-elicited emotions: comparison of three experimental methodologies. Food Qual Prefer 95:104342
119. Niedziela MM, Ambroze K (2021) The future of consumer neuroscience in food research. Food Qual Prefer 92:104124
120. Ludden GDS, Schifferstein HNJ, Hekkert P (2009) Visual e tactual incongruities in products as sources of surprise. Empir Stud Arts 27:61–87
121. Warrenburg S (2002) Measurement of emotion in olfactory research. ACS Symp Ser 825: 243–260
122. Pichon AM, Coppin G, Cayeux I, Porcherot C, Sander D, Delplanque S (2015) Sensitivity of physiological emotional measures to odors depends on the product and the pleasantness ranges used. Front Psychol 6:1821
123. Kaneko D, Hogervorst M, Toet A, van Erp JBF, Kallen V, Brouwer A, M. (2019) Explicit and implicit responses to tasting drinks associated with different tasting experiences. Sensors 19:4397
124. Fernández C, Pascual JC, Soler J, Elices M, Portella MJ, Fernández-Abascal E (2012) Physiological responses induced by emotion-eliciting films. Appl Psychophysiol Biofeedback 37(2):73–79
125. de Wijk RA, Noldus LPJJ (2021) Using implicit rather than explicit measures of emotions. Food Qual Prefer 92:104125
126. Ekman P, Friesen WV (1971) Constants across cultures in the face and emotion. J Pers Soc Psychol 17:124
127. Ekman P (1993) Facial expression and emotion. Am Psychol 48:384
128. Wieser MJ, Brosch T (2012) Faces in context: a review and systematization of contextual influences on affective face processing. Front Psychol 3:471
129. Tian Y-L, Kanade T, Cohn JF (2005) Facial expression analysis. In: Handbook of face recognition. Springer, New York, pp 247–275
130. Bailey RL (2016) Modern foraging: presence of food and energy density influence motivational processing of food advertisements. Appetite 107:568–574
131. Garcia-Burgos D, Zamora MC (2013) Facial affective reactions to bitter-tasting foods and body mass index in adults. Appetite 71:178–186
132. Meiselman HL (2017) Emotion measurement: theoretically pure or practical? Food Qual Prefer 62:374–375
133. Banse R, Scherer KR (1996) Acoustic profiles in vocal emotion expression. J Pers Soc Psychol 70:614–636
134. Johnstone T, Scherer KR (2000) Vocal communication of emotion. In: Lewis M, Haviland J (eds) The handbook of emotion, 2nd edn. Guildford, New York, pp 220–235
135. Goudbeek M, Scherer KR (2010) Beyond arousal: valence and potency/control cues in the vocal expression of emotion. J Acoust Soc Am 128:1322–1336
136. Lausen A, Hammerschmidt K (2020) Emotion recognition and confidence ratings predicted by vocal stimulus type and prosodic parameters. Humanit Soc Sci Commun 7(1): 1–17

137. Gawronski B, De Houwer J (2014) Implicit measures in social and personality psychology. In: Reis HT, Judd CM (eds) Handbook of research methods in social and personality psychology. Cambridge University Press, Cambridge, pp 283–310
138. Nosek BA, Hawkins CB, Frazier RS (2011) Implicit social cognition: from measures to mechanisms. Trends Cogn Sci 15:152–159
139. Greenwald AG, Banaji MR (1995) Implicit social cognition: attitudes, self-esteem, and stereotypes. Psychol Rev 102:4–27
140. Verhulst F, Hermans D, Baeyens F, Spruyt A, Eelen P (2006) Determinants and predictive validity of direct and indirect measures of recently acquired food attitudes. Appetite 46:137–143
141. Demattè ML, Sanabria D, Spence C (2007) Olfactory-tactile compatibility effects demonstrated using a variation of the implicit association test. Acta Psychol 124:332–343
142. Lemercier-Talbot A, Coppin G, Cereghetti D, Porcherot C, Cayeux I, Delplanque S (2019) Measuring automatic associations between relaxing/energizing feelings and odors. Food Qual Prefer 77:21–31
143. Schouteten JJ (2020) Are questionnaires the best way to measure emotions for food products and beverages? Food Qual Prefer 92: 104122
144. Siegel EH, Sands MK, Van den Noortgate W, Condon P, Chang Y, Dy J, Barrett LF (2018) Emotion fingerprints or emotion populations? A meta-analytic investigation of autonomic features of emotion categories. Psychol Bull 144:343–393
145. Barrett LF, Adolphs R, Marsella S, Martinez AM, Pollak SD (2019) Emotional expressions reconsidered: challenges to inferring emotion from human facial movements. Psychol Sci Public Interest 20(1):1–68
146. Kaneko D, Toet A, Brouwer AM, Kallen V, Van Erp JB (2018) Methods for evaluating emotions evoked by food experiences: a literature review. Front Psychol 9:911
147. Plassmann H, Karmarkar UR (2015) Consumer neuroscience: revealing meaningful relationships between brain and consumer behavior
148. Larsen JT, Berntson GG, Poehlmann KM, Ito TA, Cacioppo JT (2008) The psychophysiology of emotion. Handb Emot 3:180–195
149. Dorado R, Chaya C, Tarrega A, Hort J (2016) The impact of using a written scenario when measuring emotional response to beer. Food Qual Prefer 50:38–47
150. Schouteten JJ, De Steur H, De Pelsmaeker S, Lagast S, De Bourdeaudhuij I, Gellynck X (2015) An integrated method for the emotional conceptualization and sensory characterization of food products: the emo sensory (R) wheel. Food Res Int 78:96–107
151. Scherer KR (2005) What are emotions? And how should they be measured? Soc Sci Inf 44(4):695–729
152. Desmet PMA (2005) Typology of fragrance emotions. In: Proceedings of the fragrance research conference. ESOMAR, Amsterdam, pp 133–146
153. Ferdenzi C et al (2011) Affective dimensions of odor perception: a comparison between Swiss, British, and Singaporean populations. Emotion 11(5):1168–1181
154. Porcherot C, Delplanque S, Gaudreau N, Cayeux I (2013) Seeing, smelling, feeling! Is there an influence of color on subjective affective responses to perfumed fabric softeners? Food Qual Prefer 27:161–169
155. Porcherot C, Delplanque S, Planchais A, Gaudreau N, Accolla R, Cayeux I (2012) Influence of food odorant names on the verbal measurement of emotions. Food Qual Prefer 23(2):125–133
156. Porcherot C, Petit E, Giboreau A, Gaudreau N, Cayeux I (2015) Measurement of self-reported affective feelings when an aperitif is consumed in an ecological setting. Food Qual Prefer 39:277–284
157. Pionnier Pineau E, Rytz A, Gerebtzoff D, Godinot N, Hudry J, Maier A, et al (2010) Do different flavors generate different emotions? A multidisciplinary approach to measure the emotional response related to beverage consumption. In: Poster presented at 4th European conference on sensory and consumer research: a sense of quality, Spain, Vitoria-Gasteiz
158. Nestrud MA, Meiselman HL, King SC, Lesher LL, Cardello AV (2016) Development of EsSense25, a shorter version of the EsSense Profile®. Food Qual Prefer 48(Part A):107–117
159. Ares G, Jaeger SR (2013) Check-all-that-apply questions: influence of attribute order on sensory product characterization. Food Qual Prefer 28(1):141–153
160. Ares G, Antúnez L, Giménez A, Jaeger SR (2015) List length has little impact on consumers' visual attention to CATA questions. Food Qual Prefer 42:100–109
161. Spinelli S, Dinnella C, Ares G, Abbà S, Zoboli GP, Monteleone E (2019) Global profile:

going beyond liking to better understand product experience. Food Res Int 121:205–216

162. Lakens D (2021) Sample size justification. https://doi.org/10.31234/osf.io/9d3yf
163. Schouteten JJ, Gellynck X, Slabbinck H (2019) Influence of organic labels on consumer's flavor perception and emotional profiling: comparison between a central location test and home-use-test. Food Res Int 116:1000–1009
164. Schouteten JJ, De Steur H, Sas B, De Bourdeaudhuij I, Gellynck X (2017) The effect of the research setting on the emotional and sensory profiling under blind, expected, and informed conditions: a study on premium and private label yogurt products. J Dairy Sci 100:169–186
165. Meiselman HL (2013) The future in sensory/consumer research:............. evolving to a better science. Food Qual Prefer 27(2): 208–214
166. Jaeger SR, Hort J, Porcherot C, Ares G, Pecore S, MacFie HJH (2017) Future directions in sensory and consumer science: four perspectives and audience voting. Food Qual Prefer 56:301–309
167. Jaeger SR, Porcherot C (2017) Consumption context in consumer research: methodological perspectives. Curr Opin Food Sci 15:30–37
168. Labbe D, Ferrage A, Rytz A, Pace J, Martin N (2015) Pleasantness, emotions and perceptions induced by coffee beverage experience depend on the consumption motivation (hedonic or utilitarian). Food Qual Prefer 44:56–61
169. Jaeger SR, Spinelli S, Ares G, Monteleone E (2018) Linking product-elicited emotional associations and sensory perceptions through a circumplex model based on valence and arousal: five consumer studies. Food Res Int 109:626–640
170. Jaeger SR, Swaney-Stueve M, Chheang SL, Hunter DC, Pineau B, Ares G (2018) An assessment of the CATA-variant of the EsSense Profile®. Food Qual Prefer 68:360–370
171. Schifferstein HNJ, Fenko A, Desmet PMA, Labbe D, Martin N (2013) Influence of package design in the dynamics of multisensory and emotional food experience. Food Qual Prefer 27:18–25
172. Giboreau A, Meiselman HL (2018) Emotions before and after a meal in a natural eating situation. Food Qual Prefer 65:191–193
173. Cardello AV, Schutz H, Snow C, Lesher L (2000) Predictors of food acceptance, consumption and satisfaction in specific eating situations. Food Qual Prefer 11:201–216
174. Piqueras-Fiszman B, Jaeger SR (2015) The effect of product-context appropriateness on emotion associations in evoked eating occasions. Food Qual Prefer 40:49–60
175. Piqueras-Fiszman B, Jaeger SR (2014) The impact of the means of context evocation on consumer's emotion associations towards eating occasions. Food Qual Prefer 37:61–70
176. van Zyl H, Chaya C (2021) Emotion in beverages. In: Emotion measurement. Woodhead Publishing, pp 731–771
177. Chaya C, Pacoud J, Ng M, Fenton A, Hort J (2015) Measuring the emotional response to beer and the relative impact of sensory and packaging cues. J Am Soc Brew Chem 73(1): 49–60
178. Meilgaard MC, Carr BT, Civille GV (2006) Sensory evaluation techniques. CRC Press, Hoboken
179. Worch T, Sinesio F, Moneta E, Abbà S, Dreyfuss L, McEwan JA, Porcherot-Lassallette C (2020) Influence of different test conditions on the emotional responses elicited by beers. Food Qual Prefer 83: 103895
180. Jiang Y, King JM, Prinyawiwatkul W (2014) A review of measurement and relationships between food, eating behavior and emotion. Trends Food Sci Technol 36:15–28
181. Porcherot C, Valentin D, Ducret C, Grand O, Melis C, Vergnaud H, et al (2014) Optimization of a verbal measure of emotions elicited by fragrances: comparison between CATA and a 5-point scale. In: Poster session presented at the 6th European conference on sensory and consumer research, Copenhagen, Denmark
182. Vidal L, Ares G, Hedderley DI, Meyners M, Jaeger SR (2018) Comparison of rate-all-that-apply (RATA) and check-all-that-apply (CATA) questions across seven consumer studies. Food Qual Prefer 67:49–58
183. Reinbach HC, Giacalone D, Ribeiro LM, Bredie WL, Frøst MB (2014) Comparison of three sensory profiling methods based on consumer perception: CATA, CATA with intensity and napping®. Food Qual Prefer 32:160–166
184. Jaeger SR, Xia Y, Lee P-Y, Hunter DC, Beresford MK, Ares G (2018) Emoji questionnaires can be used with a range of population segments: findings relating to

age, gender and frequency of emoji/emoticon use. Food Qual Prefer 68:397–410

185. Jaeger SR, Roigard CM, Ares G (2018) Measuring consumers' product associations with emoji and emotion word questionnaires: case studies with tasted foods and written stimuli. Food Res Int 111:732–747
186. Jaeger SR, Jin D, Hunter DC, Roigard CM, Hedderley DI (2020) Multi- response approaches in product-focused investigations: methodological variations across three case studies. Food Res Int 132:109113
187. Krosnick JA (1991) Response strategies for coping with the cognitive demands of attitude measures in surveys. Appl Cogn Psychol 5(3): 213–236
188. Krosnick JA (1999) Survey research. Annu Rev Psychol 50(1):537–567
189. Sudman S, Bradburn NM (1982) Asking questions: a practical guide to questionnaire design. Jossey-Bass
190. Jaeger G, Schlich P, Tijssen I, Yao J, Visalli M, de Graaf C, Stieger M (2014) Temporal dominance of emotions: measuring dynamics of food-related emotions during consumption. Food Qual Prefer 37:87–99
191. Jaeger SR, Beresford MK, Lo KR, Hunter DC, Chheang SL, Ares G (2020) What does it mean to check-all-that-apply? Four case studies with beverages. Food Qual Prefer 80: 103794
192. Cliff M, Heymann H (1993) Development and use of time-intensity methodology for sensory evaluation: A review. Food Res Int 26(5):375–385
193. Lee WE III, Pangborn M (1986) Time-intensity: the temporal aspects of sensory perception. Food Technol 40:71–78
194. King SC, Meiselman HL, Carr BT (2010) Measuring emotions associated with foods in consumer testing. Food Qual Prefer 21: 1114–1116
195. Laros JM, Steenkamp EM (2005) Emotions in consumer behavior: a hierarchical approach. J Bus Res 58:1437–1445
196. Ng M, Chaya C, Hort J (2013) The influence of sensory and packaging cues on both liking and emotional, abstract and functional conceptualisations. Food Qual Prefer 29:146–156
197. Gomez-Corona C, Carrillo V, Dueñas N, Lyon D (2016) The taste of emotions. Exploring the link between sensory attributes and emotions in strawberry flavors. In: Poster presented in 7th European conference on sensory and consumer research, Dijon, 11–14 Sept 2016
198. Parente ME (2010) Cosmetics and emotions: a tool for improving consumer satisfaction. In: Presented at the 26th congress of the international federation of societies of cosmetic chemists, Buenos Aires, Argentina
199. Piqueras-Fiszman B, Spence C (2015) Sensory expectations based on product-extrinsic food cues: an interdisciplinary review of the empirical evidence and theoretical accounts. Food Qual Prefer 40:165–179
200. Thomson DMH (2007) SensoEmotional optimization of food products and brands. In: MacFie H (ed) Consumer-led and food product development. Woodhead Publishing Limited, Cambridge, pp 281–303
201. Thomson DMH (2008) Liking isn't enough. A sense of innovation. In: Paper presented at the 3rd European conference on sensory and consumer research, Hamburg, Germany
202. Seo H-S, Roidl E, Müller F, Negoias S (2010) Odors enhance visual attention to congruent objects. Appetite 54:544–549
203. Cardello AV (2007) Measuring consumer expectations to improve food product development. In: MacFie HJH (ed) Consumer-led food product development. Woodhead, Cambridge, pp 223–261
204. Anderson RE (1973) Consumer dissatisfaction: the effect of disconfirmed expectancy on perceived product performance. J Mark Res 10(1):38–44
205. Kähkönen P, Tuorila H, Hyvönen L (1995) Dairy fat content and serving temperature as determinants of sensory and hedonic characteristics in cheese soup. Food Qual Prefer 6(2):127–133
206. Tuorila H, Cardello AV, Lesher LL (1994) Antecedents and consequences of expectations related to fat-free and regular-fat foods. Appetite 23(3):247–263
207. Westcombe A, Wardle J (1997) Influence of relative fat content information on responses to three foods. Appetite 28(1):49–62
208. Caporale G, Monteleone E (2001) Effect of expectations induced by information on origin and its guarantee on the acceptability of a traditional food: olive oil. Sci Aliment 21(3): 243–254
209. Lange C, Rousseau F, Issanchou S (1999) Expectation, linking and purchase behavior under economical constraint. Food Qual Prefer 10(1):31–39
210. Siret F, Issanchou S (2000) Traditional process: influence on sensory properties and on consumers' expectation and liking application

to 'pâté de campagne'. Food Qual Prefer 11(3):217–228

211. Deliza R, MacFie HJ (1996) The generation of sensory expectation by external cues and its effect on sensory perception and hedonic ratings: a review. J Sens Stud 11(2):103–128

212. Dalton P (2002) Olfaction. In: Yantis S, Pashler H (eds) Sensation and perception, Steven's handbook of experimental psychology, vol 1, 3rd edn. Wiley, New York, pp 691–746

213. Schifferstein HN, Verlegh PW (1996) The role of congruency and pleasantness in odor-induced taste enhancement. Acta Psychol 94(1):87–105

214. Walsh G, Shiu E, Hassan LM, Michaelidou N, Beatty SE (2011) Emotions, store-environmental cues, store-choice criteria, and marketing outcomes. J Bus Res 64(7): 737–744

215. Dantec M, Mantel M, Lafraire J, Rouby C, Bensafi M (2021) On the contribution of the senses to food emotional experience. Food Qual Prefer 92:104120

216. Sinesio F, Moneta E, Porcherot C, Abbà S, Dreyfuss L, Guillamet K et al (2019) Do immersive techniques help to capture consumer reality? Food Qual Prefer 77:123–134

Chapter 9

Implicit Measures of Food "Wanting"

Donato Cereghetti, Isabelle Cayeux, Sylvain Delplanque, and Kathrin Ohla

Abstract

Many foods are inherently salient and highly rewarding stimuli that can trigger the desire for consumption even in the absence of liking. Together with the sheer constant availability of energy-dense food, this has led to a rising prevalence of obesity and associated diseases. Increasing effort to investigate the workings of food reward in different contexts and populations is pertinent to understand consumer behavior and helps develop behavioral strategies and intervention to foster healthy food choices. Here, we present an overview of implicit behavioral paradigms that have been implemented to measure the motivational facet of reward processing that precedes food intake: wanting. We discuss their strengths and weaknesses and provide an assessment of their applicability to consumer research.

Key words IAT, Association tests, Consumer behavior, Liking vs. wanting

1 Introduction

The omnipresence of calorie-dense, rewarding foods is undoubtedly contributing to the obesity pandemic, requiring consumers to repeatedly make "healthy" choices and to limit food intake. As food is a potent rewarding stimulus that captures attention and is even preferentially processed by the human brain, food-related decision making remains a challenge for many [1]. Understanding the mechanisms underlying food reward is thus pertinent to develop behavioral strategies to foster healthy choices and to tackle obesity and associated diseases. The Incentive Salience Theory (IST) [2–6] differentiates between two main components of reward processing: wanting and liking. Wanting is considered the motivational component of reward processing that is predominant during the so-called appetitive phase, before food consumption. Liking, in turn, refers to the hedonic component of reward processing and captures the enjoyment of a food during the subsequent consummatory phase.

Carlos Gómez-Corona and Heber Rodrigues (eds.), *Consumer Research Methods in Food Science*,
Methods and Protocols in Food Science, https://doi.org/10.1007/978-1-0716-3000-6_9,

Wanting and liking can both be explicit or implicit [5, 7, 8]. Implicit wanting, also referred to as incentive salience, describes the automatic capture of consumers' attention and triggering of approach and consummatory behavior; here, the food itself but also food cues are viewed as a "motivational magnet" [5]. Implicit liking refers to hedonic reactions elicited by the consumption of a food. In contrast to implicit wanting and liking, their explicit counterparts require awareness. In that sense, explicit wanting refers to the conscious experience of *desire* before consumption and explicit liking to the conscious experience of *pleasure* during consumption.

The two components can also be differentiated based on their brain activation patterns. While wanting is mediated by mesolimbic and mesocortical dopaminergic systems, liking is mediated by a network of opioid and endocannabinoid "hedonic hotspots" in the nucleus accumbens and ventral pallidum [9]. Behaviorally, wanting and liking responses are often correlated and the effort engaged to gain a particular reward is directly proportional to the pleasure felt during its consumption. Nevertheless, wanting and liking can be dissociated behaviorally whenever an amplified wanting for a reward is observed without a corresponding increase in liking during consumption. Wanting and liking can also be separately assessed with explicit measures such as self-reports, where participants provide subjective ratings. While self-reports add value, they are insufficient to capture reward processing, partly because self-reported wanting may be contaminated by expected pleasantness of a food rather than reflecting pure desire [10]. Consequently, self-reports alone might be insufficient to accurately predict food choice and food behaviors.

Understanding the mechanisms of wanting independent of liking could open new avenues to develop behavioral strategies to control food seeking and craving and thus support healthy eating behavior. In the following, we present an overview of behavioral paradigms to measure implicit wanting. While such an overview cannot be exhaustive, we focus on tasks with high potential for applied consumer as well as fundamental research on reward processing. We discuss their advantages and disadvantages to measure wanting.

2 Leeds Food Preference Questionnaire

The Leeds Food Preference Questionnaire (LFPQ) is a computer-based procedure to measure explicit and implicit food wanting [11, 12]. Visual analogue scales (VAS) are used to assess explicit wanting, whereas implicit wanting is measured in a forced choice task during which participants are asked to choose from two foods the one they "want to eat more now" (*see* Fig. 1). Participants are asked to respond as quickly as possible by pressing either of two

Fig. 1 Illustration of a typical trial of the LFPQ [11]. Pictures 0060 and 0465 from the food-pics database [13, 14]

keys of a computer keyboard or response pad. Choices and response times (RT) are recorded and used to compute an index of implicit wanting, in which faster choices are assumed to be an indicator of increased wanting.

The original version of the LFPQ [11] comprises 16 pictures of foods, which vary in fat content (low fat versus high fat) and taste (savory versus sweet) and they are assigned to four food categories (low-fat savory, low-fat sweet, high-fat savory, and high-fat sweet), which are combined to create 96 unique picture pairs. In its first applications (e.g., [15, 16]), implicit wanting was assessed from untransformed RT or by computing the d-score, which was originally developed to analyze RT in Implicit Association Tests (IAT) [17]. More recently, a frequency-weighted algorithm has been proposed, in which RTs are adjusted for frequency of choice [18, 19]:

$$I_{\mathrm{A}} = \sum_{i=1}^{N_{\mathrm{win}}} \frac{\overline{t}}{t_i} - \sum_{j=1}^{N_{\mathrm{lose}}} \frac{\overline{t}}{t_j}$$

where I_{A} equals to the implicit wanting score for foods belonging to category A of a given participant; N_{win} equals to the number of times foods belonging to category A were chosen by the participant; N_{lose} equals to the number of times foods belonging to the category A were not chosen by the participant; $\overline{t}$ equals to the mean of RT computed for an individual; t_i and t_j are individual RT observations for *win* and *lose* trials, respectively. Positive I_A scores indicate faster choices of category A as compared to the other food categories. An I_A score of zero indicates that category A is equally wanted as the others.

The LFPQ has been applied in different contexts, such as exercise training [20, 21], high-altitude trekking [22], sleep restriction [23]; and used with different populations, including older adults [24], participants with obesity [25], and patients suffering from eating disorders [18, 26]. While the ability of the LFPQ to capture implicit wanting, as defined and conceptualized by the IST, has been questioned [27, 28], we deem it useful, particularly for applied research, because it is easily implemented and efficient. A practical guide for the implementation of the LFPQ has previously been published [12].

3 Implicit Association Tests

Developed by Greenwald et al. in 1998 [29], the IAT is a computer-based procedure to measure the association strength between two targets (e.g., *high-fat food* and *low-fat food*) and two attributes (e.g., *pleasant* and *unpleasant*). In the classical IAT, participants are to sort targets and attributes as quickly and accurately as possible by pressing either of two response keys. Accuracy and RTs are typically recorded; faster and more accurate responses are observed when two highly associated concepts share the same response key, as compared to when they would share different keys.

The IAT typically consists of five successive blocks (*see* Fig. 2). In the first block, participants are trained to sort targets (e.g., *brownie*, *apple*) into their categories (e.g., *high-fat food* and *low-fat food*). In the second block, participants are trained to sort attributes (e.g., *delicious*, *disgusting*) into their respective categories (e.g., *pleasant*, *unpleasant*). In the third block, participants are required to sort targets and attributes by using the sorting rules learned in blocks one and two (e.g., *high-fat foods* + *pleasant* and *low-fat foods* + *unpleasant*). In the fourth block, participants are to sort attributes again, but this time the response keys are switched. In the fifth and final block, participants are to sort targets and attributes by combining the sorting rules learned in blocks one and four (e.g., *high-fat foods* + *unpleasant* and *low-fat foods* + *pleasant*). Accuracy and RTs are measured during blocks three and five. Positive implicit attitudes toward high-fat foods are indicated by faster and more accurate responses during block three as compared to block five. In contrast, positive implicit attitudes toward low-fat foods are indicated by faster and more accurate responses during block five as compared to block three.

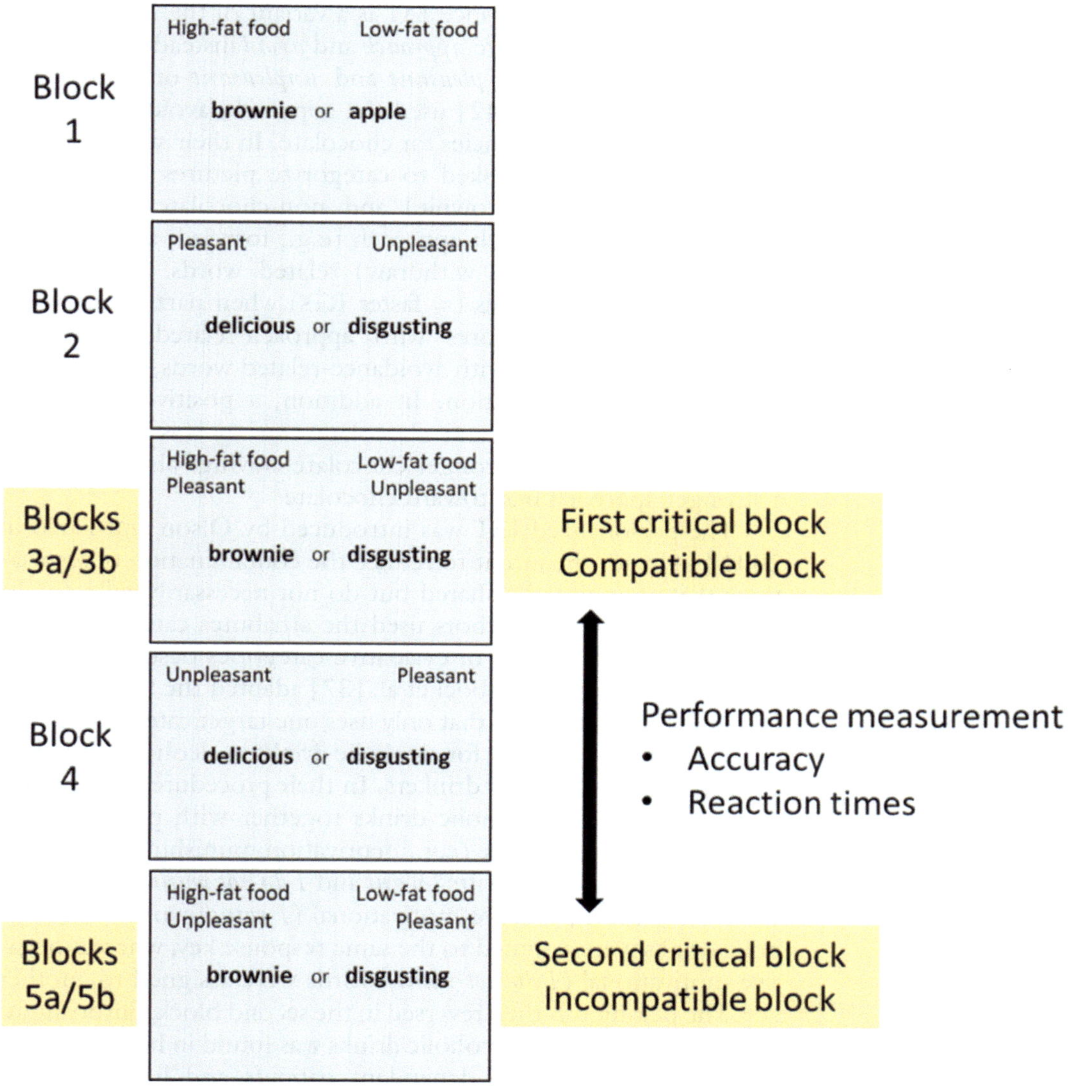

Fig. 2 Illustration of a typical evaluative IAT block structure

Several adaptations of the canonical IAT have been proposed as a measure of implicit wanting for different rewards including smoking cues [30, 31], sex [32], attractive persons [33], alcohol [34–39], beverages [40], and foods [41–43]. These studies used different procedures such as an approach/avoidance IAT [34–36, 38, 39, 42], a personalized wanting IAT [31, 32, 37, 41], a recoding-free IAT (IAT-RF) [43], or the more recent Wanting-IAT (W-IAT) [30, 33, 40]. In the following, we will describe some of these procedures in more detail.

The approach/avoidance IAT is a variant of the IAT in which the attribute categories are *approach* and *avoid* instead of the evaluative categories such as *pleasant* and *unpleasant* or *positive* and *negative*. Kemps et al. [42] used the approach/avoidance IAT to measure approach tendencies for chocolate. In their study, a group of young women was asked to categorize pictures of chocolate (e.g., chocolate bar, brownie) and non-chocolate (e.g., cake, pizza) food together with approach (e.g., forward, advance) and avoidance (e.g., escape, withdraw) related words. They found shorter response latencies (= faster RTs) when participants were to sort chocolate pictures with approach-related words and non-chocolate pictures with avoidance-related words, as compared to the inverse combination. In addition, a positive correlation between approach bias and chocolate craving was found; with participants reporting stronger chocolate cravings also presenting a stronger approach bias toward chocolate.

The personalized IAT was introduced by Olson and Fazio in 2004 [44] with the intent to reduce the contamination of associations that are culturally shared but do not necessarily reflect individual attitudes. The authors used the attributes categories *I like* and *I do not like* instead of evaluative categories described above. Inspired by this task, Tibboel et al. [37] adapted the Single Target IAT, a variant of the IAT that only uses one target category [45] to measure implicit wanting for alcoholic drinks in alcohol-dependent patients, heavy, and light drinkers. In their procedure, participants were to categorize alcoholic drinks together with positive (e.g., gain, praise) and negative (e.g., deprivation, punishment) motivational words associated with *I want* and *I do not want*, respectively. In the first block, positive motivational (*I want)* words and alcoholic drinks were assigned to the same response key, whereas negative motivational (*I do not want*) words were assigned to another key. The pairing was then reversed in the second block. Surprisingly, a stronger wanting for alcoholic drinks was found in heavy drinkers as compared to alcohol-dependent patients and light drinkers, suggesting that the relationship between wanting and alcohol consumption could not be linear. It should moreover be mentioned that this study failed to dissociate between liking and wanting, thereby casting doubts as to the validity of the procedure, at least in the context of alcohol dependency.

Rothermund et al. introduced the recoding-free IAT (IAT-RF) to avoid sorting strategies that are not based on category membership [46]. Composed of three blocks, the IAT-RF merges the tasks from blocks three and five of the IAT into a single block, in which sorting rules randomly switch between trials (i.e., after each response). Kraus and Piqueras-Fiszman proposed a variant of the IAT-RF to measure the impact of hunger state on motivational tendencies towards food [43]. First, participants watched a video clip for 10 min while they were either not permitted to eat at all (hunger group) or while they had a sandwich but not sweets (reduced hunger group). After this hunger manipulation, participants performed the motivational IAT-RF, where they sorted four classes of stimuli into their respective categories by following two sorting rules that randomly switched between trials. In trials subjected to the first rule, participants were asked to sort sandwich pictures with approach-related words (e.g., reach, toward) and pictures of sweets with avoidance-related words (e.g., escape, move away); whereas in trials subjected to the second rule, participants were asked to sort pictures of sweets with approach-related words and sandwich pictures with avoidance-related words. In the hunger group, shorter latencies were observed when participants sorted sandwich pictures with approach-related words and pictures of sweets with avoidance-related words. Inversely, in the reduced hunger group, shorter latencies were observed when participants sorted pictures of sweets with approach words and sandwich pictures with avoidance words. An increased wanting for sandwiches was found in the hunger group and for sweets in the reduced hunger group. Together the results highlight the importance of the physiological hunger state in approach and avoidance tendencies as well as in wanting.

More recently, the validity of IAT procedures that used motivational or approach/avoidance-related words (e.g., [37, 42, 43]) has been challenged. Such procedures are believed to measure semantic associations, hedonicity or, at best, a cognitive form of wanting rather than implicit wanting [30, 33, 40, 47]. Consequently, the Wanting-IAT (W-IAT) was introduced as an alternative having the peculiarity of including a motivational component. In a first study, Koranyi et al. applied the W-IAT to measure implicit preferences for attractive versus unattractive faces [33]. In this study, the motivational component was achieved by inducing thirst as a physiological need state in participants and by using need-relevant and need-irrelevant stimuli as wanting attributes in the subsequent W-IAT. Before the W-IAT, participants were to consume salty snacks to increase thirst. Then, during the W-IAT, participants were to sort (1) pictures of attractive and unattractive faces into their relevant target category, (2) pictures of need-relevant drinks (e.g., bottle of water) into the *I want* category, and (3) pictures of need-irrelevant objects (e.g., sunglasses) into the *I do not want* category. Importantly, as incentive, each correct response to pictures of drinks was rewarded with 10 mL of water for later consumption. As expected, shorter latencies were observed when heterosexual men sorted attractive female faces with need-relevant drinks (*I want* category), as compared to need-irrelevant objects (*I do not want* category), providing thus a first validation of their approach. In a more recent study, a new variant of the W-IAT was implemented to measure implicit wanting for coffee (vs. juice) in heavy and in low/non-consumers of coffee [40]. Since drinks were already used as target stimuli, they could not be used as need-relevant stimuli, and thirst induction was not an option either. Alternatively, the wanting response was incentivized using monetary rewards. For this, participants were to sort numbers in the *I want* category and letters in the *I do not want* category. Each correct response to numbers was then rewarded with a small monetary gain (0.05€) that was paid at the end of the experiment. Compared to low-coffee consumers, heavy coffee drinkers were found to be faster in sorting *I want* stimuli (numbers) together with coffee and *I do not want* stimuli (letters) together with juice, confirming thus the ability of the W-IAT procedure to measure implicit wanting for drinks. The W-IAT seems to permit to dissociate liking from wanting (e.g., [33, 40]), and thus have a higher construct validity as compared to IAT procedures using synonyms of wanting words. Taken together, the W-IAT is a new and promising procedure and its potential in applied consumer research clearly deserves to be further explored.

4 Effort-Based Tasks

Effort-based tasks are procedures in which participants are required to work to obtain or to see rewarding stimuli. In human studies, willingness to work is often measured with a handgrip dynamometer, in which frequency of squeeze (i.e., number of squeezes performed on the handgrip) or grip strength (i.e., amount of muscular force exerted on the handgrip) are measures of interest (*see* Fig. 3). Participants are often provided with real-time feedback reflecting the amount of effort being exerted on the handgrip at each instant during task trials, for example, through visualization of a fluid level moving up and down in a thermometer (*see* [48–50]). Among the most promising tasks to measure willingness to work for foods and related cues are the Pavlovian-to-Instrumental Transfer (PIT) [49–52], the LikeWant [48], the Effort Picture Selection (EPST) [53, 54], the Subliminal Motivation [54, 55], and the Grab-to-Eat [56] tasks. Some of these paradigms are detailed below.

The Pavlovian-to-Instrumental Transfer (PIT) paradigm measures the effect of reward-associated stimuli on operant behavior. Originating from animal research (e.g., [57]), the PIT has been successfully applied in human research to measure implicit wanting for monetary rewards [50, 58], sexual stimuli [59, 60], nicotine [61–63], and also for alcohol [51, 64–66], food and non-food odors [48, 49, 67], and foods [52, 68]. The PIT has three phases: the Pavlovian training, the instrumental training, and the transfer test. While the order of the two training phases can vary, the transfer test is always performed last [69]. In the Pavlovian training phase, a neutral stimulus, which initially does not elicit a specific

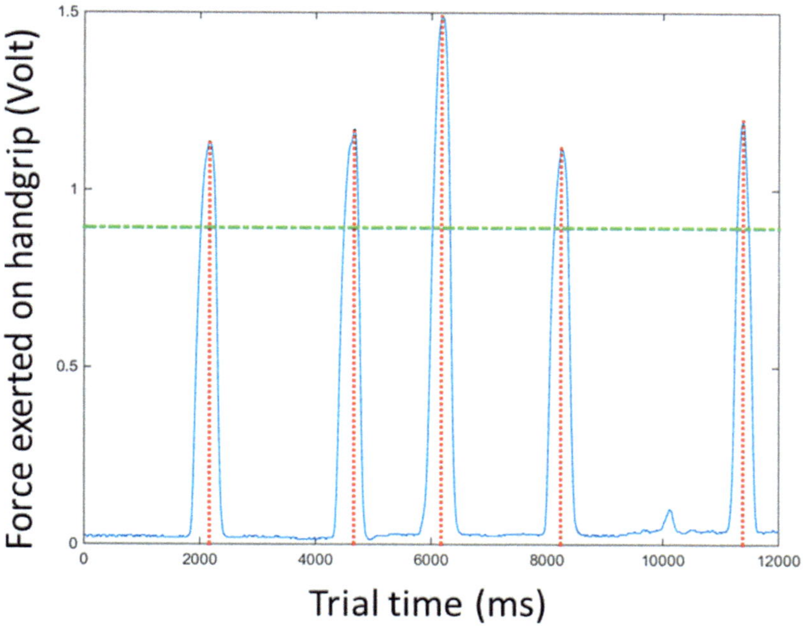

Fig. 3 Illustration of a single trial of the Likewant paradigm (from Ref. [48]), in which the participant squeezed the handgrip five times. In green, the threshold used to detect the number of squeezes (i.e., 50% of participant maximal grip strength). The total amount of muscular force exerted by the participant can also be estimated by calculating the area under the curve using the trapezoidal rule

response, is paired with a rewarding unconditioned stimulus, the UCS, a biologically relevant stimulus that naturally elicits a response. After repeated pairing, the formerly neutral stimulus acquires the ability to elicit a response that was previously only elicited by the UCS. It becomes, therefore, a conditioned stimulus (CS). In the instrumental training phase, participants are trained to squeeze the dynamometer to receive the rewarding UCS. During the transfer test, implicit wanting is measured. Here, participants are to perform the instrumental task again, this time in the presence of the CS. The ability of the CS to enhance instrumental responding is then assessed, which is known as PIT effect. A CS associated with highly rewarding UCS is expected to elicit more instrumental responding than a CS associated with a less rewarding UCS. The transfer test is usually performed under extinction, when no UCS are delivered after instrumental actions. This is crucial to obtain a wanting measure that is not contaminated by the hedonic experience from the consumption of the rewarding UCS. An example of a human PIT study comes from Pool et al., who investigated the impact of stress on wanting for chocolate [49]. In this study, participants were first trained to squeeze a dynamometer to receive a rewarding chocolate odor (UCS) delivered by a computer-controlled olfactometer. During the Pavlovian training phase, participants were exposed to a series of odor-picture pairings, where either of two neutral pictures was associated with the presence (CS+) and the absence (CS−) of chocolate odor. After both training phases, participants' stress level was manipulated using a socially evaluated cold pressor task. Half of the participants were to immerse their hand in cold water (0–2 °C; stress group), whereas the other half was to immerse their hand in warm water (35–37 °C; control group). Both groups were told that they were being videotaped for later analysis of their facial expressions and the experimenter observed them the entire time. In the transfer test, participants had to squeeze the dynamometer, while CS+ and CS− were presented on the screen. Results revealed an effect of the CS on squeeze frequency, with an increased effort during CS+ trials as compared to CS− trials (PIT effect). The PIT effect was larger in the stress than in the control group resulting in an interaction between group and CS. This result highlights that stress can increase the wanting for foods.

Since the PIT has contributed to the emergence of the IST, it should be considered as a "gold standard" to measure implicit wanting in humans. The task is, however, not readily adoptable for consumer research because it is time-consuming (approx. 40 min for an olfactory PIT) and it requires naive participants who are not yet experienced with the study protocol for the extinction phase to work. The latter is an important limitation for consumer studies which typically use within-subject designs and a series of evaluations of different products. We developed *LikeWant*, an efficient modification of the PIT to measure wanting for flavors and fragrances in consumer research [48]. For this, we have adapted the PIT to minimize the testing time through (1) the reduction of repetitions and (2) the assimilation of the instrumental training phase with the transfer test in the so-called "wanting task." To enable repeated measures, we reduced the number of extinction trials, in which implicit wanting was measured, and we introduced a reconditioning phase immediately after the extinction phase to deliver the rewards that participants expected to receive after squeezing the dynamometer. The *LikeWant* paradigm thus consists of two phases: the Pavlovian training phase and the wanting task. In the Pavlovian training phase, a rewarding odor (UCS) and an odorless air puff were each associated with a previously neutral picture, the CS+ and the CS−, respectively. In the wanting task, participants were to squeeze the dynamometer and aim to bring the liquid level in a thermometer display up and receive the odor associated to the picture presented on the screen. The wanting task consisted of four brief sub-phases: the instrumental learning phase (ILP), the partial extinction phase (PEP), the extinction phase (EP), and the reconditioning phase (RP). In the ILP and PEP, an unfamiliar picture, that has not been used in the previous Pavlovian training phase, was presented on the screen and

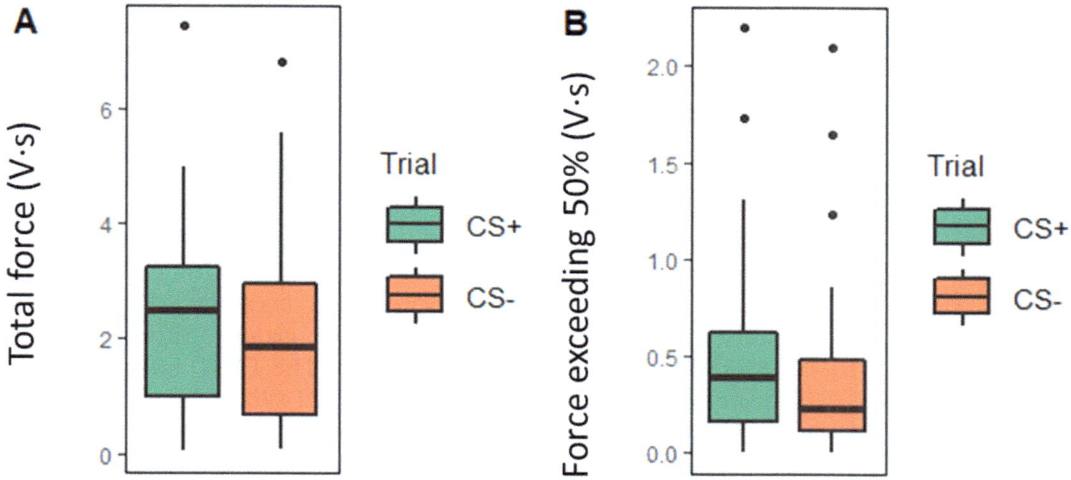

Fig. 4 Total amount of (**a**) muscular force and (**b**) unnecessary muscular force (both expressed in Volt-seconds, V·s) exerted by participants in EP [48]. Participants significatively exerted more effort and, respectively, more unnecessary effort, when CS + was presented on the screen

participants learned to squeeze the dynamometer to receive the rewarding UCS. The EP consisted of two critical trials performed under extinction, in which implicit wanting was measured. Here, the CS instead of the unfamiliar picture used in previous ILP and PEP was presented. The RP was similar, with the difference that UCS was delivered after instrumental actions: UCS for CS+ and odorless air for CS−. The results show that wanting can indeed be measured in only two trials under extinction, with participants exerting more effort in the CS+ trial as compared to the CS− trial (Fig. 4a). As an additional indicator of implicit wanting, we measured the amount of unnecessary muscular force exerted by participants as the effort was not required to obtain the rewarding UCS. During ILP and PEP trials, participants implicitly learn that 50% of their maximal force is sufficient to "fill the thermometer" and receive the rewarding UCS. In the subsequent EP, any effort exceeding the 50% threshold should thus be considered a waste of effort. Interestingly, an effect of the CS was also found on force exceeding 50%, with participants exerting more unnecessary effort in CS+ condition as compared to CS− condition (Fig. 4b). We speculate that unnecessary effort under extinction could be an indicator of implicit wanting. The sensitivity of this indicator needs to be further explored.

In a technical report, Ziauddeen et al. presented two interesting, and still under-investigated, handgrip paradigms to measure different aspects of the motivational component of food reward [54]. The Effort Picture Selection Task (EPST) measures the relative effort that participants are willing to exert to see different food pictures presented on a computer screen. In this task, participants are presented pairs of food pictures, the first of which is large and clearly visible on the screen (the default picture), while the other one is small and unrecognizable (the nondefault picture). Participants are to squeeze the dynamometer to simultaneously increase the size of the non-default picture and decrease the size of the default picture. This task was successfully used in a randomized controlled study [53] to assess the effect of the μ-opioid receptor antagonist GSK1521498 to reduce food wanting in patients suffering from binge eating disorders. In this study, 63 compulsive eaters were randomly assigned to two groups, either receiving GSK1521498 or placebo for 4 weeks. The EPST was performed before and after intervention. GSK1521498 reduced the effort exerted by patients to view high-calorie foods as compared to low-calorie foods. The Subliminal Motivation Task is an incentive force task designed to measure participants' conscious and unconscious motivation for foods. Each trial starts with the presentation of a food picture that is preceded and followed by a mask. In conscious trials, food pictures are presented for 200 ms, whereas in subliminal trials, food pictures are presented for 33 ms. After the masked food picture presentation, participants are instructed to squeeze the handgrip to win points toward the food presented on the screen; the harder they squeezed, the more points they could win and exchange for food. This procedure was used in a study carried out by Ziauddeen et al., in which participants had to work for savory foods, sweet foods and neutral non-food stimuli, as control [55]. Participants performed this task in two identical blocks of 150 trials each. Between the two blocks, sensory-specific satiety was manipulated, with half of participants eating pizza (savory) and the other half eating cake (sweet). Participants exerted less effort to win the food they had just consumed compared to the other food. Interestingly, this pattern was found in both conscious and unconscious trials, suggesting that this procedure can capture motivational processes that operate beyond awareness.

More recently, the Grab-to-Eat Task (GET) was developed to measure participants' eating motivational dynamics during an ad-libitum food consumption task [56]. In this study, participants were instructed to freely squeeze a dynamometer to receive chocolate milk, administered using a computer-controlled peristaltic pump. Each squeeze was rewarded with a sip of chocolate milk. Prior to performing the GET, participants were randomly assigned to two groups, in which hunger was manipulated. As expected, the effort exerted throughout the task was influenced by hunger state, with fasted participants exerting more effort to receive chocolate milk as compared to satiated participants. Interestingly, the effort exerted on the handgrip decreased throughout the task, showing that food value gradually decreases throughout consumption.

5 From "Wanting" to Behavior: Approach/Avoidance Tasks

Approach/avoidance procedures (AAP) were developed to assess whether environmental stimuli activate automatic behavioral tendencies based on the assumptions that appetitive stimuli should elicit approach and aversive stimuli elicit avoidance reactions. In these tasks, accuracy and RT are recorded and participants are expected to respond faster and more accurately when they approach positive (or "wanted") stimuli and avoid negative (or "unwanted") ones, as compared to the reverse. This is known as the approach/avoidance compatibility effect. A number of different AAP variants has been proposed including joystick-based tasks, in which participants are required to pull (for approaching) and push (for avoiding) a joystick in response to stimuli (e.g., [70–72]), manikin tasks, in which participants are asked to move an abstract manikin toward or away from targets (e.g., [73]), mobile-based tasks, in which participants are asked to pull and push a smartphone toward their face (e.g., [74, 75]), and tasks based on an embodied approach to cognition, in which whole-body approach/avoidance reactions are either measured (e.g., [76, 77]) or simulated (e.g., [78, 79]). Notably, AAP also vary in the instructions provided to participants [80]. In procedures using explicit instructions, participants are to evaluate the valence of target stimuli, and thus to approach or avoid targets based on the positive–negative dimension. Conversely, in procedures using implicit instructions, participants are engaged in an orthogonal task where valence is irrelevant. This can be achieved, for example, when targets are presented in one sensory modality (e.g., visual) and emotional stimuli in another (e.g., olfactory; experiment #2 in [78]). While implicit instructions are thought to capture automatic cognitive processes, they are usually found to yield smaller effect sizes as compared to explicit instructions [80].

Piqueras-Fiszman et al. used a joystick-based AAT to investigate behavioral tendencies toward appealing and disgusting foods [71]. They presented food pictures with positive (e.g., chocolate, savory crackers, fresh strawberries) and negative (e.g., half-eaten foods, rotten or spoiled foods, food with insects) valance on a computer screen together with an upward pointing triangle above or a downward pointing triangle below the picture. Implicit instructions were used, and participants were to move the joystick in the direction pointed to by the triangle regardless of the food or its valence. As expected, a compatibility effect was found, with participants moving appealing foods faster toward and disgusting foods faster away from themselves, as compared to the reverse.

Going beyond valence effects, other studies showed the feasibility to measure behavioral tendencies for different food categories. For example, the Affective Simon Task Manikin version (AST-manikin) investigates the effect of therapy on behavioral tendencies toward low- versus high-caloric foods in patients suffering from anorexia nervosa (AN) [73]. Patients were to move a manikin toward and away from three categories of stimuli: high-calorie foods (e.g., pizza, croissant, chocolate), low-calorie foods (e.g., strawberries, popcorn, carrots), and neutral pictures (e.g., bowls, mugs). The pictures were presented in top- and side-view. Pictures were sequentially presented together with a black manikin below or above the picture. Instructions were implicit. Participants were to move the manikin toward or away from the picture depending on its perspective (top- or side-view) and, thus, regardless of the food's category. The AST-manikin task was performed before and after a 1 year long therapy. An effect of the treatment on patients' behavioral tendencies was observed: patients showed a higher approach bias toward low-caloric foods as compared to high-caloric foods before but not after therapy, indicating that the therapy improved behavioral tendencies.

Mobile technologies (tablets and smartphones) represent a new way to measure approach/avoidance tendencies outside the lab, thereby reaching consumers in their ecological environment. In the mobile AAP introduced by Zech et al., visual stimuli are presented on a smartphone screen and participants are to approach them by pulling the phone toward their face and avoid them by pushing the phone away [75]. The direction of the movement, reaction times, reaction force, and motion distance are inferred using the built-in accelerometry sensors. A mobile AAP task was recently applied to measure behavioral tendencies toward Dutch and Asian, palatable and unpalatable foods in a group of Dutch consumers [74]. Throughout the task, participants were *explicitly instructed* to pull the phone toward them upon presentation of a specific food category (e.g., palatable food) and push the phone away upon the presentation of the other one (e.g., unpalatable food), the instructions being subsequently reversed. As expected, participants were found to be faster to approach familiar (Dutch) and palatable foods compared to unfamiliar (Asian) and unpalatable foods.

Paradigms based on embodied cognition theories also represent an alternative to measure behavioral reactions in ecological conditions. Here, whole-body approach/avoidance reactions are measured with motion-tracking technologies (e.g., [76, 77]) or simulated through a visual flow presented on a computer screen (e.g., [78, 79]). Tasks simulating whole-body approach/avoidance reactions have the advantage that they do not require neither hardware and software for motion tracking (e.g., depth-sensing camera or VR headsets) nor advanced expertise in signal processing to analyze motion data. They may therefore be more readily adopted by applied consumer research. In the Visual Approach/Avoidance by the Self Task (VAAST), a visual scene presented on the screen is zoomed in or out depending on the button pressed by

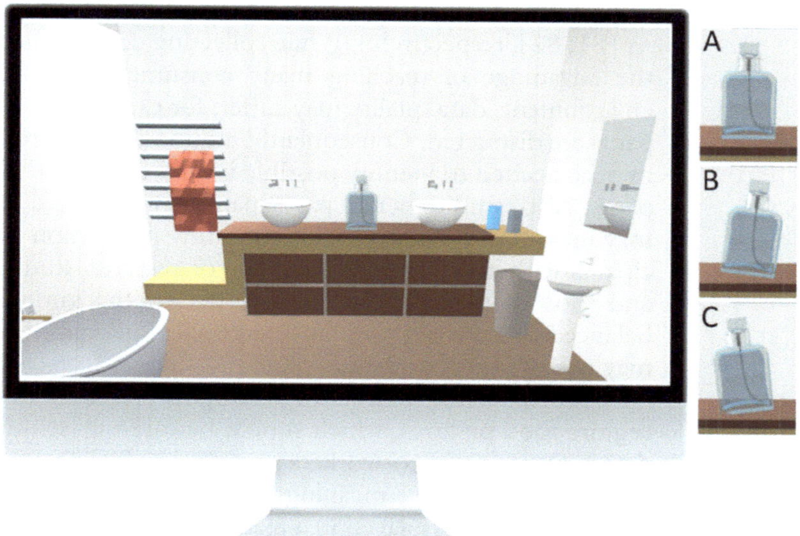

Fig. 5 Visual scene of the olfactory visual approach/avoidance by the self task [78]. (**a**) Perfume bottle presented in the vertical position (neutral stimulus). Perfume bottle tilted to the (**b**) left or (**c**) right (visual targets of the task). Half of the participants were asked to approach the left-tilted perfume bottles and avoid the right-tilted ones, whereas the remaining half were asked to approach the right-tilted and avoid the left-tilted perfume bottles

the participant, giving the impression of walking toward or away from the scene, respectively [79]. The VAAST was adapted to assess the ability of olfactory cues to produce automatic approach/avoidance reactions toward their source [78]. In each trial, participants received a positive or negative odor before they were to move toward or away from a perfume bottle presented in a congruent environment such as a bathroom (*see* Fig. 5). Notably, implicit instructions were used, with participants responding to the inclination of the bottle perfume regardless of odor pleasantness. Consistently with the approach/avoidance compatibility effect, pleasant odors were found to induce faster approach responses as compared to unpleasant odors. While the VAAST is believed to have high-ecological validity (*see* [78] for a discussion), its application in food research, with food odors or flavors, still needs to be investigated.

6 Recommendations and Best Practices

In the following subchapter, we provide general, practical recommendations intended to guide sensory and consumer scientists in their implementation of studies on food reward. Whenever possible, we recommend conducting the research in a controlled environment, such as a laboratory or sensory booths. It is worth mentioning that some of the paradigms described above can also be performed online using web applications (e.g., IAT and VAAST; *see* [81, 82], respectively). While collecting data on the internet has the advantage of reaching many consumers in their ecological environment, data quality may suffer, for example, because participants are distracted. Consequently, additional data processing steps may be needed to identify possibly invalid trials and "fake" participants. Also, studies where participants can self-elect to participate may introduce a bias and skew the study population toward participants with an interest in the topic or research studies in general and those with access to the internet. If the sample should be balanced or even representative, agencies specialized in recruiting may need to be consulted.

A considerable number of paradigms presented in this chapter require food pictures, which should be carefully selected and validated. Oustric and co-workers proposed guidelines to assist investigators to assemble a meaningful set of stimuli [12]. Accordingly, foods should be displayed in a ready-to-eat form and in their usual portion size. Foods should be familiar, culturally relevant, and suitable to the time of testing. In addition, food pictures should display a limited variety of items, while food categories should comprise a representative variety of foods. Importantly, the homogeneity of selected pictures should be ensured by using, for instance, the same image background. Foods should be selected based on macronutrient content such that high-fat foods contain more than 40% of their energy from fat and low-fat foods should contain less than 20% of their energy from fat.

Investigators can assemble their own set of pictures or use pictures from existing databases such as FRIDa [83], Food-Pics [13, 14], OLAF [84, 85], CROCUFID [86], and Full4Health (F4H) Image Collection [87]. These databases may contain pictures from the internet (e.g., FRIDa) or the authors followed standardized photographing protocol (e.g., CROCUFID), which often leads to higher consistency across images. Food-pics combined different approaches and complemented images from an existing image database with custom photographies of foods that were not readily available (e.g., steamed vegetables). Food picture databases should include objective measures that quantify relevant physical image features (e.g., brightness, contrast, complexity, size, etc.) as well as normative subjective ratings such as liking, arousal, familiarity, or perceived healthiness, desire-to-eat, and food craving, each of which could be used to select images of interest. Several food picture databases (e.g., Food-Pics, F4H) also provide an estimation of energy and macronutrient content (i.e., fat, carbohydrate, sugar, and protein content) for each food. This information can be used to objectively select and assign food pictures to different food categories (e.g., low- and high-fat foods). Before beginning the main research, a preliminary survey or study should be conducted to ensure that the selected food pictures are recognized, culturally relevant, habitually eaten, and perceived as palatable by the population of interest [12]. Finally, for the main research study, one could consider providing participants with the opportunity to select food pictures from relevant categories if the research questions profits from stimulus materials that is relevant at individual level [12].

Because reward processing is modulated by physiological states [5], special attention should be paid to either control or systematically modulate states such as hunger, satiety, thirst, and stress. Several measures can be taken to decrease inter-subject variability of physiological states. To reduce inter-subject variability of stress, the study should be conducted in the same conditions for all participants (e.g., same location, same experimenter, etc.). To reduce inter-subject variability of hunger, a standardized meal or controlled fasting period should apply to all participants. Depending on the research objectives, hunger or thirst could be manipulated within the experiment using a group design where participants are randomly assigned to either of two experimental groups that follow a fasting regimen (aka, hunger group) or that receive selected foods in a controlled manner (aka, satiety group). Finally, it is good practice to include self-reports of aspects relating to the research question, specifically hunger and thirst but also stress, in food studies. While subjective ratings can be collected at various times during the experiment (e.g., at the beginning vs. end of the experiment, before and after hunger manipulation), their timing should be carefully selected as it may unintentionally direct participants' attention toward selected aspects of the task.

7 Conclusions

In this chapter, we present a series of paradigms and tasks suitable to measure food reward. Identifying the best paradigm can be challenging given the plethora of available tasks. Now, it is worth remembering that different paradigms have been developed to answer different research questions and to capture different cognitive processes, spanning from core implicit wanting to behavioral tendencies. Each paradigm has its strengths and weaknesses, and might be more or less suitable than others in a specific context. Choosing the optimal method is not trivial and a good balance between construct validity, applicability, and sensitivity should be aspired.

High-construct validity, which is the extent to which a task captures the cognitive process of interest, is an important criterion. In principle, tasks that have been shown to dissociate liking and wanting will likely have a high-construct validity. In this regard, we recommend the PIT to measure implicit wanting as defined by the IST as it has been proven to disentangle wanting from liking in humans not only at a behavioral but also at a neural level (e.g., [50, 88]). Other tasks, such as LFPQ and IAT procedures using synonyms of wanting words, failed to dissociate wanting and liking (e.g., [31, 37]), or raised doubts about their capacity to do so [27, 28]. While these tasks might be unsuitable to significantly advance fundamental research on reward processing, we advocate that they have value for applied and consumer research given their high efficiency. It is important to recognize that in circumstances where (implicit) liking and (implicit) wanting overlap at a behavioral level, they cannot be dissociated. For example, it is unlikely that wanting and liking can be dissociated in most product tests, unless the study protocol involves controlled state manipulations (e.g., stress, hunger) and/or a clinical population that is known to exhibit wanting without liking such as patients with binge eating disorder (e.g., [89, 90]).

From a practical point of view, paradigms differ on the maximum number of food categories that can be investigated. As an example, a classical IAT measures implicit attitudes toward two opposing food categories (e.g., high-fat foods vs. low-fat foods or meat vs. meat-analogue), whereas the LFPQ can measure wanting for four (i.e., low-fat savory, low-fat sweet, high-fat savory, and high-fat sweet), or potentially more, food categories. The different tasks also vary in costs and technical requirements; some require specific hardware and/or software so that skilled personnel is required to implement and run the study and/or to analyze the data collected. Others require an olfactometer and/or gustometer to control the delivery of smell and taste stimuli over time. While hand-held olfactometry and gustometry solutions are starting to emerge, it is hard to predict when those could be routinely used in product tests. Finally, sensitivity of the procedures, defined as the ability of a task to discriminate between different products, should also be taken into consideration. In this regard, different tasks vary in sensitivity and because incentive salience is driven by macronutrient (especially fat and sugar) differences in foods, one should expect that paradigms presented in this chapter are more sensitive to capture consumers' wants for food categories that clearly differ (e.g., brownie vs. apple), than for more similar food products (e.g., brownie vs. chocolate bar).

To conclude, food reward is a relatively new research area and understanding its mechanisms will help industrials to develop "wanted" foods that can support healthy eating behaviors. Given the considerable number of approaches, further research should focus on exploring the external validity of the tasks through systematic comparisons. Identifying neural correlates of these different tasks might also be beneficial to uncover the underlying cognitive and emotional processes, and thus, further our understanding of what the different tasks measure.

References

1. Harrar V et al (2011) Food's visually perceived fat content affects discrimination speed in an orthogonal spatial task. Exp Brain Res 214(3): 351–356
2. Berridge KC (2007) The debate over dopamine's role in reward: the case for incentive salience. Psychopharmacology 191(3): 391–431
3. Berridge KC (2012) From prediction error to incentive salience: mesolimbic computation of reward motivation. Eur J Neurosci 35(7): 1124–1143
4. Berridge KC, Robinson TE (1998) What is the role of dopamine in reward: hedonic impact, reward learning, or incentive salience? Brain Res Rev 28(3):309–369
5. Berridge KC, Robinson TE (2003) Parsing reward. Trends Neurosci 26(9):507–513
6. Berridge KC, Robinson TE, Aldridge JW (2009) Dissecting components of reward: 'liking', 'wanting', and learning. Curr Opin Pharmacol 9(1):65–73
7. Anselme P, Robinson MJ (2016) "Wanting," "liking," and their relation to consciousness. J Exp Psychol Anim Learn Cogn 42(2):123–140
8. Berridge KC, Kringelbach ML (2008) Affective neuroscience of pleasure: reward in humans and animals. Psychopharmacology 199(3):457–480
9. Berridge KC, Kringelbach ML (2011) Building a neuroscience of pleasure and well-being. Psychol Well Being 1(1):1–3
10. Pool E et al (2016) Measuring wanting and liking from animals to humans: a systematic review. Neurosci Biobehav Rev 63:124–142
11. Finlayson G, King N, Blundell JE (2007) Liking vs. wanting food: importance for human appetite control and weight regulation. Neurosci Biobehav Rev 31(7):987–1002
12. Oustric P et al (2020) Measuring food preference and reward: application and cross-cultural adaptation of the leeds food preference questionnaire in human experimental research. Food Qual Prefer 80:103824
13. Blechert J et al (2019) Food-Pics_Extended-an image database for experimental research on eating and appetite: additional images, normative ratings and an updated review. Front Psychol 10:307
14. Blechert J et al (2014) Food-pics: an image database for experimental research on eating and appetite. Front Psychol 5:617
15. Finlayson G, King N, Blundell J (2008) The role of implicit wanting in relation to explicit liking and wanting for food: implications for appetite control. Appetite 50(1):120–127
16. Griffioen-Roose S et al (2010) Measuring food reward and the transfer effect of sensory specific satiety. Appetite 55(3):648–655
17. Greenwald AG, Nosek BA, Banaji MR (2003) Understanding and using the implicit association test: I. An improved scoring algorithm. J Pers Soc Psychol 85(2):197–216
18. Dalton M, Finlayson G (2014) Psychobiological examination of liking and wanting for fat and sweet taste in trait binge eating females. Physiol Behav 136:128–134
19. French SA et al (2014) Questionnaire and laboratory measures of eating behavior. Associations with energy intake and BMI in a community sample of working adults. Appetite 72:50–58
20. Martins C et al (2017) High-intensity interval training, appetite, and reward value of food in the obese. Med Sci Sports Exerc 49(9): 1851–1858
21. Zhou Y et al (2019) Effects of moderate- and high-intensity acute aerobic exercise on food reward and appetite in individuals with methamphetamine dependence. Physiol Behav 211: 112649
22. Aeberli I et al (2013) Disturbed eating at high altitude: influence of food preferences, acute mountain sickness and satiation hormones. Eur J Nutr 52(2):625–635
23. McNeil J et al (2017) The effects of partial sleep restriction and altered sleep timing on appetite and food reward. Appetite 109:48–56
24. van der Meij BS et al (2015) Specific food preferences of older adults with a poor appetite. A forced-choice test conducted in various care settings. Appetite 90:168–175
25. Blundell J et al (2017) Effects of once-weekly semaglutide on appetite, energy intake, control of eating, food preference and body weight in subjects with obesity. Diabetes Obes Metab 19(9):1242–1251
26. Cowdrey FA, Finlayson G, Park RJ (2013) Liking compared with wanting for high- and low-calorie foods in anorexia nervosa: aberrant food reward even after weight restoration. Am J Clin Nutr 97(3):463–470
27. Havermans RC (2011) "You say it's liking, I say it's wanting . . .". On the difficulty of disentangling food reward in man. Appetite 57(1): 286–294
28. Havermans RC (2012) How to tell where 'liking' ends and 'wanting' begins. Appetite 58(1):252–255

29. Greenwald AG, McGhee DE, Schwartz JLK (1998) Measuring individual differences in implicit cognition: the implicit association test. J Pers Soc Psychol 74(6):1464–1480
30. Grigutsch LA et al (2019) Implicit 'wanting' without implicit 'liking': a test of incentive-sensitization-theory in the context of smoking addiction using the wanting-implicit-association-test (W-IAT). J Behav Ther Exp Psychiatry 64:9–14
31. Tibboel H et al (2011) Testing the validity of implicit measures of wanting and liking. J Behav Ther Exp Psychiatry 42(3):284–292
32. Dewitte M (2015) Gender differences in liking and wanting sex: examining the role of motivational context and implicit versus explicit processing. Arch Sex Behav 44(6):1663–1674
33. Koranyi N et al (2017) Dissociating implicit wanting from implicit liking: development and validation of the wanting implicit association test (W-IAT). J Behav Ther Exp Psychiatry 54:165–169
34. Farris SR, Ostafin BD, Palfai TP (2010) Distractibility moderates the relation between automatic alcohol motivation and drinking behavior. Psychol Addict Behav 24(1): 151–156
35. Ostafin BD, Marlatt GA, Greenwald AG (2008) Drinking without thinking: an implicit measure of alcohol motivation predicts failure to control alcohol use. Behav Res Ther 46(11): 1210–1219
36. Ostafin BD, Palfai TP, Wechsler CE (2003) The accessibility of motivational tendencies toward alcohol: approach, avoidance, and disinhibited drinking. Exp Clin Psychopharmacol 11(4):294–301
37. Tibboel H et al (2015) Heavy social drinkers score higher on implicit wanting and liking for alcohol than alcohol-dependent patients and light social drinkers. J Behav Ther Exp Psychiatry 48:185–191
38. Wiers RW et al (2011) Retraining automatic action tendencies changes alcoholic patients' approach bias for alcohol and improves treatment outcome. Psychol Sci 22(4):490–497
39. Wiers RW et al (2010) Retraining automatic action-tendencies to approach alcohol in hazardous drinkers. Addiction 105(2):279–287
40. Koranyi N et al (2020) Dissociation between wanting and liking for coffee in heavy drinkers. J Psychopharmacol 34(12):1350–1356
41. Ashby CR, Stritzke WG (2013) Is sensitivity to reward associated with the malleability of implicit inclinations toward high-fat food? Emotion 13(4):711–723
42. Kemps E et al (2013) Implicit approach-avoidance associations for craved food cues. J Exp Psychol Appl 19(1):30–38
43. Kraus AA, Piqueras-Fiszman B (2016) Sandwich or sweets? An assessment of two novel implicit association tasks to capture dynamic motivational tendencies and stable evaluations towards foods. Food Qual Prefer 49:11–19
44. Olson MA, Fazio RH (2004) Reducing the influence of extrapersonal associations on the implicit association test: personalizing the IAT. J Pers Soc Psychol 86(5):653–667
45. Karpinski A, Steinman RB (2006) The single category implicit association test as a measure of implicit social cognition. J Pers Soc Psychol 91(1):16–32
46. Rothermund K et al (2009) Minimizing the influence of recoding in the implicit association test: the recoding-free implicit association test (IAT-RF). Q J Exp Psychol (Hove) 62(1):84–98
47. Meissner F et al (2019) Predicting behavior with implicit measures: disillusioning findings, reasonable explanations, and sophisticated solutions. Front Psychol 10:2483
48. Cereghetti D et al (2020) LikeWant: a new methodology to measure implicit wanting for flavors and fragrances. Food Qual Prefer 80: 103829
49. Pool E et al (2015) Stress increases cue-triggered "wanting" for sweet reward in humans. J Exp Psychol Anim Learn Cogn 41(2):128–136
50. Talmi D et al (2008) Human pavlovian-instrumental transfer. J Neurosci 28(2): 360–368
51. Sommer C et al (2017) Strong seduction: impulsivity and the impact of contextual cues on instrumental behavior in alcohol dependence. Transl Psychiatry 7(8):e1183
52. van Steenbergen H et al (2017) Dissociable corticostriatal circuits underlie goal-directed- vs. cue-elicited habitual food seeking after satiation: evidence from a multimodal MRI study. Eur J Neurosci 46(2):1815–1827
53. Cambridge VC et al (2013) Neural and behavioral effects of a novel mu opioid receptor antagonist in binge-eating obese people. Biol Psychiatry 73(9):887–894
54. Ziauddeen H et al (2014) Studying food reward and motivation in humans. J Vis Exp 85:e51281
55. Ziauddeen H et al (2012) Food images engage subliminal motivation to seek food. Int J Obes 36(9):1245–1247

56. Pirc M et al (2019) Grab to eat! Eating motivation dynamics measured by effort exertion depend on hunger state. Food Qual Prefer 78:103741
57. Wyvell CL, Berridge KC (2000) Intra-accumbens amphetamine increases the conditioned incentive salience of sucrose reward: enhancement of reward "wanting" without enhanced "liking" or response reinforcement. J Neurosci 20(21):8122–8130
58. da Costa RQM et al (2020) An adaptation of Pavlovian-to-instrumental transfer (PIT) methodology to examine the energizing effects of reward-predicting cues on behavior in young adults. Front Psychol 11:195
59. Sennwald V et al (2020) Outcome-specific and general Pavlovian-to-instrumental transfers involving sexual rewards. Motiv Sci 6(1):79–83
60. Stussi Y et al (2021) Individual concerns modulate reward-related learning and behaviors involving sexual outcomes. Motiv Sci 7(4): 424–438
61. Hogarth L (2012) Goal-directed and transfer-cue-elicited drug-seeking are dissociated by pharmacotherapy: evidence for independent additive controllers. J Exp Psychol Anim Behav Process 38(3):266–278
62. Hogarth L, Maynard OM, Munafo MR (2015) Plain cigarette packs do not exert Pavlovian to instrumental transfer of control over tobacco-seeking. Addiction 110(1):174–182
63. Manglani HR et al (2017) Pavlovian-to-instrumental transfer of nicotine and food cues in deprived cigarette smokers. Nicotine Tob Res 19(6):670–676
64. Garbusow M et al (2016) Pavlovian-to-instrumental transfer effects in the nucleus accumbens relate to relapse in alcohol dependence. Addict Biol 21(3):719–731
65. Hardy L et al (2017) Drug cue reactivity involves hierarchical instrumental learning: evidence from a biconditional Pavlovian to instrumental transfer task. Psychopharmacology 234(13):1977–1984
66. Martinovic J et al (2014) Electrophysiological responses to alcohol cues are not associated with Pavlovian-to-instrumental transfer in social drinkers. PLoS One 9(4):e94605
67. Chillà C et al (2019) Measuring wanting without asking: the Pavlovian-to-instrumental transfer paradigm under test. Food Qual Prefer 78:103720
68. Watson P et al (2014) Working for food you don't desire. Cues interfere with goal-directed food-seeking. Appetite 79:139–148
69. Cartoni E, Balleine B, Baldassarre G (2016) Appetitive Pavlovian-instrumental transfer: a review. Neurosci Biobehav Rev 71:829–848
70. Chen M, Bargh JA (2016) Consequences of automatic evaluation: immediate behavioral predispositions to approach or avoid the stimulus. Personal Soc Psychol Bull 25(2):215–224
71. Piqueras-Fiszman B, Kraus AA, Spence C (2014) "Yummy" versus "Yucky"! Explicit and implicit approach-avoidance motivations towards appealing and disgusting foods. Appetite 78:193–202
72. Rinck M, Becker ES (2007) Approach and avoidance in fear of spiders. J Behav Ther Exp Psychiatry 38(2):105–120
73. Neimeijer RA, de Jong PJ, Roefs A (2015) Automatic approach/avoidance tendencies towards food and the course of anorexia nervosa. Appetite 91:28–34
74. Brouwer AM et al (2021) Measuring implicit approach-avoidance tendencies towards food using a mobile phone outside the Lab. Foods 10(7):1440
75. Zech HG et al (2020) A mobile approach-avoidance task. Behav Res Methods 52(5): 2085–2097
76. Eder AB et al (2021) Embodiment of approach-avoidance behavior: motivational priming of whole-body movements in a virtual world. Motiv Sci 7(2):133–144
77. Lebowitz ER, Francois B (2018) Using motion tracking to measure avoidance in children and adults: psychometric properties, associations with clinical characteristics, and treatment-related change. Behav Ther 49(6):853–865
78. Cereghetti D et al (2021) Validation of new methods of using simulated whole-body movements as implicit indicators of sound and odor preferences. Front Psychol 12:659269
79. Rougier M et al (2018) A new look at sensorimotor aspects in approach/avoidance tendencies: the role of visual whole-body movement information. J Exp Soc Psychol 76:42–53
80. Phaf RH et al (2014) Approach, avoidance, and affect: a meta-analysis of approach-avoidance tendencies in manual reaction time tasks. Front Psychol 5:378
81. Carpenter TP et al (2019) Survey-software implicit association tests: a methodological and empirical analysis. Behav Res Methods 51(5):2194–2208
82. Aube B et al (2019) The online-VAAST: a short and online tool to measure spontaneous approach and avoidance tendencies. Acta Psychol 201:102942

83. Foroni F et al (2013) The FoodCast research image database (FRIDa). Front Hum Neurosci 7:51
84. Miccoli L et al (2016) Affective pictures and the open library of affective foods (OLAF): tools to investigate emotions toward food in adults. PLoS One 11(8):e0158991
85. Miccoli L et al (2014) Meet OLAF, a good friend of the IAPS! The open library of affective foods: a tool to investigate the emotional impact of food in adolescents. PLoS One 9(12):e114515
86. Toet A et al (2019) CROCUFID: a cross-cultural food image database for research on food elicited affective responses. Front Psychol 10:58
87. Charbonnier L et al (2016) Standardized food images: a photographing protocol and image database. Appetite 96:166–173
88. Pool ER et al (2022) Differential contributions of ventral striatum subregions to the motivational and hedonic components of the affective processing of reward. J Neurosci 42(13): 2716–2728
89. Morales I, Berridge KC (2020) 'Liking' and 'wanting' in eating and food reward: brain mechanisms and clinical implications. Physiol Behav 227:113152
90. Berridge KC (2009) 'Liking' and 'wanting' food rewards: brain substrates and roles in eating disorders. Physiol Behav 97(5):537–550

Chapter 10

Methods for Studying Multisensory Perception in Consumer Research

Fabiana M. Carvalho and Charles Spence

Abstract

The consumption of food and drink are amongst life's most multisensory experiences, and go beyond the sensory stimulation taking place in the mouth. The latest research reveals the complex multisensory interactions that influence consumers' sensory and hedonic judgments, as well as their food preferences and choices. This chapter explores the concept of crossmodal correspondences in order to understand the means by which anticipatory food cues can affect the perceptual outcome of a flavor for the consumer. It also reviews the procedures and techniques that have been most commonly used to investigate multisensory stimulation and how anticipatory sensory cues ultimately affect the overall consumption experience. The remainder of this chapter summarizes a couple of recently conducted studies that were designed to assess crossmodal effects on consumers' sensory and hedonic responses to coffee under realistic tasting conditions (i.e., real-life settings). The use of contexts that are sufficiently realistic and thus relevant to the consumer is of great utility in consumer research.

Key words Multisensory perception, Crossmodality, Extrinsic cues, Consumption, Food

1 Introduction

The consumption of food and drink are amongst life's most multisensory experiences. We effortlessly perceive flavors as multisensory experiences, rather than as a series of discrete unisensory events. Such multisensory percepts emerge as a result of the brain automatically combining the information processed by the different senses, and integrating it (where appropriate) into a unified multisensory experience of flavor that we localize to the oral cavity [1–5].

The term "multisensory" refers to the fact that a multitude of different inputs reaches our perceptual systems. These inputs can be separated as belonging to one or another sensory modality on the basis of the anatomical structures that receive and process the incoming stimulation, or the phenomenology associated with the subjective experience of stimulation. Taking this into account, flavor can be regarded as several distinct neurophysiological systems

Carlos Gómez-Corona and Heber Rodrigues (eds.), *Consumer Research Methods in Food Science*, Methods and Protocols in Food Science, https://doi.org/10.1007/978-1-0716-3000-6_10,

that are integrated effortlessly [6, 7] or else as a "functionally distinct sense that is cognitively 'constructed' from the integration of distinct physiologically defined sensory systems" ([8, p. 1007]; *see also* [4]). The research demonstrates how these distinct sensory systems interact, meaning that the functioning of a unisensory system, say olfaction, can influence (and sometimes enhance) the functioning of another unisensory system, say gustation [9, 10].

The experience of eating, as a multisensory construct, involves the senses that are typically stimulated when food and drink are consumed (i.e., gustation, retronasal olfaction, oral-somatosensation, and possibly also audition), but it is important to note that it goes well beyond the simplistic act of biting, chewing (i.e., masticating), and swallowing a food product. The anticipatory cues from the food that is about to be eaten may well stimulate some combination of the vision, audition, somatosensation (including active touch or haptics), and orthonasal olfaction [11, 12]. Those anticipatory cues concern everything that the consumer can learn about a food product prior to tasting it. They range all the way from the sensory aspects of the food itself (e.g., its visual appearance and orthonasal aroma) to the multisensory aspects of the tableware, the food packaging, and the context in which the food is presented. These sources of sensory information can be classified as "product-intrinsic" and "product-extrinsic." Product-intrinsic sources of information are those that are a physical part of the product itself, such as the color or the smell/aroma of the food itself. On the other hand, product-extrinsic sources of information are those that are somehow related or associated with the product, but are not physically a part of that which is tasted or consumed, such as, for example, a product's labelling, packaging, any tableware, and the location in which it happens to be served or consumed [13] (*see* Fig. 1). Importantly, both product-intrinsic and -extrinsic anticipatory cues can help to set up strong expectations in the mind of the consumer about what they are about to taste/experience [12] and can ultimately affect how a product's flavor is perceived, both sensorially and hedonically [12, 14–17].

Investigating how multisensory stimulation influences the consumers' tasting experience has a huge potential to deliver a better understanding of people's food preferences and choices. How can one measure the influence of anticipatory sensory cues, be they intrinsic or extrinsic to the product, on consumers' flavor perception and food choice? This chapter aims to provide the reader with an overview of the most important neural mechanisms underlying multisensory integration focusing, in particular, on crossmodal correspondences and the means by which anticipatory cues affect the perceptual outcome of a flavor for the consumer. We will then address the procedures and techniques that have been most commonly used to investigate multisensory stimulation and how anticipatory sensory cues can ultimately affect the overall multisensory

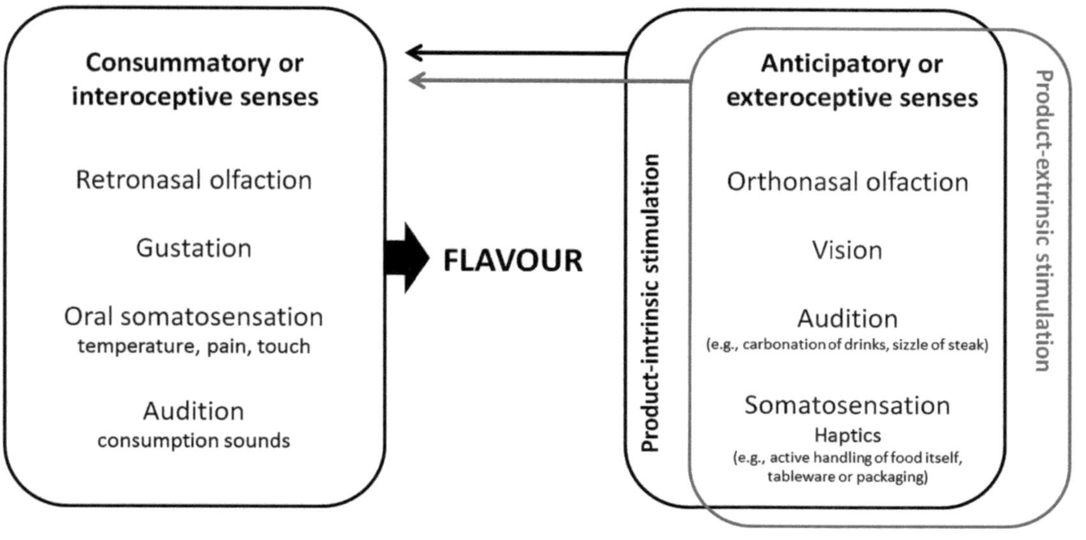

Fig. 1 The role of the various senses contributing to, or influencing, multisensory flavor perception can be conceptualized in terms of an important distinction between consummatory (or interoceptive) and anticipatory (or exteroceptive) sources of sensory information [11]. In turn, the anticipatory sensory information can be further classified as "product-intrinsic" and "product-extrinsic" [1, 4, 13]. The sensory inputs involved in flavor perception appear on the left and those that are involved in setting flavor expectations appear on the right

flavor percept as well as the experience of consumption. Finally, we close this chapter by presenting a pair of recently conducted experiments that were designed to assess crossmodal effects on consumers' sensory and hedonic responses to coffee under realistic conditions (i.e., real-life settings). The use of contexts that are realistic enough and relevant to consumers is an important aspect of consumer research. Ecologically valid paradigms (i.e., data acquisition under more naturalistic testing conditions) generate results more readily applicable to real-world situations—compared to traditional laboratory methods [18].

2 Underlying Neural Mechanisms

As has been mentioned already, multiple sensory systems are activated whenever we eat and drink. In recent years, a multitude of cognitive neuroscience studies have focused on elucidating those factors that modulate multisensory integration, including those giving rise to flavor perception or flavor binding [19, 20]. It is crucial to understand the driving features behind multisensory integration, namely, the factors that determine what is bound together and what is not. In other words, given the myriad different unisensory signals our senses transduce at any given time [21], how does the brain determine which stimuli to should be combined and which should be kept separate?

Starting from the multisensory events that occur in the mouth, several studies have shown that certain low-level sensory neurons located in the gustatory cortex (i.e., anterior insula) are multisensory, being responsive to oral somatosensory, olfactory (retronasal), and also to overall food processing [e.g., 22–24]. In addition, there is growing evidence that the olfactory primary (i.e., piriform) cortex is activated during gustatory stimulation as taste-selective information is transmitted from the gustatory to the piriform cortex [25, 26]. These intertwined neurophysiological processes suggest that there may be an automatic form of flavor binding of the oral (i.e., chemical) senses. At least as far as the senses of taste, olfaction, and oral-somatosensation are concerned, the propensity to bind the component stimuli would seem to be largely hard-wired [27], but modulated by experience. For instance, the attribution of taste qualities to odors as a result of associative learning has been linked to the activity of odor-responsive neurons in the orbitofrontal cortex (OFC). These neurons become responsive to the associated taste stimuli after taste–odor associative learning [28]. Moreover, the OFC is thought to be the main neural substrate underlying the rapid learning of visual, olfactory, and taste associations [29]. In fact, the OFC is one of the key brain areas involved in the brain's anticipatory response to the visual presentation of appetizing foods [30, 31].

In addition to this largely automatic form of binding, there are further issues concerning the factors that modulate multisensory flavor integration. First, spatial and temporal factors are important in modulating multisensory flavor binding. Generally speaking, concurrent temporal stimulation tends to favor multisensory integration [32]. That is, the closer in time the stimuli in different sensory modalities are presented, the more likely multisensory integration is to occur [33, 34]. Similarly, spatial coincidence also appears to facilitate multisensory integration, at least amongst the spatial senses of vision, audition, and touch [35] (though *see* [36]). Stimulation from the senses of taste, retronasal olfaction, audition, and oral-somatosensation can be considered to overlap, to some extent, during consumption (i.e., mastication and swallowing). In addition to spatiotemporal factors, cognitive factors such as prior knowledge greatly contribute to multisensory integration. In fact, it has been argued that learned associations between different sensory attributes may well play a major role on the multisensory integration related to flavor perception [2, 6, 9, 10, 37, 38].

In fact, one factor that provides an important means of constraining multisensory integration relates to the crossmodal correspondences—that is, to the compatibility, judged by an observer, between unisensory features or attributes of crossmodal stimulation [38, 39]. Despite audiovisual correspondences being by far the best documented of the correspondences in the literature to date, a number of studies have demonstrated that they also exist amongst the oral (i.e., consummatory) senses as well as between the oral and extra-oral (i.e., anticipatory) senses [40–45]. The crossmodal correspondences are likely to be of four qualitatively different, though not necessarily mutually exclusive, kinds—statistical, structural, semantically, and emotionally mediated [38, 46, 47].

While certain crossmodal correspondences may result from specific characteristics of the neural systems coding sensory information [38, 48–50], others may come from the internalization of either the statistical regularities or the semantic identity that are present in the observer's environment [46]. The former type (i.e., structural) is likely to involve internal determinants of the functional organization of the perceptual system (such as increased neural firing associated with increased stimulus intensity for each of the senses), whereas the latter two types of crossmodal associations (i.e., statistical and semantic) are learned.

The statistical crossmodal correspondences depend upon the observer picking-up (not necessarily consciously) on two or more of stimulus dimensions or features that happen to be correlated in nature. For instance, it has been shown that certain colors are systematically associated with particular colors, such as the association between sour tasting foods with green and yellow, and the sweet taste with red and pink [43, 51]. One mechanism that has been suggested by several researchers is that these associations reflect the internalized or learned correspondences from environmental regularity that many fruits do, for example, transition from green, unripe, and sour through to ripe, red, and sweet [52–54].

Semantic crossmodal correspondences emerge following language development as certain terms come to be associated with more than one perceptual quality or continuum [38, 55]. Early studies have already reported psychophysical associations between geometrical shapes and tastes [56]. For instance, the word "sharp" is a common linguistic term that can be used to describe geometrical shapes as well as taste and mouthfeel attributes (such as sour or pungent, or the taste of certain cheeses) [57]. A number of studies have demonstrated that the shape of the food itself or a shape associated to the food can affect participants' judgments. For instance, Gal et al. [58] reported that angular shapes presented to participants in a pre-task involving comparing the surface area of geometrical figures was associated to a cheese sample being evaluated as sharper in flavor than when round shapes had to be evaluated instead.

One may wonder what the means are by which crossmodal correspondences influence people's perception and/or behavior. It has been argued that crossmodal correspondences are not automatic in the same way that the concurrents induced in synesthesia are, but they can nevertheless affect people's behavior without their conscious control in a relatively rapid, if not entirely automatic manner [46]. The literature on crossmodal correspondences, especially when considering the role of anticipatory cues in setting our flavor expectations [13], can be interpreted in the context of the literature on crossmodal priming. Priming is the implicit memory effect in which exposure to a stimulus (i.e., the prime) influences the response to another, subsequently presented, stimulus [59]. Priming often reflects the operation of implicit or non-conscious processes, and has been shown to occur in those tasks that do not require explicit recollection of past experiences by the participant [60]. In the present context, expectations can be considered as top–down effects, based on prior knowledge, about certain aspects of food-related stimuli. Food expectations could be interpreted as a case of priming in the sense that the exposure to several anticipatory cues may potentially prime the consumer's perception, judgment, and/or behavior towards food (*see* [61] for a review).

In addition to memory-related (top-down) effects such as priming, attentional effects have also been shown to modulate people's performance in those experiments that have investigated crossmodal congruency. It had been proposed that, in the context of crossmodal correspondences, attention might operate in a bottom-up, stimulus-driven manner, depending upon a cue that can sufficiently drive the participant's attention towards the crossmodally corresponding feature, attribute, or dimension [46]. Selective attention to a sensory modality has been shown to modulate (or bias) the processing of flavor stimuli [62]. In this way, stimuli such as colors or sounds can be used to draw the consumers' attention to the corresponding sensory aspects of their tasting experience, which can, in turn, make them more salient [43, 63] (*see* [45] for a review).

Both priming and saliency effects contribute to a phenomenon named "sensation transference" [64], otherwise known as "affective ventriloquism" [65]. This refers to the observable effect of having the consumer's (conscious or otherwise) expectations or feelings about the food receptacle or packaging being transferred to their ratings of the food itself. In other words, relevant product-extrinsic cues can influence the consumers' evaluation, and presumably also their experience, of the properties of the product as well. These concepts are closely related to biasing effects such as the "halo" effect, originally described by Thorndike [66], in which the general impression of a stimulus positively biases its evaluation on (supposedly) unrelated characteristics. Thus, if the visual, acoustic. and/or haptic properties of the container are manipulated, it is plausible to infer some kind of effect on the ratings of taste and/or flavor attributes, as well as hedonic evaluation of the food product tested [67, 68].

3 Stimuli

3.1 Anticipatory Cues

Multisensory studies on consumption experience focusing on the role of anticipatory cues have tended to investigate whether exposure to multisensory stimulation prior to consumption affects the consumers' responses to, and/or evaluation of, the food or beverage being evaluated. Basically, such studies assess whether a prior encounter with certain stimuli can somehow be associated with a change (i.e., modulation) in the participant's response to another subsequently presented flavor stimulus [19].

The anticipatory cue can be a simple fully controlled stimulus, it can be a pair or a set of (controlled) stimuli, or it can be a complex mixture of stimuli, such as a whole context or environment. For instance, some studies have measured the effect of the color of foods or beverages (e.g., [69, 70], *see* [51] for a recent review) or the color of the container in which a beverage happens to be served ([71]; *see* [72] for a review) on the ratings of certain sensory attributes of the products by the participants. In these cases, the experimental design had only one varying sensory attribute (i.e., color). There are other studies that have assessed the influence of two or more anticipatory cues on aspects of consumption, such as the effect of the color of the beverage and the shape of the glass in which the beverage happened to be served on participants' flavor expectations [73], the effect of the soundtrack and the label on the evaluation of beers [74], the effect of shapes and/or speech sounds on participants' evaluation of chocolate varying in cocoa content [75], or even the effect of aroma, color, and soundtrack on taste perception and liking of fruit juices [76].

Other studies, meanwhile, have investigated the effect of a multisensory consumption environment on the participants' product judgments. For example, Velasco et al. [16] tested the effect of different environments varying in terms of their ambient colors, sounds, and smells on sensory and hedonic ratings of a whisky sample. Their result showed that the carefully constructed environments in which the whisky was evaluated were capable of modulating the grassiness on the nose, the sweetness on the palate, and the textured aftertaste of the whisky by c. 10–15%. More recently, Motoki et al. [77] reported that the color parameters of the interiors of coffee shops tended to be associated with consumers' expected sensory properties of the coffees served as well as the self-reported likelihood of visiting the establishment.

As mentioned already, anticipatory cues (i.e., the exteroceptive stimulation prior to consumption) falls into two main categories, namely, "product-intrinsic" and "product-extrinsic" cues. The intrinsic factors are those that are a physical part of the food product itself, whereas extrinsic factors are associated to the food product in some way, but are not part of what is consumed [13, 17, 78].

A multisensory experiment can make use of exteroceptive stimuli belonging to either one or both categories of anticipatory cues. Depending on the research hypotheses, the experimenter can choose to manipulate only intrinsic factors such as the color, the shape, and/or the aroma of a product (e.g., [79]), or to focus instead on varying only extrinsic factors such as visual and/or haptic aspects of the packaging or servingware associated to a certain food product [80–82]. Alternatively, the researcher could manipulate both intrinsic and extrinsic factors in a single experiment by adding factorial levels to the product itself as well as to the servingware [73, 83, 84], packaging [85], and soundtracks [74], for instance.

Depending on the experimental hypothesis, the researcher should manipulate the level of crossmodal congruency between anticipatory and consummatory information accordingly. The anticipatory stimulus could either be crossmodally congruent or incongruent with the food target, so that the experimental design should include both congruent and incongruent conditions. For instance, the color of the packaging and the flavor of the product inside can be presented to the participants in congruent versus incongruent pairs [86]. In addition, packaging and servingware congruency may be defined not only in relation to a product's sensory attributes (e.g., taste, flavor) but also to the product category [87, 88] and to the appropriateness of the pairing according to the consumer's everyday experiences [89, 90]. It is important to be aware that congruency operates at multiple levels [38], despite only some levels typically being manipulated by researchers.

3.2 Task Instructions

Anticipatory cues can act as explicit or implicit factors depending on several issues, including how the participants are debriefed or instructed regarding the experimental task. Different verbal instructions given to the participants can make them aware (or fail to make them aware) that they are under the influence of an anticipatory stimulus, such as a prime. This could potentially dissociate implicit from explicit feature processing [91, 92].

Explicit task instructions could draw attention to the prime, or else might make the dimensions on which the crossmodal correspondence operates more perceptually salient to the participant. The perceptual salience of the relevant dimension could influence the participant's performance on the subsequent evaluation (*see* [43] for a review). On the other hand, implicit instructions might distract the participant from finding out what the relevant dimension is, and what the effect could be, and feature processing would take place outside the awareness of the participant. This could also alter the participant's responses to the subsequent stimuli [61].

4 Delivering the Stimuli

The anticipatory stimulus (or cue) can be part of the food product that is being tested (i.e., intrinsic factor) or can be sensory and/or semantic aspects of another stimulus related to the food product (i.e., extrinsic factor). In the former case, since they cannot be separated from the food product that is being evaluated (i.e., the target stimulus), their presentation occurs simultaneously with the presentation of the food. An example would be the manipulation of the color, shape, or texture of the food being tested. In the latter case, the presentation of the anticipatory cue to participants could follow either a simultaneous or sequential order to the food stimulus depending on the study hypotheses (e.g., [93, 94]).

The extrinsic anticipatory stimulus can be delivered to the participants as: (1) a real/physical object or environment; (2) a computerized two-dimensional (2D) or three-dimensional (3D) image; or (3) an evoked or realistic virtual setting.[1] In the first case, the experimenter can make use of actual extrinsic stimulation such as real packaging, labelling, or servingware that will be perceived visually as a real object and, depending on the experimental design, will also be handled/touched by the participants [71, 74, 81, 91, 95]. In addition, a whole physical environment can be manipulated by the experimenter in order to create distinct multisensory conditions for consumption/tasting [16, 96]. In the second case, participants are presented with computerized 2D or 3D (i.e., includes rotation tools) images of extrinsic stimuli, such as photographs or graphic mock-ups of packaging [80, 92, 97, 98]. In the third case, in order to mimic the real-life experience of eating, the researchers can use visual (e.g., 2D video displays), auditory (e.g., soundtracks), haptic as well as olfactory stimuli in the laboratory [69, 99]. In addition, the development of new, 3D immersive technologies offer further possibilities to "place" the consumer into ever-more realistic environments, and thus potentially greatly improve the extrinsic validity of lab-based research [100–103].

It is, though, worth mentioning a few limitations of augmented reality studies, mainly regarding the use of head-mounted 3D virtual reality goggles to create immersive experience in lab-based contexts. While creating better imagery and spatial immersion, there are a number of commonly reported unwanted induced effects such as out-of-focus and motion blurring [104] as well as disorientation and even, in some cases, nausea [105]. Augmented

[1] In some cases, the anticipatory cue might be simply a written or verbal description. For instance, it has been shown that priming words such as "very sweet" vs. "less sweet" can modulate the self-report perceived sweetness as well as the activity of the gustatory cortex of participants [147]. Moreover, the literature of "evoked context" has long used written scenarios to evoke a consumption context in a sensory laboratory. Consumers are presented with brief texts or verbal descriptions meant to evoke a sense of presence in a real situation (*see* [170] for a comprehensive study).

and virtual reality methods mostly only capture visual attributes, and considering that multisensory stimulation is central in increasing consumer engagement (e.g., [106]), it would be necessary to add touch (e.g., temperature, *see* [107]), sound, taste, and olfactory dimensions to immersive virtual experiences. However, as the 3D immersive technology improves, the validity of the data obtained using these methods will likely also increase. In addition, when comparing traditional lab testing data with data acquired under immersive and more realistic contexts, it is important to keep in mind that external elements to the food or beverage, such as the servingware, should be controlled wherever possible (e.g., given the realization of how important the drinking vessel can be to the consumer's sensory and hedonic judgements of several beverages; [108]). *See* Bangcuyo et al. [100] for one such study where the environmental context was confounded with the servingware, thus making an unambiguous interpretation of the study's results in terms of the influence of the various factors difficult.

Regarding the target (food) stimulus, multisensory studies have relied on either real or 2D/3D photographs of food items in a computer screen. In these experiments, the food products that are going to be evaluated are presented to the participants usually following a side-by-side or sequential monadic presentation (e.g., [71, 79, 109]) or a two-alternative forced-choice scheme (e.g., [110]). Moreover, several multisensory studies, instead of presenting pictorial or real food stimuli, have adopt the procedure of presenting only the anticipatory cues to the participants and then measure the evoked or expected flavor attributes. For instance, it is common to ask the participants, following the presentation of the anticipatory cue, to rate the likelihood that a food would have a certain taste or flavor attribute (e.g., "how sweet does this look?") [98, 111], to perform verbal association tasks [83], or to select flavor attributes from a list with pre-defined terms (e.g., Check-All-That-Apply, CATA: [112]; or mention the first item that comes to mind: [73]).

5 Measuring the Responses

The second part of a study investigating crossmodal effects on the perception of flavor always entails some measurement of the hypothesized effect of the anticipatory cue(s), be they product-intrinsic or product-extrinsic, on participants' perception, judgment, and/or behavior. This response can be physiological (such as heart rate or skin conductance) or, more often, is a conscious perceptual or verbal response by the participant. Moreover, different types of responses can be addressed by several physiological, behavioral, and neuroimaging methods, which can be considered under the umbrella term of psychophysiological methods

[113]. These methods can be used to assess sensory, hedonic as well as affective/emotional responses by the participants.

Starting from physiological methods, studies investigating the influence of multisensory stimulation as well as crossmodal congruency on flavor perception have included measures of the autonomic nervous system, such as skin conductance, salivary flow rate, and heart rate, as well as attentional responses such as people's gaze behavior.

Galvanic skin responses and heart rate have long been considered sensitive markers for emotional arousal and valence [114], including when resulting from stimulation of olfaction and gustation [115, 116]. Regarding the multisensory aspects of consumption, it has been shown that the perceived flavor of ice cream by consumers varies as a function of the perceived pleasantness of the music that happens to be playing in the background [117] as well as the environment in which the product happens to be tasted/consumed [118]. Both of the just-mentioned studies reported that skin conductance and heart rate recordings were correlated with changes in multisensory flavor perception as a function of music liking ratings [117] and of crossmodal correspondences between eating environments and ice-cream flavor perception [118].

In another study, the participants were served fruit or vegetable juice. However, their expectations were manipulated prior to tasting by presenting an anticipatory cue which was an image of an ingredient that could either match or mismatch the target juice (i.e., congruent vs. incongruent trials, respectively). The fruit and vegetable juices were perceived as having different intensities of the basic tastes sweet, salty, and bitter in the different conditions. Moreover, skin conductance and heart rate recordings were related to attention and the orientation response, indicating increased attention for the processing of information related to the disconfirmation of sensory and hedonic expectations (i.e., on incongruent trials) [119]. Regarding measures of salivary flow rate, only a few studies have directly assessed the association between product-extrinsic cues, such as environmental factors, and changes in salivation. It has been shown that both the total light intensity (illumination level) and the color temperature (warm vs. cool) can influence salivary rate ([120]; *see* [121] for a review).

There has been a great deal of interest in characterizing eye-movements in sensory and consumer research, and the most successful technique to assess visual behavior and visual attention currently is eye-tracking [122]. Eye-tracking measures saccades and fixations, and uses these measures to characterize gaze behavior. In turn, these measures can be used to investigate the visual processing of crossmodal congruency in the context of multisensory flavor perception [123]. So, for instance, Seo et al. [124] used eye-tracking to investigate how the presentation of odors influences overt visual attention to congruent food products. The participants in this particular study were presented with four odors (orange, lavender, coffee, and liquorice) before and during the presentation of pictures containing one congruent and three incongruent food stimuli with the presented odors. Participants' visual attention, measured by the total number and length of fixations, was enhanced by congruent odors, suggesting an olfactory priming effect on overt visual selective attention.

It has been demonstrated that taste–sound crossmodal correspondences can also guide visual attention. That is, participants fixate for longer on sweet foods while listening to music that is congruent with sweetness and salty foods when listening to music that is congruent with saltiness [125]. In this case, congruent music primes relevant information about the product [126]. Additionally, the use of auditory stimuli associated with products (e.g., usage sounds or product-related jingles) has also been shown to crossmodally facilitate consumers' visual search for products, even for newly learnt associations between unfamiliar brands and sonic logos [127].

In a recent study using eye-tracking to test how packaging color affected the search for different flavor of crisps, participants were cued by a flavor word before searching for crisps packaging having this particular flavor label. The pattern of eye-movement results revealed that participants were slower to find a color–flavor incongruent target than a color–flavor congruent one [128]. Another recent study used three different visual designs for waffles—a stone plate, street-food style, and a classic white plate presentation—to investigate consumers' preferences. The results of eye-tracking revealed that the length of time looking at the waffles in the different conditions differed, as did the participants' attention to individual elements on the waffles (e.g., strawberry in the street food condition and white chocolate in the stone plate condition; [129]).[2]

Moreover, recent research on automated facial expression analysis has shown that consumers' facial expressions may be most sensitive to extrinsic properties of products. A study using soy sauces with and without brand information showed that liking scores are primarily driven by specific tastes, whereas facial expressions were affected by branding and familiarity, and to a lesser degree by specific tastes [130].

[2] Eye-tracking latency, or duration, measures, such as the fixation duration (i.e., length of time looking at an AOI) can imply different characteristics of gaze behaviour, and should be carefully interpreted based on the experimental hypothesis and design. For instance, the nature (e.g., attention, memory, problem-solving) as well as the difficulty level of the task can significantly influence the mean fixation duration [122]. Regarding foods, it has, for example, been shown that the relation between fixation duration and the level of food attractiveness is U-shaped: disgusting food is gazed at much longer than good food, and also longer than very pleasant-looking food [171]. Moreover, studies have demonstrated that gazing behaviour, including measures of fixation duration, can predict food choice (e.g., [172, 173]).

Regarding behavioral measures, studies on multisensory flavor perception make use of tools that convert the participants' internal experience into a verbal description (i.e., qualitative measures) or a number (i.e., quantitative measures). Qualitative behavioral methods such as CATA and word association (WA) tasks have been used to better understand the consumers' verbal associations concerning the multisensory stimulus, such as the expectations towards a product elicited by different packaging [112, 131, 132].

Indirect quantitative measures such as provided by the implicit association test (IAT; [133]), which provide an index of automatic preferences and attitudes based on the difference of reaction times, and performance, in different conditions (or categorization tasks), have also been used to assess the associations between specific combinations of features/attributes (*see* [134] for a review). For instance, Crisinel and Spence [135, 136] used a modified version of the IAT in order to investigate the strength of the association between different auditory pitches (high to low) and basic taste qualities (sour, bitter, and sweet). The results revealed that both sour and sweet were associated with high-pitched sounds, whereas bitter was more strongly associated with low-pitched sounds instead. Other studies have used the IAT to assess color–flavor correspondences in the packaging of potato chips and showed that regular consumers of a certain brand displayed a stronger association between the color–flavor pairing used by that brand than that used by other brands [85, 92].

Additionally, there is growing interest in the implicit approach–avoidance procedure (AAP, or approach–avoidance task), which assumes that a positive stimulus facilitates an approach orientation, whereas avoidance orientation should be evoked by a negative stimulus. For example, a study evaluated people's implicit reactions to both positive (appealing) and negative (disgusting) foods as a function of their hunger state. The authors found that the participants in the no-hunger group performed avoidance (vs. approach) movements significantly faster; and their approach movements towards positive (vs. negative) foods were significantly faster [137].

As traditional quantitative measures, labelled scales such as category scales (e.g., Likert) and visual analog scales (VAS) have been widely used to quantify the influence of crossmodal associations on sensory and hedonic responses to various foods. Category scales divide intensity into bins that usually have a numeric and/or semantic label (e.g., 1 = very weak, 5 = medium, 9 = very strong), whereas VAS are typically an unstructured line scale anchored at the ends with the minimum and maximum ratings for a particular attribute (e.g., "not bitter" to "extremely bitter"). These category and line scales can be easily and rapidly understood by participants [138]. For instance, after presenting some type of anticipatory cue to participants, several studies have used these scales to measure participants' sensory and hedonic expectation and/or perception towards various food items, as well as various aspects of likely purchase behavior (e.g., purchase intention, willingness-to-pay) [71, 74, 79, 81, 139, 140].

Finally, neuroimaging techniques have been used to shed light on the neural substrates and mechanisms underlying multisensory integration and crossmodal perceptual phenomena as they apply to food and drink. It is important to recognize that behavioral responses represent the final outcome of numerous different neural processes elicited by both external (e.g., food product) and internal (e.g., hunger) stimuli. Despite offering insights on ongoing cognitive processes underlying consumers' performance or choice, behavioral measures (e.g., sensory or hedonic ratings) do not provide a systematic and direct means of investigating these processes. Non-invasive neuroimaging techniques can potentially provide a valuable contribution to the study of the relation between neural processes and behavior in consumer research. Two neurophysiological measurements used in cognitive and sensory neuroscience that have been applied to neuromarketing (or consumer neuroscience) and neurogastronomy are functional magnetic resonance imaging (fMRI) and electroencephalography (EEG) (*see* [3, 141, 142] for reviews). Despite the various contributions from neuroimaging techniques to the understanding of brain processes underlying multisensory flavor integration [143], their ecological validity can be assumed to be particularly low compared to behavioral experimental studies. For instance, fMRI studies can be long in duration, the task and stimuli must be highly controlled (e.g., oral delivery of liquefied foods via tubes), they take place in a sterile environment, and the subject is required to lie horizontal with their head restrained (to avoid head movements) [144].

Several neuroimaging studies have specifically investigated the topic of crossmodal influences of anticipatory cues on brain regions associated with flavor perception, with or without the delivery of an actual taste, flavor or odor. fMRI studies have demonstrated that higher-order inferences derived from visual stimuli with semantic content (i.e., taste/smell words or pictures of food) are represented in those brain regions traditionally considered to represent only low-level information about taste or smell. For instance, simply reading words with strong gustatory (e.g., salt, grape, and honey) or olfactory semantic associations (e.g., cinnamon and camphor) have been shown to activate areas that process taste and olfactory sensations, respectively [145, 146]. Moreover, Woods et al. [147] reported that the expectation of different taste intensities triggered by text cues ("very sweet" vs. "less sweet") modulates the activity of participants' gustatory cortex. Expecting a very sweet drink, but tasting a less sweet one, enhanced the reported sweetness and increased activity in the gustatory cortex, as compared to tasting a less sweet drink without this expectation. Similarly, expecting that a highly aversive bitter taste but tasting a not so distasteful one led participants to report the taste as less aversive than when they had accurate expectations about the taste. In addition, the primary taste cortex was less strongly activated [148].

A more recent study using EEG showed very similar results; in particular, participants were cued to expect a low- vs. high-sweet stimulus (sucrose solutions in this case). The anticipation of low-sweetness decreased the sweetness ratings of the high-sweet solution, and likewise the anticipation of a high-sweet increased the sweetness ratings of the low-sweet solution. The expectation of taste intensity modulated activity in gustatory insular cortex [149]. More recently, a study using 7-Tesla fMRI showed that viewing pictures of foods associated with a specific taste (e.g., sweet, salty, and sour) evokes taste quality-specific activity in gustatory insular cortex [150].

A few neuroimaging studies have also assessed the crossmodal influences of color on odor perception. For instance, a superadditive neural response in the orbitofrontal and insular cortices have been reported following the presentation of congruent (as compared with incongruent) color-odor pairs (e.g., a strawberry aroma being paired with the color red vs. with turquoise) ([151]; *see* [152] for a recent review). Another study used EEG to investigate whether congruent abstract shapes could modulate olfactory perception and olfactory event-related potentials (ERPs). The participants were presented either matching or mismatching odor-shape pairs and were asked to judge odor pleasantness and intensity. Compared to the incongruent shape, the congruent shape increased the pleasantness of the matching odor and the unpleasantness of the mismatching one. Moreover, it was associated to significantly higher amplitudes and shorter latencies in the olfactory ERP (N1 peak) [153].

6 Case Studies

6.1 Introduction

This section summarizes two recent studies that investigated the effect of the physical properties of coffee cups on specialty coffee consumers' sensory and hedonic judgment of specialty coffees [95, 140]. The first study [140] assessed the effect of the color of the cup (henceforth Study 1), while the second [95] assessed the effect of the cup surface texture (henceforth Study 2) on consumers' sensory and/or hedonic judgments of specialty coffee. The experiments in both studies were carried out in real-life settings—that is, the data were obtained in a specialty coffee shop in Sao Paulo, Brazil. Relevant here, several studies have reported significant differences in hedonic and consumption behavior measures obtained in actual or virtually realistic consumption location (e.g., wine bar or coffee shop) when compared to those obtained in traditional sensory booths (e.g., [100, 102]). Highly controlled environments can all too easily end-up stripping away meaningful contextual (i.e., visual, auditory, and olfactory) information that may affect consumers' perceptions under everyday conditions, and that may be restored, at least to a certain extent, in immersive environments. Nonetheless, it is important to consider keeping control over physical contextual factors. In our studies, consumers were tested in a quiet, well-lit air-conditioned room in the second floor of a coffee shop that was isolated from the regular consumption area (i.e., the first floor). Thus, despite bringing the coffee shop context of consumption to the experiment, several environmental/contextual factors, such as room lighting, noise, and temperature, as well as the distance between participants during the assessment could be carefully controlled. The participants were led, in groups of 12, to the testing room and were then seated around six two-seater tables, with at least 1 m spacing between adjacent tasters. Designing more realistic testing conditions is a central concern of much of contemporary consumer research. However, it is important to search for intermediate situations that fall between real-life and laboratory conditions in order to improve the ecological validity of experimental environments without compromising on the results, due to lack of control over experimental testing conditions (*see* [154] for a review).

The main rationale behind these experiments concerns the fact that our drinking experience is necessarily tied to the receptacle from which a drink is consumed, be it, for instance, a glass, a cup, a mug, a bottle, or can. In particular, converging evidence demonstrates strong associations between the texture, as well as the color properties, of the drinking receptacle and the experience of taste (or flavor) in several beverages ([155, 156]; *see* [108] for a review). In addition, the correspondence between the physical attributes of the cup (i.e., its color and texture) and flavor attributes in specialty coffee were essentially unknown at the time these investigations commenced.

Our main hypothesis for both studies was based on previous evidence of the existence of sensation transference between visual (i.e., color) and haptic (i.e., texture) aspects of the drinking vessel to the sensory and hedonic evaluation of the beverage itself. In Study 1, three experiments were conducted. In the present section, we will focus only on the experiment in which both congruent and incongruent pairings of cup color and coffee flavor profile were served to the participants. In Study 2, three experiments were conducted as well. Given the aim of the present chapter, we will focus on the experiment which was conducted with amateur coffee consumers.

6.2 Materials and Methods

Participants It is important to consider the selection of a test population appropriately screened to ensure they are representative participants for the test being conducted. If the emphasis is given to a specific premium or high-quality product, such as specialty coffee, a short screening survey on demographics and consumption profile should probably be included. In our studies, we adopted the following inclusion criteria: (1) amateur consumers who had been drinking specialty coffee for at least a year with no sugar added (since the perception of natural sweetness in coffee would be assessed) and (2) not having a cold or any other self-reported impairment of their sense of smell or taste at the time the study was being conducted. Where possible, the sample size should be guided by power calculations based on previous measures of sensitivity and variance for a similar treatment and test population. For example, in Study 1, considering medium effect size (d) of 0.75, an intended power ≥ 0.8, and alpha value ≤ 0.05, the minimum sample size is 73 in two-way factorial designs [157]. Since our recruitment was via social media and the tasting section was conducted during a weekend, we were able to test 273 consumers who had met the inclusion criteria.

Including a relatively larger number of participants may not necessarily mean a greater incremental benefit in terms of power and effect size. It is, though, worth noting that oversampling will not compensate for the suboptimal sampling of stimuli [158], and in very large sample size analysis (e.g., "big data") it could even magnify the bias associated with error resulting from poor sampling or study design [159].

Test Design It is important to consider the outcomes and experimental comparison that will be made in the data collected. The two studies in question were designed to examine whether the physical characteristics of the coffee cup would affect sensory and/or hedonic ratings of specialty coffee by consumers.

In Study 1, there were two independent variables; coffee and cup. The coffee variable had two levels, that is, it either had high acidity (Kenyan) or high sweetness (Brazilian), meaning different flavor profiles. The cup variable had four levels, that is, the outer surface of the cup could be pink, green, yellow, or white (*see* Fig. 2a). Note that the inner color of the cups was always white, in order to minimize the impact of any change in the color contrast (between the coffee and the inner surface of the cup; cf. [72, 160]) on the perceived appearance properties of the coffee. The dependent variables were acidity, sweetness, and liking. To avoid a possible coffee-color bias, the tasting procedure followed a within-participants

Fig. 2 Cups used in the experiments conducted with specialty coffee consumers in (**a**) Study 1—cup color and (**b**) Study 2—cup texture. (Adapted from Refs. [95, 140], respectively)

experimental design,[3] and the evaluation of the samples followed a sequential monadic presentation scheme with the order of presentation balanced across participants. In this way, all of the participants were exposed to both congruent (pink/sweet Brazilian) and incongruent (green and yellow/acidic Kenyan) pairings of the color of the cup and coffee flavor profile. By opting for a within-participants experimental design, the experimenter should consider the issue that presenting different stimuli to the participant (i.e., cups whose color varies) might unintentionally draw the latter's attention to color as an important feature that they would otherwise perhaps not notice or consider.

As mentioned in Subheading 3.2, the perceptual salience of a relevant feature in the experiment could influence the participant's performance on the subsequent evaluation [46, 161]. The participants performed pre- and post-tasting evaluations. In the former

[3] In a "within-subjects" designed experiment, each individual is exposed to more than one of the treatments being tested, whereas in a "between-subjects" designed experiment, each individual is exposed to only one treatment. In the former, causal estimates can be obtained by examining how individual responses changed when the circumstances of the experiment changed. In the latter, on the other hand, causal estimates are obtained by comparing the responses of those in one experimental condition with the responses of those in another. It is important to bear in mind that both approaches have their respective (dis-)advantages. For instance, within-subject designs may lead to spurious effect due to carry-over and demand effects, but have several statistical advantages compared to between-subject designs, especially when it comes to measuring taste. On the other hand, between-subjects designs typically have no natural anchor and, consequently, have substantial noise and may miss important patterns. In fact, many crossmodal correspondences are relative phenomena, meaning that without comparison at within-subject level, this factor may be missing. Nevertheless, in several real situations, an individual is likely to only face a single decision, and a between design might have more external validity. Anyhow, the choice of designs should be carefully considered in the context of the question being studied, and the practical implementation of the research study should also be taken into account (*see* [174] for a review on the topic, and *see* [175] for a review on the use of incomplete block designs in consumer research).

case, they were instructed to first rate, only by looking at the cup of coffee, its expected sweetness and acidity using the rating scales provided (with a counterbalanced order of attributes across participants). Next, in the post-tasting evaluation, they moved on to sample the coffee and repeat the same ratings after having tasted it, as well as judging how much they liked it. Ratings were performed using a 10-cmVASanchored at 0 ("not at all") and 10 ("very"). Again, assessing expectations prior to tasting can draw the participants' attention to color and make them explicit to the subject (instead of keeping them implicit).

In Study 2, there was only one independent variable; cup texture, having two levels, namely smooth or rough finish (*see* Fig. 2b). Only one coffee was served in this study. The dependent variables were aroma, sweetness, acidity, body, and aftertaste. The tasting procedure followed a between-participants design and 117 consumers took part. They were instructed to evaluate the aroma before tasting the coffee using the aroma rating scale. Next, they moved on to tasting and rating the coffee's basic sweetness and acidity, with the order of presentation counterbalanced across participants, followed by ratings of body. After they had swallowed, they rated the coffee's aftertaste. Ratings were performed using a 10-cm VAS anchored at 0 ("not at all") and 10 ("very") for aroma, sweetness, and acidity. For body and aftertaste, a continuous semantic differential scale was used, anchored with the polar terms "totally smooth" (0) and "totally dry" (10). The range of attributes to be measured as dependent variables should be carefully considered to avoid halo-dumping effects [162]. The latter response bias may occur when participants are provided only one intensity scale (e.g., sweetness) to rate a mixture of two sensations (e.g., a sweet tasting strawberry flavor). It has been suggested that limiting the response alternatives may result in participants "dump"-ing the second sensation onto the only scale that they have available to them, thus artificially enhancing the ratings for sweetness (*see* Spence, in press-a, for a recent review). When the conceptual boundaries for sweetness and fruitiness become more defined (i.e., by the presentation of separate response scales for each dimension), there is less potential overlap and this dumping effect disappears (or at least is minimized).

It should be noted that (1) color and texture were the only factors that were varied in Studies 1 and 2, respectively, as all the other cup characteristics were kept as constant as possible, and (2) participants were given no information about the coffee (e.g., country of origin, processing method, varietal, etc.) they were going to taste. When testing the effect of the different drinking vessels on consumer perception, it is important to ensure that the physical characteristics that are not being assessed should vary as little as possible. For example, studies have demonstrated that virtually all physical characteristics of the container such as its shape, color, size, texture, and weight have been shown to influence the consumer's sensory and hedonic ratings of a range of different beverages ([67]; *see* [108] for a review).

6.3 Data Analysis

To assess multiple dependent variables simultaneously, the data from Study 1 were first analyzed using repeated measures multivariate analysis of variance (RM-MANOVA) on the dependent variables "sweetness" and "acidity" for expected coffee ratings (pre-tasting). A two-way RM-ANOVA was performed on the post-tasting data in order to assess the main effects of cup color and coffee type, as well as their interaction, on participants' perception of sweetness and acidity, and on their "liking" ratings. Finally, the same statistical test was used to compare the expected (pre-tasting) coffee ratings with the post-tasting ratings on the dependent variables "sweetness," "acidity," and "liking," where rating occasion (pre- or post-tasting) was included as an additional within-participants factor. All post-hoc pairwise comparisons were Bonferroni corrected, and differences were considered significant at $p \leq 0.05$.

In order to further explore the relationship between pre- and post-tasting scores, and hedonic judgments, a Pearson's correlation analysis was carried out. First, a difference index was generated by subtracting the post- from the pre-tasting rating values of sweetness and acidity for each participant (i.e., "expectation" minus "perception" ratings), for both congruent (i.e., pink cup/Brazilian coffee; green cup/Kenyan coffee) and incongruent (i.e., pink cup/Kenyan coffee; green cup/Brazilian coffee) conditions. Next, the correlations were calculated between the difference indices and the liking scores for each of these four conditions.

A between-participants MANOVA was conducted on the data from Study 2 with texture of the cup (smooth vs. rough) as the between-participant factor, and aroma, sweetness, acidity, body, and aftertaste were the dependent variables. Once again, all of the post-hoc pairwise comparisons were Bonferroni corrected.

6.4 Results

The results of Study 1 revealed that the color of the cup exerted a significant influence over both pre- and post-tasting ratings for all of the measured attributes. Prior to tasting, the colors pink and green were associated with sweetness and acidity, respectively, whereas yellow was associated with both tastes (*see* Fig. 3a). In the post-tasting phase of the study, the participants were exposed to both congruent (pink/sweet) and incongruent (green and yellow/acidic) pairings of the color of the cup and coffee flavor profile. Interestingly, liking ratings decreased significantly in the incongruent conditions (*see* Fig. 3b). The Pearson correlation analysis revealed a negative correlation between the acidity difference index and liking ratings in the incongruent condition "green

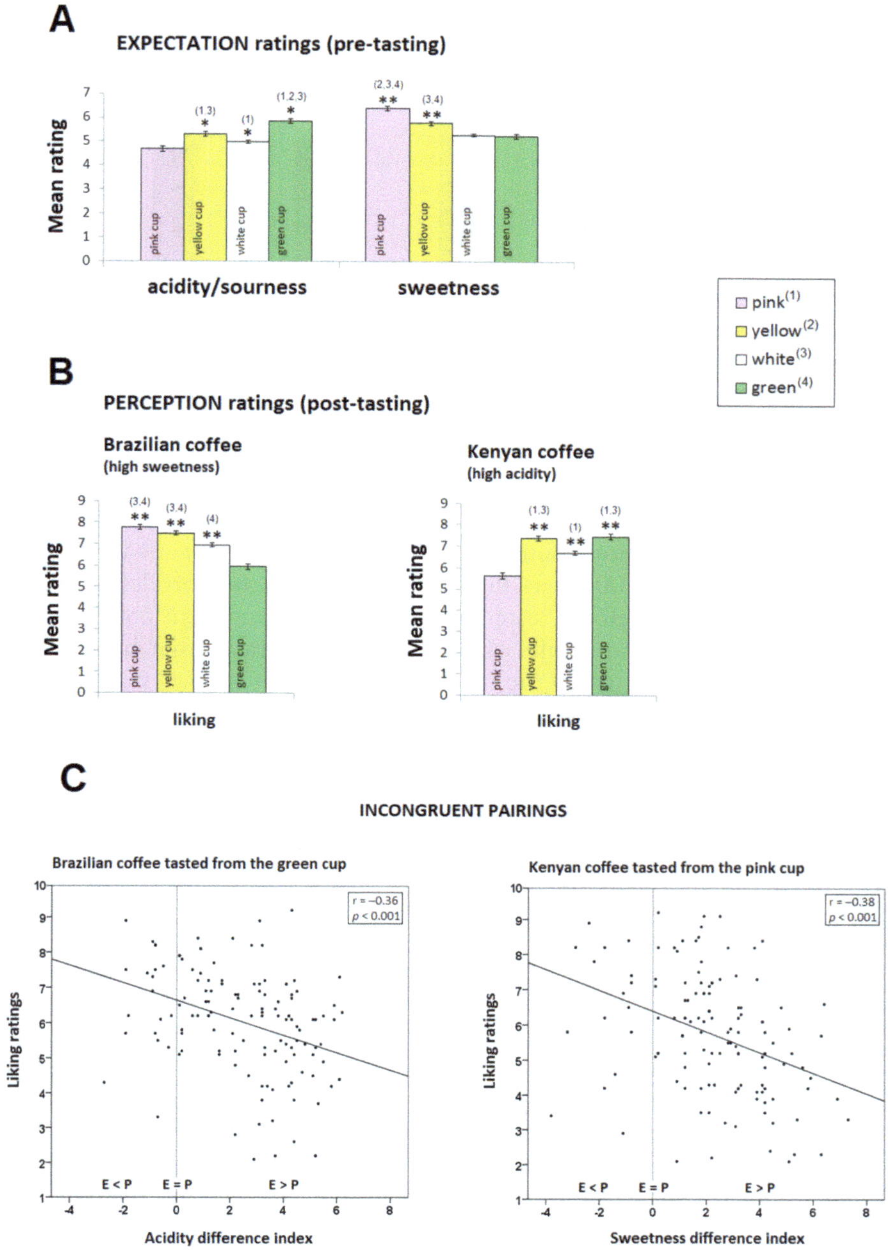

Fig. 3 Mean ratings (±SE) of (**a**) pre-tasting (expectation) and (**b**) post-tasting (perception) as a function of rating type (sweetness, acidity, and liking) for all four cup colors tested (pink, yellow, white, and green). Panel (**b**) shows separate plots for Brazilian and Kenyan coffees. Asterisks indicate statistical significance at $p < 0.05$ (*) or $p < 0.001$ (**). (**c**) Scatter-plots highlighting the correlation between the difference indices and liking ratings in both incongruent conditions. The difference indices were created by subtracting the post-tasting from the pre-tasting rating values of sweetness and acidity, respectively. E = Expectation (i.e., pre-tasting scores); P = Perception (i.e., post-tasting scores). (Adapted from Ref. [140])

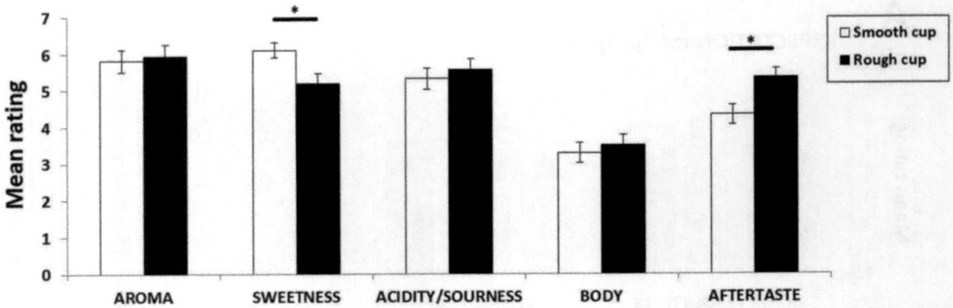

Fig. 4 Mean ratings (±SE) of aroma, sweetness, acidity, body, and aftertaste for the smooth (white bars) and rough (grey bars) cups. The attributes aroma, sweetness, and acidity were rated in terms of 'intensity' (0—"not at all"; 10—"very"), whereas the attributes of body and aftertaste were judged on a continuous "smooth-to-dry" bipolar scale. Asterisks indicate statistical significance at $p < 0.01$. (Adapted from Ref. [95])

cup/Brazilian coffee" ($r = 0.36$, $n = 133$, $p < 0.001$), and between the sweetness difference index and liking ratings in the incongruent condition pink cup/Kenyan coffee ($r = -0.38$, $n = 134$, $p < 0.001$) (*see* Fig. 3c). By contrast, no correlation was observed between the tested variables in the congruent conditions.

The results of Study 2 demonstrated that haptic information presented to the hand and/or felt on the lips from the coffee cup with the rough surface did indeed influence the tasting experience, both in terms of sweetness and mouthfeel. The coffee was also rated as tasting sweeter when sampled from the smooth (as compared to the rough) cup and was also perceived as being significantly drier, or rougher, in aftertaste when tasted from the rough ceramic (as compared to the smooth) cup (*see* Fig. 4).

6.5 Discussion

The two studies reviewed here demonstrated a significant impact that the color (Study 1) and texture (Study 2) of the coffee cup can have on both sensory and hedonic judgments of specialty coffee by regular consumers. Given that the drinking receptacle is a necessary component of the multisensory coffee drinking experience, it is crucial to recognize just how much of an influence it can actually have on the consumer's perception of the contents. Ultimately, drinking receptacles should most likely be designed taking cross-modal influences into account, in order to optimize the consumer's multisensory tasting experience of the contents [82, 108].

This shift in emphasis from previous studies on specialty coffee that have tended to focus solely on the intrinsic qualities of the beverage is essential to better understand consumer perception/behavior in the real-world. Despite the small body of empirical research that has been published on this topic to date, there has undoubtedly been a growing interest in investigating the effect of a range of product-extrinsic factors (e.g., such as the cup, packaging, context of consumption) on the sensory and hedonic aspects of coffee consumption [81, 82, 163]. In fact, the specialty coffee consumer, in addition to different coffee flavor profiles, is likely also interested in brewing methods, packaging, and ambience [163–165]. Therefore, it is important to understand the relevant factors that might help to enhance their multisensory experience in the coffee shop.

7 Conclusions

Product-extrinsic factors are a necessary part of the multisensory experience of consumption. Those experiments designed to test the influence of such product-extrinsic factors on consumer perception need to consider the crossmodal associations between the different dimensions of multisensory stimulation. These can be understood as a multitude of pieces of product information, which may or may not fit well together [166]. Generally speaking, consumers tend to prefer those products for which different pieces of sensory information match, as the congruency enhances the consumer's processing fluency (e.g., [167]) meaning that product configurations matching the associations that consumers have internalized (i.e., associations that are based on the crossmodal correspondences) are likely to be processed more fluently and, consequently, liked more (*see* [168]). In these cases, the product as a whole, including the anticipatory aspects of consumption, communicates a coherent multisensory message.

One interesting topic still to be further examined within the area of multisensory flavor perception is the impact of cross-cultural similarities and differences on consumption experiences. Cross-cultural research suggests that we learn to combine the intrinsic attributes (e.g., tastes, smells, colors, textures) as well as extrinsic elements (e.g., attributes of servingware or packaging) that regularly co-occur in foods and beverages that we grow accustomed to. In fact, a key part of figuring out why color sometimes does, and sometimes does not influence taste/flavor perception comes down to the question of exposure and the learning of specific crossmodal associations (e.g., [53, 169]). Thus, consumers with different cultural backgrounds may have very different flavor, hedonic and/or semantic expectations when exposed to exactly the same visual or auditory anticipatory cues/stimuli [73, 125].

References

1. Small DM (2012) Flavor is in the brain. Physiol Behav 107:540–552
2. Spence C (2016) Oral referral: on the mislocalization of odours to the mouth. Food Qual Prefer 50:117–128
3. Spence C (2020) Multisensory flavor perception: a cognitive neuroscience perspective. In: Sathian K, Ramachandran VS (eds) Multisensory perception. Academic, Cambridge, MA, pp 221–237
4. Stevenson RJ (2009) The psychology of flavour. Oxford University Press, Oxford
5. Stevenson RJ (2014) Flavor binding: its nature and cause. Psychol Bull 140:487–510

6. Auvray M, Spence C (2008) The multisensory perception of flavour. Conscious Cogn 17:1016–1031
7. Spence C (2019) Crossmodal contributions to the perception of piquancy/spiciness. J Sens Stud 34(1):e12476
8. Prescott J, Stevenson RJ (2015) Chemosensory integration and the perception of flavor. In: Doty RL (ed) Handbook of olfaction and gustation. Wiley Blackwell, Hoboken, pp 1008–1028
9. Spence C (2022a) Factors affecting odour-induced taste enhancement. Food Qual Prefer 96:104393
10. Stevenson RJ, Boakes RA (2004) Sweet and sour smells: learned synaesthesia between the senses of taste and smell. In: Calvert GA, Spence C, Stein BE (eds) The handbook of multisensory processing. MIT Press, Cambridge, MA, pp 69–83
11. Small DM, Veldhuizen MG, Felsted J, Mak YE, McGlone F (2008) Separable substrates for anticipatory and consummatory food chemosensation. Neuron 57:786–797
12. Spence C, Piqueras-Fiszman B (2014) The perfect meal: the multisensory science of food and dining. Wiley, Chichester
13. Piqueras-Fiszman B, Spence C (2015) Sensory expectations based on product-extrinsic food cues: an interdisciplinary review of the empirical evidence and theoretical accounts. Food Qual Prefer 40:165–179
14. Woods AT, Poliakoff E, Lloyd DM, Dijksterhuis GB, Thomas A (2010) Flavor expectation: the effects of assuming homogeneity on drink perception. Chemosens Percept 3:174–181
15. Yeomans MR, Chambers L, Blumenthal H, Blake A (2008) The role of expectancy in sensory and hedonic evaluation: the case of smoked salmon ice-cream. Food Qual Prefer 19(6):565–573
16. Velasco C, Jones R, King S, Spence C (2013) Assessing the influence of the multisensory environment on the whisky drinking experience. Flavour 2:1
17. Wang QJ, Mielby LA, Junge JY, Bertelsen AS, Kidmose U, Spence C, Byrne DV (2019) The role of intrinsic and extrinsic sensory factors in sweetness perception of food and beverages: a review. Foods 8(6):211
18. Stelick A, Dando R (2018) Thinking outside the booth—the eating environment, context and ecological validity in sensory and consumer research. Curr Opin Food Sci 21:26–31
19. Spence C (2015) Multisensory flavour perception. Cell 161:24–35
20. Stevenson RJ (2016) Attention and flavor binding. In: Piqueras-Fiszman B, Spence C (eds) Multisensory flavor perception. Woodhead Publishing, Cambridge, UK, pp 15–35
21. Spence C, Squire SB (2003) Multisensory integration: maintaining the perception of synchrony. Curr Biol 13:R519–R521
22. De Araujo IE, Simon SA (2009) The gustatory cortex and multisensory integration. Int J Obes 33(2):S34–S43
23. Uddin LQ, Nomi JS, Hebert-Seropian B, Ghaziri J, Boucher O (2017) Structure and function of the human insula. J Clin Neurophysiol 34(4):300–306
24. Huang Y, Kakusa BW, Feng A, Gattas S, Shivacharan RS, Lee EB et al (2021) The insulo-opercular cortex encodes food-specific content under controlled and naturalistic conditions. Nat Commun 12(1):1–12
25. Fortis-Santiago Y, Rodwin BA, Neseliler S, Piette CE, Katz DB (2010) State dependence of olfactory perception as a function of taste cortical inactivation. Nat Neurosci 13(2):158–159
26. Maier JX, Blankenship ML, Li JX, Katz DB (2015) A multisensory network for olfactory processing. Curr Biol 25(20):2642–2650
27. Spence C (in press) Multisensory feature integration. In: Mroczko-Wąsowicz A (ed) Sensory individuals. Oxford University Press, Oxford
28. Rolls ET (2011) Chemosensory learning in the cortex. Front Syst Neurosci 5:78
29. Rolls ET, Baylis LL (1994) Gustatory, olfactory, and visual convergence within the primate orbitofrontal cortex. J Neurosci 14:5437–5452
30. van der Laan LN, de Ridder DTD, Viergever MA, Smeets PAM (2011) The first taste is always with the eyes: a meta-analysis on the neural correlates of processing visual food cues. NeuroImage 55:296–303
31. Wang GJ, Volkow ND, Telang F, Jayne M, Ma J, Rao M et al (2004) Exposure to appetitive food stimuli markedly activates the human brain. NeuroImage 21(4):1790–1797
32. Isogai T, Wise PM (2016) The effects of odor quality and temporal asynchrony on modulation of taste intensity by retronasal odor. Chem Senses 41(7):557–566
33. Pfeiffer JC, Hollowood TA, Hort J, Taylor AJ (2005) Temporal synchrony and integration of sub-threshold taste and smell signals. Chem Senses 30(7):539–545
34. Shore DI, Barnes ME, Spence C (2006) The temporal evolution of the crossmodal congruency effect. Neurosci Lett 392:96–100

35. Green BG (2003) Studying taste as a cutaneous sense. Food Qual Prefer 14:99–109
36. Spence C (2013) Just how important is spatial coincidence to multisensory integration? Evaluating the spatial rule. Ann N Y Acad Sci 1296(1):31–49
37. Small DM (2004) Crossmodal integration–insights from the chemical senses. Trends Neurosci 27(3):120–123
38. Spence C (2011) Crossmodal correspondences: a tutorial review. Atten Percept Psychophys 73(4):971–995
39. Chen Y-C, Spence C (2017) Assessing the role of the 'unity assumption' on multisensory integration: a review. Front Psychol 8:445
40. Delwiche JF, Heffelfinger AL (2005) Cross-modal additivity of taste and smell. J Sens Stud 20(6):512–525
41. Deroy O, Crisinel A-S, Spence C (2013) Crossmodal correspondences between odors and contingent features: odors, musical notes, and geometrical shapes. Psychon Bull Rev 20(5):878–896
42. Simner J, Cuskley C, Kirby S (2010) What sound does that taste? Cross-modal mapping across gustation and audition. Perception 39: 553–569
43. Spence C, Wan X, Woods A, Velasco C, Deng J, Youssef J, Deroy O (2015) On tasty colours and colourful tastes? Assessing, explaining, and utilizing crossmodal correspondences between colours and basic tastes. Flavour 4:23
44. White TL, Prescott J (2007) Chemosensory cross-modal Stroop effects: congruent odors facilitate taste identification. Chem Senses 32(4):337–341
45. Spence C (2019) Attending to the chemical senses. Multisens Res 32(7):635–664
46. Spence C, Deroy O (2013) How automatic are crossmodal correspondences? Conscious Cogn 22(1):245–260
47. Spence C (2020) Assessing the role of emotional mediation in explaining crossmodal correspondences involving musical stimuli. Multisens Res 33:1–29
48. Mondloch CJ, Maurer D (2004) Do small white balls squeak? Pitch–object correspondences in your children. Cogn Affect Behav Neurosci 4:133–136
49. Stevens SS (1957) On the psychophysical law. Psychol Rev 64:153–181
50. Walsh V (2003) A theory of magnitude: common cortical metrices of time, space and quality. Trends Cogn Sci 7:483–488
51. Spence C, Levitan CA (2021) Explaining crossmodal correspondences between colours and tastes. I-Perception 12(2):1–28
52. Foroni F, Pergola G, Rumiati RI (2016) Food color is in the eye of the beholder: the role of human trichromatic vision in food evaluation. Sci Rep 6:37034
53. Higgins MJ, Hayes JE (2019) Learned color taste associations in a repeated brief exposure paradigm. Food Qual Prefer 71:354–365
54. Maga JA (1974) Influence of color on taste thresholds. Chem Senses 1(1):115–119
55. Gallace A, Spence C (2006) Multisensory synesthetic interactions in the speeded classification of visual size. Percept Psychophys 68: 1191–1203
56. Cytowic RE, Wood FB (1982) Synaesthesia II: psychophysical relations in the synaesthesia of geometrically shaped taste and colored hearing. Brain Cogn 1:36–49
57. Deroy O, Valentin D (2011) Tasting liquid shapes: investigating the sensory basis of cross-modal correspondences. Chemosens Percept 4(3):80–90
58. Gal D, Wheeler SC, Shiv B (2007) Cross-modal influences on gustatory perception. Available at SSRN 1030197. http://ssrn.com/abstract=1030197. Accessed 2 Aug 2021
59. Schacter DL (1992) Priming and multiple memory systems: perceptual mechanisms of implicit memory. J Cogn Neurosci 4(3): 244–256
60. Stevens WD, Wig GS, Schacter DL (2008) Implicit memory and priming. In: Roediger IHL (ed) Cognitive psychology of memory: Vol 2 of Learning and memory: a comprehensive reference. Elsevier, Oxford, pp 623–644
61. Dijksterhuis G (2016) Multisensory flavor priming. In: Piqueras-Fiszman B, Spence C (eds) Multisensory flavor perception. Woodhead Publishing, Cambridge, UK, pp 133–153
62. Ashkenazi A, Marks LE (2004) Effect of endogenous attention on detection of weak gustatory and olfactory flavors. Percept Psychophys 66:596–608
63. Spence C, Wang QJ (2015) Wine & music (II): can you taste the music? Modulating the experience of wine through music and sound. Flavour 4:33
64. Cheskin L (1972) Marketing success: how to achieve it. Cahners Books, Boston
65. Spence C, Gallace A (2011) Multisensory design: reaching out to touch the consumer.

Psychol Mark 28:267–308. https://doi.org/10.1002/mar.20392
66. Thorndike EL (1920) A constant error in psychological ratings. J Appl Psychol 4(1): 25–29
67. Kampfer K, Leischnig A, Ivens BS, Spence C (2017) Touch-flavor transference: assessing the effect of packaging weight on gustatory evaluations, desire for food and beverages, and willingness to pay. PLoS One 12(10): e0186121
68. Wang QJ, Spence C (2019) Sonic packaging: how packaging sounds influence multisensory product evaluation. In: Velasco C, Spence C (eds) Multisensory packaging: designing new product experiences. Palgrave Macmillan, Cham, pp 103–125
69. Ammann J, Stucki M, Siegrist M (2020) True colours: advantages and challenges of virtual reality in a sensory science experiment on the influence of colour on flavour identification. Food Qual Prefer 86:103998
70. Zampini M, Sanabria D, Phillips N, Spence C (2007) The multisensory perception of flavor: assessing the influence of color cues on flavor discrimination responses. Food Qual Prefer 18:975–984
71. Piqueras-Fiszman B, Spence C (2012) The influence of the color of the cup on consumers' perception of a hot beverage. J Sens Stud 27(5):324–331
72. Spence C (2018) Background colour & its impact on food perception & behaviour. Food Qual Prefer 68:156–166
73. Wan X, Velasco C, Michel C, Mu B, Woods AT, Spence C (2014) Does the type of receptacle influence the crossmodal association between colour and flavour? A cross-cultural comparison. Flavour 3:1
74. Reinoso-Carvalho F, Velasco C, Van Ee R, Leboeuf Y, Spence C (2016) Music influences hedonic and taste ratings in beer. Front Psychol 7:636
75. Ngo MK, Misra R, Spence C (2011) Assessing the shapes and speech sounds that people associate with chocolate samples varying in cocoa content. Food Qual Prefer 22(6): 567–572
76. Wang QJ, Mielby LA, Thybo AK, Bertelsen AS, Kidmose U, Spence C, Byrne DV (2019) Sweeter together: assessing the combined influence of product-related and contextual factors on perceived sweetness of fruit beverages. J Sens Stud 2019:e12492
77. Motoki K, Takahashi A, Spence C (2021) Tasting atmospherics: taste associations with colour parameters of coffee shop interiors. Food Qual Prefer 54:104315
78. Betancur MI, Motoki K, Spence C, Velasco C (2020) Assessing the relative contribution of product-intrinsic and product-extrinsic variables influencing people's choice of beer: a review. Food Res Int 137:109367
79. Wang QJ, Carvalho FR, Persoone D, Spence C (2017) Assessing the effect of shape on the evaluation of expected and actual chocolate flavour. Flavour 6:1
80. Becker L, van Rompay TJ, Schifferstein HN, Galetzka M (2011) Tough package, strong taste: the influence of packaging design on taste impressions and product evaluations. Food Qual Prefer 22(1):17–23
81. de Sousa MM, Carvalho FM, Pereira RG (2020) Colour and shape of design elements of the packaging labels influence consumer expectations and hedonic judgments of specialty coffee. Food Qual Prefer 83:103902
82. Spence C, Carvalho FM (2019) Assessing the influence of the coffee cup on the multisensory tasting experience. Food Qual Prefer 75: 239–248
83. Piqueras-Fiszman B, Alcaide J, Roura E, Spence C (2012) Is it the plate or is it the food? Assessing the influence of the color (black or white) and shape of the plate on the perception of the food placed on it. Food Qual Prefer 24:205–208
84. Fairhurst MT, Pritchard D, Ospina D, Deroy O (2015) Bouba-Kiki in the plate: combining crossmodal correspondences to change flavour experience. Flavour 4(1):1–5
85. Piqueras-Fiszman B, Spence C (2011) Crossmodal correspondences in product packaging. Assessing color–flavor correspondences for potato chips (crisps). Appetite 57:753–757
86. Velasco C, Wan X, Knoeferle K, Zhou X, Salgado-Montejo A, Spence C (2015) Searching for flavor labels in food products: the influence of color-flavor congruence and association strength. Front Psychol 6:301
87. Lyons SJ, Wien AH (2018) Evoking premiumness: how color-product congruency influences premium evaluations. Food Qual Prefer 64:103–110
88. Spence C, Velasco C (2018) On the multiple effects of packaging colour on consumer behaviour and product experience in the 'food and beverage' and 'home and personal care' categories. Food Qual Prefer 68:226–237
89. Raudenbush B, Meyer B, Eppich W, Corley N, Petterson S (2002) Ratings of

pleasantness and intensity for beverages served in containers congruent and incongruent with expectancy. Percept Mot Skills 94: 671–674

90. Spence C (2020) Multisensory flavour perception: blending, mixing, fusion, and pairing within and between the senses. Foods 9(4): 407
91. Blackmore H, Hidrio C, Godineau P, Yeomans MR (2020) The effect of implicit and explicit extrinsic cues on hedonic and sensory expectations in the context of beer. Food Qual Prefer 81:103855
92. Piqueras-Fiszman B, Velasco C, Spence C (2012) Exploring implicit and explicit crossmodal colour–flavour correspondences in product packaging. Food Qual Prefer 25(2): 148–155
93. Biswas D, Labrecque LI, Lehmann DR (2021) Effects of sequential sensory cues on food taste perception: cross-modal interplay between visual and olfactory stimuli. J Consum Psychol 31(4):746–764. https://doi.org/10.1002/jcpy.1231
94. Wang QJ, Spence C, Knoeferle K (2020) Timing is everything: onset timing moderates the crossmodal influence of background sound on taste perception. J Exp Psychol Hum Percept Perform 46(10):1118–1126
95. Carvalho FM, Moksunova V, Spence C (2020) Cup texture influences taste and tactile judgments in the evaluation of specialty coffee. Food Qual Prefer 81:103841
96. van Rompay TJ, van Hoof JJ, Rorink J, Folsche M (2019) Served straight up: effects of verticality cues on taste evaluations and luxury perceptions. Appetite 135:72–78
97. Simmonds G, Spence C (2019) Food imagery and transparency in product packaging. In: Velasco C, Spence C (eds) Multisensory packaging: designing new product experiences. Palgrave Macmillan, Cham, pp 49–77
98. Velasco C, Salgado-Montejo A, Marmolejo-Ramos F, Spence C (2014) Predictive packaging design: tasting shapes, typographies, names, and sounds. Food Qual Prefer 34: 88–95
99. Flavián C, Ibáñez-Sánchez S, Orús C (2021) The influence of scent on virtual reality experiences: the role of aroma-content congruence. J Bus Res 123:289–301
100. Bangcuyo RG, Smith KJ, Zumach JL, Pierce AM, Guttman GA, Simons CT (2015) The use of immersive technologies to improve consumer testing: the role of ecological validity, context and engagement in evaluating coffee. Food Qual Prefer 41:84–95
101. Chen Y, Huang AX, Faber I, Makransky G, Perez-Cueto FJ (2020) Assessing the influence of visual-taste congruency on perceived sweetness and product liking in immersive VR. Foods 9(4):465
102. Hannum M, Forzley S, Popper R, Simons CT (2019) Does environment matter? Assessments of wine in traditional booths compared to an immersive and actual wine bar. Food Qual Prefer 76:100–108
103. Oliver JH, Hollis JH (2021) Virtual reality as a tool to study the influence of the eating environment on eating behavior: a feasibility study. Foods 10:89
104. Kruijff E, Swan JE, Feiner S (2010) Perceptual issues in augmented reality revisited. In: 2010 9th IEEE international symposium mixed and augmented reality (ISMAR), 3–12 October
105. Saredakis D, Szpak A, Birckhead B, Keage HA, Rizzo A, Loetscher T (2020) Factors associated with virtual reality sickness in head-mounted displays: a systematic review and meta-analysis. Front Hum Neurosci 14: 96
106. Bravo-Moncayo L, Reinoso-Carvalho F, Velasco C (2020) The effects of noise control in coffee tasting experiences. Food Qual Prefer 86:104020
107. Heschong L (1979) Thermal delight in architecture. MIT Press, Cambridge, MA
108. Spence C, Wan X (2015) Beverage perception and consumption: the influence of the container on the perception of the contents. Food Qual Prefer 39:131–140
109. Maggioni E, Risso P, Olivero N, Gallace A (2015) The effect of a container's weight on the perception of mineral water. J Sens Stud 30(5):395–403
110. Pich J, Chuquichambi EG, Blay NT, Corradi GB, Munar E (2020) Sweet and bitter near-threshold solutions activate cross-modal correspondence between taste and shapes of cups. Food Qual Prefer 83:103891
111. Lick E, König B, Kpossa MR, Buller V (2017) Sensory expectations generated by colours of red wine labels. J Retail Consum Serv 37: 146–158
112. Riofrio-Grijalva R, Lago M, Fabregat-Amich-P, Guerrero J, Cuesta A, Vázquez-Araújo L (2020) Relationship between tactile stimuli and basic tastes: CATA with consumers with visual disability. J Sens Stud 35(1):e12549
113. Koc E, Boz H (2018) How can consumer science be used for gaining information about consumers and the market?: the role of psychophysiological and neuromarketing

research. In: Cavicchi A, Santini C (eds) Case studies in the traditional food sector. Woodhead Publishing, Cambridge, UK, pp 129–152

114. Greenwald MK, Cook EW, Lang PJ (1989) Affective judgment and psychophysiological response: dimensional covariation in the evaluation of pictorial stimuli. J Psychophysiol 3: 51–64

115. Bensafi M, Rouby C, Farget V, Bertrand B, Vigouroux M, Holley A (2002) Autonomic nervous system responses to odours: the role of pleasantness and arousal. Chem Senses 27(8):703–709

116. Rousmans S, Robin O, Dittmar A, Vernet-Maury E (2000) Autonomic nervous system responses associated with primary tastes. Chem Senses 25(6):709–718

117. Kantono K, Hamid N, Shepherd D, Lin YHT, Skiredj S, Carr BT (2019) Emotional and electrophysiological measures correlate to flavour perception in the presence of music. Physiol Behav 199:154–164

118. Xu Y, Hamid N, Shepherd D, Kantono K, Reay S, Martinez G, Spence C (2019) Background soundscapes influence the perception of ice-cream as indexed by electrophysiological measures. Food Res Int 125:108564

119. Verastegui-Tena L, van Trijp H, Piqueras-Fiszman B (2019) Heart rate, skin conductance, and explicit responses to juice samples with varying levels of expectation (dis) confirmation. Food Qual Prefer 71:320–331

120. Katsuura T, Jin X, Baba Y, Shimomura Y, Iwanaga K (2005) Effects of color temperature of illumination on physiological functions. J Physiol Anthropol Appl Hum Sci 24(4):321–325

121. Spence C (2011) Mouth-watering: the influence of environmental and cognitive factors on salivation and gustatory/flavor perception. J Texture Stud 42(2):157–171

122. Duerrschmid K, Danner L (2018) Eye tracking in consumer research. In: Ares G, Varela P (eds) Methods in consumer research, vol 2. Woodhead Publishing, Cambridge, UK, pp 279–318

123. Motoki K, Saito T, Onuma T (2021) Eye-tracking research on sensory and consumer science: a review, pitfalls and future directions. Food Res Int 145:110389

124. Seo H-S, Roidl E, Müller F, Negoias S (2010) Odors enhance visual attention to congruent objects. Appetite 54(3):544–549

125. Peng-Li D, Byrne DV, Chan RC, Wang QJ (2020) The influence of taste-congruent soundtracks on visual attention and food choice: a cross-cultural eye-tracking study in Chinese and Danish consumers. Food Qual Prefer 85:103962

126. Yeoh JP, North AC (2010) The effect of musical fit on consumers' preferences between competing foods limitation. Music Sci 14:165–180

127. Knoeferle K, Knoeferle P, Velasco C, Spence C (2016) Multisensory brand search: how the meaning of sounds guides consumers' visual attention. J Exp Psychol Appl 22:196–210

128. Huang J, Peng Y, Wan X (2021) The color-flavor incongruency effect in visual search for food labels: an eye-tracking study. Food Qual Prefer 88:104078

129. Bercik J, Paluchova J, Neomaniova K (2021) Neurogastronomy as a tool for evaluating emotions and visual preferences of selected food served in different ways. Foods 10(2): 354

130. de Wijk RA, Ushiama S, Ummels M, Zimmerman P, Kaneko D, Vingerhoeds MH (2021) Reading food experiences from the face: effects of familiarity and branding of soy sauce on facial expressions and video-based RPPG heart rate. Foods 10(6):1345

131. Ares G, Deliza R (2010) Studying the influence of package shape and colour on consumer expectations of milk desserts using word association and conjoint analysis. Food Qual Prefer 21(8):930–937

132. Piqueras-Fiszman B, Velasco C, Salgado-Montejo A, Spence C (2013) Using combined eye tracking and word association in order to assess novel packaging solutions: a case study involving jam jars. Food Qual Prefer 28(1):328–338

133. Greenwald AG, Nosek BA, Banaji MR (2003) Understanding and using the implicit association test: I. An improved scoring algorithm. J Pers Soc Psychol 85(2):197–216

134. Kraus AA, Piqueras-Fiszman B (2018) Measuring implicit associations in food-related consumer research. In: Ares G, Varela P (eds) Methods in consumer research, vol 2. Woodhead Publishing, Cambridge, UK, pp 203–230

135. Crisinel A-S, Spence C (2009) Implicit association between basic tastes and pitch. Neurosci Lett 464(1):39–42

136. Crisinel A-S, Spence C (2010) A sweet sound? Food names reveal implicit associations between taste and pitch. Perception 39(3): 417–425

137. Piqueras-Fiszman B, Kraus A, Spence C (2014) "Yummy" versus "yucky"! Explicit and implicit approach—avoidance

motivations toward appealing and disgusting foods in normal eaters. Appetite 78:193–202
138. Hayes JE, Allen AL, Bennett SM (2013) Direct comparison of the generalized visual analog scale (gVAS) and general labeled magnitude scale (gLMS). Food Qual Prefer 28(1): 36–44
139. Carvalho FM, Spence C (2018) The shape of the cup influences aroma, taste, and hedonic judgements of specialty coffee. Food Qual Prefer 68:315–321
140. Carvalho FM, Spence C (2019) Cup colour influences consumers' expectations and experience on tasting specialty coffee. Food Qual Prefer 75:157–169
141. Dalenberg JR, Hoogeveen HR, Lorist MM (2018) Physiological measurements: EEG and fMRI. In: Ares G, Varela P (eds) Methods in consumer research, vol 2. Woodhead Publishing, Cambridge, UK, pp 253–277
142. Spence C (2019) Neuroscience-inspired design: from academic neuromarketing to commercially relevant research. Organ Res Methods 22(1):275–298
143. Shepherd GM (2012) Neurogastronomy: how the brain creates flavor and why it matters. Columbia University Press, New York
144. van Atteveldt N, van Kesteren MT, Braams B, Krabbendam L (2018) Neuroimaging of learning and development: improving ecological validity. Frontline Learn Res 6(3): 186–203
145. Barros-Loscertales A, González J, Pulvermüller F, Ventura-Campos N, Bustamante JC, Costumero V et al (2012) Reading salt activates gustatory brain regions: fMRI evidence for semantic grounding in a novel sensory modality. Cereb Cortex 22(11): 2554–2563
146. González J, Barrós-Loscertales A, Pulvermüller F, Meseguer V, Sanjuán A, Belloch V, Ávila C (2006) Reading cinnamon activates olfactory brain regions. NeuroImage 15:906–912
147. Woods AT, Lloyd DM, Kuenzel J, Poliakoff E, Dijksterhuis GB, Thomas A (2011) Expected taste intensity affects response to sweet drinks in primary taste cortex. Neuroreport 22(8):365–369
148. Nitschke JB, Dixon GE, Sarinopoulos I, Short SJ, Cohen JD, Smith EE, Kosslyn SM, Rose RM, Davidson RJ (2006) Altering expectancy dampens neural response to aversive taste in primary taste cortex. Nat Neurosci 9:435–442
149. Wilton M, Stancak A, Giesbrecht T, Thomas A, Kirkham T (2018) Intensity expectation modifies gustatory evoked potentials to sweet taste: evidence for bidirectional assimilation in early perceptual processing. Psychophysiology 56(3):e13299
150. Avery JA, Liu AG, Ingeholm JE, Gotts SJ, Martin A (2021) Viewing images of foods evokes taste quality-specific activity in gustatory insular cortex. Proc Natl Acad Sci 118(2):e2010932118
151. Österbauer RA, Matthews PM, Jenkinson M, Beckmann CF, Hansen PC, Calvert GA (2005) Color of scents: chromatic stimuli modulate odor responses in the human brain. J Neurophysiol 93:3434–3441
152. Spence C (2020) Olfactory-colour crossmodal correspondences in art, science, & design. Cogn Res Princ Implic 5:52
153. Seo H-S, Arshamian A, Schemmer K, Scheer I, Sander T, Ritter G, Hummel T (2010) Cross-modal integration between odors and abstract symbols. Neurosci Lett 478(3):175–178
154. Giboreau A (2018) Situational factors and the design of in situ evaluations. In: Ares G, Varela P (eds) Methods in consumer research, vol 2. Woodhead Publishing, Cambridge, UK, pp 109–116
155. Risso P, Maggioni E, Etzi R, Gallace A (2019) The effect of the tactile attributes of a container on mineral water perception. Beverages 5(1):23
156. van Rompay TJL, Finger F, Saakes D, Fenko A (2017) "See me, feel me": effects of 3D-printed surface patterns on beverage evaluation. Food Qual Prefer 62:332–339
157. Gacula M Jr, Rutenbeck S (2006) Sample size in consumer test and descriptive analysis. J Sens Stud 21(2):129–145
158. Westfall J, Kenny DA, Judd CM (2014) Statistical power and optimal design in experiments in which samples of participants respond to samples of stimuli. J Exp Psychol 143(5):2020
159. Kaplan RM, Chambers DA, Glasgow RE (2014) Big data and large sample size: a cautionary note on the potential for bias. Clin Transl Sci 7(4):342–346
160. Lyman B (1989) A psychology of food, more than a matter of taste. Avi, van Nostrand Reinhold, New York
161. Wilson TD, Schooler JW (1991) Thinking too much: introspection can reduce the quality of preferences and decisions. J Pers Soc Psychol 60:181–192
162. Clark CC, Lawless HT (1994) Limiting response alternatives in time-intensity scaling:

an examination of the halo-dumping effect. Chem Senses 19(6):583–594

163. Spence C, Carvalho FM (2020) The coffee drinking experience: product extrinsic (atmospheric) influences on taste and choice. Food Qual Prefer 80:103802
164. de Luca P, Pegan G (2014) The coffee shop and customer experience: a study of the US market. In: Musso F, Druica E (eds) Handbook of research on retailer-consumer relationship development. IGI Global, Hershey, pp 173–196
165. Spence C (2021) Sonic seasoning and other multisensory influences on the coffee drinking experience. Fron Comp Sci 3:644054. https://doi.org/10.3389/fcomp.2021.644054
166. Schifferstein HN (2021) Designing food experiences: a multisensory approach. In: Massari S (ed) Transdisciplinary case studies on design for food and sustainability. Woodhead Publishing, Cambridge, UK, pp 115–130
167. Reber R, Winkielman P, Schwarz N (1998) Effects of perceptual fluency on affective judgments. Psychol Sci 9(1):45–48
168. Spence C, Velasco C (2019) Packaging colour and its multiple roles. In: Velasco C, Spence C (eds) Multisensory packaging. Springer, Cham, pp 21–48
169. Shankar MU, Levitan C, Spence C (2010) Grape expectations: the role of cognitive influences in color-flavor interactions. Conscious Cogn 19:380–390
170. Jaeger SR, Fiszman S, Reis F, Chheang SL, Kam K, Pineau B et al (2017) Influence of evoked contexts on hedonic product discrimination and sensory characterizations using CATA questions. Food Qual Prefer 56:138–148
171. Haindl M, Duerrschmid K (2010) Food attractiveness and gazing behaviour. In: Spink AJ, Grieco F, Krips OE, Loijens LWS, Noldus LPJJ, Zimmerman PH (eds) Proceeding of measuring behaviour. Eindhoven, The Netherlands, pp 419–421
172. Jantathai S, Danner L, Joechl M, Dürrschmid K (2013) Gazing behavior, choice and color of food: does gazing behavior predict choice? Food Res Int 54(2):1621–1626
173. van der Laan LN, Hooge ITC, de Ridder DTD, Viergever MA, Smeets PAM (2015) Do you like what you see? The role of first fixation and total fixation duration in consumer choice. Food Qual Prefer 39:46–55
174. Charness G, Gneezy U, Kuhn MA (2012) Experimental methods: between-subject and within-subject design. J Econ Behav Organ 81(1):1–8
175. Wakeling IN, Buck D (2001) Balanced incomplete block designs useful for consumer experimentation. Food Qual Prefer 12(4): 265–268

Chapter 11

Doing (Consumer) Research with Children: What's New?

Sandrine Monnery-Patris and Kaat Philippe

Abstract

This chapter focuses on the use of alternative methods to more properly allow children's voices to be heard in consumer studies as well as in behavioral or social sciences. It is still a challenge to sensorialists and behavioral scientists to consider different ways of engaging children in their protocol and to involve them as active participants. The present chapter aims to provide examples for both academics and practitioners wishing to engage children in their studies. By describing the different steps of the creation of a protocol designed to target attitudes towards foods in children and by highlighting the added value of engaging them as experiment designers, this chapter ought to offer a cornerstone for new approaches in the understanding of the determinants of eating behaviors and food-related decision-making.

Key words Consumer research, Children, Method, Attitudes, Engaging protocol, Co-creation

1 Parents' and/or Children's Prism

With growth, children's cognitive abilities as well as linguistic capacities are constantly changing. So, involving the child in traditional sensory and consumer science while the child is growing is quite challenging, both from scientific and methodological points of view.

A classical way to challenge these difficulties is to involve the parents instead of the children in studies and to directly ask them to fill in questionnaires, diaries, or scales about their child's everyday food consumption practices and preferences. Since most research involving parents does in fact address mothers [1], the results may in turn not reflect the child's answers but the maternal perceptions of children's behaviors. Moreover, it has been shown that agreement between parent and child reports may vary according to the tools used. On the one hand, previous studies have shown that parents may quite accurately predict their children's lunch choices from actual menus [2], making their predictions by using specific knowledge about their child's likes but also by projecting their own

Carlos Gómez-Corona and Heber Rodrigues (eds.), *Consumer Research Methods in Food Science*,
Methods and Protocols in Food Science, https://doi.org/10.1007/978-1-0716-3000-6_11,

preferences. On the other hand, some studies have reported that parent and child reports may not completely overlap: for instance, it has been shown that young girls' perceptions of parental pressure-to-eat showed a good concordance with parents' reports, but parent and child reports of restriction were not significantly related [3]. One could argue that a social desirability bias could operate differently for these two practices and that parents may have been more reluctant to report frequent use of restrictive feeding practices, whereas they may have interpreted the pressure-to-eat practices as an opportunity to encourage child intake of healthy foods. In other words, feeding a child is a very sensitive issue, so parents may be reluctant to report socially unfavorable practices or behaviors and, hence, may answer to strongly conform to social expectations by overreporting more desirable answers [4].

Interestingly, some authors have specifically assessed the links between parental feeding practices and children's eating behaviors through the prism of the child. To do so, adapted versions of the Child Feeding Questionnaire have been developed [5–7]. Originally, this is a parent-report questionnaire. In order to reflect children's experiences, child-report ("Kids") versions have been developed for school-aged children. In a study involving French children aged 9–11 years [8], the authors confirmed that children are aware of parental feeding practices aimed at restricting and/or pressuring their intake. Based on their results, they assumed that considering child perceptions could be a window of opportunity for limiting social desirability when reporting on the use of parental feeding practices.

However, children's responses may also reflect biases. A way to challenge or limit these biases is to favor indirect protocols with children, using, for instance, priming methods or assessing attitudes towards foods (*see,* for instance, [9]). It is well known that food choice decision-making in adults as well as in children is impacted by both conscious (explicit) and non-conscious (implicit) factors [10–12]. Since direct tasks are assumed to be more sensitive to the participants' normative responses and to social norms [13, 14], indirect protocols constitute interesting alternatives to more properly target the influence of underlying implicit knowledge and attitudes on eating behaviors and thus to limit the impact of social desirability. For example, a study examined whether overweight youngsters like food more than lean peers by measuring their implicit attitudes with an Implicit Association Task (IAT) [15]. In this study, the authors have adapted the IAT initially developed by Greenwald et al. [16]. The IAT may be defined as a response time-based computer task in which items/pictures/words have to be categorized according to one of four categories: two target categories (e.g., flowers and insects) and two attribute categories (e.g., positive and negative nouns). In this task, sorting has to be realized as quickly and accurately as possible, and each response is attributed to one target and one attribute category. It is assumed that when the same response is used for targets and attributes that are associated in memory (e.g., flowers/positive versus insects/negative), response time is faster than when the same response is used for targets and attributes that are unrelated (e.g., flowers/negative versus insects/positive). According to the authors, this facilitation effect is thought to reflect an implicit and automatic processing. Using a personalized version of the IAT (with 'I like' and 'I dislike' instead of 'positive' and 'negative' as attributes), Craeynest et al. [15] have namely tested whether a preference for palatable unhealthy food over palatable healthy food was revealed in youngsters with overweight than in lean controls. This test was selected as it is assumed to more properly assess personal implicit attitudes than societal views in children. The results finally did not reveal the group's differences. This result has been confirmed recently in a study involving French children [17]: the authors observed that French children with overweight have implicit hedonic-based attitudes towards food similar to those of normal-weight French children, but they have more explicit nutrition-based attitudes. Moreover, combining implicit and explicit attitudes, they revealed a discrepancy between hedonic-based implicit and nutrition-based explicit attitudes more common in children with overweight than in those with a normal weight.

Using child-report measures and including children as participants in studies is a first step for involving children and obtaining insight into their point of view, their preferences, and behaviors. However, more can be done and achieved, for example, by involving children very early in the creation of the design of an experiment or food-product development [18, 19]. In the following paragraphs, we will detail two studies (an experimental study and an interventional one) relying on implicit and explicit tasks involving children in participatory co-creation initiatives with gamification items to more properly engage them in the research and to gain valuable insight into the predictors of eating behaviors, as well as to favor a pleasurable experience for the child.

2 Children as Experiment Designers

Attitudes are important precursors of eating behaviors and food choices. In order to assess children's attitudes (i.e., hedonic- and nutrition-based) towards foods, a new protocol has been developed relying on two tasks: an implicit task and explicit task [20], following different steps as described below.

2.1 Co-creation of the Implicit Pairing Task

The implicit pairing task was inspired by Rozin et al. [21]'s task, in which the adult participant associates a food term preferentially with a culinary term (e.g., bread–butter) or with a nutritional term (e.g., bread–carbohydrate) by circling the word that he/she believes is the most different from the other two. To adapt this task to the cognitive abilities of 5- to 11-year-old children, individual co-creation sessions have taken place, including two different steps.

In the first step, children were invited to describe what they saw on pictures of foods (instead of words as in Rozin's study) and to share their feelings towards them. They were invited to freely categorize and/or classify pictures if they wanted. No instructions were given in this step except "don't hesitate to say what comes to mind while seeing these pictures." This step, named *"children's voices,"* was designed to mainly select the highly familiar and inspiring foods' pictures for the child whatever her/his age.

Then, in a second step, *"through the eyes of the child,"* the children were invited to help the researchers to select triplets of pictures that made sense to them and also to express their understanding of the pairing task. It appeared that it was easier for them to choose the two foods that *"go best together"* among the different triplets rather than to point at the food they believed to be the most different, as in the Rozin et al.'s task. The spontaneous verbatims of the children as well as their preference for the way in which they would like to carry out the task were selected in the instruction of the final version of the task. Besides, their spontaneous verbatims were recorded after each pairing to check the motivations underlying each association and to make sure that the associations such as Nutella-butter or peas-beans (coded as nutritional association) were justified by nutritional consideration (for instance, "it is fat," "there are vegetables"); or that the pairings such as bread-jam or pasta-sauce (coded as hedonic association) were motivated by culinary consideration (e.g., "I like to put jam on bread," "I prefer pasta with sauce"). These co-creation steps finally allowed us to select 11 triplets of three pictures of foods for the pairing task (*see* Fig. 1) and to adapt to the instructions.

Fig. 1 Example of a triplet selected in the pairing task [20]

2.2 Co-creation of the Explicit Categorization Task

An explicit categorization task inspired by Berger's analyses of children's categorization tools [22] has also been adapted. The pictures of highly identifiable foods were selected during the first step described above (i.e., *"children's voices"*). Then, "*through the eyes of the child*," the children were invited to help the researchers to select different pictures to co-create the categories spontaneously made and named by them. This resulted in the spontaneous emergence of five categories (e.g., *"yummy," "yucky," "makes you strong," "makes you fat,"* and *"I don't know this food"*). Interestingly, the researchers involved hoped that a category like *"good for health" or "good for the planet"* would spontaneously emerge during this step but this was not the case, so these categories were simply not retained in the final version of the task. Notably, each dimension (hedonic, nutritional) paralleled two a priori opposite sides (respectively, yummy/yucky; strong/fat). This co-creation protocol allowed the researchers to finally select 49 pictures of foods to create a new task with the children and to adapt the instructions to the children's age and living environment (Fig. 2).

To sum up, this experience of co-creation of two adapted tasks allowed the researchers to validate new implicit and explicit tasks (respectively, the implicit pairing task and the explicit categorization task) suitable to children's abilities and to understand more properly the developmental trajectory of hedonic and nutritional attitudes in children. Besides, complementary researches allowed them to disentangle how these attitudes may predict food choices in children, showing that children with more hedonic-based implicit/explicit attitudes toward food were more likely to choose healthy food options from a buffet than children with nutrition-based attitudes [11]. Finally, these results suggest that pleasure from eating might not be an enemy but an ally with regard to healthy eating among children. Promising insights have emerged in children's attitudes and values, demonstrating that emphasizing healthiness per se may have counterproductive effects on food choices for the child. Given the increase in childhood obesity around the world [23], interventions relying on the pleasure to eat (healthy) foods appeared warranted.

Fig. 2 Illustration of the pictures for the categories selected in the categorization task (e.g., "yummy," "yucky," for the hedonic categorization, "makes you strong," "makes you fat," for the nutritional categorization, and "I don't know this food") from Ref. [20]

3 Children as Intervention Designers

A recent study assessed the impact of an in-home pleasure-oriented intervention on the nutritional quality of mid-afternoon snacks chosen by mother-child dyads [24]. To develop the intervention, four separate focus groups (two with five mothers—not described here—and two with five children) had been conducted to co-create the content of the intervention. The main idea of the intervention was to stimulate the different dimensions (sensory, cognitive, and social) of pleasure of eating healthy foods [25] without emphasizing healthiness. To do so, children were actively involved as co-creators. During the focus groups, different foods, pictures of foods, games, kitchen utensils, and infographics were placed on the table. Children were invited to "*help the researchers to find a good way to help children to eat fruits/cereals/dairy products for the mid-afternoon snack.*" They were told to "*imagine that your friend does not like these foods, what would you suggest to help her/him to taste them?*" Here the children were encouraged to play, taste, manipulate, and evaluate, for instance, the convenient and fun aspects of the utensils, to share their opinions regarding the foods and how they appreciate (or not) to eat them (recipe, context, etc.).

Based on the results of the focus groups, the authors created an intervention relying on three "yummy boxes" containing different tools to stimulate the pleasure of consuming healthy foods. Specifically, the box contained a card about the five senses, which targeted the sensory dimension by describing with specific vocabulary the sensations and feelings experienced through the different senses when consuming fruits, cereal products, or dairy products, mainly based on the verbatim of the children. The box also contained a kitchen utensil (an apple peeler, a set of bamboo picks to make fruit skewers, molds for cereal bars, etc.), a recipe card (for instance, a banana-apple pie recipe), and a culinary challenge highly appreciated by the children. The recipes targeted the sensory dimension of pleasure from eating, the culinary challenge invited the participants to make a recipe for their mid-afternoon snacks with the kitchen utensils present in the box and to post a picture of the result on a dedicated blog. By involving the child/parent in a common activity, which acted as a source of social interaction, the culinary challenge focused on the social dimension of pleasure from eating. Finally, the box also contained infographics about the history and origin of fruits/cereals/dairy products, quiz, and crossword puzzle or labyrinth inspired by the children. This targeted the cognitive dimension of pleasure from eating, with the aim to modify their representations and thus to increase their attraction to healthy products.

Interestingly, the results of this interventional study revealed a significant increase in positive attitudes towards healthy foods, which may constitute the first step in a behavioral change in favor of healthier choices in children [24, 25].

4 Conclusions and Recommendations for Future Research

Knowing that one in five children is living with overweight or obesity in the world, the use of participative sciences makes sense to more properly assess the determinants of food choices in children and to really allow children's voices to be heard. Alternative protocols engaging children as co-designers may offer promising insights, both in terms of methods as well as concepts actually relevant for them. Co-creation with children can also be used to develop actionable ideas and prototypes of new food products that are healthy and well liked [19].

Overall, it is highly recommended to involve children from the start when conceptualizing an experiment or developing new food products for children. Children have different food preferences than adults (e.g., [26]) and their food choices are more strongly determined by food pleasure than adults [27, 28]. From their own experiences with food and their preferences, they can help researchers to use methods and terminology that match these experiences and perceptions. They can also help to generate original ideas for new, healthier food products or interventions to stimulate healthy and happy eating in children in the short and long term.

When co-creating with children, it is important to take into account their developmental and linguistic skills. More limited skills may call for the need of more non-verbal and simple co-creation activities and a stronger involvement of a researcher or moderator. The older and more skilled the children are, the more autonomously they can contribute (e.g., writing down their ideas, using more advanced cooking skills) [29]. Gamification of the tasks (e.g., using a story, rewards, fun images, . . .) is also highly recommended to actively engage and motivate children to participate [30–32]. It also helps to make them experience the tasks as pleasurable.

Depending on the aims of the co-creation activities with children, it can be preferable to interact with children one-on-one or to involve them in group sessions with multiple children. If opting for group sessions, it may be interesting to select more heterogenous groups of children in order to generate more diverse ideas [33]. However, it may also be beneficial to select children that know each other, because their group cohesion can facilitate the group dynamics and their active participation in the co-creation sessions [19].

Since research is still limited with regard to these alternative protocols engaging children as co-designers in the food domain, future research is recommended for developing and evaluating productive techniques for co-creation with children.

References

1. Philippe K, Chabanet C, Issanchou S, Monnery-Patris S (2021) Are food parenting practices gendered? Impact of mothers' and fathers' practices on their child's eating behaviors. Appetite 166:105433. https://doi.org/10.1016/j.appet.2021.105433
2. Mata J, Scheibehenne B, Todd PM (2008) Predicting children's meal preferences: how much do parents know? Appetite 50(2–3): 367–375, ISSN 0195-6663. https://doi.org/10.1016/j.appet.2007.09.001
3. Carper JL, Orlet Fischer J, Birch LL (2000) Young girls' emerging dietary restraint and disinhibition are related to parental control in child feeding. Appetite 35(2):121–129. https://doi.org/10.1006/appe.2000.0343
4. Kröller K, Warschburger P (2008) Associations between maternal feeding style and food intake of children with a higher risk for overweight. Appetite 51(1):166–172. https://doi.org/10.1016/j.appet.2008.01.012
5. Birch LL, Fisher JO (2000) Mothers' child-feeding practices influence daughters' eating and weight. Am J Clin Nutr 71(5): 1054–1061. https://doi.org/10.1093/ajcn/71.5.1054
6. Birch LL, Fisher JO, Davison KK (2003) Learning to overeat. Maternal use of restrictive feeding practices promotes girls' eating in the absence of hunger. Am J Clin Nutr 78(2): 215–220. https://doi.org/10.1093/ajcn/78.2.215
7. Carper JL, Fisher JO, Birch LL (2000) Young girls' emerging dietary restraint and disinhibition are related to parental control in child feeding. Appetite 35(2):121–129. https://doi.org/10.1006/appe.2000.0343
8. Monnery-Patris S, Rigal N, Chabanet C, Boggio V, Lange C, Cassuto DA, Issanchou S (2011) Parental practices perceived by children using a French version of the Kids' Child Feeding Questionnaire. Appetite 57(1):161–166. https://doi.org/10.1016/j.appet.2011.04.014
9. Monnery-Patris S, Chambaron S (2020) Added-value of indirect methods to assess the relationship between implicit memory and food choices in adult consumers as well as in children. Curr Opin Food Sci 33:14–20. https://doi.org/10.1016/j.cofs.2019.10.004
10. Craeynest M, Crombez G, Koster EH, Haerens L, De Bourdeaudhuij I (2008) Cognitive-motivational determinants of fat food consumption in overweight and obese youngsters: the implicit association between fat food and arousal. J Behav Ther Exp Psychiatry 39(3):354–368. https://doi.org/10.1016/j.jbtep.2007.09.002
11. Marty L, Miguet M, Bournez M, Nicklaus S, Chambaron S, Monnery-Patris S (2017) Do hedonic-versus nutrition-based attitudes toward food predict food choices? A cross-sectional study of 6-to 11-year-olds. Int J Behav Nutr Phys Act 14(1):1–10. https://doi.org/10.1186/s12966-017-0618-4
12. Papies EK, Potjes I, Keesman M, Schwinghammer S, Van Koningsbruggen GM (2014) Using health primes to reduce unhealthy snack purchases among overweight consumers in a grocery store. Int J Obes 38(4): 597–602. https://doi.org/10.1038/ijo.2013.136
13. Higgs S (2015) Social norms and their influence on eating behaviours. Appetite 86:38–44. https://doi.org/10.1016/j.appet.2014.10.021
14. Robinson E, Blissett J, Higgs S (2013) Social influences on eating: implications for nutritional interventions. Nutr Res Rev 26(2): 166–176. https://doi.org/10.1017/S0954422413000127
15. Craeynest M, Crombez G, Haerens L, De Bourdeaudhuij I (2007) Do overweight youngsters like food more than lean peers? Assessing their implicit attitudes with a personalized implicit association task. Food Qual Prefer 18(8):1077–1084
16. Greenwald AG, McGhee DE, J.L. (1998) Schwartz measuring individual differences in implicit cognition: the implicit association test. J Pers Soc Psychol 74:1464–1480
17. Marty L, Chambaron S, Bournez M, Nicklaus S, Monnery-Patris S (2017) Comparison of implicit and explicit attitudes towards food between normal-and overweight French children. Food Qual Prefer 60:145–153.

https://doi.org/10.1016/j.foodqual.2017.04.013

18. Galler M, Gonera A, Varela P (2020) Children as food designers: the potential of co-creation to make the healthy choice the preferred one. Int J Food Des 5(1–2):125–131(7). https://doi.org/10.1386/ijfd_00015_3
19. Velázquez AL, Galler M, Vidal L, Varela P, Ares G (2022) Co-creation of a healthy dairy product with and for children. Food Qual Prefer 96:104414. https://doi.org/10.1016/j.foodqual.2021.104414
20. Monnery-Patris S, Marty L, Bayer F, Nicklaus S, Chambaron S (2016) Explicit and implicit tasks for assessing hedonic-versus nutrition-based attitudes towards food in French children. Appetite 96:580–587. https://doi.org/10.1016/j.appet.2015.10.026
21. Rozin P, Fischler C, Imada S, Sarubin A, Wrzesniewski A (1999) Attitudes to food and the role of food in life in the USA, Japan, Flemish Belgium and France: possible implications for the diet–health debate. Appetite 33(2):163–180. https://doi.org/10.1006/appe.1999.0244
22. Berger C (1997) Catégorisation, Formation de concepts et induction: Rôle des informations perceptives et conceptuelles chez le jeune enfant. Annee Psychol 97:495–517
23. World Health Organization (2018) Taking action on childhood obesity (No. WHO/NMH/PND/ECHO/18.1). World Health Organization
24. Poquet D, Ginon E, Sénécal C, Chabanet C, Marette S, Issanchou S, Monnery-Patris S (2020) Effect of a pleasure-oriented intervention on the nutritional quality of midafternoon snacks and on the relationship between food liking and perceived healthiness within mother-child dyads. Food Qual Prefer 84:103947. https://doi.org/10.1016/j.foodqual.2020.103947
25. Poquet D, Ginon E, Marty S, Franzon C, Chabanet C, Issanchou S, Monnery-Patris S (2022) Effect of a pleasure-oriented intervention conducted at home on the energy intake of mid-afternoon snacks consumed by children. https://doi.org/10.1086/720452
26. Forestell CA, Mennella JA (2015) The ontogeny of taste perception and preference throughout childhood. In: Doty RL (ed) Handbook of olfaction and gustation. Wiley, Hoboken, pp 795–828
27. Marty L, Chambaron S, Nicklaus S, Monnery-Patris S (2018) Learned pleasure from eating: an opportunity to promote healthy eating in children? Appetite 120:265–274. https://doi.org/10.1016/j.appet.2017.09.006
28. Nicklaus S (2016) The role of food experiences during early childhood in food pleasure learning. Appetite 104:3–9. https://doi.org/10.1016/j.appet.2015.08.022
29. Dean M, O'Kane C, Issartel J, McCloat A, Mooney E, Gaul D et al (2021) Guidelines for designing age-appropriate cooking interventions for children: the development of evidence-based cooking skill recommendations for children, using a multidisciplinary approach. Appetite 161:105125. https://doi.org/10.1016/j.appet.2021.105125
30. Brewer R, Anthony L, Brown Q, Irwin G, Nias J, Tate B (2013) Using gamification to motivate children to complete empirical studies in lab environments. In: Proceedings of the 12th international conference on interaction design and children, pp 388–391. https://doi.org/10.1145/2485760.2485816
31. Chitakunye P (2012) Recovering children's voices in consumer research. Qual Mark Res 15(2):206–224. https://doi.org/10.1108/13522751211215903
32. Chou YK (2019) Actionable gamification: beyond points, badges, and leaderboards. Packt Publishing Ltd
33. Van Mechelen M, Gielen M, vanden Abeele V, Laenen A, Zaman B (2014) Exploring challenging group dynamics in participatory design with children. In: Proceedings of the 2014 conference on interaction design and children, pp 269–272. https://doi.org/10.1145/2593968.2610469

Chapter 12

Creativity on Our Plates: Measuring Creativity in Cooking

Nicolas Pichot and Eric Bonetto

> Some people like to paint pictures, or do gardening, or build a boat in the basement. Other people get a tremendous pleasure out of the kitchen, because cooking is just as creative and imaginative an activity as drawing, or wood carving, or music.
> Julia Child, cooking teacher and cookbook author

Abstract

Cooking is considered a creative activity. We observe creativity not only in the kitchens of Michelin- starred chefs known for their innovations but also in the kitchens of ordinary people in everyday life. The purpose of this chapter is to discuss ways to measure creativity in cooking. Three main questions will be addressed. First, the "*What?*" question, that is the question of what we intend to measure when it comes to investigating creativity (i.e., what is creativity?). Second, we will address the "*Why?*" question, focusing on why it is interesting to measure creativity in cooking. Third, and finally, we will address the "*How?*" question by presenting a method for measuring creativity in the kitchen. Thus, the evaluation of the creative character of culinary productions can be carried out in a more precise way by drawing on theories and methodologies that have been developed in the scientific field of creativity.

Key words Creativity, Innovation, Food, Measure

Creativity research has been particularly flourishing over the last decades [1, 2] and has shown important applications in various fields (e.g., education, arts, sciences, and technology) and society as a whole [3]. Numerous works have thus been conducted on creativity in many fields of research, among which the domain of cooking at different levels, from everyday cooking [4] to haute cuisine [5, 6].

There is no doubt that cooking is considered a creative activity. Indeed, this activity is often considered as a way to express one's own creativity [7]. An example of creativity in cooking is provided by Beghetto, Kaufman, and Hatcher: a portrait of Kevin Bacon made out of bacon [8]. Creativity in the kitchen can be found in haute cuisine and televised cooking shows (e.g., Iron Chef; [6]),

Carlos Gómez-Corona and Heber Rodrigues (eds.), *Consumer Research Methods in Food Science*,
Methods and Protocols in Food Science, https://doi.org/10.1007/978-1-0716-3000-6_12,

but also in everyday cooking. Indeed, even home chefs can use unusual combinations of ingredients to create unique dishes or new ways to cook classic dishes (e.g., cooking random foodstuffs found in the fridge; [8]). We thus observe creativity in the kitchens of Michelin-starred chefs known for their innovations, but also in the kitchen of ordinary people. As a consequence, various self-reported measures of creative behaviors include behaviors such as cooking an original dish or creating a new recipe, along with other creative behaviors in other domains (writing a poem or creating a sculpture; [9]).

The purpose of this chapter is to discuss ways to measure creativity in cooking. In order to do this, three main questions will be addressed. First, the "*What?*" question, that is the question of what we intend to measure when it comes to investigating creativity. This first question will bring us to define what we mean by "*creativity.*" These points related to the definition of creativity will illustrate the links between theory and method, as they have important consequences for the selection of variables to include in food studies when it comes to measuring creativity. Second, we will address the "*Why?*" question, focusing on why it is interesting to measure creativity in cooking. We will present some studies showing the relevance of evaluating creativity in cooking, especially when it comes to better understanding consumer reactions. Third, and finally, we will address the "*How?*" question by presenting the Consensual Assessment Technique, a method for measuring creativity in the kitchen.

1 What?

1.1 What Is This Thing Called Creativity?

A first question that arises concerns the nature of what we intend to measure: What is creativity? Researchers have struggled to reach a consensus on the definition of the concept of *creativity* and of what can (or cannot) be considered *creative* [10–12]. The "*standard definition*" of creativity [11]—the most consensual definition of creativity—postulates that to be creative a production has to be both original (novel, unusual, unique) and valuable (effective, feasible, appropriate to the task). If the *novelty* criterion may seem intuitively very related to creativity, the *value* criterion may seem questionable. The choice of this last criterion reflects the desire of researchers to draw up two distinct categories between what is simply new and what would be creative. Runco and Jaeger express this idea by pointing out that many random processes (e.g., automatic word or sentence generator, monkey typing) can be the source of very new things that have no use and/or are not efficient in fulfilling their intended functions.

The logical argument for including value in the standard definition has also been questioned in the scientific literature. Weisberg considers that the criterion of utility/value, because of its variable character in time (e.g., what was useful or had value in 1950 could have lost or gained value today), cannot be a constitutive criterion of creativity in that it creates logical paradoxes [13, 14]. Indeed, Weisberg indicates that in accepting the standard definition of creativity, we should also accept the idea that some artists, whose work was never recognized or even boycotted during their lifetime, might have become creative after their deaths, their work having found an audience and recognition posthumously (e.g., Van Gogh). Still with the objective of distinguishing what is new from what is creative without including a criterion of value, Weisberg proposes an alternative definition of creativity based on a criterion of intentionality [12, 13]. He states that to be considered creative, the production must be original and intentionally produced.

In the light of these definitional debates, and in line with Weisberg or Wimmer proposals [15, 16], creativity will be considered here as a special case of problem-solving activity [15]. In their most formal form, problem solving activities refer to the passage from an initial state to a final state that represents a goal to be achieved (i.e., X must be effective for Y; X must be desirable for Y). There can be as many goals as one wishes, and the value criterion simply represents the adequacy of the solution in view of the goals to be reached.

Creativity would thus be a specific type of problem in which the solutions must satisfy a certain number of criteria defined a priori by the researcher. This definition allows a certain flexibility in the definition (and thus the measurement) of creativity: while it invariably retains the criterion of novelty (the main criterion linked to creativity; [10]), the value can be defined in different ways according to the requirements of the considered domain. In other words, the value of a production will not have the same meaning (and will not be measured in the same way) in the domain of painting, in science, in engineering…, or in cooking. For instance, in a cooking competition in which the competitors have been asked to create a *trompe l'oeil*, the creativity of the dishes created will be defined according to their originality and their capacity to deceive the consumers on what they eat. In the same way, if asked to cook for children, the dishes will need to be original and appealing to children (e.g., have a fun character, do not include ingredients usually rejected by children).

It should be noted that it follows from these defining elements that a production process can be qualified as creative on the basis of the same criteria. In the same way, an individual will be judged creative if he or she has produced a creation that meets the defining criteria of creativity.

1.2 Defining Creativity in Cooking

Creativity thus refers to a special case of problem-solving situations in which solutions (i.e., creations) need to be new and valuable [11]. Consequently, in the domain of cooking, a dish must be both *new* and *valuable* to be considered creative. As we said previously, the value criterion needs to be defined according to the nature and constraints of the domain considered, in this case, the domain of cooking. In the cooking context, this dimension could be understood as the evaluation of the taste of the dish, as the quality of the execution of the dish, or as the respect of particular constraints such as the dietary aspect of the dish or the respect of particular food restrictions.

In addition, one might question the relative importance of novelty and value when it comes to defining—and thus measuring—creativity in cooking. In this regard, the distinction between functional and aesthetic creativity [17] offers such a weighting [8]. Indeed, aesthetic creativity has beauty or entertainment as its primary goal. This creativity emphasizes the originality of creations. This corresponds in particular to creativity in artistic fields, in which there is greater latitude in terms of value (despite specific rules defining an artistic genre, for example). Functional creativity, for its part, places the emphasis on value. In this type of creativity, the creation must of course be new, but it is crucial that its value be very high (i.e., that it effectively responds to the problem at hand). This is notably the case with eminent technological advances. This distinction is important to remember in the culinary field because the creator can emphasize aesthetics at the expense of value (a very beautiful dish, very well presented, but with an almost banal taste; i.e., aesthetic creativity), emphasize value at the expense of novelty (a very good dish presented in a banal way; i.e., functional creativity), or satisfy both types of creativity in an equivalent way (a very nice and very good dish; i.e., both aesthetic and functional creativity). Although in most cases, creative cooking seeks to satisfy both functional and aesthetic creativity, understanding this distinction and the balance between the two components of creativity can be very helpful when it comes to understanding the different ways that cooking can be creative [8]. This distinction will also be important in the construction of experimental designs and in the measurement of creativity in cooking (*see* point 3).

In a nutshell, all these points related to the definition of creativity—and more specifically, creativity in cooking—are of great importance when it comes to measuring creativity in cooking. Indeed, they indicate that before attempting to measure creativity in cooking, we must first define the criteria that will allow us to distinguish what is creative from what is not (what answers the creative problem and what does not). More specifically, researchers need to define (1) what will allow them to consider a dish or a process of creating a dish as new (new compared to what?) and valuable (according to what criteria? Taste? Compliance with specific instructions?) within the context of the study (haute cuisine or everyday cooking?). Moreover, researchers may also need to define (2) the expected balance between novelty and value, between pure aesthetic creativity and pure functional creativity. It should be noted here that these defining points apply not only to the dishes themselves (e.g., a new dish proposed on the menu of a restaurant) but also to the processes of their creation (e.g., a new cooking method) or to the tasting experience (tasting in a particular context, with staging, music, stimulation of senses).

2 Why?

Though research on the creative aspects of cooking remains relatively limited [7, 18], measuring creativity in this domain involves numerous issues.

The restaurant sector is a highly competitive environment: succeeding in this sector is a challenge, and business failures are frequent [19]. In this context, creativity and innovation represent increasingly important competitive advantages, playing an important role in the financial success of restaurants [18, 20]. Thus, the fierce competition in the restaurant industry is fueling the need for unique and creative products and experiences [21]. Yet, if creativity and innovation are generally considered as important for business success and consumer satisfaction, researchers need to know how to measure creativity in order to be able to promote it. This need to better understand creativity in cooking motivated—and still motivates—research on the consequences of creative food-related experiences on consumers.

For instance, Leong et al. diffused a questionnaire among the customers of a Michelin-starred restaurant in order to test the influence of cuisine creativity on consumers' pleasure (feeling relaxed, satisfied, happy), arousal (feeling excited, interested, surprised), and dominance emotions (feeling important, superior, autonomous) [21]. The authors used three components of cuisine creativity, namely "value" (whether the cuisine is worthwhile), "uniqueness/rarity" (including novelty, originality, thoughtfulness, difficult to imitate or substitute), and "aesthetics" (including elaboration, beauty, and elegance; [21]). Results showed that aesthetics (linked to value in the standard definition) emerged as the main contributor to pleasure, arousal, and dominance. For their part, uniqueness/rarity (i.e., novelty in the standard definition of creativity) and value only impacted pleasure and dominance. Thus, the different components of creativity showed specific influences on consumers' emotional states. It should be noted that other contributions showed that food's creative aspects tend to influence positive emotions but not negative emotions [19].

Research also investigated the effect of perceived product creativity on individuals' intentions to consume this product. For instance, Rubera, Ordanini, and Griffith showed that perceived novelty and meaningfulness (i.e., value) of drinks positively predicted consumers' intentions to buy these drinks in two different cultural contexts (Italy and the USA) [22]. Conversely, previous research also showed that when a product is considered "*too new,*" it can be rejected by users and consumers or be considered less useful [23–26].

These few examples of the implications of the research carried out in the field of creativity show the fundamental and applied interest of this research, which has important repercussions for the food industry. A better understanding of the impact of the creativity of food products on the evaluations and consumption choices of individuals makes it possible, for example, to identify determining factors of the success of chefs and restaurants, and of the development of the food industry as a whole.

3 How?

Measuring creativity is a real challenge. In its general sense, *measuring* refers to the "*assignment of numerals to objects or events according to rules*" [27] (p. 677). This definition was later clarified to indicate that it is not the events, objects, or persons that are measured but some of their properties. The question of the measurement of creativity thus simply refers to (1) determining the properties to which a number is to be assigned and (2) the way in which this number is to be assigned.

The consensual evaluation technique (CAT) proposed by Amabile allows us to meet these two constraints [28]. CAT is based on the idea that the best measure of the creative character of a production (an artistic work, a chef's dish, a scientific theory, or any other artifact) is using the combined evaluation of different experts in the domain in which the production occurs [29]. Indeed, unlike concepts such as weight or size, no device or tool can approximate and provide a totally objective measure of the originality, effectiveness or feasibility of a production. Thus, using different experts to assign a value to the different observable properties of the production is an appealing solution. This method is very similar to what is currently done in most prize committees (i.e., Fields Medal in mathematics) or in the research field through peer-reviewing.

However, there are some differences between the assessment procedure put forward by the CAT and the way that most creative activities are assessed in the scientific literature. The most important difference is that CAT is based on the so-called "*Consensual Definition of Creativity*" that considers creativity as a directly observable variable. It is thus somewhat opposed to the *Standard*

Definition of Creativity, which implicitly considers that creativity is only accessible through its components: novelty and value [11]. Thus, in its strict use, CAT implies *not* giving specific criteria to evaluate creativity and evaluating concepts such as technical quality or aesthetics as differentiated variables (for more details, *see* [28]). In fact, it is noteworthy that most of the measurement scales developed in recent years broadly follow the CAT procedure without, however, respecting the latter's instructions concerning the criteria (e.g., Dean et al., Horn and Salvendy, Pichot et al. [25, 30, 31]. Indeed, if one adheres to the hypothesis that creativity, as a property of a production, is an unobservable variable, it becomes necessary to develop a set of criteria that is sufficiently distinctive and measurable to capture this concept.

As we have seen in the section on "*What*," the first two components to be evaluated are novelty and value, and it is possible to add other supposedly independent components such as aesthetics, desirability, or even, in some contexts, feasibility [31]. Moreover, it is important to find different items that allow us to evaluate these components (i.e., the multiplication of measures allows us to approximate the real score of the production on the considered component). For example, it is possible to use synonymous items (i.e., original, unusual, unconventional) when evaluating the novelty component or to evaluate the novelty with respect to different aspects of the production. An example is proposed by Gero and Kannengiesser with the distinction between function, behavior, and structure [32]. Indeed, in the kitchen, a chef could very well try to do new things in terms of structure, "*What it is*" (e.g., trompe-l'oeil dishes; destructured—molecular cuisine), in terms of behavior, "*What it does*" (e.g., some desserts involve a procedure where the consumer interacts with a part of the dish to reveal a new part; consuming the dish in particular situations: in the dark, standing up, with the hands), or in terms of function, "*What it is for*" (e.g., satisfy the tastes of the people who taste a dish, be aesthetically pleasing). Long story short, the evaluation of the creativity of the productions will be all the more precise and reliable if the framework of expression and the criteria of evaluation have been thought out beforehand.

The last important point to develop concerning the evaluation of creative productions concerns the judges and their levels of expertise. This aspect of evaluation is at the very heart of the CAT whose use is based on the assumption that: "*it is possible to obtain reliable judgements of creativity of products, given an appropriate group of judges*" ([28], p. 1001). The CAT recommends the use of judges with significant knowledge of the domain, which is perfectly justified to meet some criteria such as novelty, technical quality, or feasibility. An accurate evaluation of these criteria calls for a comparison with the existing, established norms and rules in a specific field (i.e., functional approach; [17]). It is therefore necessary that the person carrying out the measurement has a thorough knowledge of the domain in which the production is expressed. However, certain criteria that call for a subjective appreciation (i.e., aesthetic approach; [17]) can also be studied and call for less or more restrictive categories of judges. Criteria related to *desirability*, *aesthetic appeal*, *beauty*, and *pleasantness* can also be part of the objectives to be reached for the production to be considered as creative. The assignment of a precise measure to these criteria does not require the use of objective knowledge of the domain. Their subjective nature leads judges to assign a score according to their own subjective experience. Thus they can provide additional information on the creative character of the production insofar as it represents a goal to be achieved).

It is definitely possible to evaluate what is on our plates based on the scores given by the judges on each criterion. However, there are several rules of thumb to consider in order to properly manipulate the data from the judges' evaluations. For example, the aggregation of the scores of each judge for the same item can only be done if their evaluations are generally consistent. It is therefore important to have a good inter-judge agreement if we want to average the scores obtained from the evaluations made. Therefore, a set of items aiming at evaluating the same dimension can be averaged in order to obtain a global score. However, it is necessary to ensure that all the items measure the same dimension by studying their reliability (i.e., Cronbach's alpha; McDonald's omega coefficient). We can thus measure and compare culinary productions on the basis of the scores obtained on the different dimensions studied separately.

But the most important questions that some readers might ask are: *Is it possible to have a general score of creativity of the production? How can it be obtained? What are the implications?* It is possible to create a creativity index by combining the different scores obtained on each dimension into an overall creativity score [31]. However, this approach requires a specific ontological position. Indeed, the creation of a creativity index only makes sense through a constructivist stance. Thus, creativity would represent a social construct or a pragmatic concept resulting from the aggregation of the scores obtained on each dimension. By adopting this ontological position, the general level of creativity is considered to be the cause of its underlying dimensions and does not constitute a real phenomenon. Thus, on the ontological level, we have here approached creativity from a constructivist point of view, considering that creativity is caused by the observed levels of effectiveness, novelty, desirability, etc. A competing approach, the so-called realist approach, would consider that it is creativity that causes the observed levels of the manifest variables (e.g., novelty, appropriateness, etc.). In this approach, it would only be possible to obtain general creativity scores and to consider creativity as a variable if there are substantial correlations between each supposed dimension of creativity. The interpretation of these general creativity scores and the comparison of items on the new variable thus created will depend essentially on the intensity and direction of the correlations obtained between the items.

To conclude, the evaluation of the creative character of culinary productions can be carried out in a more precise way by drawing on methodologies and research tools that have been developed in the scientific field of creativity. However, it should be noted that there are important conceptual, methodological, and ontological limitations. These limitations are not specific to the field of cooking but to the field of creativity as a whole. Indeed, as we mentioned at the beginning of this chapter, there are still fundamental debates about the very criteria defining the concept of creativity and consequently about the dimensions to be taken into consideration for its evaluation [10, 12].

References

1. Bonetto E, Pichot N, Girandola F, Bonnardel N (2021) The normative features of creativity: creative individuals are judged to be warmer and more competent. J Creat Behav 55:649–660
2. Qian M, Plucker JA, Yang X (2019) Is creativity domain specific or domain general? Evidence from multilevel explanatory item response theory models. Think Skills Creat 33:100571
3. Runco MA, Albert RS (2010) Creativity research: a historical view. Cambridge University Press, New York
4. McCabe M, de Waal Malefyt T (2015) Creativity and cooking: motherhood, agency and social change in everyday life. J Consum Cult 15:48–65
5. Albors-Garrigos J, Barreto V, García-Segovia P, Martínez-Monzó J, Hervás-Oliver JL (2013) Creativity and innovation patterns of haute cuisine chefs. J Culin Sci Technol 11:19–35
6. Bouty I, Gomez ML (2013) Creativity in haute cuisine: strategic knowledge and practice in gourmet kitchens. J Culin Sci Technol 11:80–95
7. Mosko JE, Delach MJ (2021) Cooking, creativity, and well-being: an integration of quantitative and qualitative methods. J Creat Behav 55:348–361
8. Beghetto RA, Kaufman JC, Hatcher R (2016) Applying creativity research to cooking. J Creat Behav 50:171–177
9. Diedrich J, Jauk E, Silvia PJ, Gredlein JM, Neubauer AC, Benedek M (2018) Assessment of real-life creativity: the inventory of creative activities and achievements (ICAA). Psychol Aesthet Creat Arts 12:304–316
10. Pichot N, Bonetto E, Pavani JB, Arciszewski T, Bonnardel N, Weisberg RW (2022) The construct validity of creativity: empirical arguments in favor of novelty as the basis for creativity. Creat Res J 34:2–13
11. Runco MA, Jaeger GJ (2012) The standard definition of creativity. Creat Res J 24:92–96
12. Weisberg R, Pichot N, Bonetto E, Pavani JB, Arciszewski T, Bonnardel N (2021) From explicit to implicit theories of creativity and back: the relevance of naive criteria in defining creativity. J Creat Behav 55:839–856
13. Weisberg RW (2015) On the usefulness of "value" in the definition of creativity. Creat Res J 27:111–124
14. Harrington DM (2018) On the usefulness of "value" in the definition of creativity: a commentary. Creat Res J 30:118–121
15. Weisberg RW (2017) Problem solving. In: Rodrigues P (ed) The Routledge international handbook of thinking and reasoning. Routledge, London. pp 607–623
16. Wimmer L (2016) Problem solving as a sufficient condition of the creative process: a case for closer cooperation of creativity research and problem solving research. Front Psychol 7:488
17. Cropley D, Cropley A (2010) Functional creativity. In: Kaufman JC, Sternberg RJ (eds) Cambridge handbook of creativity. Cambridge University Press, pp 301–318
18. Horng JS, Hu ML (2008) The mystery in the kitchen: culinary creativity. Creat Res J 20: 221–230

19. Lee C, Hallak R, Sardeshmukh SR (2019) Creativity and innovation in the restaurant sector: supply-side processes and barriers to implementation. Tour Manag Perspect 31:54–62
20. Tosh M (1998) Who is cooking? Progress Groc 77:77–82
21. Leong MWA, Yeh SS, Fan YL, Huan TC (2020) The effect of cuisine creativity on customer emotions. Int J Hosp Manag 85:102346
22. Rubera G, Ordanini A, Griffith DA (2011) Incorporating cultural values for understanding the influence of perceived product creativity on intention to buy: an examination in Italy and the US. J Int Bus Stud 42:459–476
23. Boudreau KJ, Guinan EC, Lakhani KR, Riedl C (2016) Looking across and looking beyond the knowledge frontier: intellectual distance, novelty, and resource allocation in science. Manag Sci 62:2765–2783
24. Hoeffler S (2003) Preferences for really new products. J Mark Res 40:406–420
25. Pichot N, Bonetto E, Arciszewski T, Bonnardel N, Pavani JB (2021) Toward a general factor of disruptiveness: the most novel creative objects tend to be the least valuable and feasible ones
26. Zhou J, Wang XM, Bavato D, Tasselli S, Wu J (2019) Understanding the receiving side of creativity: a multidisciplinary review and implications for management research. J Manag 45:2570–2595
27. Stevens SS (1951) Mathematics, measurement, and psychophysics. In: Stevens SS (ed) Handbook of experimental psychology. Wiley, New York
28. Amabile TM (1982) Social psychology of creativity: a consensual assessment technique. J Pers Soc Psychol 43:997–1013
29. Baer J, McKool SS (2009) Assessing creativity using the consensual assessment technique. In: Schreiner CS (ed) Handbook of research on assessment technologies, methods, and applications in higher education. IGI Global, pp 65–77
30. Horn D, Salvendy G (2009) Measuring consumer perception of product creativ-ity: impact on satisfaction and purchasability. Hum Factors Ergon Manuf Serv Ind 19:223–240
31. Dean DL, Hender J, Rodgers T, Santanen E (2006) Identifying good ideas: constructs and scales for idea evaluation. J Assoc Inf Syst 7: 646–699
32. Gero JS, Kannengiesser UDO (2006) A function-behaviour-structure ontology of processes. In: Gero JS (ed) Design computing and cognition'06. Springer, Dordrecht. pp 407–422

Part III

Methods from Marketing

Chapter 13

Qualitative Data Analysis: Challenges and Opportunities for Food Researchers

Jorge Herman Behrens

Abstract

Qualitative research is traditional in humanities and health sciences, and has played a significant role in marketing research, especially in consumer behavior studies. Regarding sensory science, a field where experimentation and the hypothetico-deductive scientific model predominate, qualitative methods have been relatively underutilized—and even misused—though qualitative research methods might be of choice to gain a broader understanding of a myriad of food issues such as food choices, culinary practices, sensory preferences, and consumer experience, just to mention a few. Sensory studies primarily focus on the product and how it is perceived by means of the human senses, and qualitative research, on the other hand, can be of choice when the voice of the consumer must be heard. This chapter aims to present and discuss the mainstream qualitative techniques based on individual or group interviews, including key elements for designing and conducting sound research: defining appropriate strategies to research questions; selecting participants, interviewers, and suitable interview techniques; performing content analysis; and, finally, reporting. At the end of this chapter, Internet-based studies and mixed methods (qualitative combined with quantitative methods) are addressed for they bring up a promising and exciting horizon of consumer studies.

Key words Qualitative research, Interviews, Focus group, Content analysis, Mixed methods, Internet

1 Introduction

Food choices are a complex phenomenon that involves a set of diverse variables, from intrinsic cues such as color, flavor, and texture, to extrinsic factors like brand, price, convenience, and usage context, just to mention a few [1–3].

In the domain of food studies, quantitative methods predominate. Food science and technology, as well as sensory and consumer science, are intrinsically related fields, sharing a common scientific paradigm based on quantitative measures subjected to statistical analyses [4]. Food researchers are usually interested in making inferences, seeking the most likely explanation from their data. Regarding sensory and consumer data, questionnaires (a set of

Carlos Gómez-Corona and Heber Rodrigues (eds.), *Consumer Research Methods in Food Science*,
Methods and Protocols in Food Science, https://doi.org/10.1007/978-1-0716-3000-6_13,

scales combined to measure psychological constructs and/or percepts) and large numbers of respondents have been long the mainstream, though it is uncertain that all the effort in statistical analysis and interpretation will shed light on the underlying aspects of the research problem. In this sense, qualitative methods based on asking consumers to state their opinions, feelings, attitudes, and preferences might not only complement but also direct further quantitative studies on a food-related research topic [5, 6].

Qualitative research involves interviews or observations in a less structured way. As new information comes out, the flow and strategy of the investigation may change [7, 8], and, perhaps, this is the key point that makes many food researchers feel unease about qualitative studies. Guerrero [9] argues that the relatively low use of qualitative methods in the food science domain might be due to two main interrelated factors: firstly, the difficulty in analyzing the information; secondly, publishing scientific papers based solely on qualitative techniques is relatively difficult, as usually they are not supported by statistical analysis that provides them with a more objective and less interpretive character.

A variety of qualitative techniques have been used in consumer studies, both in academia and companies, where consumers can be thought of as "co-designers" of products [8, 10] and services as well [11, 12]. Common methods include personal or one-on-one interviews, group interviews (more frequently referred to as focus groups), and observational methods (ethnography) [13].

Despite the recent advances in qualitative research, one-on-one and group interviews have been the most traditional and popular techniques in consumer research and, therefore, this chapter will primarily focus on the analysis of textual information. As for the way in which the data are collected, one-on-one interviews are of choice when the research focuses on individual behavior, such as eating habits, or when the issue is very personal, emotionally charged, or involves very specific types of respondents [8]. On the other hand, focus group is a group dynamic set to explore the processes of construction of reality common to a social group. Focus groups are very popular in marketing research [7, 14].

Ethnography is a naturalistic observation process of registering the behavior of a selected group of people in the everyday life. Ethnographic research has been applied to issues that involve food-related behaviors, such as culinary practices, purchase choices, product use in context, etc. Ethnography as a qualitative research strategy, in spite of its suitability to food and consumer studies, is beyond the scope of this chapter and the reader can delve into ethnography in Outtrey et al. [15].

Depending on the research goals, a qualitative study may solely be exploratory, a pilot of a larger (e.g., a mixed methods) study, or even the very heart of an academic project. Harris et al. [16] describe some contexts in which qualitative research is preferable:

- Determine causal explanations of phenomena in their natural settings
- Study the process or natural history of a phenomenon
- Understand the culture, traditions, symbols, perception, emotions, language, and meaning of phenomena from the participants' perspective
- Describe the context of phenomena
- Describe an unfamiliar community or culture
- Generate tentative theories and hypotheses
- Validate theory
- Complement quantitative research

The following sections will cover the key points in a qualitative study, from the design to the final report. Recent advances in the field will also be discussed.

2 Methodology

The overall purpose of qualitative research is to discern the meanings people give to their experiences and gain understanding of a given phenomenon from the perspective of research participants. When designing a qualitative study, the researcher must consider four elements to ensure validity, reliability, and relevance of the findings: the research strategy, the method of data collection, the questions that will be made, the selection of the participants, and how the data will be analyzed.

2.1 Research Strategies

Three major established approaches to qualitative research are widely discussed in the literature: grounded theory, phenomenology, and ethnography.

Grounded theory is an inductive, theory discovery methodology that allows the researcher to develop a theoretical account of the general features of a topic while simultaneously grounding the account in empirical observations or data [17]. This approach is suitable to identify and explain social processes and behaviors in specific situations or contexts. Grounded theory involves research setting and asking general questions such as "What is going on here?" or "What are the problems here?" but sometimes asking more specific questions such as, "How do these people cope with such issue?" Using interviews, focus groups, and/or document analysis, the researcher can find out the answers and propose situation-specific theory and hypotheses for further testing. Moreover, grounded theory is applicable to compare groups within the same theme and assess within and across-group saturation [16, 18, 19].

Phenomenology, in its turn, examines the meaning of human experience from the individual's perspective in terms of people's emotions, attitudes, thoughts, meanings, perceptions, and bodily experiences—as or after they have experienced a phenomenon. This approach mainly uses one-on-one interviews, for there is a special concern for the meaning, and the way in which it influences the experience [13]. Classical content analysis, discourse analysis, and keywords-in-context are suitable analytical strategies to gain understanding on how individuals explain and justify their personal experiences and their reasoning process. The analysis can be performed across groups (i.e., between-group analysis), within one group (i.e., within-group analysis), or for an individual in a group (i.e., intramember analysis) [17].

Ethnography studies social systems, cultures, social life, and activities of everyday life, through observation in naturalistic settings. Data is collected while conducting field work and the researcher takes part in the daily life of the participants, usually for a given period, using personal experience, informal conversations, and interviews, in addition to other sources of information for analysis (e.g., images, documents, etc.) [17, 20]. According to Gomez-Corona [21] contemporary ethnography in consumer research provides a cultural understanding of consumers by sharing day-to-day activities such as cooking, shopping, eating, etc., capturing the meaning that participants associate with those actions. Although physical environments are typical in ethnographic research, netnography or the ethnography online have gained interest of consumer researchers [44, 49].

Harris and colleagues [16] present other research strategies based on symbolic interaction, narrative, participatory action research, and case study. Kahlke [22] discussed the generic qualitative approach as an alternative that is neither focused on culture as ethnography nor in building a theory like the grounded theory. A generic qualitative approach seeks to investigate and understand a phenomenon, a process or perspectives and worldviews of the people studied, possibly blending established methodological approaches to create something new or claiming no formal methodological framework at all [23]. Generic qualitative studies might be suitable for "seeking to understand" exploratory studies due to the relative freedom from a theoretical framework [24]. However, there is still some criticism about this approach, especially when incompatible elements of more than one established methodology are blended.

2.2 Data Collection

Interviews in consumer research can be conducted individually or with a group of selected individuals. Both techniques have their strengths and drawbacks, and the selection will largely depend on the objectives of the study.

Interviews are audio and/or video recorded upon consent of the participants given prior to the beginning. In this sense, ethical considerations and the protection of participants are fundamental aspects of interviewing, particularly in the academic field where the research protocols must be submitted to and approved by a research ethics committee prior to the data collection process. The issue of ethics in qualitative research is suitably discussed by Sanjari et al. [25] and Aguirre and Hyman [26].

2.2.1 One-on-One Interviews

The one-to-one interviews are quite common in health and social research, being a valuable method of gaining insight into people's perceptions and understandings and experiences of a given phenomenon, especially when the issue under investigation is somewhat personal or sensitive. Interviews in qualitative research can be structured (yielding more straightforward answers) or semi-structured (argumentative, expository answers), but the recommended technique in the one-on-one scenario is the in-depth interviewing, which consists of asking very general questions to prompt unbounded exploration of an issue. Then follow-up questions are spontaneously created based on the answers of the participant, producing a more detailed, comprehensive, and representative picture of the phenomenon [16]. The researcher must bear in mind that an interview is more than a conversational interaction between two people, and it requires considerable preparation of the interviewer. The nature of the questions, the questioning technique, listening skills, and the rapport interviewer–interviewee are crucial to obtaining a successful outcome [11].

One-on-one interviews generally last between 1 and 2 h and they are always carried out one at a time. There is no scientifically accurate number of interviews in a study, which may range from three to ten interviewees, or as high as tenths [13], to as many as tenths [27]. However, the number of interviews will depend on time and budget restrictions, as well as the moment when information saturation is reached.

2.2.2 Focus Groups

Focus groups take the form of an open group discussion to get insights and ideas on a certain issue [4]. Traditionally, focus groups have been used in the marketing field, but have gained in popularity in other areas such as sensory and consumer research [6, 28, 29].

Typically, a focus group comprises 6–10 individuals representing a specific target of interest, who are asked questions associated with the research interest (eventually with the help of complementary materials such as videos, mock-ups, samples, activities, etc.) and prompted to respond freely. Video recordings are taken to gain insights from body language, whereas audios are transcribed verbatim to have a written record of responses [16]. As a general recommendation, 3–4 focus group sessions are expected to be sufficient [7] for the researcher to check if the information saturation point was reached, or if more sessions are needed. Fewer than six participants may not provide an effective group dynamic, and more participants (>10) can make it difficult to control the group and generate multiple distractions.

Like one-on-one interviews, a question guide is necessary to conduct the session by a moderator. Multiple sessions are often conducted to obtain data and draw meaningful conclusions. If group dynamics is the main advantage over one-on-one interviews, concerns are related to the risk of biasing the discussion to obtain a desired response or even failing in promoting the group dynamics [30]. The role and importance of the moderator in focus group is discussed in the following section.

2.3 Conducting the Interview

Both one-on-one and focus group interviews need an interview guide phrased in simple and objective language. The best questions are short and easy to read, not double-barreled (e.g., "Do you think plant-based and organic food are healthful and sustainable?" where one thing is plant-based, and the other is organic food) and avoid technical jargon [7, 8]. Questions should be open-ended to evoke conversation instead of single responses or just yes/no answers, permitting those beliefs, feelings, or ideas emerge during the discussion [18]. Before starting, a pretest of the guide must be taken and for this purpose, one can run a few interviews [3–5] or discuss the questions with specialists in the theme under research [27]. The idea is to check (1) whether the questions work as we expected, (2) there is something unexpected or even contradictory in the answers obtained, (3) the duration of the interview is adequate, and (4) the depth of the responses seem adequate [18]. Examples of question guides are available in the work of Ribeiro et al. [27] who investigated consumer's views on genetically modified food, and Behrens et al. [31] on consumer purchase habits and views on food safety.

One critical issue in conducting interviews is the skill and characteristics of the interviewer/moderator, for they have a significant impact on the data produced. Thus, the following cornerstones must be considered [4, 7, 13, 18].

- Gender of the interviewer—which may influence or even bias the discussion
- Sociability—the way they behave in different contexts involving different kinds of people (e.g., children, adolescents, seniors, illiterate, LGBTQIA+, etc.)
- Self-control—capacity for concentration, memory, and patience
- Nationality and ethnicity—it can make a difference when interviewing people with diverse ethnicity, language, religion, and cultural background

The interviewer or moderator should try not to be influenced by their experience, previous knowledge, or direct participation in the design of a study, etc., which might bias the data ("You only see what your eyes want to see"). Depending on the research design, and when there are no budget restrictions, it is worth considering hiring a professional, skilled moderator.

2.4 Participant Selection

There is no specific qualitative sampling criterion, but the individuals in the sample must come from the target population, characterized by age, gender, socioeconomic status, education level, culture, lifestyle, etc. Moreover, there should not be any conflict of interest between the participants—including the interviewer [13].

Regarding focus group specifically, internal homogeneity groups are preferable, that is, participants of the same gender, socioeconomic level, ethnicity, similar attitudes toward a product, brand, person, etc. Behrens and colleagues [32], exploring consumer views on food irradiation, conducted three independent groups, being two feminine and one masculine. The main reason is that men usually seem to be more interested in the features of technology while women show concern about the risks and benefits that it poses. Moreover, differences in socioeconomic status led to the decision of conducting a group with more affluent women and another with lower income women, which showed to be efficient to reveal differences toward food choices and behaviors.

Sometimes, it can be interesting to mix people, for example, young people and old people, men and women, etc., when they share a fundamental trait of interest, and the group dynamics can provide a broader view of the population under research.

3 Data Analysis

Qualitative data is usually extensive, greatly diverse, and carries underlying meaning and symbolism. When handling with interview datasets, the first step consists of the transcription of the verbal information, which can be made (1) verbatim—transcribing the exact words as they were said by the interviewees, and (2) transcription with registration of time (pauses made during an interview), intonation (high or low tones during conversations), body language (rejection, acceptance, boredom, etc.) [16]. Transcription can be performed by the researcher or professional transcription services. Commercial software for qualitative research like NVivo (www.qsrinternational.com/) and MAXQDA (www.maxqda.com), besides supporting all the process of data analysis, also enable transcription of audio and video.

With the transcripts available, the researcher gets acquainted with the content and looks at the entire set of interviews in search for patterns and peculiarities. Such exploration increases the understanding of the meaning that the participants attached to the topics discussed in the interview. The results of the data exploration should be recorded and available throughout the course of the content analysis and the notes taken can be varied: general remarks, hypotheses, or ideas for analysis; initial answers to research questions; highlights of individual text passages, formulations, metaphors, etc.; notes on the interview situation, and text passages that need in-depth examination [33].

The most basic way of analyzing the outcome of interviews consists of examining transcripts for each interviewee in one-on-one interviews, or within a group in a focus group session [34], when the consensus opinions and interactions are of major interest. However, caution must be taken not to simply disregard the dissenters at the risk of losing some relevant information [35].

Subjectivity is inherent to qualitative data analysis and the researcher must be aware of the tendency to pay attention to only those pieces of information that fit in with their personal expectations [4]. To overcome this drawback, an alternative to make the analysis of qualitative data more objective is the researcher triangulation [9, 18], a method used to increase the credibility (trustworthiness and how believable a study is) and validity (the extent to which a study accurately reflects or evaluates the concept or ideas being investigated) of research findings. Triangulation can help ensure that fundamental biases arising from a single researcher are overcome [36], and usually 2–3 researchers take part in the triangulation process [37].

3.1 Analytical Techniques

Despite the long history of qualitative research, there is no specific framework on how to analyze the data obtained. Onwuegbuzie et al. [19] described four different analytical techniques suitable for the research strategies.

Constant comparison analysis is commonly applied to focus group data and when ground theory is the research approach. It is usually conducted as per the following steps: firstly, the data are chunked into small units; the researcher attaches a descriptor, or code, to each of the units; then, the codes are grouped into categories (axial coding), and the final step is the construction of a core category, that is, when all the concepts are pulled together to offer and explanation of the phenomenon [17]. Commonly, a study comprises multiple focus groups, and a specific concern is referred to as saturation or the point at which information appears repetitive and further data collection becomes redundant [38]. So, the constant comparison of multiple focus groups indicates the need for more data collection to assess the meaningfulness of the themes (or categories) and/or to refine them, until data saturation and/or theoretical saturation is reached [19]. In this sense, researchers are encouraged to design studies with multiple focus groups so that themes can be properly tested. Hennink and colleagues [38] identified five parameters influencing saturation in focus group data: study purpose, type of codes, group stratification, number of groups per stratum, and type and degree of saturation to be reached.

In the classical content analysis small chunks of the data are selected and then a code is placed with each chunk. Instead of creating a category from the codes, the researcher places them into similar groupings and counts them. The analysis can be made to identify whether each participant or each group used a given code, and all the occurrences of the code. This permits the calculation of frequencies of categories and then the use of correspondence analysis (CA), a multivariate analysis technique applicable to contingency tables [6]. CA yields maps depicting the association of categories and objects (e.g., segments of respondents either within a group or from different groups) which summarizes the most salient information. Although counts normally provide very useful information (e.g., level of consensus, response patterns among the interviewees, etc.), they should never be used to reduce the essence of qualitative information [19].

Discourse or thematic analysis goes beyond the strict meaning of the words, and its main objective is to gain understanding on how speaking builds different versions of reality, at different times or contexts, and what are the consequences of what is said [13]. In other words, the way an individual or group of people speaks does not have a neutral form by itself but is based on a social representation with a symbolic and cultural load.

Discourse analysis in social research addresses behaviors, mental states, and mental processes that are anchored in social contexts. Through a social psychology perspective three fundamental entities should be taken into account [39]: (1) the self, the individual person as the object of its own reflective consciousness, characterized on the biological, affective, emotional, motivational, and cognitive levels; (2) the others, or an individual or a group of them, who may or may not be present, and who have a level of influence in the social context of the person, and (3) the context: the individual participates in a social situation that implies a social position occupied between the self and the others, also characterized by the culture, the language, the environment, etc.

Given the richness of empirical data available for the discourse analysis, it is worth having an analytical perspective that not only reveals people's thoughts but also permits the selection of what is prominent in the speech of the interviewees. In this sense, the social representations framework has a potential in studies focused on food choices. As defined by Moscovici [40], social representations are "sets of values, beliefs and metaphors expressed by the members of a social group as a code for social exchange, so they can name and classify unambiguously the various aspects of their world and their personal and group history." Social representations underlie the construction of practical knowledge—or the common sense—as a product of all information disseminated by the media, cultural diffusers, academy, etc., debated in the various environments such as religious associations, workplaces, family, and closer relations. The Collective Subject Discourse (CSD) technique, based on the Theory of Social Representations, was set as the framework for the content analysis of studies on customer's perceptions of safety, knowledge of food-borne diseases and self-involvement in the food safety chain in restaurants [41], consumer views on genetically modified foods [27], and to analyze and understand the knowledge, interpretations, and practices of education officers and nutritionists in the management of school feeding programs in Brazilian counties [42].

Like discourse analysis, keywords-in-context aims to determine how words are used in context with other words. Keywords-in-context involves a contextualization of words that are considered central to the development of themes and theory by analyzing words that appear before and after each keyword, leading to an analysis of the culture of the use of the word [19]. People use words in different ways reflecting their background, social status, education, etc., hence it is necessary to examine how words are used in context, which is likely to happen in a focus group due to its interactive nature. Thus, each word voiced by a participant not only should be interpreted as linked to all the other words voiced during the discussion, but it should be interpreted with respect to the words voiced by all other participants of the focus group. This analysis can be performed across groups, within one group, or for specific participants.

3.2 Coding

Coding is the process of labeling and organizing data form transcripts, audios, or images/videos to identify different categories and the relationships between them [18, 43].

At the beginning of the content analysis the researchers have already familiarized themselves with the data, marked important text passages, registered ideas for the analysis, and perhaps even written initial case summaries. Then the transcript excerpts are marked with codes, which are the building blocks for theory or model building. A code is most often a word or short phrase that symbolically assigns a summative, salient, essence-capturing, and/or evocative attribute for a specific section of a transcript [37].

When clustered together according to similarity and regularity, codes actively facilitate the development of categories [33]. Normally, multiple analysts perform the coding process so that the reliability and validity of the coded data through inter-analyst agreement measures are assessed [36] and, for some studies, as the only reasonable way to handle the sheer amount of data.

Table 1
Categorization methods used in content analysis of qualitative data

Type of category	Definition	Examples
Fact	A piece of unambiguous information presented as having objective reality utilized to classify information.	"Consumer of a particular brand"; "users of food delivery apps"
Thematic	Indicates themes and are exceptionally common in practice.	"Sustainability," "veganism"
Evaluative or scaling	Used to assess a phenomenon's degree or qualities in both dichotomous and ordinal ways.	"Beer consumption" (low, average, and high); "Food intolerances," "present/absent"
Analytical	Characterized by a better degree of reflection and elucidation of a phenomenon.	"Compulsion"; "conspicuous consumption"
Theoretical	Allude to an existing hypothesis, or it can be deducted straightforwardly from the hypothesis.	"Cognitive dissonance," "Social representation"
Natural	Words and phrases used by the interviewees usually called in vivo categories, particularly in the grounded theory approach.	Type of language (e.g., formal/ colloquial), aphorisms, slang, etc.
Formal	Formal traits of interview subjects.	The word count in the transcript or the quantity of interviewees' discourse commitments.

MacQueen and colleagues [37] and Rädiker and Kuckartz [33] provide detailed directions to coding and categorization of qualitative data. Categories are used for classification, reduction, abstraction, and attribution of meaning and can be used for indexing, describing, and explaining the data. Table 1 presents examples of different types of categories that can be identified during the content analysis.

4 Reporting

Rädiker and Kuckartz [33] make a valuable recommendation for reporting qualitative research: "write continuously during the entire analysis phase, that is, from the very beginning."

For company reports, a bulleted style is generally recommended detailing objectives, summary, the research process—details of method, characteristics of the participants, locations, dates, etc.—key findings, interpretations (if needed), and recommendations. The question guide may be appended [8].

As previously stated, the development of coding and the categories are essential for the analysis of interviews, but simply counting the category frequencies would be a rather simplistic form of reporting. Instead, it is important to explain the concepts and describe the relationships between categories and, in this sense, quotations from the participants' speeches can illustrate the categories and make them more plausible. It is preferable to use short (and no more than three) quotations, which are cleverly interwoven into the development of the arguments. A maximum of a quarter of the text in the results section should consist of quotations [33]. This would give the reader a holistic view of the phenomenon studied, especially if the research aims to characterize target group and propose theories, hypotheses, and complex models to be further tested.

Regarding dissertations and PhD thesis, these must be thoroughly described, including the accurate description of all the steps in content analysis, from the transcription to the development of categories. Appendices should contain relevant documents of the study (approval of a research ethics committee, letters of invitation to the participants, consent forms), the question guide or questionnaire, the rules by which the interviews were transcribed, coded, and the documentation of the category system.

Visual representations of the information can summarize and illustrate different results from different groups of individuals. Concept mapping is a method to frame a research project, reduce qualitative data, analyze themes and interconnections in a study, and present findings whereas word clouds display the most prominent or frequent words in a body of text. Examples of the use of concept mapping can be found in Lawless and Heymann [8] (Chapter 16—Qualitative Consumer Research Methods) and in Brown et al. [29].

5 Online Qualitative Research

The Internet has been the new frontier of qualitative research. New technologies for accessing and interacting with people in different parts of the world are bringing new possibilities and shaping the future of qualitative research and the data collection process. In this sense classical qualitative methods can be adapted to Internet-based methods such as online focus groups, personal interviews, and "netnography"—an ethnographic research approach to studying and understanding consumption-related aspects of consumers' lives online [44]. Both synchronous (real-time) and asynchronous (not-real-time) methods have been used in several research fields.

Although online qualitative research is not exactly a novelty, it has been greatly boosted by the restrictions imposed by the Covid-19 pandemic. Since traditional face-to-face investigations of all kinds were restricted, videoconferencing services turned possible to perform "socially distant" studies. As people have become familiar with various Internet-based communication platforms, one might assume that their digital skills and competences have accordingly grown, consequently making their participation in online research data collection easier [45].

Online interviews allow researchers to collect qualitative data from a geographically dispersed population (e.g., cross- or intercultural studies) and/or difficult-to-access individuals, being especially useful when having neither the time nor the resources to arrange in-person meetings [4]. However, there are also concerns such as the security of the platform, confidentiality (for participants at home or other environments where they can be overheard), and the logistical needs (e.g., computer or mobile phone, bandwidth, lighting, and the quality of the participant's video camera and audio).

Basically, online qualitative studies are versions of traditional ones. When performed synchronously, which is particularly appropriate for group dynamics, there are concerns such as participants' need to think, type (at their own speed), look at the screen, read the text, and maintain a logical thread of answering [46]. To assure smooth participation, especially in focus groups, Lobe et al. [45] recommend having participants engage in a short one-on-one pre-session with the researcher, in order to prevent unforeseen technical issues and for the participants to familiarize themselves with the use of a particular platform's features. Even though the platforms usually allow for a large number of people in a single session, videoconferencing works best with a relatively small number of participants, ideally 3–5, as focus groups with a larger number of participants are particularly difficult to hold on mobile devices with small screens that make it difficult to see the windows with the other participants.

As for asynchronous data collection, participants log in the platform and answer discussion topics on their own time, through questionnaires or discussion groups (e.g., blogs, comments on social media, etc.). This procedure has several advantages: it allows the participants to overcome global time differences and lower typing skills as they have more time to answer; it is more appropriate to sensitive topics and permits collecting more answers than in a focus group. On the other side, the group dynamics is complex and may create conflict among the participants (e.g., discussion lists or social media comments), and the spontaneity of responses, as well as non-verbal inputs, are compromised. Besides, the rationality bias is a concern in asynchronous studies, once the participants have more time to think about their answers and even to look for information on the topic discussed. The greater rationality in the answers implies a greater effect of possible stereotypes, social norms, group effect, influence of the moderator, and possible leaders within the group [4].

A number of video conferencing services are available for remote qualitative research: Microsoft Teams (www.microsoft.com), Google Meet (meet.google.com), Zoom (www.zoom.us), Skype (www.skype.com), just to mention some. These platforms have some common features but also peculiarities that can be accessed in the services websites. Free use is possible but limited, and more functionalities (e.g., larger groups of participants, interaction devices, or even audio transcription) can be purchased according to the research needs.

Many are the studies and revisions recently published on the use of video-conference devices in qualitative research and the reader is encourage do dive into this topic as it will certainly be one of the major outcomes of the challenge imposed to qualitative researchers during the Covid-19 times.

As the number of Internet users has increased around the world, and more parts of people' lives began occurring online, social medias open new possibilities as they permit one to capture information from target populations, from almost anywhere in the world, in an efficient and inexpensive way to better understand consumer habits and preferences, opinions about products and services, brands, etc. [47]. Social media comes in many different forms, for example, blogs and publishing networks (WordPress, Tumblr, Medium), collaborative projects (Wikipedia), social networks (Facebook, LinkedIn, Twitter), media-sharing networks (Instagram, Snapchat, YouTube, TikTok), discussion forums (Reddit, Quora, Digg), bookmarking and content curation networks (Pinterest, Flipboard), and consumer review networks (Yelp, Google reviews, TripAdvisor) [48].

To gain a deeper understanding of marketplace relationships and actors, netnography, aka virtual ethnography, online ethnography, or digital ethnography is and adaptation of traditional ethnographic research techniques to the study of cultures and communities manifesting through computer-mediated communications [49]. Netnographic studies involve observing what people are doing in social media as they are constantly connected to the Internet through computers and other mobile devices, especially the younger generations that share many of their opinions, experiences, and everyday activities online. Thus, the evolution of netnographic research is closely linked to the evolution of the Internet. The plethora of traces left by customers on various online platforms is valuable data for consumer researchers. See Heinonin and Medberg [44] for an extensive review on netnographic studies.

Last but not least, ethical issues are highlighted when it comes to Internet-based interviews and social media data mining. Although most of the fundamental ethics in online interviewing are the same as in face-to-face contexts, some critical points must not be overlooked such as privacy of participants, confidentiality of data and security. In this sense, it is essential to investigate the data security policies of all platforms and services [45]. Guidelines on ethical issues in online research are available from the Association of Internet Researchers [50].

6 Mixed Methods

Tashakkori and Creswell [51] defined mixed methods as research in which the investigator collects and analyzes data, integrates the findings, and draws inferences using both qualitative and quantitative methods in a single study. Mixed methods research should be used when the research question suggests that combining quantitative and qualitative approaches is likely to provide superior findings.

Molina-Azorin [5] argues that the answers to quantitative research questions could be made easier and more adequately if, prior to the quantitative inquiry, a qualitative phase were carried out with the aim of acquiring a deeper understanding of the research problem. However, there is no rule indicating that the qualitative phase must always happen before the quantitative phase. Indeed, the methods can be mixed in a way that has complementary strengths and non-overlapping weaknesses.

Mixed method studies have several purposes that can be considered as advantages of this approach: (1) triangulation—to seek convergence and corroboration of findings from different methods that examine the same phenomenon, (2) development—using the results from one method to help develop or inform the other method, (3) expansion—using different methods to assess different facets of a phenomenon, yielding an enriched, elaborated understanding of that phenomenon, and (4) complementarity—clarifying, enhancing, or illustrating the results from one method with the results from the other method [52].

Projective techniques (e.g., free word association, completion tasks, etc.) or sensory evaluation methods (e.g., projective mapping and sorting tasks) can be either used together or complement interviews and focus group, since supplementing qualitative data with counts yields a form of mixed methods data analysis. In fact, moderators and interviews use projective stimuli to improve the questioning process quite frequently. Projective techniques may be introduced at different stages of the interview once the participants feel comfortable enough, being a good way of "breaking the ice" in a focus group discussion at the start [53], or to mitigate the rationality bias and capture more emotional and less stereotyped responses [4].

The study of Brown and colleagues [29] is an interesting example of mixed methods. To understand American premium chocolate consumer perception of craft chocolate and desirable chocolate product attributes, they conducted a mixed-method study using focus groups and projective mapping. Firstly, 27 selected participants performed a projective mapping (PM) task, through which they had to plot on a sheet of paper 47 stickers of various chocolate products, from mainstream to premium and craft chocolates according to similarities and dissimilarities between the brands and note down appropriate descriptors for the characteristics of the chocolates. In a second stage, focus groups were held and the participants brought their maps to use them as a visual aid for describing how they perceive chocolate quality. During the sessions, the participants explained their projective maps and tasted five pieces of chocolate bars (all of them included among the stickers used in PM) representing the range of commercial chocolates in the US market. Afterward, they described the product in terms of flavor, packaging, certification labels, and other elements and attributes they found appealing and unappealing. Finally, factors that encourage them to purchase a new food product and shared words they would use to describe the chocolates tasted were discussed. Grounded theory was considered appropriate to analyze the focus group data because craft chocolate is a novel food product and there was no prior research from which to preconceive codes. PM data was analyzed by multifactor analysis (MFA).

The study demonstrated the utility of using a mixed-method approach to uncover consumer perception of a novel product such as craft chocolate. Premium chocolate consumers expressed excitement, were intrigued by the flavor of craft chocolate, and had difficulty contextualizing it compared to the premium and mainstream chocolates. Besides, they paid close attention to search attributes such as a handcrafted aesthetic, the color gold, cocoa percentage, and thick foil in all the products they mapped and sampled. Grounded theory analysis of focus groups revealed that consumers find trust to be the primary desirable credence attribute, while utility and/or joy are primary desirable experience attributes when consuming chocolate.

7 Final Remarks

Qualitative research is essential to understanding people's thinking and behavior and much of the knowledge in the area has come from health and social research where qualitative methods are widely used. Researchers in the areas of food, nutrition, and sensory science might benefit from qualitative research methods to design studies aimed at understanding the myriad of food-related issues in today's world. Qualitative studies can reveal information that numbers are not able to provide. Moreover, contact restrictions during the Covid-19 pandemic have boosted Internet-assisted communication technologies including social media interaction. In this sense, food researchers now have the possibility of studying consumers everywhere, real-time or not. Traditional qualitative techniques such as interviews and focus groups can now be conveniently applied online to accessing diverse targets. In this scenario, netnography opens up the possibility of investigating consumer habits, preferences, opinions, and feelings through the analysis of not only verbal but also a rich imagery language.

References

1. Rozin P (2006) The integration of biological, social, cultural and psychological influences on food choices. In: Shepherd R, Raats M (eds) The psychology of food choice. CABI, Cambridge
2. Prescott J (2012) Taste matters – why we like the foods we do. Reaktion, London
3. Cardello AV, Meiselman H (2018) Contextual influences on consumer responses to food products. In: Ares G, Varela P (eds) Methods in consumer research, vol 2. Elsevier, Oxford
4. Guerrero L, Xicola J (2018) New approaches to focus groups. In: Ares G, Varela P (eds) Methods in consumer research, vol 1. Elsevier, Oxford
5. Molina-Azorin JF (2012) Mixed methods research in strategic management: impact and applications. Organ Res Methods 15(1):33–56
6. Warde A (2014) Food studies and the integration of multiple methods. Política y Sociedad 51(1):51–57
7. Krueger RA, Casey MA (2008) Focus groups: a practical guide for applied research. SAGE, Thousand Oaks
8. Lawless HT, Heymann H (2010) Qualitative consumer research methods. In: Sensory evaluation of food. Food Science Text Series. Springer, New York, NY. https://doi.org/10.1007/978-1-4419-6488-5_16
9. Guerrero L, Guàrdia MD, Xicola J, Verbeke W, Vanhonacker F, Zakowska S (2009) Consumer-driven definition of traditional food products and innovation in traditional foods. A qualitative cross-cultural study. Appetite 52:345–354
10. Ares G, Giménez A, Gámbaro A (2008) Understanding consumers' perception of conventional and functional yogurts using word association and hard laddering. Food Qual Prefer 19:636–643
11. Hancock B, Ockleford E, Windridge K (2007) An introduction to qualitative research. The NIHR RDS EM/YH. Available at: https://www.rds-yh.nihr.ac.uk/wp-content/uploads/2013/05/5_Introduction-to-qualitative-research-2009.pdf
12. Trischler J, Zehrer A (2012) Service design: suggesting a qualitative multistep approach for analyzing and examining theme park experiences. J Vacat Mark 18(1):57–71
13. Goméz-Corona C (2019) Estudios qualitativos tradicionales. In: Goméz-Corona C, Lelièvre-Desmas M (eds) Introducción al análisis sensorial y estudios con consumidores: Nuevas perspectivas. XOC Estudio, Ciudad de Mexico
14. Saint-Denis C (2018) Consumer and sensory evaluation techniques: how to sense successful products. Wiley, Hoboken
15. Ottrey E, Jong J, Porter J (2018) Ethnography in nutrition and dietetics research: a systematic review. J Acad Nutr Diet 118(10):1903–1942. https://doi.org/10.1016/j.jand.2018.06.002
16. Harris JE, Gleason PM, Sheen PM, Boushey C, Beto JA, Bruemmer B (2009) An introduction to qualitative research for food and nutrition professionals. J Am Diet Assoc 109:80–90
17. Goulding C (2005) Grounded theory, ethnography and phenomenology – a comparative analysis of three qualitative strategies for marketing research. Eur J Mark 39(4):294–308
18. Flick U (2004) Uma introdução à pesquisa qualitativa, 2nd edn. Bookman, Porto Alegre
19. Onwuegbuzie AJ, Dickinson WB, Leech NL, Zoran AG (2009) A qualitative framework for collecting and analyzing data in focus group research. Int J Qual Methods 8(3):1–21
20. Freedman I (2016) Cultural specificity in food choice: the case of ethnography in Japan. Appetite 96:138–146
21. Gómez-Corona C, Escalona-Buendía HB, García M, Chollet S, Valentin D (2016) Craft vs. industrial: habits, attitudes and motivations towards beer consumption in Mexico. Appetite 96:358–367
22. Kalhke RM (2014) Generic qualitative approaches: pitfalls and benefits of methodological mixology. Int J Qual Methods 13(1):37–52
23. Caelli K, Ray L, Mill J (2003) "Clear as mud": toward greater clarity in generic qualitative research. Int J Qual Methods 2(2):1–24
24. Bellamy K, Ostini R, Martini N, Kairuz T (2016) Seeking to understand: using generic qualitative research to explore access to medicines and pharmacy services among resettled refugees. Int J Clin Pharm 38:671–675
25. Sanjari M, Bahramnezhad F, Fomani FK, Shoghi M, Cheraghi MA (2014) Ethical challenges of researchers in qualitative studies: the

necessity to develop a specific guideline. J Med Ethics Hist Med 7:14

26. Aguirre GC, Hyman MR (2016) Ethics in qualitative consumer research. In: Hackett PMW (ed) Qualitative research methods in consumer psychology: ethnography and culture. Routledge, New York
27. Ribeiro TG, Barone B, Behrens JH (2016) Genetically modified foods and their social representation. Food Res Int 84:120–127
28. van Kleef E, van Trijp HCM, Luning P (2005) Consumer research in the early stages of new product development: a critical review of methods and techniques. Food Qual Prefer 16:181–201
29. Brown AL, Bakke AJ, Hopfer H (2020) Understanding American premium chocolate consumer perception of craft chocolate and desirable product attributes using focus groups and projective mapping. PLoS One 15(11): e0240177. https://doi.org/10.1371/journal.pone.0240177
30. Stewart DW, Shamdasani PN (2007) Focus groups: theory and practice, 2nd edn. SAGE, Thousand Oaks
31. Behrens JH, Barcellos MN, Frewer J, Nunes TP, Franco BDGM, Destro MT, Landgraf M (2010) Consumer purchase habits and views on food safety: a brazilian study. Food Control 21:963–969
32. Behrens J, Barcellos MN, Frewer LJ, Nunes TP, Landgraf M (2009) Brazilian consumer views on food irradiation. Innovative Food Sci Emerg Technol 10:383–389
33. Rädiker S, Kuckartz U (2020) Focused analysis of qualitative interviews with MAXQDA – step by step. MAXQDA Press, Berlin
34. Duggleby W (2005) What about focus group interaction data? Qual Health Res 15(6): 832–840
35. Maxwell JA (2005) Qualitative research design: an interactive approach, 2nd edn. SAGE, Newbury Park
36. Noble H, Heale R (2019) Triangulation in research, with examples. Evid Based Nurs 22(3):67–68
37. MacQueen K, McLellan E, Kay K, Milstein B (1998) Codebook development for team-based qualitative analysis. Cult Anthropol 10(2):31–36
38. Hennink MM, Kaiser BN, Weber MB (2019) What influences saturation? Estimating sample sizes in focus group research. Qual Health Res 29(10):1483–1496. https://doi.org/10.1177/1049732318821692
39. Jodelet D (1984) Représentation sociale: Phénomènes, concept et théorie. In: Moscovici S (ed) Psychologie Sociale. PUF, Paris
40. Moscovici S (1988) Notes towards a description of social representations. Eur J Soc Psychol 18:211–250
41. Behrens JH, Vedovato GM, Cervato-Mancuso AM, Bastos DM (2015) Social representations of safety in food services. Food Res Int 74: 324–328
42. Nogueira RM (2016) The National School Feeding Programme (PNAE) in the interpretation of the managers of the Metropolitan Region of Campinas/SP. PhD Thesis, University of Campinas
43. Barbour RS (2001) Checklists for improving rigour in qualitative research: a case of the tail wagging the dog? BMJ 322:1115–1117
44. Heinonen K, Medberg G (2018) Netnography as a tool for understanding customers: implications for service research and practice. J Serv Mark 32(6):657–679
45. Lobe B, Morgan D, Hoffman KA (2020) Qualitative data collection in an era of social distancing. Int J Qual Methods 19:1–8
46. Tuttas CA (2015) Lessons learned using web conference technology for online focus group interviews. Qual Health Res 25(1):122–133. https://doi.org/10.1177/1049732314549602
47. Vidal L, Ares G, Jaeger S (2018) Application of social media for consumer research. In: Ares G, Varela P (eds) Methods in consumer research, vol 1. Elsevier, Oxford
48. Digital Vydia. Types of social media. Available at: https://www.digitalvidya.com/blog/types-of-social-media/. Accessed 15 Oct 2021
49. Kozinets RV (2002) The field behind the screen: using netnography for marketing research in online communities. J Mark Res 39:61–72
50. Franzke A, Bechmann A, Zimmer M, Ess C (2020) Internet research: Ethical guidelines 3.0. https://aoir.org/reports/ethics3.pdf. Accessed 15 Oct 2021
51. Tashakkori A, Creswell J (2007) The new era of mixed methods. J Mixed Methods Res 1:3–7
52. Aguinis H, Werner S, Abbott J, Angert C, Park J, Kohlhausen D (2010) Customer-centric science: reporting significant research results with rigor, relevance, and practical impact in mind. Organ Res Methods 13:515–539
53. Donoghue S (2000) Projective techniques in consumer research. J Fam Ecol Consum Sci 28: 47–53

Chapter 14

Consumer Research in Social Media: Guidelines and Recommendations

Carlos Gómez-Corona and Hoa Nguyen

Abstract

This chapter focuses on understanding the general practices and some applications of social media research in the food and beverage domain.

The first part of the chapter is intended to give a general introduction and define what social media is versus other types of media (mainstream media, reviews, etc.). It also explores the human relationship with social media as an extended "digital self," and why the use of social media has exploded. An additional point is covered with the legal and ethical considerations required to perform research by using social media monitoring tools.

The second and third parts of this chapter focus on general applications of social media research (exploratory studies, trend watch, netnography, and other applications) and then provide a detailed view of how to perform social media studies step-by-step: from the basic Boolean search formula to natural language processing and human analysis.

Finally, two case studies are shared to compare the results of social media research versus online quantitative tests. The objective of this comparison is to explain the strengths and limitations of each research and how they can be complementary, as they are usually compared in both academic and industrial applications.

Key words Social media, Consumer research, Digital tools, Image, Emojis

1 Introduction to Social Media Research

The exponential growth of social media in recent decades has not gone unnoticed by consumer researchers. Media such as Facebook, Twitter, and Instagram are increasingly being used to study consumers' perception toward products such as wine [1], beer [2], different types of consumption (e.g., sustainable consumption: [3]), or attitudes toward services branding [4]. But despite being relatively new to consumer researchers, the roots of social media research can be traced back to the study of social networks [5]. In fact, one of the first major publications of social media research intended to better understand social networks in a University [6].

Carlos Gómez-Corona and Heber Rodrigues (eds.), *Consumer Research Methods in Food Science*,
Methods and Protocols in Food Science, https://doi.org/10.1007/978-1-0716-3000-6_14,

There are several factors that can explain the growing interest in the use of social media as a tool for consumer research. The first factor is the digital content, as social media allows any person to publish what they want, with little effort, at a low cost, and for a potential audience of millions of people around the world [7]. This led to a second factor, which is the large amount of data available on social media. Billions of users around the world share valuable information that can be "easily" analyzed. Consumer researchers code and analyze this data to understand human behavior [8]. The third factor is the advances in digital technologies that continue to make communication channels and platforms more ubiquitous and effortless; human beings are more "connected" to each other than ever before [9]. The fourth factor is that, as the research material is published on the Internet, it involves no direct contact between the subject and the researcher. It avoids one of the problems facing much qualitative research, namely that of interviewer bias, whereby what is said is influenced by the researcher [10]. The fifth factor can be the variety of social media platforms, which enables the study of different aspects of digital content, such as text, images, or videos.

Figure 1 summarizes the diversity of social media platforms, categorizing them by their main feature. The first category of social media is those focusing on networking: Facebook, Vkontakte, Sina Weibo, and LinkedIn (focus on business networking). These networking platforms allow the sharing of text, video, and images. Facebook is still the platform with the highest number of users (2740 million, in late 2021), followed by 740 million for LinkedIn, 500 million for Sina Weibo, and 400 million for VK. The second category is focused on text sharing with Twitter, with 353 million, and Tumblr, with a similar number of users. It is important to highlight that although Twitter is usually considered as a text-based platform, it allows the use of pictures and short videos (like Facebook or LinkedIn). One important aspect of social media studies in consumer research is that many of the studies focus their research on Twitter, as it is still the social media channel with the highest number of open accounts, meaning that the users restrict their Twitter accounts less compared to Facebook. The video content platforms may be the least explored in consumer research due to the complexity of analyzing videos; however, their millions of users represent an interesting source of research, and like Twitter, the video content platforms usually have less restrictive filters defined by their own users. While YouTube is the most popular with 2291 million users, other platforms such as TikTok (700 million users) are captivating different audiences (especially teenagers, Generation Z). From the image content platforms, the most used in consumer research is Instagram (1221 million users). The multiple-purpose platforms of WeChat and Xiaohongshu mix social media from China with chat and e-commerce functionalities, half private–half public content. However, the specific case of social

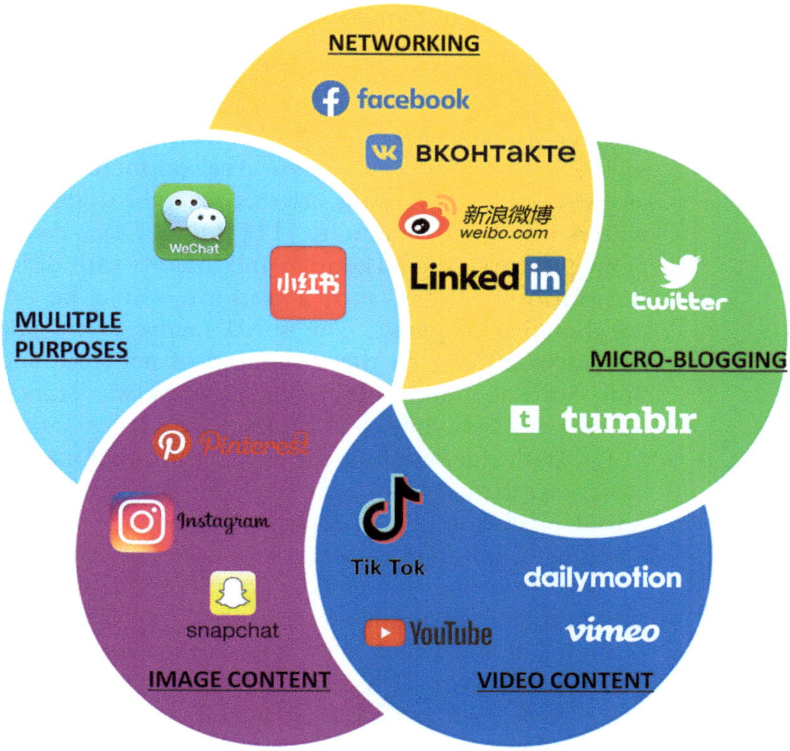

Fig. 1 Social media categorization depending on their main feature: networking, micro-blogging, video content, image content, and multiple purposes

media in China is far more complex. There are currently several restrictions to both accessing sites from China, and to China. Additionally, there are hundreds of firewalls that limit the content that can be posted on social media, for example, certain sensitive content cannot be posted by a Chinese user, nor can it be accessed outside China (e.g., a LinkedIn post from France cannot be accessed in China). And finally, the social media landscape in China is constantly changing, as there are multiple new social media channels that emerge every month. It may be worth including an additional chapter on how to perform social media research in China, but that subject will not be covered here.

Some social media platforms are used almost globally (e.g., LinkedIn but which has stopped operations in China at the moment of writing this Chapter), others are not able to be used in certain countries (e.g., Facebook, Twitter, or Instagram in China), and there are also regional applications such as Vkontakte (mainly used in Russia, Ukraine, and Belarus), Sina Weibo, and WeChat (mainly used in China). From a generational perspective, some applications are more popular with younger people (e.g., TikTok) or older consumers (e.g., Facebook). These differences across platforms are important not only at the moment of the study definition but also when drawing a line to delimit what social media is.

1.1 What Is Social Media... and What Is Not?

The era of social media as we understand it today probably started in 1998, when Bruce and Susan Abelson founded "Open Diary," an early social networking site that brought together online diary writers into one community [11]. Open Diary enabled different people to create digital content and interact with different users of the platform. Today, such a platform would be categorized as collaborative blogging. The social media as we know it today started in 2005 with the beginning of networks such as Yahoo 360, YouTube, and Facebook (Facebook was launched in 2004 but was available only to Harvard students. *See*: Edosomwan et al. [12]). With the development of different platforms specialized in either text sharing (e.g., Twitter), video sharing (e.g., YouTube), or mixed platforms (e.g., Facebook), it became important for the scientific community to define what social media is.

We can define social media as a group of Internet-based applications that build on the ideological and technological foundations of Web 2.0 and which allow the creation and exchange of user-generated content. This content needs to fulfill three basic requirements to be considered as such: first, it needs to be published either on a publicly accessible website or on a social networking site accessible to a selected group of people; second, it needs to show a certain amount of creative effort; and finally, it needs to have been created outside of professional routines and practices [11]. To better illustrate this definition, Fig. 2 separates social media into two groups: social networks and forums and illustrates the different types of mainstream media that can also be susceptible to consumer research: reviews, news and press, and institutional media.

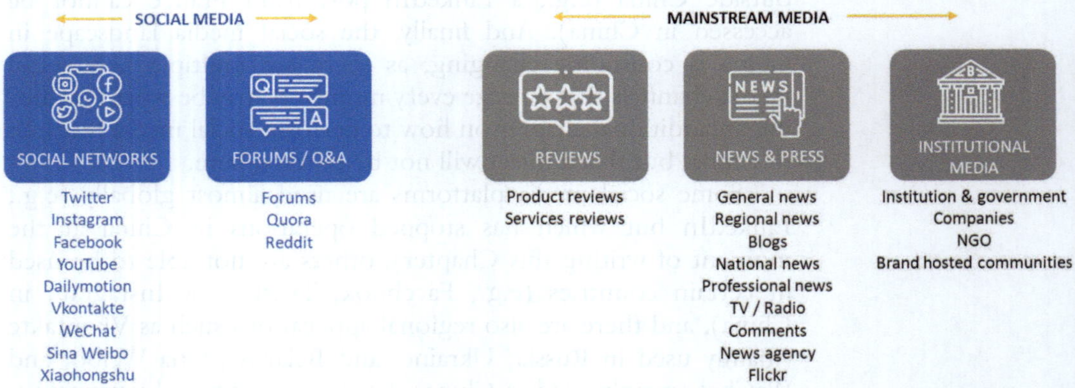

Fig. 2 Media categorization in social and mainstream

1.2 Social Media as a Digital Self

The rise of cell phones, as well as Internet-based chat applications, has changed both our forms of written communication and our expectations about public behavior. Cell phones allow individuals to be linked to two places at the same time, where they are physically present in one place while in connection with the people on the other phone in a different location [13]. When a person is selecting the information to be shared on social media, they are building what the authors of this chapter call a "digital self." This digital self usually represents one part of the person that is posting social media content, a self that is carefully selected to be at the regard of hundreds or thousands of persons. This digital self has a story and a defined space and time. It has a physical date of birth and an account creation date. The digital self can be gendered or genderless (e.g., in Instagram, a user can be gendered as a man, woman, non-binary, or any other gender defined by the user). The digital self usually geo-localizes itself with an account of origin (e.g., an Instagram account from France) but can also geo-localize where they are at a specific time and space (e.g., in Singapore for a business trip). When the content is seen on the social media, their social network can interact with them, and they can be digitally present while being physically absent. As the social media user is posting what he/she decides, a social media researcher can only analyze what is publicly accessible in the media. Despite this, a priori selection of the social media user, the amount of data is so big that it helps consumer researchers to understand different phenomena. But what drives a person to post so much content on social media?

Prada [14] suggests that social media reflects one important characteristic of our society, which is undoubtedly a confessional one. In the current communicative excess typical of the network, everything accesses the condition of language, everything is publicized and externalized. For many Internet users, no impulse or idea seems unacceptable for it to access a social and public dimension; every occurrence, thought, or desire manifests itself; everything seems to find satisfaction in the field of representing the performance for the camera of a mobile device. In the words of Sibilia [7], the contemporaneous consumer becomes the protagonist of his own story. A story in which the person is at the same time the producer, the protagonist, and the diffuser of their own story. At least in the United States, it is estimated that more than half of young people publish their biographical data and images on the Internet without any concern regarding the defense of their own privacy or that of their friends, relatives, and colleagues, who also tend to inhabit their audio-visual confessions.

In a similar direction, Toubia and Stephen [15] suggest that there are two main motivations for consumers to post content on social media: intrinsic utility and image-related utility. Intrinsic utility assumes that users receive direct utility from posting content and leads to "the doing of an activity for its inherent satisfactions rather than for some separable consequence." Image-related utility, on the other hand, assumes that users are motivated by the perceptions of others. Image-related utility is also related to status seeking or prestige motivation. However, not all social media posting is about one's experiences with the social life, as people also use social media to talk about material or services experiences. According to Balbi et al. [16] many consumers express their opinions about a product or service without any censorship by simply posting or commenting on various word-of-mouth channels such as Twitter, Facebook, and even Amazon. Many consumers or social media influencers are also paid or persuaded to make positive or negative comments on digital platforms, either in social or in mainstream media, and this is called "paid review" [17]. As commented by Choi et al. [18], while most of the customers simply write about their satisfaction or feelings about a product or service on social media, some influencers provide a detailed review of key functions of a product or service and/or frank criticisms of their shortcomings.

1.3 Dos and Don'ts: Legal and Ethical Considerations

Some authors mention that there are several challenges in determining how to transfer traditional ethical principles to the world of social media, addressing the duties and obligations of the researcher [10]. There are several examples of serious ethical mistakes in social media research. The first one dates back to the early years of social media. In September 2008, a group of researchers publicly released data collected from the Facebook accounts of an entire group of college students. The dataset comprised machine-readable files of virtually all the information posted on approximately 1700 Facebook profiles by an entire cohort of students at an anonymous, northeastern American university. Profiles were sampled at one-year intervals (with permission from Facebook and the university in question) beginning in 2006. This first wave covers first-year profiles, and three additional waves of data will be added over time, one for each year of the cohort's college career [19]. The researchers took various steps to protect the identity of the subjects, including the removal of student names and identification numbers from the dataset. However, despite these steps, the name of the university was easily identified within days of its public release as Harvard College [20]. This re-identification of the source institution of the dataset reveals the fragility of the presumed privacy of the subjects under study [21].

The second and more recent example involves a data research company. In March 2018, news emerged that Cambridge Analytica was involved in the collection of identifiable personal data from more than 87 million Facebook users to influence voter views on behalf of politicians. This instance of social media research by Cambridge Analytica was "grossly unethical," according to Liptak [22], with Facebook losing millions in market value in the space of a few hours of the scandal being announced [23], and Cambridge

Table 1
Four privacy violations to avoid in social media research

What	Why
Amount of personal information collected	Privacy violations can occur when "extensive amounts of personally identifiable data are being collected and stored in databases"
Improper access to personal information	Privacy violations might occur when information about individuals might be readily available to people not properly or specifically authorized to have access to the data
Unauthorized secondary use	The use of personal information raises the concern that information collected from individuals for one purpose might be used for another secondary purpose, without authorization from the individual, and thus the subject loses control of their information
Errors in personal information	The impact of possible errors within datasets
Research publication	The improper publication of social media research, showing personal details of the people involved in the study, may also be a privacy violation

Analytica closing its doors to further operations a few months after the scandal. How to avoid this series of ethical challenges? Table 1 shows five different situations that may occur in privacy violation when conducting social media research. For further understanding, the readers can see the Internet Research: Ethical Guidelines, from the Association of Internet Researchers—AOIR [24], as well as the General Data Protection Regulation—GDPR, which is the European Data Protection Regulation and is applicable as of May 25th, 2018 in all European Union member states to harmonize data privacy laws [25].

2 Different Applications of Social Media Research

Today, the use of social media as a research tool can be used to cover a wide range of objectives. From its early uses to understand human networks [20] to the new applications, the number of publications is growing and growing. Figure 3 shows an attempt to summarize the multiple uses of social media into four big areas: (1) Exploratory research—to understand general attitudes, perceptions, emotions, and expressions of consumers; (2) Trend-watch—to discover new products, new flavor combinations, innovation, etc.; (3) Netnography—Ethnographic approaches online to follow the online content associated with specific social media accounts; and (4) Community management—which is usually more corporate, where companies, brands, and social media influencers analyze the impact of their content in social media. These four applications are explored in more detail in the following sections.

Fig. 3 Four different uses of social media research: exploratory research (study of Mexican ingredients through chefs), trend watch (Instagram account dedicated to trends in flavors), netnography (online ethnography study of targeted Brazilian consumers' food consumption), and community management. (Images from authors: first three images with permission of users, last image adapted from pixabay.com, royalty free)

2.1 Exploratory Consumer Research: Emotion, Perception, and Attitudes

The most common use of social media research in the food industry (in consumer research departments), and in academic research, might be the exploration of consumer attitudes [1, 26], habits [27], perception [4] (of brands, products, ingredients, flavors, etc.), emotions evoked [28] (of products or brands), and general opinions (e.g., of food legislation, labeling, etc.) [29]. Examples of these studies usually use the number and word counts in tables, word clouds, histograms, and sometimes more complex analysis. For example, Fig. 4 shows a study on mint perception in Latin America carried out in 2019 by the authors. In this study, the social media mentions of the use and perception of mint and mint products is crossed with Google searches. The scatterplot is divided into four zones, when it is crossed by the mean of social media mentions and of Google searches. The Prioritize zone highlights the areas that have both high social media mentions and high searches on Google. The Watch zone highlights a high number of social media mentions but lower searches in Google. The Investigate zone is the area that has a high number of Google searches but low social media content, which means that the consumers are interested but not sharing any information, or very little. And finally, the Weak signals zone is where both the social media mentions and Google searches are very low.

The use of this type of matrix is illustrated in Fig. 5, where the Prioritize and Watch zones are qualitatively clustered to identify the main uses and application of mint in different products and benefits. The clusterization is more evident for some products/uses than others. For example, the first dimension identified is at the far left of the graph, where cocktail is classified in the dimension of Alcoholic Drinks. At the far right, Tea is also kept as a unique dimension. The dimensions in the middle of the graphs can be confused between

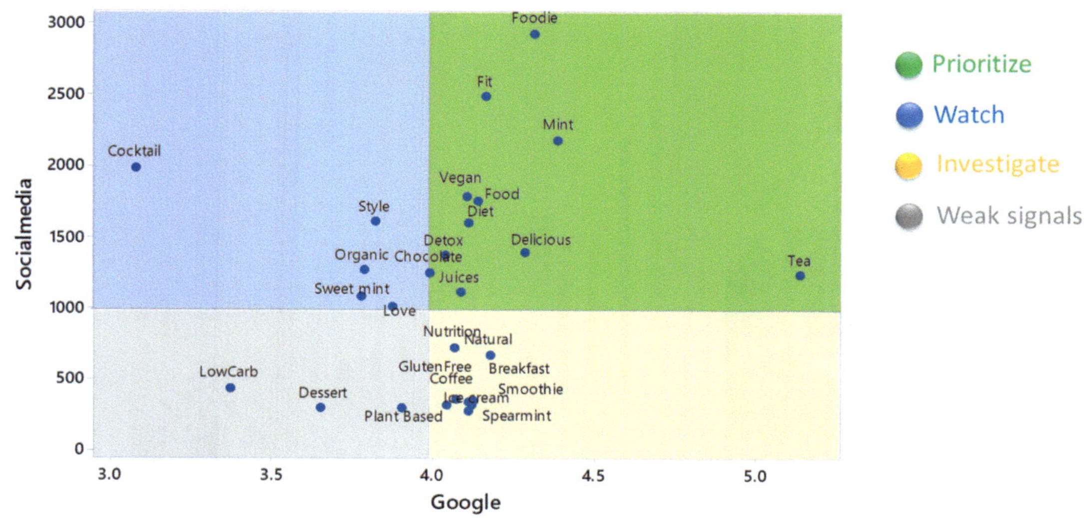

Fig. 4 Matrix crossing the Google search in the horizontal axes in the logarithm scale and the number of social media mentions in the vertical axes for different uses and perception of mint. (Data from Google Keyword Planner (searches) and Synthesio (social media). Data collection: Latin American countries, from August 2017–2019. Languages for data search: Spanish, Portuguese, and English)

Fig. 5 Prioritize and Watch zones were taken to perform a qualitative classification of all possible mint uses and product perception. Data was then structured in five main uses: alcoholic drinks, organic & natural, healthy-fit-detox & vegan, foodie, and tea. (Images: Royalty free from pixabay.com)

Fig. 6 Graph showing the number of mentions for the emotions associated with mint and mint products. Green bars are those that have a higher number of mentions than the average of all mentions (average = 332). Gray bars are below the average of mentions. (Data from Social Media study August 2017–2019 (same as Figs. 4 and 5))

them, but they were separated into three independent dimensions of Organic & Natural, Healthy-Fit-Detox-Vegan, and Foodie. This basic analysis can be of great support in explaining the main different uses and benefits of mint.

In addition to the uses and perception of the mint products, a more specific analysis can be done by simply counting the main emotions associated with mint. Figure 6 shows the number of emotions; the search is done manually in the same Mints study in Latin America. The emotions were filtered and counted manually. To facilitate the search for emotions, the list of terms was determined by 25 emotions from the ScentMove model [30]. The graph shows all the emotions in a histogram, where the blue bars indicate that the emotions were used above the average of emotions and the gray bars are emotions that were elicited below the average score.

2.2 Trend Watch: Identifying Emerging Conversations and Influencers

Given the Internet's scalability and speed of diffusion, some contributors attract a mass audience, build a fan base, and become a source of advice for their followers, thus developing into social media influencers (SMIs). Who these SMIs are and how they influence consumer behavior is now a significant and inevitable topic for both academics and practitioners [31]. Some of these SMIs are considered trend-setters, and therefore their study can be a source of inspiration for new product trends, ingredients, opinions, etc. They can be studied as a group of persons or as individuals (for more on this, refer to the next section on netnography). Other trends can be studied as a function of a global phenomenon such as the COVID pandemic.

Fig. 7 Word cloud of touch-related mentions. Total mentions in study = 416,000. (Data from Social Media through Synthesio platform. Global reach in English, period January–November 2020. Social media posts from users were rebuilt using main words from the original post ("...") indicates that original text continued) to avoid social media user identification. Name and pictures were also removed from the original post)

Figure 7 shows an example of a study performed during the COVID pandemic in 2020 with the objective of understanding the effect of lack of touch, especially during COVID lockdown. In this study, a first exploration of the words related to touch is shown in the form of a word cloud, and at the right of the figure, there are some examples of the actual posts these words come from. After performing a textual content analysis, the words were divided into three Re-Touch dimensions.

The Re-Touch dimensions are an example of the analysis of a consumer trend, where consumers are searching for a way to replace the sensation of touch with another sensory stimulus. The first dimension is physical touch; an image at the left illustrates the way in which the people are replacing the touch of a mug (with hot chocolate) and connecting it to a physical touch. The second dimension is connecting (social) touch, where the persons search for a connection between food and digital connection (e.g., Sharing food in virtual parties with friends via digital platforms during the lockdown period). Finally, the third dimension is spiritual touch and is associated with a connection between food and other persons that cannot be physically present; the example shown talks about a memory of grandparents linked to a specific food odor or taste (Fig. 8).

Fig. 8 Social media dimensions of Re-Touch through flavor experiences: physical touch, spiritual touch, and connecting (social) touch. One example per dimension is shown. Social media posts from users were rebuilt using main words from the original post ("..." indicates that original text continued) to avoid social media user identification, images were replaced with royalty-free ones from freepik.com

2.3 Longitudinal Monitoring: Netnography

Netnography is a relatively new research method, which adapts research techniques of ethnography to study cultures and communities through computer-mediated communications [32]. Initially defined by Kozinets [33], based on the developments of research methods in the field of marketing, it is also referred to, but less frequently as cyber-ethnography, digital ethnography, or online ethnography. As for classical ethnography (please *see* Chapter 3), the netnographic process should include enough time to observe the phenomena being studied, and it also follows the standard qualitative techniques of data analysis [34]. In the case of netnography applied to social media, it has previously been published how longitudinal monitoring of behavior, on Twitter, for example, can provide sufficient detail for researchers to understand specific phenomena [35].

In the area of consumer research applied to food and beverages, there are also publications of netnography, most of them in the area of tourism, like wine tourism [36], food tourism [37], or Indonesian gastronomy [38]. Therefore, this is clearly an opportunity for consumer researchers working in the field of food and beverages and is also a methodology that can provide additional information for consumer researchers. For example, Fig. 9 shows a selection of four images from a netnography performed in Brazil with "foodies." The study was performed during a one-year period in specific Instagram accounts and with the permission of the social media users. The objective was to understand attitudes toward food,

Fig. 9 Netnography of a Brazilian "foodie." Pictures and hashtags are from the user and show the person's food consumption in their local city (Sao Paolo, Brazil), and in different cities visited between 2018 and 2019. (Images from Instagram published with user's permission)

perception of different ingredients, and more versus less common ingredients and flavor combinations. The analysis, not presented here, was done following the traditional aspects of qualitative analysis (*see* Chapter 5).

2.4 Other Applications: Community Management and Communication

Maybe the most common use of social media, and one which is usually out of the scope of consumer researchers, is community management. The objective of community management is to listen and measure the digital social impact of social media activities and how a specific audience or users are reacting (like, comment, or share a post) toward a specific social media content posted by an influencer or a brand content writer. According to Ang [39], a better term could be community relationship management, because it more accurately reflects what people do in online communities— connect, converse, create and collaborate.

This tracking activity is crucial for the community manager for the following:

(a) Handle brand e-reputation survey, quickly detect and address online crisis in real time, improve brand image
(b) Benchmark brand visibility and liking, compared to competition
(c) Refine the content creation strategy and set the best posting timeline (frequency of posts, best days in the week, and best hours of the day) in order to maximize the reach, retain targeted audience, and develop follower base
(d) Brand, service, or product engagement to help increase relationship, trust, commitment, or loyalty [40]

3 Social Media Research Step by Step

Designing and executing social media research can be quite challenging for someone outside the discipline of consumer research. It will be easier for a consumer researcher to start social media projects versus someone outside the discipline, for example, a marketeer or a food scientist. Overall, all the good practices of consumer research (ethics, objectives definition, data extraction and analysis, etc.) are applicable to social media research. The new aspects that emerge when doing social media research are the data extraction principles (different options can be used, *see* Fig. 11) as thousands of persons can be included in the study, compared to the usual hundreds in traditional consumer research. To better explain the usual process of social media, the reader can see Fig. 10 as a general framework of a social media study. However, as in every consumer research study, some studies can be rather unique, and other principles should be taken into consideration.

Step one of the social media study is to set up the objectives, which should be accomplished through social media listening. As simple as it may sound, it is actually not that obvious. The first mistake made in social media research is thinking that everything can be solved through social media and going back to Subheading 1.2, the researcher needs to think and sometimes do an initial pilot study to understand if the objective can be accomplished. For example, the study of the needs of low-income consumers cannot be easily studied through social media, as low-income people may not have access to social media in certain countries, and when these consumers have access to social media, their specific first-hand needs can be difficult to study through social media posts. Therefore, we recommend researchers do a first social media check and see if the objectives of the study can be accomplished through social media research. Once the objectives are clear, the selection of the social media channel should follow (use of Twitter, Instagram, YouTube, etc.), together with the desired country, language, and time frame. It is common to have studies that cover a wide range of

STEP 1 **OBJECTIVES**	STEP 2 **DATA PROCESSING**	STEP 3 **DATA ANALYSIS**	STEP 4 **FINAL INTERPRETATION**
• Fix clear objectives for study that can be solved through social media research. • Define social media channel & target users (country, time, etc.).	• Select data collector tool. • Define Boolean search with key words & hashtags. • Perform a pilot run to check the accuracy of the dataset. • Remove noise after pilot run and define final social media query. • Apply final query and run the search in social media monitoring tool.	• Define filters (e.g., emotions, conversation drivers, etc.). • Word count + Textual analysis (statistical analysis, thematic analysis). • Image analysis, emojis & hashtag. • Perform data clustering. • Make additional analysis (e.g., correspondence analysis, comparing themes) with human review.	• Extract key dimensions, semantic and sentiment analysis to illustrate the findings. • Assure that analysis responds to the defined study objectives. • Create final data visualization.

Fig. 10 Framework of the main steps in social media research

time, going from 1 month to years. Today, certain social media tools enable the study of social media content in a 3-year frame (in other words, we can go back 3 years in social media listening).

The second step to follow is the selection of the data extraction tool or data collector tool. Open tools can be found, and multiple academic publications use open sources such as **R** (twitte**R**) or simply the advanced search function of Twitter. Tools such as twitte**R** embedded in statistical software have only limited functions, and the data extraction does not allow a highly specific search in terms of theme (e.g., complex Boolean queries), geographical localization, use of multiple languages, or restriction of specific words. However, these open sources like Twitter are easy to use. On the other hand, private tools (e.g., Synthesio or Twitter Analytics) have a higher range of options to improve the extraction of the social media content but licenses need to be paid, which sometimes exceed the budget of certain researchers. Nevertheless, the data extraction tool used by the researcher must allow the definition of the key words to use in the extraction through Boolean search (*see* Subheading 3.1) and remove the noise after performing a pilot search. Noise is defined as the social media content that appears in the data extraction that is not related to the study objective, and if the noise is included in the data analysis, it will heavily bias the results of the study. Examples of noise can be retweets (if the researcher is only interested in primary social media content) or unrelated content which is confused in the extraction (e.g., political content). Once this noise is identified, it should be automatically removed in a subsequent extraction by adding filters (*see* operators NOT in Subheading 3.1). It is quite common to perform a first pilot run of the data and make a top-level analysis of that initial data. Additional noise can be removed and then the final query should be defined to run the extraction and have the final data. A query is the equivalent of a questionnaire in a traditional consumer test; the query contains all the words used to extract the data; these words should be in a special format, using Boolean operators (also *see* Subheading 3.1).

The third step is the data analysis, and in this part, we work with the data extracted but we also define additional filters. For example, if we work on the perception of fresh flavors, we can define specific filters of the data to extract the emotions associated with fresh flavors, their sensory attributes, and the main words associated with that fresh flavor's conversation. The filters are a subset of key words related to that theme. In the case of emotions, we can filter what the emotions used were, and to do that, we can use a set of emotional terms from a specific methodology (e.g., ScentMove [30]). Then the emotions are counted (how many times nostalgic is mentioned versus joy, happiness, etc.), and we can then make sub-filters, for example, the emotions associated with fresh flavors by different categories of food (savory or sweet goods); and then this wordcount can be re-analyzed into a correspondence analysis, or simply a percent of the emotions column chart.

The fourth and last step is the final data interpretation. It consists of putting together all the information to extract the key dimensions of the study. Going back to the fresh flavors project example, we could understand what dimensions are associated, such as: product categories, sensory attributes, and emotions associated with masculine flavors. With all this data, we need to go back to the objectives (as always) to be sure that the data, the quality, extension, and level of analysis and details match our objectives. And we then want to make additional data visualization to explain what the social media users comment around fresh flavors (for example).

3.1 Source's Definition and Boolean Search

Defining clear objectives of the research is the first important step. This helps define the right coverage (country, language, timeline, etc.) and the right keyword list to be set in the query; these keywords used are the heart of the social media study. Most of the current social media monitoring tools use Boolean formulas to associate keywords, in order to select the right content and the right topics, from different sources. Boolean search or Boolean formulas are technically called Boolean IR (information retrieval) systems [41]; and they are a set of operators that enable a logical search in different information systems such as Internet, intranet, or social media. The most common Boolean operators are "OR," "AND."

A query sentence is built up using two types of wordings:

(a) *Operators*: OR, AND, NEAR, NOT. Operators are always written in English, and in uppercase, and are not to be translated to other languages even if the research coverage is other than English-speaking countries. Other common Boolean operators are quotation marks ("·") to search for an exact phrase and brackets (), to group multiple search strings and set priorities [42].

The case of NEAR is specific and can be used simply as NEAR, or some social media tools enable the distance function in the NEAR operator. For example, if we want to look for words near five words of "fresh," we can use the operator NEAR/5 (flavoNEAR/5 fresh). The objective of this distance operator is to reduce the noise and identify words or concepts that are related. It will not be the same to find: "I love fresh air from mountains while hiking and I love the flavor of a dry hoppy Indian pale ale beer" versus "Dry hops give such a fresh flavor to Indian pale ale beers" (the first sentence is not accurate if we look for fresh flavor, while the second sentence will do).

(b) *Keywords*: a list of keywords should be defined corresponding to the themes and objectives of the research. Those are different terms and their synonyms that social media would use to talk about the topic of your interest. These terms should be written in the local language of the geographic coverage

In a social media query, different units of keywords can be (or should be) associated to each other by operators:

- Operator "OR" is used when there are many synonyms or many possibilities of writing for one concept.
- Operator "AND" helps narrow down too broad a topic to a more specific one by linking it to another set of keywords. Operator "AND" could be replaced by operator "NEAR" to collect a smaller but more accurate dataset.
- Operator "NOT" helps exclude noise and unwanted content.
- Asterisk (*) have been recently used in some social media tools to search for words having the same root or, its invariable nucleus, together with its derivations (conjugation, quantity, adjectival, adverb, etc.). For example, for the word emotion: emotional, emotions, emotionally, emotionalize, etc. In other words, the asterisk will cover the derivations of emotion to: -al, -s, -ally, -ize.

Let's analyze an example of a basic Boolean search query:

If your objectives are to understand how people in the UK are talking about strawberry ice-cream, but you are not interested in the conversation talking about strawberry ice-cream promotion or ice-cream packaging, nor people complaining about a delivery issue of strawberry ice-cream; your query should be written as:

(("ice cream" OR "icecream") AND (strawberry OR strawberries)) NOT (delivery OR packaging OR promotion)

Now... **Let's analyze an example of a more complex Boolean search query:**

If your objectives are to understand the perception of mint, spearmint, or peppermint in premium sweet goods categories, the dashboard should be written as:

(((mint OR minty OR menthol OR spearmint OR peppermint) NEAR/7 (baked OR baking OR biscuit OR brioche OR brownies OR bun OR butter OR cake OR candies OR candy OR cereals OR cheesecake OR "chewing gum" OR chewinggum OR chocolate OR confit* OR cookies OR cracker OR croissant OR dessert OR dough OR doughnut OR "ginger bread" OR gingerbread OR gum OR gummy OR jam OR jellies OR jelly OR licorice OR lollipop OR macaroon OR marshmallow OR muesli OR oats OR panettone OR "panna cotta" OR pannacotta OR pie OR shortbread OR sorbet OR sugar OR tart OR yogurt OR yoghurt)) NEAR/7 (fancy OR extraordinary OR rare OR phenomenal OR exception* OR incredible OR premium* OR unique* OR styl* OR classy OR sophisticat* OR complex* OR luxe OR elegan* OR outstanding OR posh OR fashion* OR prestige OR distinct* OR refinement OR delicacy OR quality OR "high quality" OR highquality OR superior* OR special OR perfect* OR excellent* OR ideal OR sublime OR great* OR expensive)) NOT (amazon OR buy OR amzn OR offer* OR voucher* OR lenovo OR nike OR google OR iphone OR android OR candle* OR adidas OR gin OR deal* OR offer* OR coupon OR shipping OR set OR price OR pack* OR "hot deal" OR coupons OR job* OR "cell phone" OR ebay OR pope OR marriage OR "blue tooth" OR ovarian OR jowl OR pet OR dog OR todayDealsSteal OR mydealdotcom OR 2mydeal)

3.2 Text Analysis and Natural Language Processing (NLP)

The text analysis step is key in the social media monitoring process. Clustering, filtering, counting frequencies, detecting conversation drivers, and observing relationships between different subtopics are the main actions that lead to insightful information in social media research.

Some social media monitoring tools have developed a text analysis technique called NLP or Natural Language Processing. This method considers the syntax structure, places or order of nouns, verbs, adjectives, adverbs, etc., in a sentence and can run an automatic text analysis by detecting natural clusters of words in a specific dataset. This technology helps analyze recurrent types of datasets faster compared to manual filtering methods. Usually, NLP technology can be run by paid analytics tools, but also in open sources such as IRaMuTeQ (http://www.iramuteq.org/), which is an **R** interphase of multidimensional text analyses, based on Reinert textual clustering algorithm [43].

To give an example, when the NLP is used in a social media tool, it can regroup all the verbs of the "strawberry ice-cream" dataset in one cluster and show the most frequently used verbs that the social media users are posting, such as: love, buy, order, eat, pair, make (at home, myself), hate, etc. In the same way, this function could be used to highlight nouns, such as brands distributing strawberry ice cream that the consumers are talking about.

The automatic processes are changing every day and becoming more efficient; however, the human understanding and analysis of the text processing remains particularly important after performing any type of automatic text process (such as NPL). Only humans and human analysis will have the ability to understand humor, language subtleties, local language-specific word variations (e.g., Spanish from Mexico versus Spanish from Argentina), emotional connections, and any type of hidden meaning when using words that machine learning cannot process or achieve yet.

3.3 Social Media Monitoring Tools

We can count dozens of major social media monitoring tools that can help the consumer researcher to extract the data from social media efficiently. There are open tools such as twitte**R** (from **R** project, as mentioned before in the Chapter), or even Twitter search engine. In the world of academia, most of the published articles might use twitte**R** (e.g., Vidal et al. [44]), but more and more articles use paid tools that have access to specific functions that will enable specific objectives to be achieved, such as cross-cultural research (Twitter cannot easily filter mentions from different languages in specified countries). There are, however, some academic articles that mention the use of specific social media paid tools (e.g., Arellano-Covarrubias et al. [2]), search engines of social media platforms (e.g., YouTube [45]), or even a mix of social media and online surveys [46].

Regarding paid tools for social media research, Fig. 11 shows an attempt to categorize different options based on their current offering (or expected share of market), and their strategy shown in their own Internet sites. All paid platforms have different strengths and functionalities, and they also have a big variation in terms of costs. It is not the scope of this chapter to make a review of social media tools or a recommendation. We leave Fig.11 as an inspiration and source of information for the readers.

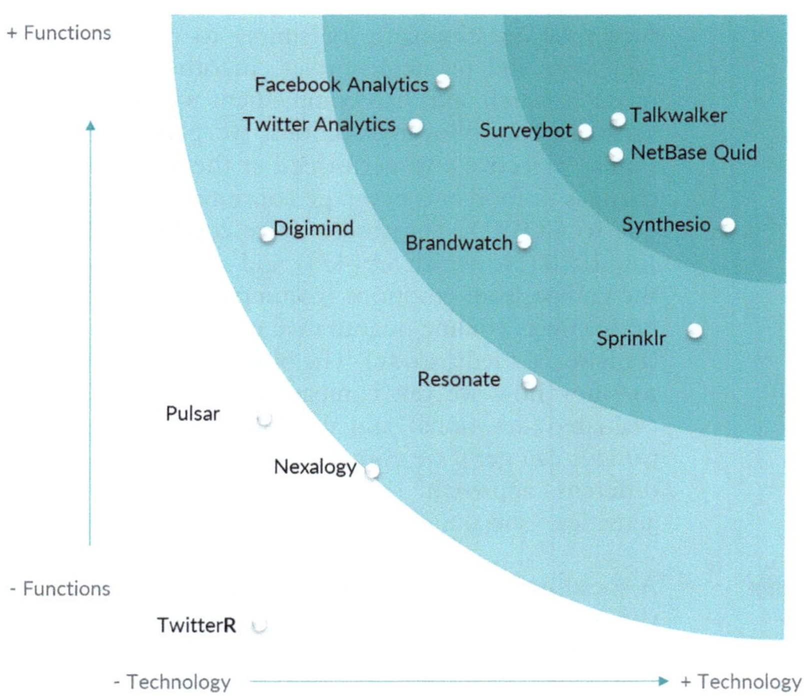

Fig. 11 Social media tools map. A qualitative positioning based on the authors' knowledge and perception of different tools. (Note: The image does not represent market share or actual offering or formal strategy. The image is built to support readers in understanding the competitive landscape of social media monitoring tools via a qualitative categorization)

4 A Case Study: COVID-19 and Its Impact on Food and Beverages

According to Vidal et al. [8] one of the prime advantages of conducting social media research is the volume of data that is available and continuously being generated by users in a spontaneous way. Compared to more traditional methods, a vast amount of information on a wide range of topics can be gathered more efficiently and at relatively low financial cost. But despite the richness of social media research, it is still questioned both in academia and the industry due to the potential biases that social media may have compared to other more "traditional methods," such as face-to-face or online surveys. The question might not be which approach is better (social media vs. online surveys), but how they can contribute to understanding consumption habits, and there are several authors that already acknowledge it. For example, Spiro [35] mentions that survey and social media data methods complement each other by alleviating weaknesses in the other approach. Moreover, linking social media to survey data affords the ability to associate self-reported behaviors with those expressed (directly or indirectly) online.

In practice, combining social media data with surveys is not without challenges. Study participants can be asked to volunteer their social media identities as part of contact information on a survey or questionnaire, or simply to combine the results as if they were independent studies. In other words, conduct social media research and then complement with online surveys, or vice versa. In the following sections, we present two studies with a similar objective and performed at the same time: to understand changes in food and beverage consumption and general behaviors due to COVID. The studies were conducted between April and July 2020 (Firmenich SA [47]), and only two questions are used for the comparison: emotions (common in both studies) and human body effects (online quantitative questionnaire)—human activities (only social media study). The objective of this small comparison is to show how we can compare the results of the same variable evaluated (emotions), and also to see the complementary information we can get if we compare similar questions but with a slightly different approach (body effect —online questionnaire and activities—social media).

4.1 Social Media Results

A social media study was performed, collecting social media conversation between the months of April to July 2020 in seven English-speaking countries: USA, Canada, United Kingdom, South Africa, Nigeria, India, and Australia. The study was conducted through Synthesio platform, using a query of COVID-19 linked to flavors and perfumery (*see* Annex 1 for complete query).

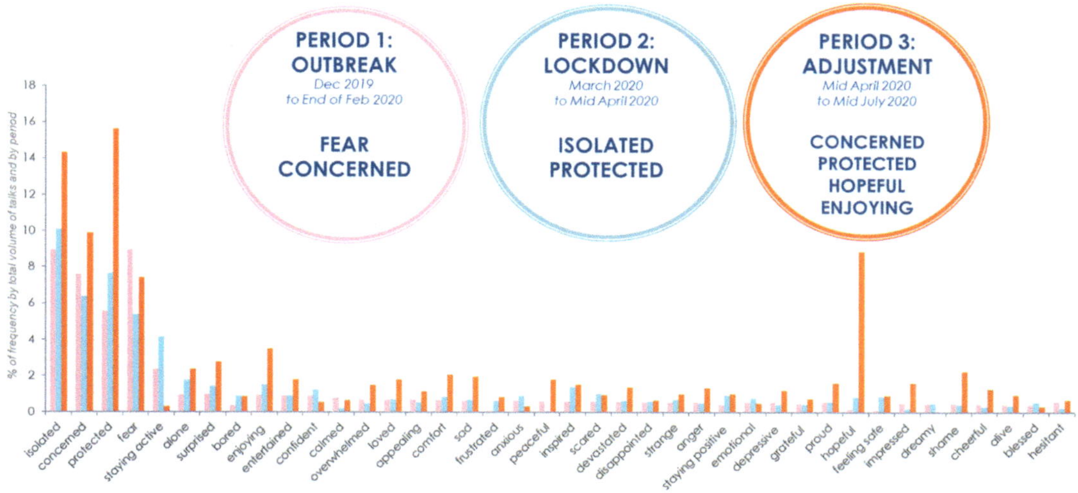

Fig. 12 List of the main emotions elicited in the social media study. Data represent the percent of total mentions

4.1.1 Emotions

The results of the emotions are shown in Fig. 12; 26 different emotions could be identified. The emotion with the highest percent of elicitation (percent of total number of mentions) is protected (15.6%), followed by isolated (14.3%), and concerned (9.9%). The rest of the emotions have a lower percent of elicitation; however, it is possible to analyze a wide variety of emotions, almost any emotion can be identified, yet they might have a very low number of mentions, compared to the top emotions of Fig. 14 (an online survey study with invariable number of emotions in the questionnaire). The list of emotions identified in social media are both positive and negative.

4.1.2 Human Activities

Figure 13 shows a graph of activity ranking from the three different studies performed in social media analysis in three different periods. The most important human activities identified in social media conversation were: work from home, use of face mask, going out to restaurants, and seeing friends. The chart makes the comparison between different periods of time easy.

4.2 Online Quantitative Results

An online study was conducted between June 19 and 25, 2020, in 22 countries: USA, Mexico, Colombia, Brazil, Argentina, United Kingdom, Germany, France, Italy, Russia, Turkey, Nigeria, South Africa, Saudi Arabia, Pakistan, India, Indonesia, Vietnam, Thailand, the Philippines, China, and Japan. A total of 6000 consumers took part in the study through an online platform. Demographic characteristics were 60/40 Women and Men, and ages 40% 18–34, 40% 35–54, 20% 55+. The questionnaire consisted of

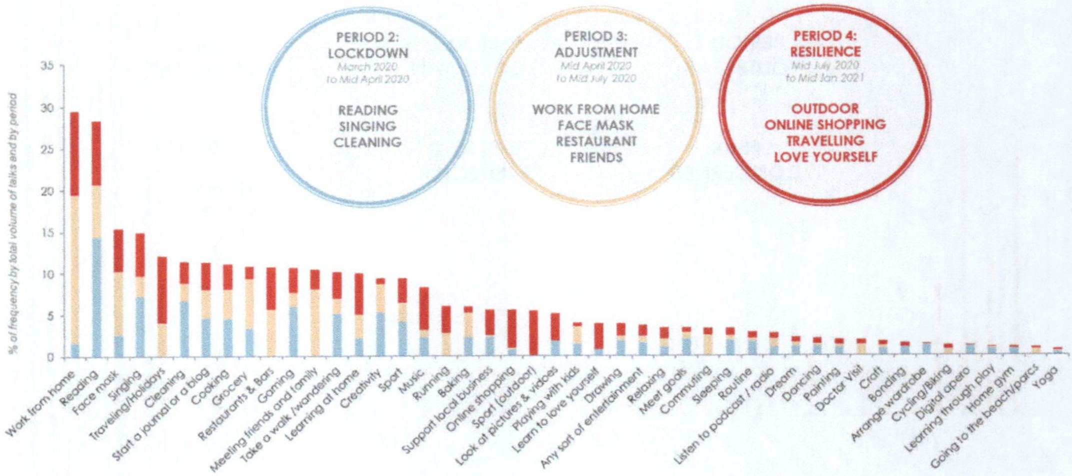

Fig. 13 List of activities in the social media study. The data is arranged by percent of elicitation compared to the total number of mentions. Orange columns from period 3 correspond to the study from April to July 2020

19 questions, ranging from CATA, single option, and scale types of questions. The questionnaire was translated into local languages and applied equally in all the countries [48].

4.2.1 Emotions

For the emotions results, question five was used: "Which, if any, of the following effects have you personally experienced more of in the past 2 weeks compared to before the crisis? Select all that apply." The results are shown in Fig. 14. To facilitate the comparison with the social media study, two bars are shown: the blue one corresponds to the total percent (22 countries), while the orange bar corresponds to the English-speaking countries (almost the same as the social media study: USA, UK, South Africa, Nigeria, India, the Philippines, and Australia). The top three emotions identified were boredom (40%), stress (35%), and family bonds (39%).

4.2.2 Body Impacts

For the impact on body or body changes, the following question is used: "Which, if any, of the following effects have you personally experienced more of in the past 2 weeks compared to before the crisis? Select all that apply." Like the emotion question, in blue the bars of the global average are shown and in orange the English-speaking countries, to make a more direct comparison with the social media study (performed in seven English-speaking countries, Fig. 15). Taking into consideration the results from the English-speaking countries, the variable with the higher elicitation is "lack of movement," with 38%, followed by "gain weight," with 37%, and "lack of sleep" in third place, with 28% elicitation.

4.3 Takeaway

By just comparing two out of many questions, social media and online studies show similar information. The metrics are different (number of mentions in social media and % in the online study).

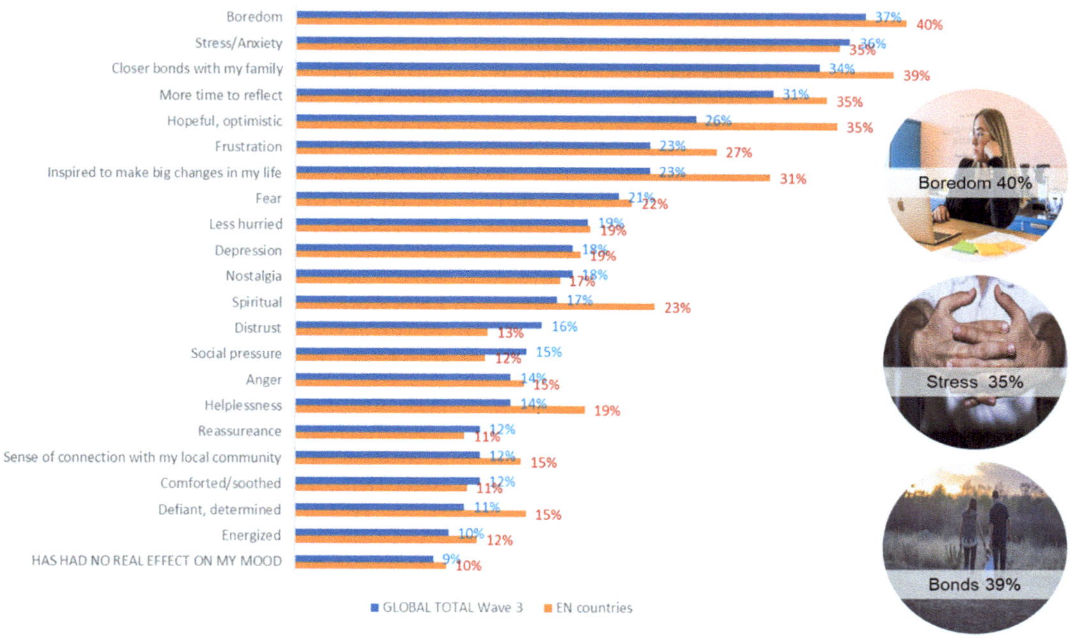

Fig. 14 Percent of elicited emotions obtained in a CATA question of emotions. Blue bars are the weighted average of the total 22 countries, and orange bars are the weighted average of the English-speaking countries

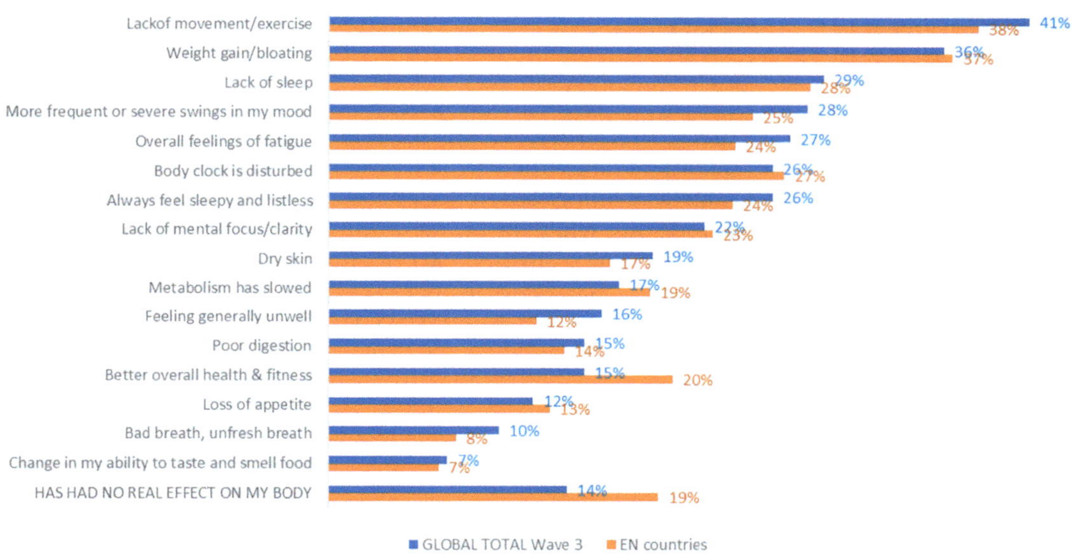

Fig. 15 Percent of body-change responses obtained in a CATA question of emotions. Blue bars are the weighted average of the total 22 countries, and orange bars are the weighted average of the English-speaking countries

However, the results can be comparable, not in terms of similar variable results, but in the overall trend. The social media study shows a greater variety of responses, and the data can easily be split by gender, age, and country (like the online study). On the other hand, the online study shows clearer trends of data with higher percentages across the variables measured, which is a result of the design per se of the study (invariable number of responses, invariable list of options in each question, etc.).

In a numerical comparison, the results can seem similar, but maybe the key question relies on the nature of the response being measured rather than the mere numerical comparison. In the case of social media, the number of mentions corresponds to a match of words in a post, which depends strongly on the context of that specific post, and the tweets and retweets may or may not count as double mentions (depending on the configuration of the social media search and the platform used for the extraction). For example, in the case of the emotions, the emotion with the highest percent of elicitation is "protected." But when analyzing the details of each Instagram post or tweet, protected can be separated into "I want to feel protected" versus "I do not feel protected," and therefore the social media results will strongly depend on the context of each post. Meanwhile, in the online survey, as it contains a specific question, with a detailed option of responses, the question can be considered as it was written in the questionnaire (e.g., I feel boredom). The results of the online study could initially be a study with a higher accurate response, however, in online studies, there is always a percentage of error due to fatigue and errors or misconceptions when reading both the questions and the options for the responses [49], socially accepted answers in surveys [50], or simply bad quality of respondents [51]. And most importantly, the list of emotions was simply different. The social media could filter a long list of emotions, while the online study was fixed to a set of 25 terms coming from the ScentMove list of emotions [30]. In the end, the online or the social media studies are taking different approaches to assess the perception of the COVID-19 changes, and the results can be taken as complementary rather than competing to be the "best" approach.

5 Conclusion: Strengths and Limitations

One of the main challenges of social media research is related to the analysis of large amounts of textual information, which ideally requires the use of advanced automated text analysis tools. Also, language and cultural skills are required to interpret social media data, especially when dealing with data from different world regions [8].

A short list of advantages:

- Unbiased access to respondents as the respondents are not directly contacted to answer a question
- Global access to consumers, any time, and for almost all possible topics.
- Large amount of data extracted in text and images
- Fast data extraction, enabling faster studies when the data is processed automatically
- Use of natural consumer language

A short list of disadvantages:

- The data are highly unstructured and need text processing and cleaning before performing the data analysis.
- Open social monitoring tools have restricted functions, as they usually do not filter by time, geo-localization, and an effective Boolean search.
- The data strongly depend on the context and have high risk of misinterpretation with unexperienced social media researchers.
- Social media give us access to what people show digitally, and not necessarily to what people normally do.
- Social media research may not apply in specific research objectives or specific communities (e.g., low-income consumers or can reach out to only the population that is very connected).

Acknowledgments

The authors wish to thank Jeffrey SCHMOYER for sharing the COVID-19 online study conducted by his team Human Insights, of the Taste & Beyond division of Firmenich, SA in 2020.
Images: Authors also wish to thank the social media users for their permission for the use of images: Itzia Rojas, Patricia Niizu, and Elke Dewolfs.

References

1. Lunardo R, Jaud DA, Corsi AM (2021) The narcissistic wine consumer: how social attractiveness associated with wine prompts narcissists to engage in wine consumption. Food Qual Prefer 88:104107
2. Arellano-Covarrubias A, Gómez-Corona C, Varela P, Escalona-Buendía HB (2019) Connecting flavors in social media: a cross cultural study with beer pairing. Food Res Int 115: 303–310
3. Simeone M, Scarpato D (2020) Sustainable consumption: how does social media affect food choices? J Clean Prod 277:124036
4. Khan S, Umer R, Umer S, Naqvi S (2021) Antecedents of trust in using social media for E-government services: an empirical study in Pakistan. Technol Soc 64:101400
5. Simmel G, Wolff KH (1964) The sociology of Georg Simmel. Free Press, Glencoe

6. Lewis K, Kaufman J, Gonzalez M, Wimmer A, Christakis N (2008) Tastes, ties, and time: a new social network dataset using Facebook.com. Soc Networks 30(4):330–342
7. Sibilia P (2017) La intimidad como espectáculo. FCE, Ciudad de Mexico
8. Vidal L, Ares G, Jaeger SR (2018) Application of social media for consumer research. In: Ares G, Varela P (eds) Methods in consumer research. New approaches to classic methods, vol 1. Woodhead Publishing, UK
9. Pittman M, Reich B (2016) Social media and loneliness: why an Instagram picture may be worth more than a thousand Twitter words. Comput Hum Behav 62:155–167
10. McKee R (2013) Ethical issues in using social media for health and health care research. Health Policy 110:298–301
11. Kaplan AM, Haenlein M (2010) Users of the world, unite! The challenges and opportunities of social media. Bus Horiz 53:59–68
12. Edosomwan S, Prakasan SK, Kouame D, Watson J, Seymour T (2011) The history of social media and its impact on business. J Appl Manag Entrep 16(3):79–91
13. Stillerman J (2015) The sociology of consumption. A global approach. POLITY, Cambridge
14. Prada JM (2018) El ver y las imágenes en el tiempo de internet. AKAL/Estudios Visuales, Madrid
15. Toubia O, Stephen AT (2013) Intrinsic vs. image-related utility in social media: why do people contribute content to Twitter? Mark Sci 32(3):368–392
16. Balbi G, Magaudda P (2018) A history of digital media: an intermedia and global perspective. Routledge, New York
17. Ko MC, Huang HH, Chen HH (2017, August) Paid review and paid writer detection. In: Proceedings of the international conference on web intelligence, pp 637–645
18. Choi J, Yoon J, Chung J, Coh BY (2020) Social media analytics and business intelligence research: a systematic review. Inf Process Manag 57:102279
19. Kaufman J (2008) Tastes, ties, and time: Facebook data release. Retrieved from: https://cyber.harvard.edu/node/94446. March 2021
20. Zimmer M (2008) More on the "anonymity" of the Facebook dataset – It's Harvard College (Updated). Retrieved from: https://michaelzimmer.org/2008/10/03/more-on-the-anonymity-of-the-facebook-dataset-its-harvard-college/. March 2021
21. Zimmer M (2010) "But the data is already public": on the ethics of research in Facebook. Ethics Inf Technol 12:313–325
22. Liptak A (2018) The US government alleges Facebook enabled housing ad discrimination. The Verge
23. Michaelidou N, Micevski M, Cadogan JHW (2021) Users' ethical perceptions of social media research: conceptualisation and measurement. J Bus Res 124:684–694
24. AOIR – Association of Internet Researchers (2019) Internet research: ethical guidelines 3.0. Retrieved from: https://aoir.org/reports/ethics3.pdf. March 2021
25. GDPR, General Data Protection Regulation (2021) General data protection regulation. Retrieved from: https://gdpr-info.eu/. November 2021
26. Hama Kareem JA, Rashid WN, Abdulla DF, Mahmood OK (2016) Social media and consumer awareness toward manufactured food. Cogent Bus Manag 3(1):1266786
27. Hawkins LK, Farrow C, Thomas JM (2020) Do perceived norms of social media users' eating habits and preferences predict our own food consumption and BMI? Appetite 149: 104611
28. Beaudoin CE, Hong T (2021) Emotions in the time of coronavirus: antecedents of digital and social media use among Millennials. Comput Hum Behav 123:106876
29. Jackson M, Brennan L, Parker L (2021) The public health community's use of social media for policy advocacy: a scoping review and suggestions to advance the field. Public Health 198:146–155
30. Ferdenzi C, Delplanque S, Barbosa P, Court K, Guinard JX, Guo T et al (2013) Affective semantic space of scents. Towards a universal scale to measure self-reported odor-related feelings. Food Qual Prefer 30(2):128–138
31. Vrontis D, Makrides A, Christofi M, Thrassou A (2020) Social media influencer marketing: a systematic review, integrative framework and future research agenda. Int J Consum Stud 00:1–28
32. Tuikka AM, Chau N, Kimppa KK (2017) Ethical questions related to using netnography as research method. ORBIT J 1(2):1–11
33. Kozinets R (1998) On netnography: initial reflections on consumer research investigations of cyberculture. Adv Consum Res 26:366–371
34. Valentin D, Gomez-Corona C (2018) Using ethnography in consumer research. In:

Ares G, Varela P (eds) Methods in consumer research. New approaches to classic methods, vol 1. Woodhead Publishing, UK
35. Spiro ES (2016) Research opportunities at the intersection of social media and survey data. Curr Opin Psychol 9:67–71
36. Thanh TV, Kirova V (2018) Wine tourism experience: a netnography study. J Bus Res 83:30–37
37. Mkono M, Markwell K, Wilson E (2013) Applying Quan and Wang's structural model of the tourist experience: a Zimbabwean netnography of food tourism. Tour Manag Perspect 5:68–74
38. Irwansyah I, Triputra P (2016) Indonesia gastronomy brand: Netnography on virtual culinary community. Soc Sci 11(19):4585–4588
39. Ang L (2011) Community relationship management and social media. J Database Mark Cust Strategy Manag 18(1):31–38
40. Dessart L (2017) Social media engagement: a model of antecedents and relational outcomes. J Mark Manag 33(5–6):375–399
41. Frants VI, Shapiro J, Taksa I, Voiskunskii VG (1999) Boolean search: current state and perspectives. J Am Soc Inf Sci 50(1):86–95
42. Aliyu MB (2017) Efficiency of Boolean search strings for information retrieval. Am J Eng Res 6(11):216–222
43. Reinert M (1990) Alceste une méthodologie d'analyse des données textuelles et une application: Aurelia De Gerard De Nerval. Bulletin de Méthodologie Sociologique 26(1):24–54
44. Vidal L, Ares G, Machín L, Jaeger SR (2015) Using Twitter data for food-related consumer research: a case study on "what people say when tweeting about different eating situations". Food Qual Prefer 45:58–69
45. Laguna L, Fiszman S, Puerta P, Chaya C, Tárrega A (2020) The impact of COVID-19 lockdown on food priorities. Results from a preliminary study using social media and an online survey with Spanish consumers. Food Qual Prefer 86:104028
46. Kuttschreuter M, Rutsaert P, Hilverda F, Regan Á, Barnett J, Verbeke W (2014) Seeking information about food-related risks: the contribution of social media. Food Qual Prefer 37: 10–18
47. Firmenich SA (2021) COVID-19 consumer behavior survey: wave 3
48. Schmoyer JR, Brain J, Koza E (2021) Never average a crisis Better understanding of COVID 19's impact on food drink behaviors through emotional clustering. In: Poster presented at 14th pangborn sensory science symposium, online, 9–12 August 2021
49. Adams MJ, Umbach PD (2012) Nonresponse and online student evaluations of teaching: understanding the influence of salience, fatigue, and academic environments. Res High Educ 53(5):576–591
50. DeMaio TJ (1984) Social desirability and survey. Surv Subj Phenom 2:257
51. Huang HM (2006) Do print and Web surveys provide the same results? Comput Hum Behav 22(3):334–350

Chapter 15

Implementing Eye Tracking Technology in Experimental Design Studies in Food and Beverage Advertising

Kostoula Margariti, Leonidas Hatzithomas, and Christina Boutsouki

Abstract

This chapter focuses on the use of eye-tracking technology in experimental studies addressing advertising effectiveness in the food and beverage sector. Experimental design prevails in advertising research, and it significantly contributes to the disclosure of the factors and conditions that direct consumers' attitudes and behaviors. According to most hierarchical models (i.e., A.I.D.A) in advertising, attention is a fundamental variable for any subsequent cognitive processing that shapes consumers' attitudes and behaviors. Traditionally, attention is recorded through self-report measures, such as recall and recognition. However, the increasingly popular technology on tracking and neuroimaging enables researchers and companies to accurately and objectively detect consumers' attention. Taking a head start among these approaches, eye tracking technology records eye movement and fixations, thus detecting specific areas of (consumers') interest (AOI's) in advertising. The present chapter aims to provide a useful guide for both academics and practitioners wishing to engage in eye-tracking experimental studies. It further provides clear guidance on the settings of the experimental process.

Key words Experimental design, Food and beverage advertising, Eye tracking, Attention

1 Introduction

Consumers are constantly exposed to numerous food and beverage advertising stimuli that influence their attitudes, purchase intentions, and dietary behaviors [90]. According to the World Health Organization (WHO), food advertising predominantly promotes the consumption of unhealthy food, especially to kids and youth, who seem to be significantly affected [10]. A great deal of food and beverage advertising targets youth and children in various ways, including outdoor advertising [6] and magazines [38]. Television, the Internet, and social media are also significant contemporary drivers of attitude in food and beverage advertising [9, 42] that mainly promote high-calorie products [43].

Carlos Gómez-Corona and Heber Rodrigues (eds.), *Consumer Research Methods in Food Science*,
Methods and Protocols in Food Science, https://doi.org/10.1007/978-1-0716-3000-6_15,

Due to an increasing number of high-calorie advertised products that target children, the link between food and beverage advertising and the prevalence of obesity among youth has been the focus of systematic research [36]. A significant body of research also addresses adults' susceptibility to food advertising stimuli. Experimental studies have focused on how advertising stimuli affect considerations, attitudes, and purchase intentions toward organic, green, or bio food products and how they promote unhealthy dietary preferences and consumption [13, 25].

According to most hierarchical advertising models (i.e., A.I.D.A), attention is a fundamental variable for any subsequent cognitive processing and behavior. The mental processes taking place during consumers' attention to advertising appeals and their association with decision-making guide research endeavors in advertising. At the same time, different theories in psychology (i.e., the limited capacity theory) indicate that consumers can absorb, elaborate, and recall only a limited amount of information. Therefore, understanding which advertising factors direct consumers' attention and interest is of utmost importance to determine advertising effectiveness. Top-down (endogenous) factors that are related to consumers' characteristics, needs, and aims, and bottom-up factors, such as the manipulation of color, size, brightness, geometry, and quantity of information in an ad, can significantly affect attention. Adults, for instance, are more skeptical than children and activate stronger defensive mechanisms toward advertising stimuli [64].

Traditionally, attention is recorded through self-report measures, such as recall and recognition, despite a growing criticism on their effectiveness to indicate actual attention. In other words, marketing researchers are questioning consumers' ability and willingness to disclose their actual thoughts and feelings when responding to experiments and surveys. Yet even though attention is partly involuntary, it may be accurately reflected in eye movements. Milosavljevic and Cerf [57] have also highlighted two problems with recall and recognition measures: (1) either a stimulus is attended to, but the awareness stage is not reached, thus making it impossible to maintain it in the memory and/or (2) even if a stimulus is maintained in the memory, individuals might forget it. To overcome the difficulties of memory evaluations, physiological responses associated with attention, such as eye movements [17], are considered more reliable indicators. As a result, an increasing number of companies in the food and beverage sector, such as Kraft Foods, Unilever, Procter & Gamble, and Frito Lay, have adopted novel technologies such as eye-tracking to measure consumers' responses toward food marketing and advertising [78].

Eye-tracking technology records eye movements (saccades), fixations, and scanpaths with great precision and unveils consumers' attention on specific areas of interest (AOIs). Yet, the greatest benefit associated with the use of eye-tracking techniques in experimental studies is the provision of accurate, objective measures of consumers' responses. Hence, most eye-tracking studies are predominantly experiments that aim to record consumers' attention. In food advertising, in particular, research that incorporates eye-tracking techniques focuses predominantly on how visual attention is formed by children and teenagers and less by adults [64, 87].

The present chapter focuses on the use of eye-tracking technology in experimental studies that address advertising effectiveness in the food and beverage sector. The development and increasing acceptance of eye-tracking techniques create new avenues in food and beverage research.

2 Background Studies

2.1 Attention

To date, there is a widespread consensus among researchers that attention is a precondition that leads to subsequent cognitive processes and, also, it is easily captured, retained, and measured via recall and recognition measures [92]. In today's competitive and cluttered advertising environment, grabbing consumers' attention is increasingly complex and challenging [12, 45] and suggests that measuring attention through recall and recognition is perhaps not totally effective.

Studies in psychology demonstrate that attention is a fundamental mechanism that significantly affects information processing [51]. Attention might also closely reflect the subsequent behavior and cognitive processes of individuals [75]. Besides, the plethora of visual stimuli constantly calls for consumers' visual attention, which might exceed their limited processing capacities [56]. Under such circumstances, improving advertising effectiveness lies on the understanding of the mechanisms that affect visual attention [15]. Moreover, even though recall and recognition are conventional and widespread measures of attention, their effectiveness is often questioned [7, 30]. For instance, a common and straightforward question used to address recognition is "have you seen brand X on TV recently?" However, not all consumers can provide precise answers, as many of them pay less attention to advertising than they report. Hence, when asked to recall the advertisement of a brand, they sometimes can recall nothing.

2.1.1 Factors Affecting Attention

The determinants of visual attention can be studied according to two main types of processes [68]. The design characteristics of an ad stimulus (bottom-up factors) [53] and individual characteristics (top-down factors), such as goals, motivation, and familiarity [89], constitute critical factors of visual attention [15, 92]. With respect to the design characteristics, a company may achieve a salient place in consumers' memories through more appealing advertising

[27]. As Kotler et al. [48, p. 668] claim, the advertiser must turn the big idea into an actual ad execution that will capture the audience's attention and interest. Prior literature shows that people understand and record some visual elements more readily than others. For instance, Valiente [85] demonstrates that color was the second (after ad size, [74]) most significant factor explaining ad recognition, while Raphel [70] claims it is the graphics that impact attention. According to Treisman [83] color increases the speed of ad processing up to three times compared to the graphic elements. With respect to the individual factors, prior research suggests that attention to the same advertisements can vary significantly among consumers, implying that these differences might be attributed to distinct personal traits. For instance, Rosbergen, Pieters, and Wedel [73] proved that participants who perceived shampoo as a relevant and gratifying product focused on the pleasurable brand information (for instance, the pictures of the female model and product). In the same vein, Marquart, Matthes, and Rapp [55] demonstrated that political motivation predicts exposure to political advertisements by the closer political party. In addition, familiarity with the ad and the brand negatively affects overall visual attention to print ads [65, 67].

2.1.2 Attention in Food and Beverage Advertising

Food and beverage advertisements aim to address consumers' innate instincts for food and drink. In static advertising (i.e., print media), the gastronomic icons (pictures) and images usually incorporated in food ads aim to provoke those instincts. Icons usually demonstrate the physical values of the advertised product or service in the best light [18]. Moreover, to emphasize quality, close-up shots are usually used. Visual metaphors and tropes are often used to display food and drink products in a more imaginative and creative way that would attract attention [59]. In video food advertising, emphasis is placed on attention-grabbing techniques such as vivid colors, animations, jingles, and special effects. Previous studies demonstrate that advertisements of products, such as pizza and fast-food, sweets, and other convenience foods, use animated content, emotional appeals, and special effects to imply quality and good taste for the advertised product and its consumption.

Advertisements that target children promote more detailed pictures of products to be better understood [50, 91]. With respect to verbal cues, taste and nutrition claims constitute the dominating types of advertising claims in food advertisements. In ads that promote functional foods (i.e., cereals and sports drinks) nutrition and health claims prevail, while hedonic food (i.e., ice cream, potato chips, cookies, etc.) ads use taste claims [46]. Additionally, the portrayal of eating behaviors or identification with the advertising characters might grab consumers' attention and lead to consumption of the advertised food [4].

2.2 Experimental Design Approaches in Food and Beverage Advertising

Experiments are a common methodological approach to test advertising effectiveness with respect to consumers' attention [5, 65, 66], attitudes, and behavior [11, 54, 86]. Attention is a critical measure of advertising effectiveness, as it can have a subsequent effect on consumers' cognitive processes, attitudes, and behavior [34, 80]. Experiments are a common methodological approach in the study of food and beverage advertising effectiveness. Puškarević et al. [69] used an experimental approach to show how the contents of food and beverage print advertisements can influence emotional responses. Folkvord, Anschütz, Wiers, and Buijzena [20] used an experimental study to indicate that individuals playing an advergame that incorporates food cues show greater total intake. Kemps, Tiggemann, and Hollitt [44] experimental studies suggest that overweight and obese individuals, when exposed to food advertisements, indicate greater desire for food consumption. Harris, Bargh, and Brownell [29] indicated a 45% increase in children's food intake when they are exposed to food advertising and a significant increase in snack consumption by adolescents exposed to snack food advertising. Subsequently, Velazquez and Pasch's [87] experimental study associated attention to food and beverage advertising with unhealthy food and beverage choices by both children and adults.

2.3 Measuring Attention with Eye-Tracking Technology

Stated recall and recognition are poor indicators of consumers' attention toward advertising, since not all the consumers can recollect what they have seen in an advertisement. Therefore, a behavioral measure of attention (i.e., eye fixations and saccades) can be a more accurate reflection of actual consumer behavior. According to Davenport and Beck [16] to know what individuals are paying attention to, one should follow what they are looking at. Eye movements are highly associated with visual attention, making them strong indicators of the visual attention process, often recorded by contemporary eye-tracking equipment [92] (Fig. 1). Eye-tracking methods provide companies and academics with the opportunity to evaluate the effectiveness of an advertisement by minimizing cognitive biases and recording moment-to-moment physiological data [15].

Prior literature in advertising has applied eye-tracking technology to examine ad effectiveness in terms of attention. The use of eye-tracking techniques in recent studies indicates that advertising is fundamental to gaining consumers' attention [79] and that the

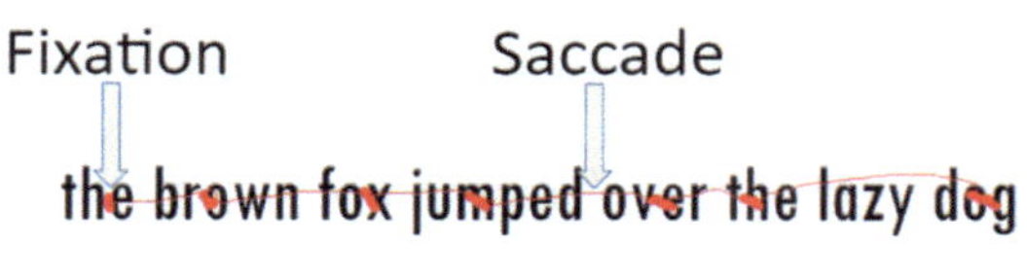

Fig. 1 Eye movements (Fixation and saccades). (Source: Al Madi & Khan [1])

size of print ads exerts a positive influence on visual attention in cluttered platforms [63]. Similarly, Hutton and Nolte [35] proved that when the model's gaze was directed at the advertised product, participants looked for a longer time at the product, the brand logo, and the rest of the advertisement compared to when the model was looking toward the participant. Furthermore, high visual complexity does not always have a negative effect on advertising effectiveness. Complex designs with a high level of creative elaboration generate greater visual attention (viewing time) [67]. According to Pieters et al. [67], although high levels of creative elaboration generate greater average fixation duration to magazine ads, advertisers should be cautious using excessive amounts of visual details. Also, Teixeira, Wedel, and Pieters [81] found that emotions (i.e., joy) influence attention and lead to increased cognitive processing (multiple eye fixations), possibly due to the interaction between bottom-up and top-down processes.

A step-by-step guide on the use of eye-tracking techniques in experimental studies addressing attention to advertising is outlined next.

3 Experimental Design: The Method

3.1 Experimental Settings

Experimental studies strive to adjust settings to the real-world environment to improve the ecological validity of their findings. Initially, participants are provided with specific instructions that guide their behavior during the experiment (i.e., they are not allowed to move aimlessly), so as not to affect eye-tracking recordings. Participants are required to provide their consent to the experimental process and the details of the experiment prior to taking their place in front of the eye-tracker for the initial calibration process. At this initial stage, researchers should be prepared to recruit more participants in case calibration to the eye-tracker is not feasible for some participants [24].

Eye-tracking systems may be either mounted to participant's head or they may be remotely placed. Head-mounted eye trackers can detect and record eye movement even through significant head movement (approximately one cubic foot). Yet, they are intrusive and costly (Fig. 2a, b). A significant drawback of this type of tracking method is that participants are constantly aware of a system recording their eye movements.

Remote eye trackers are less costly and consist of a camera lens located under or beside the computer screen used in the experiment. An infrared source is typically located in-line with the camera lens. The presence of the camera is easily overlooked by the participants, making this a very non-intrusive procedure. The remote process often requires two computers; the stimulus computer used

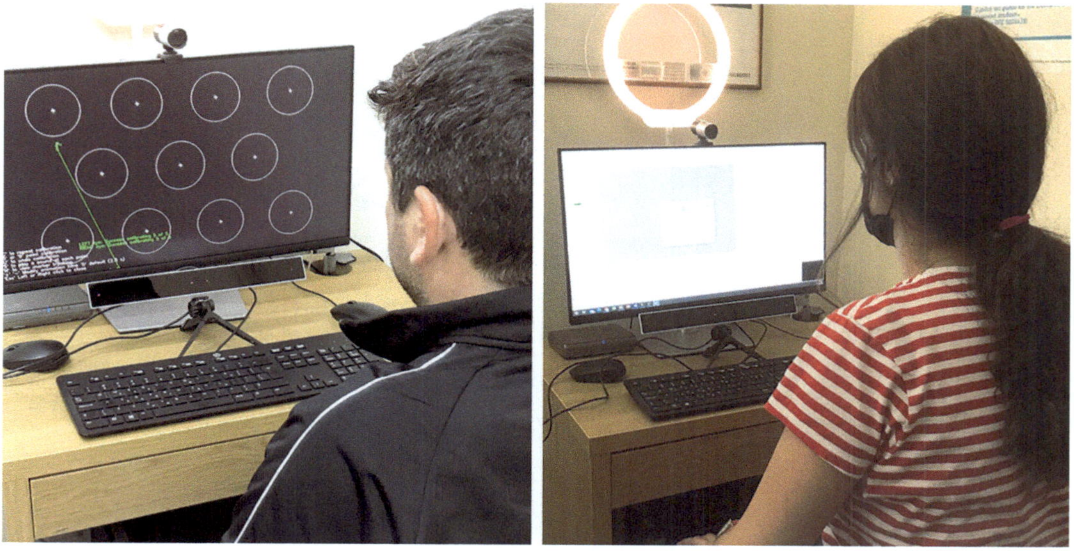

Fig. 2 (**a**, **b**) Remotely placed eye-tracking equipment. (Source: The photos were taken by the authors)

to present stimuli and to interact with the participants and the control computer that receives and processes the tracking information.

The eye-tracking equipment is adjusted to the control computer. The host/stimulus computer should have ample disk space for data storage, either from a remote or from a head-mounted eye-tracker. Some researchers, though, use a two-computer setup, an independent computer with the software under evaluation and the host computer where the researcher monitors the data collection process. The host computer sends signals to the eye tracker to start or stop tracking. Similar signals such as mouse movements or button presses can also be sent by the participants using the second computer [24].

The eye tracking system gathers x/y location and pupil size information at typical rates from 30 to 250 Hz. The video-based pupil and corneal reflex system usually record data at 60 Hz. Sampling rates that are slower do not provide enough resolution of visual attention, especially for scrolling or other tasks that involve moving visual elements, whereas faster sampling rates might lead to a massive data reduction problem. The x/y locations are translated and processed into fixations and saccades, which can be combined into scanpaths, unveiling more information about the cognitive processing during a task [24, 39]. Eye trackers usually have a tracking accuracy of about 0.5°–1° of visual angle. The screen resolution can be set, for instance at 1680 × 1050, and the average viewing distance between participants and the screen is carefully selected (i.e., 700 mm). Viewing and recording are binocular, with monocular gaze estimated separately and simultaneously for each eye (Fig. 3).

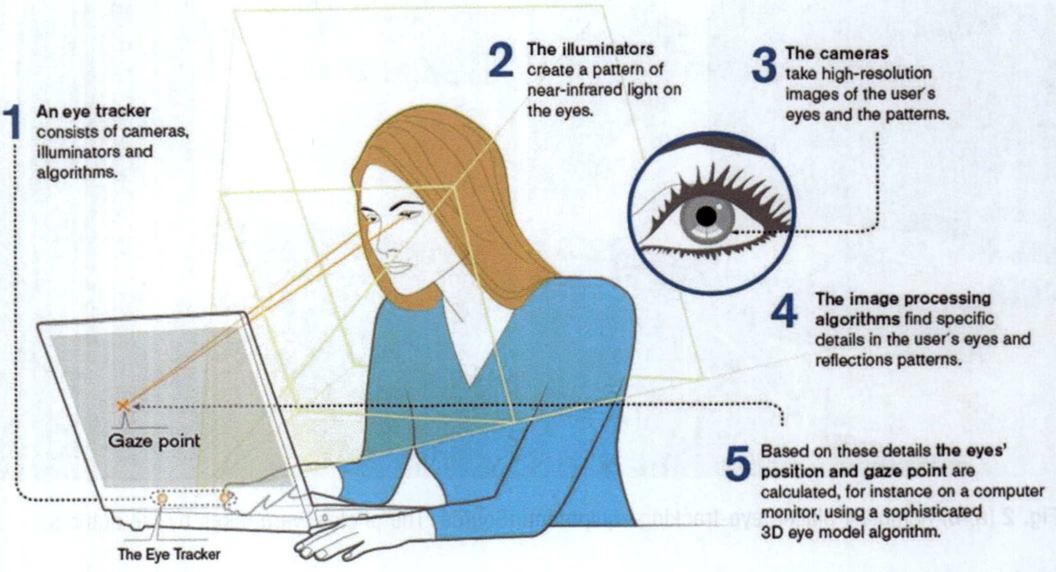

Fig. 3 An example of how the eye tracking system works. (Source: https:/www.tobiipro.com/learn-and-support/learn/eye-tracking-essentials/how-do-tobii-eye-trackers-work/)

In general, for corneal reflection, video-based systems dominate the applied eye-tracking market. These combine video imaging technology with small camera lenses and an infrared light source. Other characteristics that could also be used are light reflection from special contact lenses, video images of the pupil and/or iris, brightness differences between the iris and sclera, microscopic images of the retina, and the bulge produced by the cornea [93]. Some of these characteristics provide researchers with more accurate, higher-quality data, whereas some other are used by relatively inexpensive eye trackers yet of lower accuracy.

Small pupils generate poor eye-tracking data as they are easily hidden by the lower eyelid and are difficult to be recorded by an ineffectively positioned camera. The pupil reduces or widens depending on the luminance and contrast of an observed display. Large, bright areas might condense the pupil, whereas dark displays can expand it. Hence, it is significant to minimize ambient lighting. All other bright surfaces should be removed from the experiment room, as any potential light glints could confuse the eye-tracking system. Instead, fluorescent lighting, known to minimize infrared noise in the eye image, is preferable [32, p. 17]. Researchers should also avoid bright lighting and/or bright task screens and control ambient lighting and the eye tracker's infrared source with rheostats. If the brightness level on the test screens is highly variable, higher room luminance may negate the screen influences on the pupil size changes and enable a steadier pupil size. Researchers should similarly control the camera set up. Other variations on the pupil size may be attributed to animations.

3.2 Participants' Posture and Behavior

With respect to the physical position of the participants, researchers should ensure a minimum head motion during the calibration and recording process. For participants with dark irises, both bright and dark pupil tracking methods should be tested. Individuals with glasses or contact lenses cannot participate in the experiment.

An appropriate contact level between the eye tracker and the participant's eye should be set. The eye-tracking camera is typically set up to record eyes from a position slightly lower than the participant's horizontal view. If it is mounted lower than the required level, the lower eyelid can prevent proper pupil imaging [24].

3.3 Data Quality

Data quality is the "property of the sequence of raw data samples produced by the eye-tracker" [32, p. 29]. Data quality can be influenced by the eye-tracker, the experimental setup, such as participants, the operator, and his/her instructions, as well as the ambient environment, and in particular lighting conditions.

Quality can vary considerably across different eye trackers. Accuracy is one of the most important dimensions of data quality in eye trackers [32, pp. 41–43]. It refers to the distance between the actual gaze location and the recorded position in the eye-tracker data. The average tracker accuracy is over 0.5°, although remote systems are often reported as less accurate. Removing invalid recordings can improve the average accuracy. For instance, Komogortsev and Khan [47] used an eye-tracker with a 0.5° accuracy specification but found that the average accuracy among participants improved to 1.0° when they removed the invalid recordings. Accuracy is also strongly affected by participants' characteristics [33], such as head movements, astigmatism, and eyelid movement.

Precision is yet another significant driver of eye-tracking data quality and is defined as the ability to reliably reproduce a measurement given a fixating eye. The precision of video-based eye trackers ranges from around 0.001° to 1.03° [32, p. 40]. Lower values imply that microsaccades can be reliably detected, while higher values complicate the uncovering of fixations. High precision facilitates the detection of saccades or fixations. Poor precision might occur due to technical factors, such as the quality of the eye camera, or due to participant characteristics, such as the color of the eyes [32, p. 43].

Data quality issues imply in general that fixations can never be measured at their true location, questioning, for instance, the ability to reliably study small objects with eye-tracking technology. According to Orquin and Holmqvist [61], larger size stimuli enhance the capture rate, hence, even for excellent eye trackers (accuracy = 0.5 and precision = 0.1), stimuli must be over 5° in diameter to achieve a high capture rate. Moreover, as accuracy and precision gradually decline, the capture rate deteriorates mostly for smaller stimuli (≤2°). Large objects (≥8°) retain a high capture rate even for extremely poor levels of accuracy and precision.

As previously mentioned, factors such as glasses, contact lenses, eye color, eyelashes, and mascara may affect data quality. Furthermore, invalid samples can occur due to objects occluding the pupil or corneal reflection, poor image analysis algorithms, or additional features or reflections that resemble the pupil or corneal reflections. Specifically:

- The operator during the experimental process might affect the quality of calibration. Operators are responsible for the eye image set up, instructing the participant on posture and behavior during calibration and recording, and accepting or not the calibration and other recordings [60].
- The eye's physiology affects the stability of the eye feature data during sampling, and thus the quality of the calibration. Contact lenses, bifocal glasses [32, p. 118], and certain eye colors [40] hinder the extraction of features. For instance, a participant wearing mascara may have offsets far beyond the expected levels, which would make it difficult to record subtle effects, given the large source of error.

3.4 Calibration

Eye-tracking calibration establishes a mapping between eye image features (i.e., position of the pupil and any corneal reflection), the physical orientation of the eye, and/or gaze position of the stimulus [28]. Initially, participants are asked to look at some predefined positions in the stimulus. At each calibration target, the eye tracker detects several eye image features and associates their positions in the eye image with the position of the target. When displayed, these positions are usually near the corners and at the display's center in a 5-point calibration. A 9- or 13-point calibration locates these positions at the vertices of successively larger squares, starting from the display center [24]. Calibration stimuli are usually presented in black color on a bright background. The outer diameter of each target spanned 0.5° of the visual angles.

With eyes completely still before, during, or after looking at the gaze point, a significant problem in the calibration process is to choose the right moment to sample the coordinates of the eye image features [60]. Hence, calibration should be performed multiple times, not only inter- but also intra-user sessions [24].

Calibration can be accomplished in three ways: the operator-controlled, the system-controlled, and the participant-controlled calibration. In the operator-controlled calibration, operators accept calibration targets once they certify that participant's eye is still and directed toward the target. For successful calibration, operators need to:

1. Set up the camera for optimal recording.
2. Check for correct pupil tracking and certify that no unpredicted factors such as mascara-covered lashes might affect pupil tracking.
3. Track the corneal reflection [32].
4. Provide clear instructions to participants (i.e., "please, look at the points that appear on the screen, do not remove your eyes until the point disappears").
5. Manipulate the presentation software to present the first calibration target on the screen.
6. Evaluate the results. If calibration is valid, they can proceed with the data collection. If it is rejected, the calibration process should be repeated (perhaps using another type of calibration).

If calibration is controlled by the system, the eye-tracker software and specific algorithms specify whether participants' eye is directed toward a calibration target. Each calibration target is automatically accepted when a fixation is detected. In the participant-controlled calibration scenario, participants decide when they are fixating a calibration target and confirm this decision by clicking a mouse or keyboard button. Since each period of stillness could be considered as a valid calibration target, it is significant that participants' eyes are directed toward the correct calibration target without having to trace it. Irrespective of the calibration method used, participants are required to gaze at the center of each target and to keep staring at it until it moves [60].

3.5 Experimental Process

Following the calibration process, the advertising stimuli are displayed on the computer screen, and participants are asked to observe it as any other type of document. Different versions of advertisements are usually displayed, based on the manipulation of the variables under study. Several stimuli characteristics/elements can be manipulated to provide the independent variables in the experiment (i.e., the luminance of an image). In the case of multiple advertisements being displayed, participants can click the mouse and move to the next advertisement.

Studies with computer-loaded eye trackers require participants to comfortably sit on a chair in front of the computer screen. Infrared-type corneal reflection eye tracking systems, more commonly used in research, track the location of observers' pupils relative to a small, reflected light glint on the surface of the cornea [58, 93]. A camera lens is focused on participant's eye, while a second lens is pointed toward the visual display or scene in view. In the computer-mounted eye tracker, a scan converter is frequently used in place of the second camera. Participants' pupil is tracked based on its relative contrast with the surrounding iris. The bright pupil systems shed light on the pupil with infrared light, whereas the dark pupil systems do not illuminate the pupil at all; some systems may be switched between these modes, to find the most robust pupil imaging for a testing environment. The recorded data can be used in real-time or can be saved for subsequent statistical analysis. Contemporary eye-tracking software and hardware automatically filter out blinks and other anomalies, as they are constantly searching for unique light-dark-light signatures of reflected light usually associated with blinking [24].

Once they have finalized the eye-tracking process, participants are asked to complete a questionnaire that usually addresses ad and brand attitudes, recall and recognition, comprehension, and purchase intentions.

3.6 Data Analysis: Metrics and Visualizations

As eye tracking gains ground in a wide range of fields, different approaches to eye-tracking data analysis have been developed, such as statistical algorithms (descriptive or inferential) [32] and visualization-related techniques [2]. The objective of researchers conducting eye-tracking experiments is to quantitatively describe the attentional behavior of participants. Eye tracking data analysis typically transforms a vast number of fixations' raw data to specific metrics. Reduction of eye-tracking data into meaningful information can be a demanding process. Raw data are collected offline and are switched into meaningful units of fixations and saccades. The fixations are collected into a file that lists start/end times and locations among others [22, 23]. The analysis of fixations and saccades requires some form of fixation identification. Fixation identification is a description of tracked eye movement behaviors in a statistical manner. This signifies that raw eye-movement data points should be translated to fixation locations on the visual display. This significantly reduces the size and complexity while at the same time it retains the most important characteristics for the purposes of explaining cognitive and visual processing behavior.

Regardless of the identification algorithms' precision, the identification problem is a subjective process. A way to ensure the validity of such algorithms is to compare resultant fixations to an observer's subjective impressions. Karsh and Breitenbach [41] demonstrated how different identification algorithms produce different interpretations even when analyzing identical stimuli. Good identification algorithms ensure valid fixation and saccade locations as well as durations that in turn influence inferences such as processing complexity and visual search paths. Salvucci and Goldberg [76] provide a taxonomy according to which identification algorithms are classified based on spatial characteristics into velocity-based, dispersion-based, and area-based. Velocity-based algorithms focus on the velocity information in the eye-tracking protocols, building on the low velocities of fixation points and high velocities of saccades. Dispersion-based algorithms focus on the dispersion (i.e., spread distance) of fixation points, taking into consideration that fixation points generally occur near one another, while area-based algorithms identify points within provided areas of interest that represent relevant visual targets. Salvucci and Goldberg [76] further distinguish between these algorithms building on temporal characteristics, that is, if the algorithm uses duration information and if it is locally adaptive. The use of duration information lies on the fact that fixations are rarely less than 100 milliseconds (ms) and often range between 200 and 400 ms. The incorporation of local adaptivity enables the interpretation of a given data point to be influenced by the interpretation of temporally adjacent points. Further, researchers might use a code that corresponds to the quality of tracking. Since fixations' rate is approximately 3 Hz, the datafile length is automatically reduced. Eye tracking systems can also identify which defined objects/areas of interest are viewed by each fixation, and therefore compute the instantaneous and cumulative dwell times by screen objects [22, 23]. In the advertising context, analysis of variance (ANOVA) examines whether there are statistically significant differences in the recorded data for different objects or AOI's [77].

Even though eye tracking is a complex, highly data-driven approach, it is becoming increasingly popular among researchers across sectors. Multiple metrics analysis can be simplified by (1) developing a hypothesis from theory and "translating" in terms of eye movements, (2) getting the stimulus and drawing participants' potential eye movements, (3) considering what is most important in simulating eye movements to test the hypothesis: movement, position, latency, or numerosity measures, and (4) defining a list of measures to check the hypotheses [32].

High-precision and low-cost infrared eye-tracking equipment has been constructed to record consumers' eye movements when they experience multiple visual stimuli under quasi-natural conditions. Eye tracking equipment can indicate precisely where participants are looking or whether they are reading or scanning, thereby distinguishing between a systematic fixation on word clusters and a scanning for particular words or phrases. It further reports where the first view is directed and measures the relative intensity of participants' attention to different elements in an advertisement (areas of interest). The scan patterns of all participants in an experiment can be comparatively displayed [82].

Eye tracking equipment records the focus position of the eye's fovea 60 times per second and can discriminate between fixations and saccades. Saccades are the rapid eye movements that typically last around 20–40 ms, whereby the foveal focus is moved to a new fixation point. Saccades occur in both eyes simultaneously and range between 2 and 10 degrees of visual angle. With rotational velocities of 500–900 degrees/second, they indicate very high acceleration [14]. Saccades are typically exhibited with a latency of about 250 ms following the onset of a visual target. During a fixation, the eye is relatively still, typically for around 200–500 ms [71]. Eye tracking also allows the measurement of fixations when participants' eyes pause to examine an element in the advertisement [72]. It is an indication of the level of cognitive processing, whereby the more fixations exist, the higher the cognitive processing is.

Additional types of eye recordings are smooth pursuit eye movements (eye movements that are tracked toward slowly moving targets), compensatory eye movements (slow movements in an opposite rotational direction from the head or trunk movement), vergence eye movements (rotational movements of the two eyes in opposite directions, in response to focusing error from visual targets moving closer or further away—not usually observed in computer-oriented studies), miniature eye movements (small corrective motions, as well as tremor), and nystagmus eye movements (sawtooth-pattern of eye movements caused by a rapid back-and-forth movement of the eyes—not typically observed in usability evaluations) [24].

Each advertisement can be divided into different areas of interest (AOI's) with separate statistics provided for each. Manipulating the size of AOI could improve a model's fit [62]. A common feature among eye-tracking studies is the comparison of multiple AOIs on multiple eye-movement metrics [88]. For instance, Antúnez et al. [3] in their study on food nutrition labels, compared six AOIs in one condition and four AOIs in a second condition on five different metrics, yielding 105 significance tests. Eye-tracking software can produce visual representations such as scan paths for all respondents or a heat map, a graphic with colorful displays of the number of fixations in an area of interest.

Some of the most important measures in eye-tracking methodology are:

- Total time spent on a page or an advertisement, defined as the time spent viewing the advertisement.
- Number of fixations, defined as the times a participant pauses to examine an element.
- Fixation count, defined as the total number of gazes. A higher number indicates more gazes on a specific element.
- AGD index indicates the intensity of cognitive processing. It is one of the most common eye movement measures to date. AGD is calculated by dividing participants' fixation time by the number of fixations on a specific element/area of interest (the fixation time per fixation on the element/area) [37, 94].
- Fixation frequency which refers to the number of eye fixations on the target stimuli.
- Fixation duration which is defined as the total duration of eye fixation on the target stimuli [52].

- Fixation count within an AOI. More fixations on a specific AOI imply that it is more noticeable and thus more interesting to the viewer compared to other areas.
- Dwell count applied to compare attention distributed between targets.
- Fixation spatial density, defined as the fixations gathered in a small area. Evenly spread fixations reflect widespread and inefficient search [21].
- Gaze duration on each area of interest, defined as the cumulative duration and average spatial location of a series of consecutive fixations within an area of interest [31].
- Number of voluntary fixations (>320 ms).
- Number of involuntary fixations (<240 ms) [26].
- Time to first fixation of a specific area of interest, defined as the seconds or minutes it takes a participant to view a specific area of interest [19].
- Fixation/saccade time ratio [49].

Apart from the quantitative metrics for eye data analysis, there are visualization techniques that are divided into point-based visualization techniques, AOI-based visualization techniques, and combination visualization techniques. Additional categorizations of visualization techniques are those with information about stimulus type, the in-context visualizations, the animated visualizations, the interactive visualizations, and the active stimulus content. Statistical graphics are the most common visualization techniques for eye-tracking data, such as line charts, bar charts, box plots, or scatter plots. These visualization techniques can be used to quantitatively analyze eye-tracking data [8].

Scan paths and heat maps summarize the viewing patterns and potential differences in the way participants in an experiment view each version of an advertisement. The scan path, for instance, indicates whether participants looked longer at the headline or the image of the ad, or which was the first looked element of the ad. Saccades are represented with lines, while fixations are represented with dots. A heat map displays the overall gaze time spent on each area of interest, represented with opaqueness and color. A red opaque area represents the highest gaze time values, while the absence of color represents no gaze at all [77].

4 Conclusion

Eye tracking is evolving into a simpler and affordable technology with growing popularity and applicability in numerous research fields, such as psychology, marketing, and advertising. Eye-tracking methods in advertising provide information about the spatiotemporal dimensions of attentive behavior. For instance, Treistman and Gregg's [84] eye-tracking study demonstrated that the products displayed in an advertisement that were longer viewed, also received more sales. Eye movements are strongly linked to high-order cognitive processes, while attention contributes to communication processing and effectiveness. As such, visual marketing has much to learn from eye-tracking methods with respect to consumers' moment-to-moment processing of visual marketing stimuli [92]. Eye-tracking methods complement conventional measures of attention, such as recall and recognition, that do not always provide valid information about the unconscious, attentive behavior of the participants.

The present chapter provides academics and practitioners with a useful guide on eye-tracking experimental studies with respect to the experimental environment and conditions that enhance the validity of data, the calibration process, the experimental process, and the data analysis. The discussion suggests that eye-tracking methods in food and beverage advertising could help unveil consumers' initial impressions toward specific elements in an advertisement (i.e., the brand, the ingredients, the image, etc.) or consumers' preferences toward a specific color used in a food ad. Considering that food advertising research is highly focused on how advertising might affect youth and adult obesity and food disorders, eye-tracking methods could provide valuable information on consumers' perceptions toward the healthiness of food. Relevant approaches include, for instance, examining the fixation density on the image of a healthy or an unhealthy food or on the nutrition information included in the ad [21].

References

1. Al Madi N, Khan J (2018) Constructing semantic networks of comprehension from eye-movement during reading. In: 2018 IEEE 12th international conference on semantic computing (ICSC), IEEE, pp 49–55
2. Andrienko G, Andrienko N, Burch M, Weiskopf D (2012) Visual analytics methodology for eye movement studies. IEEE Trans Vis Comput Graph 18(12):2889–2898
3. Antunez L, Vidal L, Sapolinski A, Gimenez A, Maiche A, Ares G (2013) How do design features influence consumer attention when looking for nutritional information on food labels? Results from an eye-tracking study on pan bread labels. Int J Food Sci Nutr 64(5): 515–527
4. Bargh JA, Morsella E (2010) Unconscious behavioral guidance systems. In: Then a miracle occurs: focusing on behavior in social psychological theory and research. Oxford University Press, Oxford, pp 89–118
5. Berger R, Rauhut H, Prade S, Helbing D (2012) Bargaining over waiting time in ultimatum game experiments. Soc Sci Res 41(2): 372–379
6. Bhargava M, Donthu N (1999) Sales response to outdoor advertising. J Advert Res 39(4):7–7
7. Biel AL (1993) Converting image into equity. Brand equity and advertising: Advertising's role in building strong brands 26(10):67–81
8. Blascheck T, Kurzhals K, Raschke M, Burch M, Weiskopf D, Ertl T (2014) State-of-the-art of visualization for eye tracking data. In: EuroVis (STARs)
9. Boyland EJ, Halford JC (2013) Television advertising and branding. Effects on eating behaviour and food preferences in children. Appetite 62:236–241
10. Boyland EJ, Whalen R (2015) Food advertising to children and its effects on diet: review of recent prevalence and impact data. Pediatr Diabetes 16(5):331–337

11. Burgers C, Konijn EA, Steen GJ, Iepsma MA (2015) Making ads less complex, yet more creative and persuasive: the effects of conventional metaphors and irony in print advertising. Int J Advert 34(3):515–532
12. Burke RR, Srull TK (1988) Competitive interference and consumer memory for advertising. J Consum Res 15(1):55–68
13. Butland B, Jebb S, Kopelman P, McPherson K, Thomas S, Mardell J, Parry V (2007) Tackling obesities: future choices-project report, vol 10. Department of Innovation, Universities and Skills, London, p 17
14. Carpenter RH (1988) Movements of the eyes, 2nd rev. Pion Limited, London, UK
15. Casado-Aranda LA, Sánchez-Fernández J, Ibáñez-Zapata JÁ, Liébana-Cabanillas FJ (2020) How consumer ethnocentrism modulates neural processing of domestic and foreign products: a neuroimaging study. J Retail Consum Serv 53:101961
16. Davenport TH, Beck JC (2001) The attention economy. Ubiquity 2001(May):1-es
17. Deubel H, Schneider WX (1996) Saccade target selection and object recognition: evidence for a common attentional mechanism. Vis Res 36(12):1827–1838
18. Eco U (1973) Kultura, komunikacija, informacija
19. Ellis S, Candrea R, Misner J, Craig CS, Lankford CP, Hutchinson TE (1998) Windows to the soul? What eye movements tell us about software usability. In: Proceedings of the usability professionals' association conference, June, pp 151–178
20. Folkvord F, Anschütz DJ, Wiers RW, Buijzen M (2015) The role of attentional bias in the effect of food advertising on actual food intake among children. Appetite 84:251–258
21. Gere A, Kókai Z, Sipos L (2017) Influence of mood on gazing behavior: preliminary evidence from an eye-tracking study. Food Qual Prefer 61:1–5
22. Goldberg JH, Kotval XP (1998) Eye movement-based evaluation of the computer interface. In: Advances in occupational ergonomics and safety. IOS Press, Amsterdam, pp 529–532
23. Goldberg JH, Kotval XP (1999) Computer interface evaluation using eye movements: methods and constructs. Int J Ind Ergon 24(6):631–645
24. Goldberg JH, Stimson MJ, Lewenstein M, Scott N, Wichansky AM (2002) Eye tracking in web search tasks: design implications. In: Proceedings of the 2002 symposium on eye tracking research & applications, March, pp 51–58
25. Gorn GJ, Goldberg ME (1980) Children's responses to repetitive television commercials. J Consum Res 6(4):421–424
26. Graf W, Krueger H (1989) Ergonomic evaluation of user-interfaces by means of eye-movement data. In: Proceedings of the third international conference on human-computer interaction, (Work with computers: organizational, management, stress and health aspects), vol 1, September. pp 659–665
27. Haan MA, Moraga-González JL (2011) Advertising for attention in a consumer search model. Econ J 121(552):552–579
28. Hammoud RI (2008) U.S. Patent No. 7,331,671. Washington, DC, U.S. Patent and Trademark Office
29. Harris JL, Bargh JA, Brownell KD (2009) Priming effects of television food advertising on eating behavior. Health Psychol 28(4):404
30. Heath R, Nairn A (2005) Measuring affective advertising: implications of low attention processing on recall. J Advert Res 45(2):269–281
31. Hendrickson J (1989) Performance, preference, and visual scan patterns on a menubased system: implications for interface design. In: Proceedings of the ACM CHI'89 human factors in computing systems conference. ACM Press, pp 217–222
32. Holmqvist K, Nyström M, Andersson R, Dewhurst R, Jarodzka H, Van de Weijer J (2011) Eye tracking: a comprehensive guide to methods and measures. OUP, Oxford
33. Hornof AJ, Halverson T (2002) Cleaning up systematic error in eye-tracking data by using required fixation locations. Behav Res Methods Instrum Comput 34(4):592–604
34. Hummel G, Zerweck I, Ehret J, Winter SS, Stroebele-Benschop N (2017) The influence of the arrangement of different food images on participants' attention: an experimental eye-tracking study. Food Qual Prefer 62:111–119
35. Hutton SB, Nolte S (2011) The effect of gaze cues on attention to print advertisements. Appl Cogn Psychol 25(6):887–892
36. IOM (Institute of Medicine) (2006) Food marketing to children and youth: threat or opportunity. The National Academies Press, Washington, DC
37. Jacob RJ, Karn KS (2003) Eye tracking in human-computer interaction and usability research: ready to deliver the promises. In: The mind's eye. North-Holland, pp 573–605
38. Jones SC, Gregory P, Kervin L (2012) Branded food references in children's magazines:'advertisements' are the tip of the iceberg. Pediatr Obes 7(3):220–229

39. Just MA, Carpenter PA (1976) Eye fixations and cognitive processes. Cogn Psychol 8(4): 441–480
40. Kammerer Y, Wollny E, Gerjets P, Scheiter K (2009) How authority-related epistemological beliefs and salience of source information influence the evaluation of Web search results–an eye tracking study. In: Proceedings of the 31st annual conference of the cognitive science society, July, pp 2158–2163
41. Karsh R, Breitenbach FW (1983) Looking at looking: the amorphous fixation measure. In: Eye movements and psychological functions: international views. Routledge, London, pp 53–64
42. Kelly B, Halford JC, Boyland EJ, Chapman K, Bautista-Castaño I, Berg C et al (2010) Television food advertising to children: a global perspective. Am J Public Health 100(9): 1730–1736
43. Kelly B, King L, Baur L, Rayner M, Lobstein T, Monteiro C et al (2013) Monitoring food and non-alcoholic beverage promotions to children. Obes Rev 14:59–69
44. Kemps E, Tiggemann M, Hollitt S (2014) Biased attentional processing of food cues and modification in obese individuals. Health Psychol 33(11):1391
45. Kent RJ (1993) Competitive versus noncompetitive clutter in television advertising. J Advert Res 33(2):40–47
46. Kim K, Cheong Y, Zheng L (2009) The current practices in food advertising: the usage and effectiveness of different advertising claims. Int J Advert 28(3):527–553
47. Komogortsev OV, Khan JI (2008) Eye movement prediction by Kalman filter with integrated linear horizontal oculomotor plant mechanical model. In: Proceedings of the 2008 symposium on eye tracking research & applications, March, pp 229–236
48. Kotler P, Armstrong G, Saunders J, Wong V (1999) Priciples of marketing (Second European Edition ed.). Prentice Hall Inc., Upper Sadle River
49. Kotval XP, Goldberg JH (1998) Eye movements and interface component grouping: an evaluation method. In: Proceedings of the human factors and ergonomics society annual meeting, vol 42, no. 5, SAGE Publications, Los Angeles, October, pp 486–490
50. Kunkel D (2001) Children and television. In: Handbook of children and the media, pp 375–395
51. LaBerge D (1995) Attentional processing: the brain's art of mindfulness, vol 2. Harvard University Press
52. Lee J, Ahn JH (2012) Attention to banner ads and their effectiveness: an eye-tracking approach. Int J Electron Commer 17(1): 119–137
53. Li K, Huang G, Bente G (2016) The impacts of banner format and animation speed on banner effectiveness: evidence from eye movements. Comput Hum Behav 54:522–530
54. Margariti K, Boutsouki C, Hatzithomas L, Zotos Y (2019) Visual metaphors in food advertising: a cross-cultural study. Food Res Int 115:338–351
55. Marquart F, Matthes J, Rapp E (2016) Selective exposure in the context of political advertising: a behavioral approach using eye-tracking methodology. Int J Commun 10:20
56. Meißner M, Oll J (2019) The promise of eye-tracking methodology in organizational research: a taxonomy, review, and future avenues. Organ Res Methods 22(2):590–617
57. Milosavljevic M, Cerf M (2008) First attention then intention: insights from computational neuroscience of vision. Int J Advert 27(3): 381–398
58. Mulligan JB (1997) Image processing for improved eye-tracking accuracy. Behav Res Methods Instrum Comput 29(1):54–65
59. Nedeljković U, Puškarević I, Novaković D, Pinćjer I (2013) The effectiveness of sex appeal in print ads in relation to a visual register of the message: articulating a new framework. Mark Theory 44(3):247–263
60. Nyström M, Andersson R, Holmqvist K, Van De Weijer J (2013) The influence of calibration method and eye physiology on eyetracking data quality. Behav Res Methods 45(1):272–288
61. Orquin JL, Holmqvist K (2018) Threats to the validity of eye-movement research in psychology. Behav Res Methods 50(4):1645–1656
62. Orquin JL, Ashby NJ, Clarke AD (2016) Areas of interest as a signal detection problem in behavioral eye-tracking research. J Behav Decis Mak 29(2–3):103–115
63. Peschel AO, Orquin JL (2013) A review of the findings and theories on surface size effects on visual attention. Front Psychol 4:902
64. Pettigrew S, Tarabashkina L, Roberts M, Quester P, Chapman K, Miller C (2013) The effects of television and internet food advertising on parents and children. Public Health Nutr 16(12):2205–2212
65. Pieters R, Wedel M (2004) Attention capture and transfer in advertising: brand, pictorial, and text-size effects. J Mark 68(2):36–50
66. Pieters R, Wedel M, Zhang J (2007) Optimal feature advertising design under competitive clutter. Manag Sci 53(11):1815–1828

67. Pieters R, Wedel M, Batra R (2010) The stopping power of advertising: measures and effects of visual complexity. J Mark 74(5):48–60
68. Plassmann H, Ramsøy TZ, Milosavljevic M (2012) Branding the brain: a critical review and outlook. J Consum Psychol 22(1):18–36
69. Puskarevic I, Nedeljkovic U, Pincjer I, Franken G, Pusnik N (2018). The effects of typeface and image complexity on consumer'-visual attention and attitudinal responses in advertising. In: 8th conference on Information and graphic arts technology
70. Raphel M (1988) Ad techniques: pictures urge the reader to stop, look, listen. Bank Mark 20(8):32–33
71. Rayner K (1998) Eye movements in reading and information processing: 20 years of research. Psychol Bull 124(3):372
72. Rayner K, Foorman BR, Perfetti CA, Pesetsky D, Seidenberg MS (2001) How psychological science informs the teaching of reading. Psychol Sci Public Interest 2(2): 31–74
73. Rosbergen E, Pieters R, Wedel M (1997) Visual attention to advertising: a segment-level analysis. J Consum Res 24(3):305–314
74. Rouse RA (1991) Yellow page advertising: an empirical analysis of attributes contributing to consumer interest, liking and preference. J Prof Serv Mark 6(2):35–44
75. Russo JE (1978) Eye fixations can save the world: a critical evaluation and a comparison between eye fixations and other information processing methodologies. ACR North American Advances
76. Salvucci DD, Goldberg JH (2000) Identifying fixations and saccades in eye-tracking protocols. In: Proceedings of the 2000 symposium on eye tracking research & applications, November, pp 71–78
77. Scott N, Green C, Fairley S (2016) Investigation of the use of eye tracking to examine tourism advertising effectiveness. Curr Issue Tour 19(7):634–642
78. Shayon S (2012) Eye-tracking helping marketers boost shelf awareness. https://www.brandchannel.com/2012/07/17/eye-tracking-helping-marketers-boost-shelf-awareness/
79. Simmonds L, Bellman S, Kennedy R, Nenycz-Thiel M, Bogomolova S (2020) Moderating effects of prior brand usage on visual attention to video advertising and recall: an eye-tracking investigation. J Bus Res 111:241–248
80. Story M, Kaphingst KM, Robinson-O'Brien R, Glanz K (2008) Creating healthy food and eating environments: policy and environmental approaches. Annu Rev Public Health 29:253–272
81. Teixeira T, Wedel M, Pieters R (2012) Emotion-induced engagement in internet video advertisements. J Mark Res 49(2): 144–159
82. Theuner G, Pischke K, Bley T (2008) Analysis of advertising effectiveness with eye tracking. Meas Behav 2008:229
83. Treisman A (1982) Perceptual grouping and attention in visual search for features and for objects. J Exp Psychol Hum Percept Perform 8(2):194
84. Treistman J, Gregg JP (1979) Visual, verbal, and sales responses to print ads. J Advert Res 19(4):41–47
85. Valiente R (1973) Mechanical correlates of ad recognition. J Advert Res 13(3):13–18
86. Van Mulken M, van Hooft A, Nederstigt U (2014) Finding the tipping point: visual metaphor and conceptual complexity in advertising. J Advert 43(4):333–343
87. Velazquez CE, Pasch KE (2014) Attention to food and beverage advertisements as measured by eye-tracking technology and the food preferences and choices of youth. J Acad Nutr Diet 114(4):578–582
88. Von der Malsburg T, Angele B (2017) False positives and other statistical errors in standard analyses of eye movements in reading. J Mem Lang 94:119–133
89. Vu TMH, Tu VP, Duerrschmid K (2016) Design factors influence consumers' gazing behaviour and decision time in an eye-tracking test: a study on food images. Food Qual Prefer 47:130–138
90. Wadden TA, Brownell KD, Foster GD (2002) Obesity: responding to the global epidemic. J Consult Clin Psychol 70(3):510
91. Warren R, Wicks RH, Wicks JL, Fosu I, Chung D (2008) Food and beverage advertising on US television: a comparison of child-targeted versus general audience commercials. J Broadcast Electron Media 52(2):231–246
92. Wedel M, Pieters R (2008) A review of eye-tracking research in marketing. Rev Mark Res 4(2008):123–147
93. Young LR, Sheena D (1975) Survey of eye movement recording methods. Behav Res Methods Instrum 7(5):397–429
94. Zhang X, Yuan SM (2018) An eye tracking analysis for video advertising: relationship between advertisement elements and effectiveness. IEEE Access 6:10699–10707

Chapter 16

Field Studies in Food Settings: Lessons Learned and Concrete Cases

Tobias Otterbring

Abstract

This chapter focuses on field studies in food settings and why such studies are important for the generalizability and perceived relevance of published research findings in the food domain. After a discussion of some potential downsides with solely sticking to lab and online studies capturing self-reported responses as well as a set of salient strengths associated with fieldwork capturing real, observable behavior, five concrete cases are presented in the form of published field studies. These studies, which range from correlational, through quasi-experimental, to experimental in design, seek to showcase the methodological pluralism of research conducted in actual consumption contexts, while simultaneously exemplifying some measures commonly used in field-based investigations, such as naturalistic consumer choice, transaction data, and the amount of information search carried out by customers. The chapter ends with a series of recommendations for scholars, funding agencies, journal editors, and reviewers. Specifically, scholars interested in running future field studies and combining such fieldwork with lab or online studies are offered several suggestions regarding aspects that are advantageous to consider when doing research "in the wild." Similarly, journal editors, reviewers, and funding agencies are provided with advice and arguments for the notion that field studies cannot—and should not—be assessed through the same evaluation criteria as those used to assess controlled experiments conducted inside the walls of a lab.

Key words Field studies, Ecological validity, Generalizability, Methodological pluralism, Behavioral data, Consumer behavior

1 Introduction

Most of consumers' food-related decision-making, such as going to a restaurant, doing grocery shopping, and ordering a beer at a bar, arguably takes place outside the walls of university labs and beyond computer screens. Nevertheless, field studies in food settings designed to capture consumers' actual choice behavior, product purchases, and decision-making processes are relatively rare in the areas of food science, marketing, and consumer behavior. Instead, a

Carlos Gómez-Corona and Heber Rodrigues (eds.), *Consumer Research Methods in Food Science*, Methods and Protocols in Food Science, https://doi.org/10.1007/978-1-0716-3000-6_16,

large chunk of the scholarly studies tends to be based on self-reported attitudes and intentions or stated preferences by university students and paid online panel participants [1–5]. This is understandable for a myriad of reasons, given that lab and online experiments—when cleverly conducted—offer researchers a wide range of advantages. Some of the commonly discussed benefits of these experiment types include increased rigor and control and the chance to measure a larger number of variables, enabling a more nuanced theoretical account of the obtained results, the ability to rule out competing conceptualizations, and the possibility to capture the specific mechanisms that explain *why* a cause-effect relationship emerges as well as the boundary conditions of *when* such a relationship does and does not hold [6–9]. Moreover, the controlled conditions that are characteristic of these experiment types often enable scholars to isolate effects and hence infer causal inferences between variables with greater certainty (at least in similar settings). On top of these salient strengths, online and lab experiments are typically less time-consuming, considerably less costly, and more easily implemented than what is usually the case for field studies in actual grocery stores, restaurants, bars, and other food-related consumption contexts [7, 10, 11]. In fact, it often takes weeks, if not months, to even be able to conduct a field study and then several additional days or weeks to get a sufficiently large sample size, whereas the rapid rise in recruiting online panel participants means that scholars can conveniently collect data from thousands of participants in less than an hour.

Considering all the above-mentioned assets linked to lab and online experiments, combined with the discussed drawbacks of field studies, a motivated question is certainly why scholars should ever bother to conduct studies "in the real world." This chapter aims to provide some arguments for the advantages of studying consumers' food-related decision-making "in the wild" and hence delineate a set of favorable features linked to field studies. In what follows, I start by providing some inherent strengths associated with research from actual consumption contexts while also discussing why an exclusive emphasis on lab and online experiments risks reducing the perceived relevance of research in food science and related disciplines. Next, I present five field studies from food settings where I have acted as an author or a co-author. In so doing, I highlight a series of common measures, tools, procedures, and designs that are frequently used in field-based research. I close the chapter by presenting a few takeaway messages based on learned lessons, with the goal of providing concrete recommendations for scholars whose ambition is to conduct field studies and to remind journal editors, reviewers, and funding agencies that such studies must be assessed differently than lab and online studies.

2 Field Studies in Food Settings

So, what is the value of conducting field research and what are the aversive aspects associated with abstaining from such fieldwork? One benefit of field studies, regardless of whether they are correlational, experimental, or quasi-experimental in nature, is that they usually allow for studying *behavioral responses* in settings with a *high degree of realism* (and hence high ecological validity). This is important as only a minority of empirical studies in the areas of food science, marketing, and consumer behavior include behavioral variables as the focal outcomes (e.g., product purchases, meal choices, sales volumes of packaged foods), as collected in actual field settings. Instead, the literature is packed with studies based on self-reported perceptions, attitudes, and behavioral intentions, as well as stated preferences and other similar metrics. Such studies can surely generate profound insights and scholars often have several valid arguments for *not* conducting field studies capturing behavioral outcomes, given that certain research questions and hypotheses are simply not appropriate or feasible to address in real-life contexts before an intervention has been carefully tested under more controlled conditions [6, 7, 12]. Nevertheless, the act of scholars strategically abandoning fieldwork capturing real, observable behavior in naturally occurring settings in favor of "easier-to-run" self-report studies under more rigorous but usually less realistic settings may have some dystopic downsides.

First, behavioral outcomes have a strong selling point compared to self-report measures in that they rarely require a strong justification for their relevance to become both concrete and compelling [13, 14]. As an example, supermarket managers will likely have an easier time understanding the meaningfulness of a particular intervention if it is found to increase choice likelihood of a target product by 25% or to result in consumers spending approximately $10 more, on average, during their shopping trip compared to telling the managers that the same intervention increases students' purchase intentions from 4.2 to 4.6 on a 7-point scale.

Second, we cannot explicitly conclude that the frameworks developed and investigated through intentions, attitudes, and preferences will carry downstream effects on consumers' actual purchase or choice behavior or their shopping responses in more realistic consumption contexts (e.g., at the restaurant instead of inside the lab). As such, scholars may satisfice with developing theories that can explain how students and paid online panel participants behave under controlled conditions based on their self-reported responses, despite the risk of not knowing whether these theories will have any impact in the more complex and naturalistic environments that are symptomatic for most consumers' decision-making processes and purchase patterns [8, 10, 15, 16]. Indeed, whereas self-report can be an extremely useful tool to *explain* behavior, it is most likely less desirable that this tool replaces the study of behavior entirely in case scholars aim to draw conclusions with certainty on consumers' behavioral manifestations [6]. Such a switch from behavioral to self-report outcomes and from field to lab or online studies also risks reducing the perceived relevance of research [7], as self-reported responses do not always predict behavioral outcomes (consumers do not always teach as they preach). In fact, people's behavior is frequently at odds with what self-report suggests [17–20].

Third, relying too heavily on student samples, participant pools associated with universities, or online panel members may make a lot of the scientific claims vulnerable to overgeneralization [21, 22]. This potential problem can be mitigated when scholars conduct their studies in actual shopping settings, as these settings are more likely to include customers from a given target segment, while still including participants with more diverse demographic backgrounds (e.g., in terms of gender, education, occupation, and socioeconomic status). Thus, although online panel studies can be more representative than studies based on other common sample types [23–26] and often yield comparable or higher data quality when compared to that obtained through traditional samples, such as university students [27–29], collecting data on real consumers in naturally occurring settings can still reduce the WEIRD bias in the published literature, conceptualized as the tendency for scholars to base their research on participants from Western, educated, industrialized, rich, and democratic societies [30–32], typically in the form of university students or crowdsourced online panel samples operating in Western regions.

Having delineated a set of possible disadvantages linked to a sole focus on online and lab studies and some salient strengths associated with work conducted in naturally occurring settings, I now describe five field studies, aimed at exemplifying and showing the methodological pluralism of food-related research carried out in actual consumption contexts.

2.1 Peer Presence and Popular Choices

One relatively uncommon approach from a methodological point of view is to have observers unobtrusively watch consumers from a distance and code their consumption-relevant responses according to a predefined and carefully crafted observation scheme. As an example of a recent study relying on this observational approach, Otterbring [33] examined whether the presence (vs. absence) of peers and other consumers, respectively, could influence a target consumer's propensity to purchase popular brands (i.e., a market leader) within a given product category. Based on previous literature indicating that customers' inclination to choose more rather than less expensive brands increased in the presence of others [34] and that shopping with a friend rather than alone generally increased spending [35], the author suggested that such results could have, at least in part, emerged due to customers being more inclined to choose a market leader in the presence of others. Indeed, market leaders typically charge higher prices than an average brand within a product category [36], consistent with the price-quality link [37]. However, because others' social identity appears to be contingent on social presence effects, such that the presence of others who belong to the same in-group has a stronger impact on people's decision-making than out-group members [38], Otterbring [33] argued that the presence of peers (i.e., family members, a significant other, friends, and acquaintances) should have a stronger influence on customers' choice behavior than the presence of unknown others (i.e., other customers). Accordingly, peer presence, more so than other shoppers' presence, should promote popular brand choices in the form of a market leader.

To examine this main hypothesis, an observational field study was carried out at a large supermarket. The data were collected during 2 days after receiving permission from the supermarket. Two observers unobtrusively coded a set of behaviors stated on a structured observation scheme for customers standing in front of a supermarket shelf, including whether the observed customers were shopping alone or together with peers; whether any other customers were present or absent around a spice shelf from which the observed customers made their spice choice; and whether the brand of the spice choice corresponded to a market leader or not. Spices were used as the category of interest, given that this category had a clear market leader in Denmark when the study was conducted and because customers' demand for spices is predicted to increase materially during the next 5 years [39]. Peer presence was defined as a customer shopping together with known others during the point at which the brand choice was made (i.e., walking to a shelf as a group or socializing when selecting the spice brand), whereas the presence of other customers was operationalized as customers who spend time at or in the proximity of the spice shelf but not customers who spontaneously passed the aisle without staying or spending time close to the shelf. The supermarket offered a large selection of spices and brands, with the spice shelf being several meters wide, comprising over 25 distinct brands (e.g., private labels, well-known, less known). All customers who entered the shelf were observed, as long as the observers did not already monitor other customers in front of the shelf or pose follow-up questions to those who had just made a spice choice.

Consistent with the main hypothesis, Otterbring [33] found peer presence, but not the presence of other customers, to increase choice likelihood of popular brands, with this effect mediated by customers' stated familiarity with their chosen brand. These findings suggest that peer presence amplifies customers' accessibility of concepts linked to popularity (e.g., familiarity), resulting in congruent popular brand choices. In other words, peer presence seems to increase the saliency of popularity-linked concepts in customers' minds because peers are typically perceived as popular, thereby leading to more popular brand choices.

2.2 Supermarket Signs and Cereal Choices

A drawback of observational field studies is that some behaviors exhibited by customers may be interpreted differently by different observers. Such problems can be mitigated by having two observers code the same actions of each individual customer, allowing for computations of inter-rater agreement. However, it is important to note that a well-structured observation scheme with clear definitions of what does and does not qualify as a particular behavior is often sufficient as long as the actions to be coded are unambiguous and measurable. For example, time to purchase decision—from the point at which a customer enters a shelf until the customer has put a product in his or her shopping basket/cart—can be measured reliably through a stopwatch. Similarly, it rarely requires two independent observers to code something as basic as how many products a given customer has touched among the available alternatives in a shelf. Nevertheless, a superior (but substantially more costly) method in terms of reliability is to incorporate eye tracking as part of the study setup. Eye tracking is a process-tracing method which, through recordings of customers' eye movements, enables scholars to capture more of the decision-making process. As such, through the use of customers' visual attention, it is possible to get a more fine-grained understanding of the process preceding an actual choice or product purchase, which means that we learn more than simply studying an outcome (as available in customers' shopping baskets or stated on their receipts). An additional asset with eye-tracking methodology is that eye movements are difficult to control or suppress, meaning that such metrics are less prone to response and social desirability biases than other process-tracing techniques (e.g., think-aloud protocols) and self-report measures [40–43].

Otterbring et al. [44] conducted two field studies using eye-tracking methodology to examine the extent to which in-store signage is used during navigation and decision-making and how the viewing of such signage material influences customers' visual attention and choice behavior. Their second study sought to test whether incidental exposure to a cereal product depicted on an in-store sign would be sufficient for customers to detect that particular product more rapidly and look at it more frequently when later standing in front of a supermarket shelf with cereals, with possible downstream effects on choice behavior (Fig. 1). This would be plausible given that previously seen stimuli enhance subsequent identification of cues with similar perceptual properties [45–47] and that easily processed stimuli tend to be liked better and chosen more frequently [48–50]. Once the head-mounted eye-

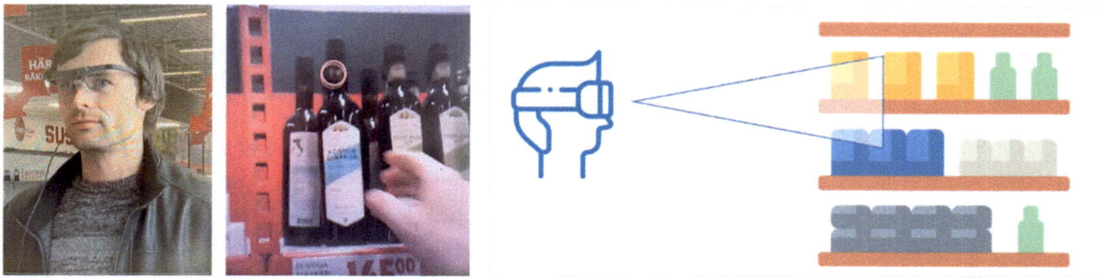

Fig. 1 Study performed in a real supermarket using a head-mounted eye tracker to record customers' Visualization featuring the author and image from the author

tracking system (similar to a pair of glasses) had been successfully calibrated, recordings of customers' visual attention began. A research assistant incidentally asked customers whether they could see an in-store sign, pointing toward it and explaining that the shelves behind the sign contained a large assortment of cereals. Customers were then instructed to go to the shelf, choose a cereal product that they could consider purchasing, and pick it up. Some customers were exposed to one cereal product, while others were exposed to another comparable product on the in-store sign (randomized between customers), with both these products available on the cereal shelves. As such, the product which customers were incidentally exposed to acted as a target, whereas the comparable product served as a control. Otterbring et al. [44] found that customers detected the target more rapidly and looked at it more frequently than the control, irrespective of which specific product was used as the target. However, only 4% of customers selected the target, whereas 45% chose a product they always reported buying, highlighting past purchases as constituting a substantially more influential factor than in-store signage in predicting consumer choice.

2.3 Hungry and Helpful?

Across one field study and a follow-up lab experiment, Gidlöf et al. [51] examined the impact of hunger on customers' prosocial responses in terms of their attitudes and behavior linked to sustainable, eco-labeled foods. This investigation was inspired by a larger stream of literature indicating potential discrepancies between hungry individuals' attitudes and behavior. Specifically, hungry individuals have been shown to behave more selfishly, act greedier, and make more immoral decisions than their satiated counterparts while still expressing more prosocial attitudes [52–54]. Gidlöf and colleagues tried to conceptually replicate prior related research in a context where the positive effects for the greater good (i.e., by purchasing eco-labeled food) are more implicit but also more frequently encountered in customers' everyday lives. Additionally, the authors tested hunger effects on prosocial attitudes and behavior in the same context (i.e., choices of and attitudes toward eco-labeled foods), which differs from previous studies, where attitudes and behavior have not always been clearly linked to the same overarching phenomenon [54, 55].

The field study took place at a supermarket and relied on eye-tracking methodology. Customers who agreed to participate and signed a consent form got a head-mounted eye tracker calibrated, after which they were asked to buy a product from the pasta category as well as two additional grocery items with their own money. Through recordings of eye moments, data were collected with respect to whether customers selected an eco-labeled pasta product or not and the amount of visual attention they devoted toward eco-labeled pasta options. This latter measure was important as it could shed light on whether hungry customers would only signal prosociality to others as a form of "cheap talk" (in terms of their attitudes toward sustainability) or, alternatively, whether such potential tendencies of prosociality would also be reflected in their decision-making process (by means of their visual attention). When customers had made their pasta choice, they completed a brief survey at the end of their shopping trip. The survey included a single-item measure of hunger and a general measure of attitudes toward environmental issues during grocery shopping, with customers also being instructed to rate the degree to which a set of attributes had influenced their pasta choice, of which organic labeling was one of the available attributes. Gidlöf et al. [51] found a significant link between customers' hunger levels and their environmental concerns, such that hungry (vs. satiated) consumers reported higher environmental concerns when shopping for groceries and expressed significantly higher importance ratings of buying eco-labeled foods. However, hunger levels were neither linked to customers' choice likelihood of eco-labeled foods nor their visual attention toward such options, illustrating a salient gap between attitudes and behavior. Thus, asking customers about their general environmental concerns and their importance ratings of buying eco-labeled products may indeed have represented a seemingly "cheap" way to signal prosociality without having to make some self-sacrifices linked to such choices (e.g., paying a higher price for sustainable foods). Interestingly, these correlational findings were largely replicated in the subsequent lab-based experiment by the authors, suggesting that the relationship between hunger and consumers' prosocial responses is generalizable across study settings, sample types, and paradigms and that this link may well be causal.

2.4 Healthy or Harmful Habits?

Machín et al. [56] conducted an eye-tracking field study in three supermarkets to explore how people make their food choices and which information they use during in-store decision-making across a wide range of different product categories. This investigation was warranted given the scarcity of former studies investigating how customers make their purchase decisions in the real store environment; for a notable exception, see [57]. Customers were approached upon store entry by one of the researchers and were told that the study aimed at understanding how consumers make their in-store purchase decisions. Customers who agreed to take part in the study were asked to make their purchases, just as they would have done during a regular shopping trip while wearing an eye tracker that recorded their eye movements. Customers received a calibration procedure of the head-mounted eye tracker and were instructed to contact a researcher as soon as they had picked up all the products they planned to purchase *before* entering the cash register area. The latter was important for privacy reasons, given that it would have been possible to detect customers' transactions (e.g., their pin codes) through the eye-tracking recordings. Finally, after having paid for their purchases, customers responded to some survey measures, after which they were debriefed about the detailed study purpose.

The results revealed that virtually all customers purchased at least one product. Those who did examined products in roughly three store sections, on average, with the vast majority investigating self-service sections containing ultra-processed products whose consumption has been linked to cardiovascular diseases and cancer risks [58–62]. For two-thirds of the chosen products, customers went immediately to a given product, grabbed it, and put it in their carrying equipment (basket or cart), seemingly without visible comparison between products. Thus, most customers engaged in limited information search. Supporting this notion, less than 7 s passed, on average, from the point at which customers grabbed a product until they put it in their carrying equipment, indicating that customers' food purchases are highly habitual in consumption contexts characterized by ample availability of ultra-processed products. Together, these findings suggest that interventions and public policies that seek to decrease customers' purchase propensity of ultra-processed products need to interfere with the habitual nature of grocery shopping in order to move purchase decisions from harmful to healthful.

2.5 Tailored Typefaces and Tipsy Transactions

An easily quantifiable data source from field settings is financial transactions. Rolschau et al. [63] conducted a field study at a bar to test whether exposure to a beer board with the beer options written either in angular or round letters would influence customers' beer selection toward sweet or sour beers using such transactional data. This examination was based on research on sensory integration effects, commonly referred to as crossmodal correspondences, with consumers often exhibiting a tendency to match stimuli across different sensory modalities. Specifically, round shapes and typefaces tend to be associated with sweetness, whereas angular shapes and typefaces are more related to sourness [64–66]. However, very few studies have tested such effects in naturalistic settings, and the vast majority of studies in this domain have focused on customers' taste expectations or taste perceptions, sometimes combined with their liking levels of drinking and dining options presented in different typefaces. Importantly, Rolschau et al. [63] could not find a single typeface-taste study that had used naturalistic choice as the key outcome variable, thus motivating such an investigation to find out whether the formerly demonstrated effects would "survive" in "the wild."

To this end, the authors conducted a quasi-experimental field study at a beer bar, with a visible blackboard containing the name, style, and alcohol content of 20 different beers representing the menu. The data consisted of transactions from the cash register system, including the amount of the transaction as well as the number, size, and style of the beers, with the conditions (round vs. angular typeface on the beer board) counterbalanced across comparable weekends and weekdays. As objective taste metrics, pH levels were used as a proxy for sourness, and Final Gravity served as a proxy for sweetness. These measures were complemented by a more subjective classification made by a bartender, who coded all beer options according to whether they were deemed to be sour or sweet, respectively. In terms of the main findings, Rolschau et al. [63] found no differences between typeface conditions regarding the sweetness of the selected beers. However, customers selected beers that were significantly sourer, by means of their pH levels, when the board menu featured beer options written in round rather than angular typeface, with comparable results emerging for the bartender's subjective sourness classification. As such, these results seem to go against an account based on crossmodal correspondences. Instead, the findings imply that customers may have been more prone to purchase beer options that, normally, would have been viewed as too sour when they were exposed to the round typeface, which theoretically should induce expectations of sweetness. Thus, the expected sweet taste induced by the round typeface could have exerted a compensatory effect on customers' choice behavior, such that they opted for beer alternatives with contrasting sour rather than congruent sweet tastes [67]. Interestingly, Rolschau et al. [63] also found that customers spent significantly more money in the round as opposed to the angular typeface condition.

3 Concluding Remarks and Recommendations

The five field studies delineated above have a set of common denominators but also certain key differences in terms of designs, measures, and other study-specific features. To end this chapter with some "hands-on" advice on things that are vital to consider when planning for and eventually running field-based investigations, the importance of the following points cannot be overstated.

Keep things as simple as possible If you are elaborating on whether to add a third control condition in a field experiment that already has two relevant conditions, a motivated question is whether this third condition is really needed. It adds complexity, requires a larger number of participants for the study to have sufficient statistical power, usually mandates additional time for data collection, and may be linked to substantially higher costs for recruiting enough participants to meet the target sample size. A simple design is also likely to be perceived more positively when pitching a field study idea to a store manager, whose acceptance may be a requirement for the study to be conducted in the first place.

Use few but good measures Consumers in actual field settings are typically visiting consumption contexts to fulfill their own goals. For that reason, they are rarely willing to answer lengthy survey instruments simply to help a researcher control for multiple potentially confounding variables or distinguish between different theoretical models [7, 64]. More often than not, field studies even necessitate the exclusion of variables that scholars wanted to capture but which had to be discarded to make it feasible for study participants in terms of time and effort. Consequently, multi-item scales are often replaced by single-item scales, which tend to be valid as long as they measure clear and unambiguous constructs [68–71]. Similarly, self-reported measures are often replaced by real, observable behavior in terms of, for example, customers' actual purchase or choice behavior. In fact, such behavioral measures constitute one of the key strengths of field studies, not least considering the many shortcomings linked to self-report [33, 72].

Pretest stimuli and use powerful manipulations As field studies often imply smaller sample sizes due to their labor-intensive and costly nature, it is generally a good idea to pretest the stimuli intended for use in the main study. Therefore, pilot testing of manipulations and stimulus material is well-advised in field research. Relatedly, if resource constraints imply that only a sample of, say, 100 customers divided into two comparable groups can realistically be recruited within the set timeframe of a field study, scholars should realize that such a sample size only allows for detecting moderate-to-large effect sizes with sufficient statistical power. Thus, if sample size restrictions constitute an issue, experimental manipulations need to be powerful to carry downstream effects on the focal variables of interest. Power calculations before conducting a field study is generally a good idea to know the minimum sample size required to detect an effect of a certain magnitude with sufficient statistical power.

Plan the study inside out before data collection takes place Field studies often require that several individuals coordinate tasks and collaborate during data collection, survey administration, experimental manipulations, etc. While researchers may vary in the extent to which they can plan for and estimate the number of individuals needed for a study to be smoothly conducted, anecdotal evidence seems to suggest that it is often a good idea to rely on more people than predicted in order to successfully execute a field study. Therefore, make sure that the entire study and all its crucial building blocks have been carefully gone through by each person involved and consider whether more manpower can be beneficial *before* proceeding with data collection. Needless to say, make sure to follow local ethical guidelines for research involving human subjects. This is perhaps even more important in field studies, as customers in actual consumption contexts are sometimes studied without their explicit knowledge, meaning that informed consent is not always collected prior to a manipulation has taken place. For that reason, some research questions are not suitable to address in field settings [6, 10, 12].

Mix methods A triangulation of study methods is generally highly desirable, meaning that field-based investigations can compellingly be combined with lab or online studies to create an entirety that is greater than the sum of its parts. By combining these study types, scholars can establish effects under controlled, rigorous conditions, exclude alternative theoretical accounts, control for confounding factors, demonstrate boundary conditions for the applicability of a given effect (as moderated by a situational factor or an individual-differences trait), and document the psychological mechanism driving the effect while also speaking about the generalizability of the effect and its ultimate behavioral consequences. As such, combining field, lab, and online studies can increase both internal and external validity.

Beyond these recommendations aimed at motivating scholars to conduct more fieldwork, it is worth raising some final remarks with respect to the assessment of field studies. First, field studies cannot be expected to have the same level of rigor and control in terms of experimental design, dependent variables, and measurement of potentially confounding factors as a controlled experiment.

Thus, journal editors, reviewers, and funding agencies need to stop asking why, say, three potentially confounding variables were not controlled for in a field study, why a third control condition was not added to the already complex study setup, why the manipulation check was not captured through a multi-item scale instead of a single-item scale, and why behavioral outcomes were not complemented with self-report measures of the presumed psychological process or the boundary conditions for the documented effect. While all such questions can be highly relevant to address in one or several follow-up studies, fieldwork should not be evaluated by the same assessment criteria used for online and lab studies conducted in more controlled and rigorous conditions. Thus, as some of the key assets of field studies include increased realism and perceived relevance, enhanced ecological validity, greater generalizability, and boosted beliefs regarding the behavioral consequences of a particular phenomenon, those evaluating field studies need to emphasize these aspects as top priorities instead of automatically applying the assessment toolbox used for evaluating studies conducted under more controlled conditions. Doing so will arguably remove some of the bitter barriers that scholars frequently encounter when trying to publish research from actual consumption contexts and hence will likely inspire further fieldwork in food settings.

References

1. Doliński D (2018) Is psychology still a science of behaviour? Soc Psychol Bull 13(2):1–14
2. Li JQ, Rusmevichientong P, Simester D, Tsitsiklis JN, Zoumpoulis SI (2015) The value of field experiments. Manag Sci 61(7): 1722–1740
3. Loebnitz N, Frank P, Otterbring T (2022) Stairway to organic heaven: the impact of social and temporal distance in print ads. J Bus Res 139:1044–1057
4. Otterbring T, Bhatnagar R, Samuelsson P, Borau S (2021) Positive gender congruency effects on shopper responses: field evidence from a gender egalitarian culture. J Retail Consum Serv 63:102738
5. Otterbring T, Wu F, Kristensson P (2021) Too close for comfort? The impact of salesperson-customer proximity on consumers' purchase behavior. Psychol Mark 38(9):1576–1590
6. Baumeister RF, Vohs KD, Funder DC (2007) Psychology as the science of self-reports and finger movements: whatever happened to actual behavior? Perspect Psychol Sci 2(4): 396–403
7. Cialdini RB (2009) We have to break up. Perspect Psychol Sci 4(1):5–6
8. Otterbring T (2021a) Evolutionary psychology in marketing: deep, debated, but fancier with fieldwork. Psychol Mark 38(2):229–238
9. Viglia G, Dolnicar S (2020) A review of experiments in tourism and hospitality. Ann Tour Res 80:102858
10. Otterbring T, Sundie J, Li YJ, Hill S (2020b) Evolutionary psychological consumer research: bold, bright, but better with behavior. J Bus Res 120:473–484
11. Pham MT (2013) The seven sins of consumer psychology. J Consum Psychol 23(4):411–423
12. Reis HT, Gosling SD (2010) Social psychological methods outside the laboratory. In: Fiske ST, Gilbert DT, Lindzey G (eds) Handbook of social psychology. John Wiley & Sons, pp 82–114
13. Baca-Motes K, Brown A, Gneezy A, Keenan EA, Nelson LD (2013) Commitment and behavior change: evidence from the field. J Consum Res 39(5):1070–1084
14. Goldstein NJ, Cialdini RB, Griskevicius V (2008) A room with a viewpoint: using social norms to motivate environmental conservation in hotels. J Consum Res 35(3):472–482

15. Gneezy A (2017) Field experimentation in marketing research. J Mark Res 54(1):140–143
16. Morales AC, Amir O, Lee L (2017) Keeping it real in experimental research—understanding when, where, and how to enhance realism and measure consumer behavior. J Consum Res 44(2):465–476
17. Kristensson P, Wästlund E, Söderlund M (2017) Influencing consumers to choose environment friendly offerings: evidence from field experiments. J Bus Res 76:89–97
18. Otterbring T, Ringler C, Sirianni NJ, Gustafsson A (2018) The Abercrombie & Fitch effect: the impact of physical dominance on male customers' status-signaling consumption. J Mark Res 55(1):69–79
19. Otterbring T, Shams P (2019) Mirror, mirror, on the menu: visual reminders of overweight stimulate healthier meal choices. J Retail Consum Serv 47:177–183
20. Townsend C, Liu W (2012) Is planning good for you? The differential impact of planning on self-regulation. J Consum Res 39(4):688–703
21. Saad G (2020) The marketing of evolutionary psychology. J Bus Res 120:485–491
22. Yarkoni T (2020) The generalizability crisis. Behav Brain Sci:1–37
23. Buhrmester M, Kwang T, Gosling SD (2016) Amazon's Mechanical Turk: a new source of inexpensive, yet high-quality data? Perspect Psychol Sci 6:3–5
24. Goodman JK, Paolacci G (2017) Crowdsourcing consumer research. J Consum Res 44(1): 196–210
25. Otterbring T, Festila A (2021) Pandemic prevention and personality psychology: gender differences in preventive health behaviors during COVID-19 and the roles of agreeableness and conscientiousness. J Safe Sci Resil 3:87–91
26. Peer E, Brandimarte L, Samat S, Acquisti A (2017) Beyond the Turk: alternative platforms for crowdsourcing behavioral research. J Exp Soc Psychol 70:153–163
27. Hauser DJ, Schwarz N (2016) Attentive Turkers: MTurk participants perform better on online attention checks than do subject pool participants. Behav Res Methods 48(1): 400–407
28. Otterbring T, Folwarczny M (2022) Firstborns buy better for the greater good: birth order differences in green consumption values. Personal Individ Differ 186:111353
29. Porter CO, Outlaw R, Gale JP, Cho TS (2019) The use of online panel data in management research: a review and recommendations. J Manag 45(1):319–344
30. Henrich J, Heine SJ, Norenzayan A (2010a) Most people are not WEIRD. Nature 466(7302):29–29
31. Henrich J, Heine SJ, Norenzayan A (2010b) The weirdest people in the world? Behav Brain Sci 33(2–3):61–83
32. Muthukrishna M, Bell AV, Henrich J, Curtin CM, Gedranovich A, McInerney J, Thue B (2020) Beyond western, educated, industrial, rich, and democratic (WEIRD) psychology: measuring and mapping scales of cultural and psychological distance. Psychol Sci 31(6): 678–701
33. Otterbring T (2021) Peer presence promotes popular choices: a "Spicy" field study on social influence and brand choice. J Retail Consum Serv 61:102594
34. Argo JJ, Dahl DW, Manchanda RV (2005) The influence of a mere social presence in a retail context. J Consum Res 32(2):207–212
35. Kurt D, Inman JJ, Argo JJ (2011) The influence of friends on consumer spending: the role of agency–communion orientation and self-monitoring. J Mark Res 48(4):741–754
36. Cao Y, Gruca TS (2004) The influence of pre-and post-purchase service on prices in the online book market. J Interact Mark 18(4): 51–62
37. Völckner F, Hofmann J (2007) The price-perceived quality relationship: a meta-analytic review and assessment of its determinants. Mark Lett 18(3):181–196
38. Chou EY, Nordgren LF (2017) Safety in numbers: why the mere physical presence of others affects risk-taking behaviors. J Behav Decis Mak 30(3):671–682
39. Fortune Business Insights (2019) Spices and seasonings market. Retrieved from: https://www.fortunebusinessinsights.com/industry-reports/spices-and-seasonings-market-101694. Accessed 20 Dec 2021
40. Glaholt MG, Reingold EM (2011) Eye movement monitoring as a process tracing methodology in decision making research. J Neurosci Psychol Econ 4(2):125–146
41. Otterbring T, Wästlund E, Gustafsson A (2016) Eye-tracking customers' visual attention in the wild: dynamic gaze behavior moderates the effect of store familiarity on navigational fluency. J Retail Consum Serv 28: 165–170
42. Otterbring T, Gidlöf K, Rolschau K, Shams P (2020a) Cereal deal: how the physical appearance of others affects attention to healthy foods. Perspect Behav Sci 43(3):451–468

43. Wästlund E, Otterbring T, Gustafsson A, Shams P (2015) Heuristics and resource depletion: eye-tracking customers' in situ gaze behavior in the field. J Bus Res 68(1):95–101
44. Otterbring T, Wästlund E, Gustafsson A, Shams P (2014) Vision (im) possible? The effects of in-store signage on customers' visual attention. J Retail Consum Serv 21(5): 676–684
45. Kristjánsson Á, Campana G (2010) Where perception meets memory: a review of repetition priming in visual search tasks. Atten Percept Psychophys 72(1):5–18
46. Reber R, Schwarz N, Winkielman P (2004) Processing fluency and aesthetic pleasure: is beauty in the perceiver's processing experience? Personal Soc Psychol Rev 8(4):364–382
47. Shah AK, Oppenheimer DM (2008) Heuristics made easy: an effort-reduction framework. Psychol Bull 134(2):207–222
48. Berger J, Fitzsimons G (2008) Dogs on the street, pumas on your feet: how cues in the environment influence product evaluation and choice. J Mark Res 45(1):1–14
49. Forster M, Leder H, Ansorge U (2013) It felt fluent, and I liked it: subjective feeling of fluency rather than objective fluency determines liking. Emotion 13(2):280–289
50. Schwarz N (2004) Metacognitive experiences in consumer judgment and decision making. J Consum Psychol 14(4):332–348
51. Gidlöf K, Lahm ES, Wallin A, Otterbring T (2021) Eco depletion: the impact of hunger on prosociality by means of environmentally friendly attitudes and behavior. J Retail Consum Serv 62:102654
52. Briers B, Pandelaere M, Dewitte S, Warlop L (2006) Hungry for money: the desire for caloric resources increases the desire for financial resources and vice versa. Psychol Sci 17(11): 939–943
53. Elbaek C, Mitkidis P, Aarøe L, Otterbring T (2021) Material scarcity and unethical economic behavior: a systematic review and meta-analysis
54. Petersen MB, Aarøe L, Jensen NH, Curry O (2014) Social welfare and the psychology of food sharing: short-term hunger increases support for social welfare. Polit Psychol 35(6): 757–773
55. Aarøe L, Petersen MB (2013) Hunger games: fluctuations in blood glucose levels influence support for social welfare. Psychol Sci 24(12): 2550–2556
56. Machín L, Curutchet MR, Gugliucci V, Vitola A, Otterbring T, de Alcantara M, Ares G (2020) The habitual nature of food purchases at the supermarket: implications for policy making. Appetite 155:104844
57. Hoyer WD (1984) An examination of consumer decision making for a common repeat purchase product. J Consum Res 11(3): 822–829
58. Ares G, Vidal L, Otterbring T, Aschemann-Witzel J, Curutchet MR, Giménez A, Bove I (2021) Communication campaigns to support the use of nutritional warnings: different messages for different people? Health Educ Behav 48:584–594
59. Ares G, Antúnez L, Otterbring T, Curutchet MR, Galicia L, Moratorio X, Bove I (2020) Sick, salient and full of salt, sugar and fat: understanding the impact of nutritional warnings on consumers' associations through the salience bias. Food Qual Prefer 86:103991
60. Ares G, Machín L, Vidal L, Otterbring T, Aschemann-Witzel J, Curutchet MR, Bove I (2020) Uruguayan citizens' perception of messages to promote healthy eating through the use of nutritional warnings. J Nutr Educ Behav 52(10):918–927
61. Fiolet T, Srour B, Sellem L, Kesse-Guyot E, Allès B, Méjean C, ... Touvier M (2018) Consumption of ultra-processed foods and cancer risk: results from NutriNet-Santé prospective cohort. BMJ 360
62. Srour B, Fezeu LK, Kesse-Guyot E, Allès B, Méjean C, Andrianasolo RM et al (2019) Ultra-processed food intake and risk of cardiovascular disease: prospective cohort study (NutriNet-Santé). BMJ 365
63. Rolschau K, Wang QJ, Otterbring T (2020) Seeing sweet and choosing sour: compensatory effects of typeface on consumers' choice behavior. Food Qual Prefer 85:103964
64. Otterbring T, Rolschau K, Furrebøe EF, Nyhus EK (2022) Crossmodal correspondences between typefaces and food preferences drive congruent choices but not among young consumers. Food Qual Prefer 96:104376
65. Spence C, Deroy O (2013) Tasting shapes: a review of four hypotheses. Theoria et Historia Scientiarum 10:207–238
66. Velasco C, Hyndman S, Spence C (2018) The role of typeface curvilinearity on taste expectations and perception. International Journal of Gastronomy & Food Science 11:63–74
67. Biswas D, Szocs C (2019) The smell of healthy choices: cross-modal sensory compensation effects of ambient scent on food purchases. J Mark Res 56(1):123–141

68. Bergkvist L (2015) Appropriate use of single-item measures is here to stay. Mark Lett 26(3): 245–255
69. Bergkvist L, Rossiter JR (2007) The predictive validity of multiple-item versus single-item measures of the same constructs. J Mark Res 44(2):175–184
70. Otterbring T (2020) Appetite for destruction: counterintuitive effects of attractive faces on people's food choices. Psychol Mark 37(11): 1451–1464
71. Otterbring T, Danielsson CB, Pareigis J (2021) Office types and workers' cognitive vs affective evaluations from a noise perspective. J Manag Psychol 36(4):415–431
72. Nisbett RE, Wilson TD (1977) Telling more than we can know: Verbal reports on mental processes. Psychol Rev 84(3):231–259

Part IV

Methods Econometrics and Sensometrics

Chapter 17

Econometric Tools for Food Science

Nicolas Depetris Chauvin and Jonas Di Vita

Abstract

In this chapter, we introduce basic econometric techniques which can be applied to food sciences. We define econometrics and Introduce linear regression models, simple and multiple. We further discuss the Ordinary Least Squares estimation procedure as well as the Gauss-Markov assumptions necessary to ensure the estimators are the most efficient and unbiased in the class of linear estimators. Then, we propose some visual inspection tools to validate the model. The use of categorical variables and interactions is also presented as a way to achieve additional understanding of the relationships among the variables. A brief discussion about inference is also provided in order to give the foundations for hypothesis testing in linear regression models. Some common caveats and violations of assumptions are discussed, such as the issue of multicollinearity and endogeneity and we provide some guidance to approach those problems. Lastly, some practical examples are presented to further illustrate the use of econometric techniques in food science.

Key words Econometrics, Statistics, Economic theory, Ordinary least squares, Linear regression, Estimators, Randomized controlled trial, Hedonic regression, Dummy variables

1 Introduction

Research methods are split broadly into quantitative and qualitative methods. Most of the literature on food science uses qualitative methods. Which type of method you choose will depend on your research questions, your underlying philosophy of research, and your preferences and skills. The good news is that you can use both at the same time. Often, the use of qualitative techniques allows the researcher to understand why certain behavior is observed, while quantitative techniques allow for a quantification of the effects.

In this chapter, we intend to provide a brief but complete overview of one of the most used quantitative methods, regression analysis, and its application with real-world data. Regression analysis is a fundamental part of econometrics models. The chapter does not cover theoretical aspects nor complete mathematical

Carlos Gómez-Corona and Heber Rodrigues (eds.), *Consumer Research Methods in Food Science*,
Methods and Protocols in Food Science, https://doi.org/10.1007/978-1-0716-3000-6_17,

derivations but is rather concerned with providing intuition on when, how, and why a given method is being used.

We will first introduce what econometrics is and its purpose and use not only in economics but also in other fields. Then, we will introduce the Ordinary Least Squares estimation method in the context of simple and multiple linear regression. Here, we will cover how to qualify a model as adequate or not and mention a few caveats to look out for such as collinearity, omitted variable bias, and simultaneity. Lastly, we will discuss on how to interpret the results from our model and conduct statistical inference.

In the last part of this chapter, we will provide three practical examples to further illustrate the use of the tools presented with real-world data. The examples will cover the use of hedonic regression, dummy variables, and difference-in-difference regressions. More precisely, example 1. will show you how to estimate the influence of specific components in the price of a bottle of wine, example 2. how to analyze data resulting from a blind tasting experiment, and example 3. how to analyze data coming from a Randomized Controlled Trial (RCT).

2 What Is Econometrics: A Brief Definition

Econometrics is a vast field of research that is concerned with different aspects, one of the main aspects being the statistical modeling of economic problems. It attempts to establish the causality between dependent and independent variables as well as to estimate quantities and effects (i.e., relationships). As a result, econometrics is at the crossroads of economic theory, statistics, and mathematics as stated by the Econometric Society [9].

Ultimately, the objective is to build a simple but sophisticated enough model to empirically verify the veracity of postulated models from economic theory. Although complicated models exist, the attempt to find the most simplistic yet powerful model is a crucial objective which aims at facilitating the interpretation of results and representing complex dynamics in a simplified way. This process makes extensive use of statistical and mathematical tools while heavily relying on theory to either confirm or disprove it. However, in this chapter, we will cover basic statistical aspects while focusing on intuition.

Econometrics deals with three types of data which are known as cross-sectional, time series, and panel data. Cross-sectional data is when we observe many different subjects at a specific time point, such as a consumers' willingness to pay for a particular food product at a given period of time. Time series data are observations for one subject collected over a time period, like the monthly level of expending on food in France over 3 years. Lastly, panel data is a combination of cross-sectional and time series as it contains many subjects over time such as a food-tasting experiment where the same assessors taste the same product over many weekends. In this chapter, we focus mostly on cross-sectional data.

Tools and techniques developed and used in econometrics can also be used in different fields such as food sciences, consumer sciences, and others. For the purpose of this chapter, we will cover the Ordinary Least Squares (OLS) estimation method as it can be viewed as the essential basis for more advanced econometric and statistical models. Additionally, OLS estimation enjoys a number of advantages that will be covered later on in this chapter.

3 Linear Regression Analysis

Linear regression analysis is a technique that attempts to establish and quantify a linear relationship between one dependent variable and one or more independent variables. Dependent variable can sometimes be referred to as response variable or outcome variable while independent variables can be referred to as explanatory variables or covariates. The case where a response variable is regressed on one covariate only is known as simple linear regression and is discussed in Subheading 3.1. On the other hand, the case where multiple covariates are used is called multiple linear regression and is discussed in Subheading 3.2.

We can view a regression model as a set of ingredients that can be formalized as follows:

$$y=f(x_1,x_2,\ldots,x_k)+\epsilon \tag{1}$$

To better understand the intuition behind linear regression analysis, let us consider the necessary ingredients to it. What we attempt to do is to explain an endogenous response variable y by one or more exogenous explanatory variables x_1, x_2, ..., x_k (k is typically not too large as we prefer parsimonious models to preserve interpretation) through a functional form f (which is a linear combination in our case). We say that y is an endogenous variable because it is determined within the model, while our covariates are called exogenous since they are determined outside the model (we will discuss caveats resulting from violation of the exogeneity conditions later on).

In other words, we want to build an econometric model that attempts to reconcile models from economic theory with what we observe in reality (empirical data). For this reason, our covariates x_j ($j = 1, \ldots, k$) are usually provided by economic theory or the hypotheses of the researcher, and the ultimate goal of linear regression analysis is to quantify the effect of those variables on the expected value of an outcome variable (technically, we model the conditional expectation of y given x: $E[y|\, x]$).

Additionally, an error term ϵ is added to the model to account for the discrepancy between theory and reality. Note that this error term is a random variable on which we impose a set of assumptions which are discussed in Subheading 3.1. This error term comes from different sources such as omitted relevant explanatory variables, measurement errors, or having wrongly specified a linear relationship between y and one of the x_j.

Lastly, in this chapter, the explanatory variables are assumed to be fixed (i.e., nonstochastic), but it is important to note that all the results which will be derived also hold if they were random, provided we condition on them. The setting where the covariates are random themselves is known as Stochastic Regression Framework.

3.1 Simple Linear Regression Model

As mentioned in Subheading 3, simple linear regression is when only one covariate is used in order to linearly explain an outcome variable. A mathematical formulation of such a model is as follows:

$$y_i = \beta_0 + \beta_1 x_{i1} + \epsilon_i \; i = 1, \ldots, n \tag{2}$$

Where y_i (a continuous variable) is the outcome variable for the *ith* observation, β_0 is the intercept, x_{i1} is the explanatory variable for the *ith* observation, β_1 is the coefficient (or slope) of the explanatory variable, and ϵ_i is the error term for the *i*th observation, which can be rewritten as follows: $\epsilon_i = y_i - \beta_0 - \beta_1 x_{i1}$. The errors represent the difference between the observed values of y and the true unobserved values. It is important to note that this model specification is abstract in nature. Indeed, we never observed the true value of the parameters or the value of the errors. These quantities come with their own interpretation. As mentioned already, β_0, the intercept, can be interpreted as the expected value of the response variable y when $x = 0$. However, the intercept may not have a sensible interpretation, especially when the value of x cannot be 0. The other parameter, β_1, is the slope associated with x_{i1} and represents the change in the expected value of y for a unit increase in x_{i1}. We can visualize each of these element in Fig. 1.

The objective of our estimation procedure is therefore to estimate the model parameters, namely β_0 and β_1, which will be represented with a hat once estimated ($\widehat{\beta}_0$ and $\widehat{\beta}_1$). These estimates will enable us to quantify the effect of an explanatory variable on the response and better understand the potential linear relationship existing between them. The formulation of the estimated model becomes:

$$\widehat{y}_i = \widehat{\beta}_0 + \widehat{\beta}_1 x_{i1} \tag{3}$$

Equation (3) is the estimated regression line where $\widehat{y}_i$ is referred to as the fitted values for the *ith* observation, $\widehat{\beta}_0$ and $\widehat{\beta}_1$ are the estimated quantities resulting from an estimation procedure called

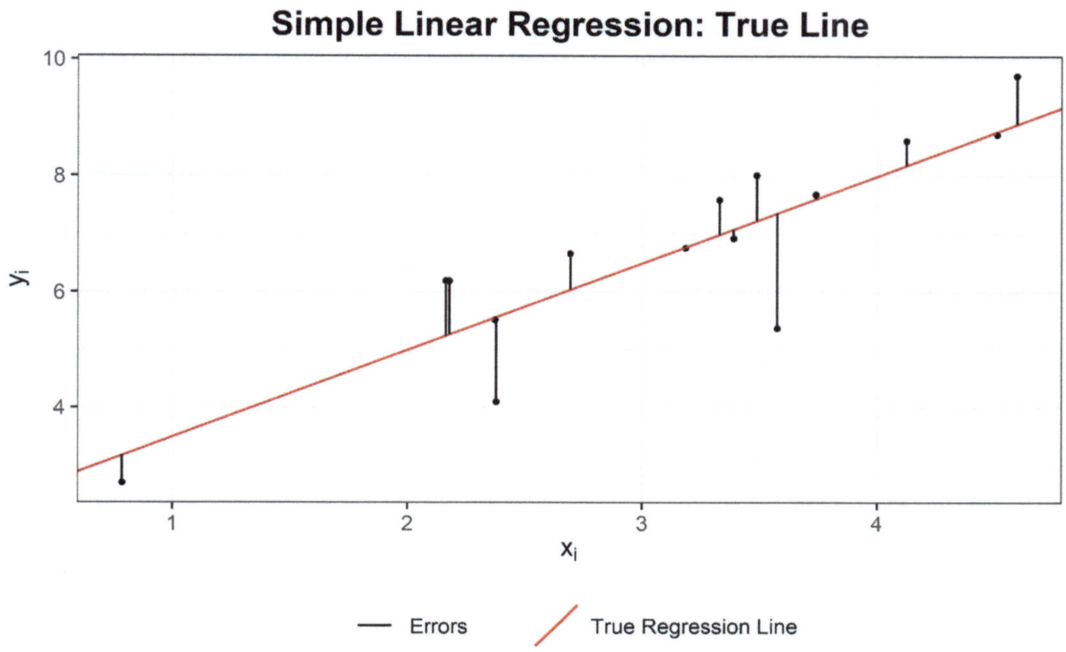

Fig. 1 Simple linear regression: true model

Ordinary Least Squares. Also, we notice that the error term ϵ is no more in the estimated regression line. It is therefore convenient and useful to introduce a new quantity called the residuals which are denoted $\widehat{\epsilon}_i = y_i - \widehat{y}_i = y_i - \widehat{\beta}_0 - \widehat{\beta}_1 x_{i1}$ for the ith observation. The residuals are the difference between the observed value and the fitted value from our model. They can be viewed as a sample estimate of the errors, although technically and theoretically they are not the same. Figure 2 illustrates all the components mentioned so far.

To derive the estimated quantities, OLS minimizes the sum of squared residuals we introduced above:

$$\min_{\beta_0,\beta_1} \sum_{i=1}^{n} \widehat{\epsilon}_i^2 = \min_{\beta_0,\beta_1} \sum_{i=1}^{n} \left(y_i - \widehat{\beta}_0 - \widehat{\beta}_1 x_{i1} \right) \tag{4}$$

From Eq. (4), we obtain a system of equations known as normal equations from which we derive the First Order Condition (FOC) for both parameters:

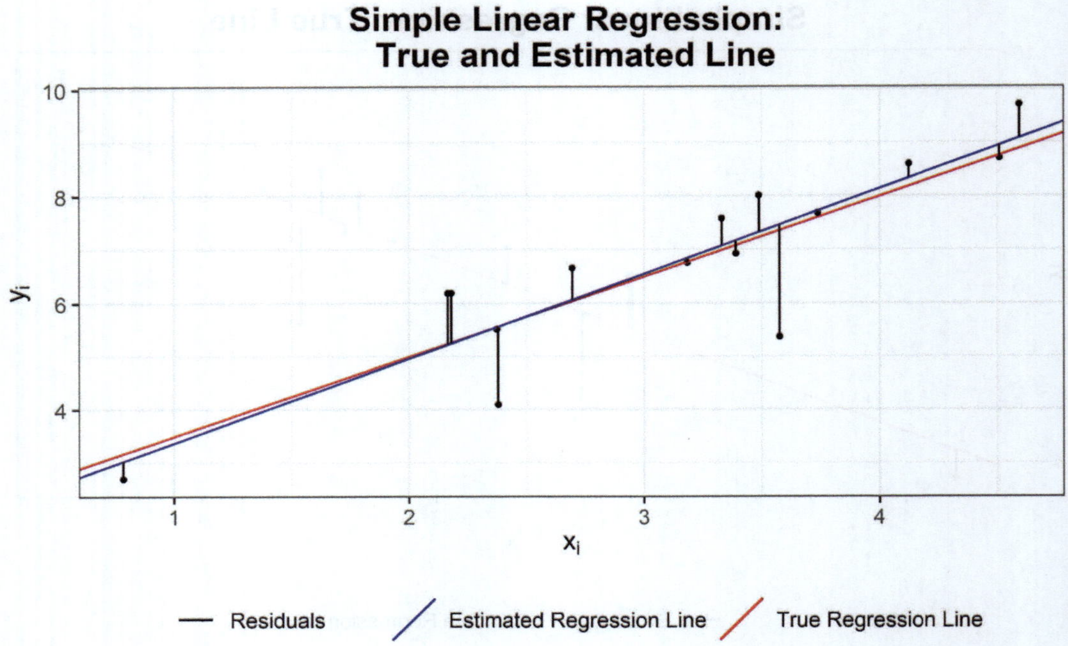

Fig. 2 Simple linear regression: true and estimated model

$$\begin{cases} \dfrac{\partial}{\partial \widehat{\beta}_0} \sum_{i=1}^{n} \left(y_i - \widehat{\beta}_0 - \widehat{\beta}_1 x_{i1}\right)^2 = 0 \\ \dfrac{\partial}{\partial \widehat{\beta}_1} \sum_{i=1}^{n} \left(y_i - \widehat{\beta}_0 - \widehat{\beta}_1 x_{i1}\right)^2 = 0 \end{cases} \tag{5}$$

By solving the normal equations, we obtain the OLS estimators:

$$\widehat{\beta}_{0\,\mathrm{OLS}} = \overline{y} - \widehat{\beta}_{1_{\mathrm{OLS}}} \overline{x}$$

$$\widehat{\beta}_{1\,\mathrm{OLS}} = \frac{\sum_{i=1}^{n} (x_i - \overline{x})(y_i - \overline{y})}{\sum_{i=1}^{n} (x_i - \overline{x})^2} \tag{6}$$

We just showed that the OLS estimators obtained in Eq. (6) are the ones that give us the line of best fit. In other words, it is the regression line that minimizes the distance between the observed value and the value given by our model, as shown in Fig. 2. We denote these estimators with an OLS subscript to specify that the estimators were obtained using Ordinary Least Squares procedure. However, for the sake of simplifying notation and as we only consider OLS estimation in this chapter, we will drop this notation. It is important to notice that nobody does this type of calculations by hand. The estimation can be done without much effort with any statistical package (i.e., Stata, SPSS, R, Python, etc.) and even with MS Excel (go to Data -> Data Analysis -> Regression).

The OLS estimators come with some properties and are known to be BLUE under some assumptions, but before laying down the necessary assumptions, let us understand what BLUE means:

- *B*est: an estimator is best when its variance is the smallest compared to other estimators.
- *L*inear: the estimators are a linear function of the response variable y.
- *U*nbiased: the expected value of the estimator is the true model parameter.
- *E*stimator: a statistic used to infer on a population parameter.

As mentioned, for this model specification to provide reliable (BLUE) estimates, a few assumptions are necessary and should always be verified before carrying on with interpretation and/or inference. The properties of the OLS estimators depend on the assumptions we are willing to impose on our model.

The following assumptions are known as the Gauss-Markov assumptions. If those are met, the OLS estimators is BLUE. Note that in most econometrics books, those assumptions are stated conditional on x (i.e., $E[\epsilon_i|x_i]$) because the assumption that x is fixed is very unlikely to hold with economic data since most data is observational and not the result of a controlled experiment.

- *Gauss-Markov Assumptions*

  1. The true underlying relationship between y and x is linear in its parameters.
  2. The expectation of the errors is 0: $E[\epsilon_i] = \mu = 0, \forall\ i$.
     - On average, the observed values lie on the true regression line.
  3. The variance of the errors is homoscedastic (equal variance): $(\epsilon_i) = \sigma^2, \forall\ i$.
     - Each observation has the same level of uncertainty.
  4. The covariance between the errors is 0: $\text{cov}(\epsilon_i, \epsilon_j) = 0$ for $i \neq j$.
     - Errors are linearly independent; knowing ϵ_i does not give us information about ϵ_j for $i \neq j$.
  5. The covariance between the covariate and the errors is 0: cov $(x_{i1}, \epsilon_i) = 0, \forall\ i$.
     - Covariate is independent of the errors (exogeneity assumption).
  6. The covariates do not exhibit perfect multicollinearity (*see* Subheading 3.2).

To verify some of those assumptions, visual inspection is extensively used and a number of plots which are usually referred to as diagnostic plots can be checked. Among those diagnostic plots, commonly used ones are Residuals vs. Fitted plot, Normal Quantile-Quantile Plot (QQ-plot), Scale-Location plot, Residuals vs. Leverage plot, and plotting the residuals against their index.

- *Residuals vs. Fitted Plot*: $\left(\widehat{y}_{i,}\widehat{\epsilon}_i\right)$
  - Plotting the residuals against the fitted values of our model allows us to verify the linearity assumption (assumption 1). We want to observe a cloud of points randomly and equally scattered around a horizontal line (i.e., lack of structure). If we observe a non-linear pattern between *y* and *x*, it means that our linear model has not captured the true underlying relationship.
  - Note that to further check linearity between *y* and *x*, it is also recommended to plot the original data in a scatter plot with the regression line added to it. If the relationship doesn't look linear, data transformation can sometimes be performed in order to linearize the relationship (e.g., logarithmic transformation).
- *Normal QQ-plot*
  - Although not specified in the assumptions above, normality of the errors gives additional beneficial properties to the OLS estimator. If this assumption is met, the OLS estimator is equal to the Maximum Likelihood Estimator and becomes the Best Unbiased Estimator (BUE) among the class of linear and non-linear estimators. The normal QQ-plot allows us to check this assumption. If the residuals are well distributed along the straight line, we can assume that the errors do follow a normal distribution.
- *Scale-Location Plot*
  - This diagnostic plot allows us to verify assumption 3 of homoscedastic variance. As for the Residuals vs. Fitted plot, we want to observe randomly and equally scattered points around a horizontal line. If we fail to observe that, statistical tests such as the Breusch-Pagan test are available to statistically check the assumption of homoscedasticity.
- *Residuals vs. Leverage Plot*
  - This plot allows us to check for influential observations. Those observations are those which alter the estimation of the regression line. This diagnostic uses Cook's distance to identify influential values. If an observation lies beyond the dashed lines representing Cook's distance, it may be an indication to look closer at this particular observation to better

understand it. Domain knowledge will tell whether this observation should be removed or treated with robust statistics techniques.

- *Residuals vs. Index*
  - Lastly, plotting the residuals in the order of measurement allows to visually identify if serial correlation exists among the residuals, which would violate assumption 4. We want to make sure that no patterns can be observed. This would typically be the case for time-ordered observations. Note that another way to check for autocorrelation would be an autocorrelation plot.

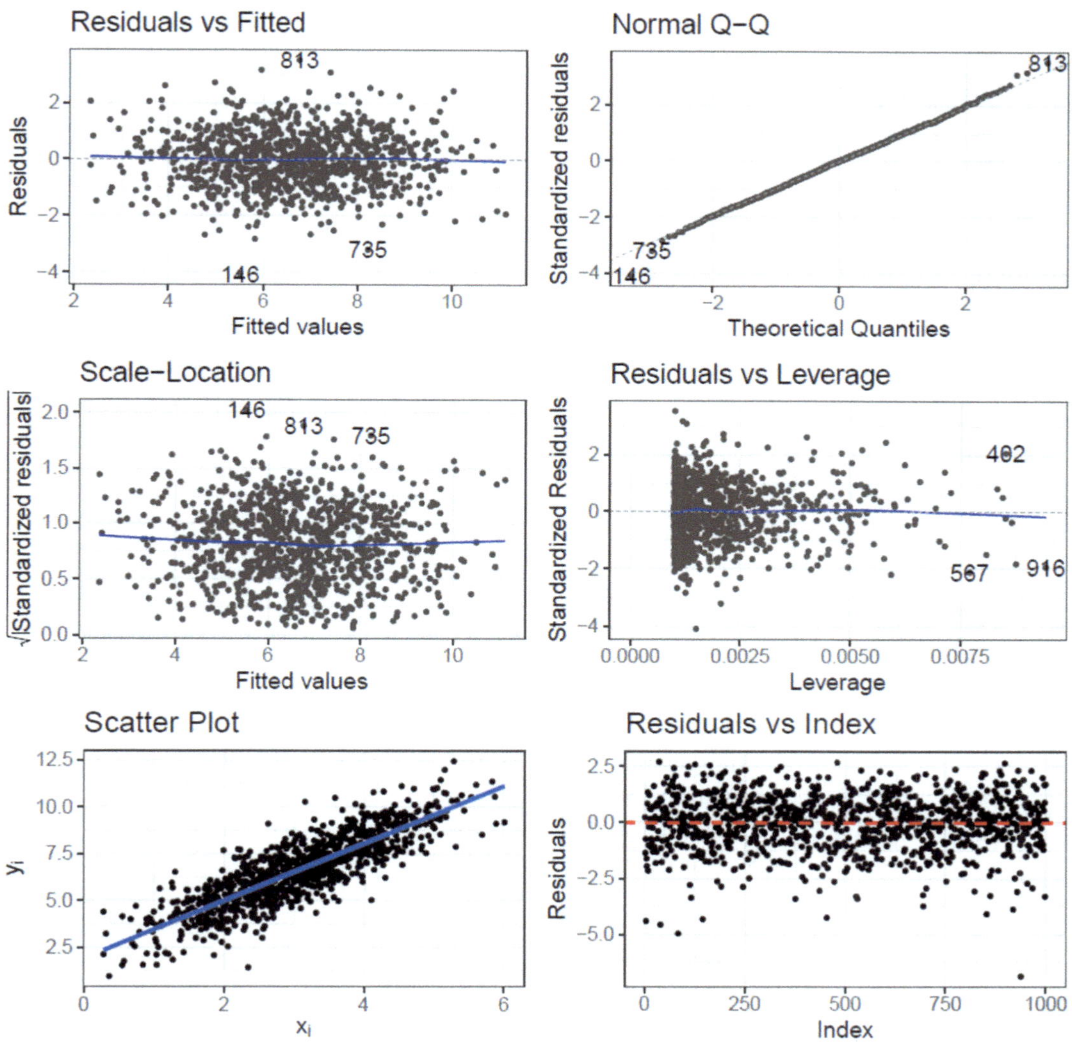

Fig. 3 Satisfactory diagnostic plots

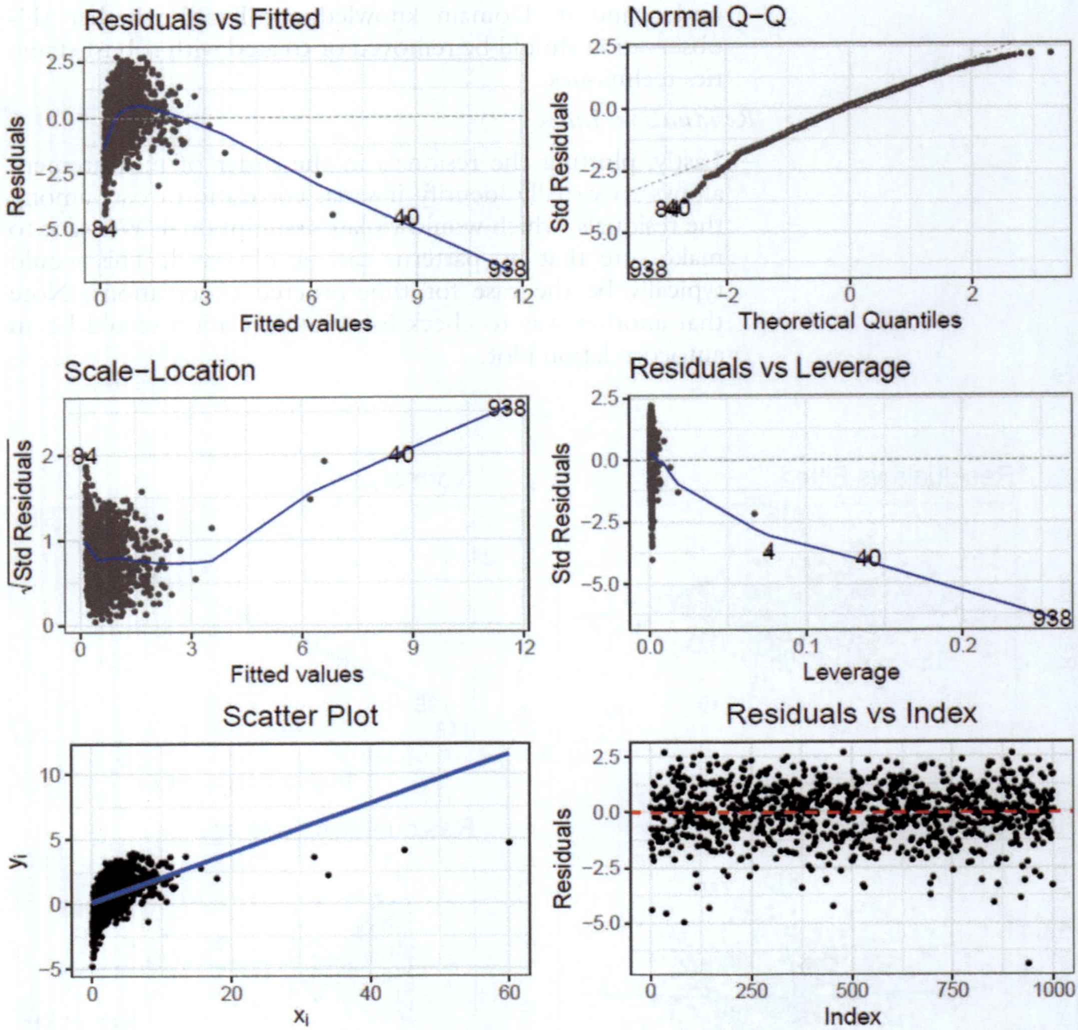

Fig. 4 Unsatisfactory diagnostic plots

If the diagnostic plots are satisfactory, one can conclude that the model seems to fit the data well. Figure 3 illustrates cases where the diagnostic plots are satisfactory, while Fig. 4 shows plot where the assumptions are clearly violated.

On top of the diagnostic we can make on the model, there exists another measure of goodness-of-fit which is commonly reported, namely the coefficient of determination R^2. This statistic expresses the amount of variability present in y which is explained by the model. Therefore, the higher the R^2, the better the model is at explaining variation in the response variable (provided the diagnostic resulted satisfactorily). R^2 takes on values between 0 and 1 and can be used to compare models built on the same dataset (note that in some cases R^2 can be negative). Mathematically, it can be expressed as follows:

$$R^2 = 1 - \frac{\text{Residual Sum of Squares (RSS)}}{\text{Total Sum of Squares (TSS)}} = 1 - \frac{\sum_{i=1}^{n} (y_i - \widehat{y}_i)^2}{\sum_{i=1}^{n} (y_i - \overline{y}_i)^2} \quad (7)$$

We will see, however, that an adjusted version of this statistic is more reliable in the case of multiple linear regression (*see* Subheading 3.2).

To summarize:

- Simple linear regression models are models that linearly associate a dependent variable with one single independent variable.
- This linear relationship is estimated in order to minimize the amount of mistakes we make between the model and the observed values.
- Ordinary Least Squares estimation provides BLUE estimates for the regression line under the Gauss-Markov assumptions.
- Model diagnostic must be carried out on the model residuals in order to verify the assumptions and ensure interpretation and inference is reliable.
- R^2 is another measure of goodness-of-fit of the model that tells us how much information is explained by the model.

3.2 Extension to Multiple Linear Regression Model

In Subheading 3.1, we covered the case of simple linear regression. However, it is rarely the case that an independent variable is well explained by one single covariate. Indeed, it is more probable that more than one covariate is required to establish a sound relationship and avoid omitted variable bias (*see* Subheading 3.3.2). For this reason, we now introduce the case of multiple linear regression, where k explanatory variables can be used to explain y. The model becomes:

$$y_i = \beta_0 + \beta_1 x_{i1} + \beta_2 x_{i2} + \ldots + \beta_k x_{ik} + \epsilon_i \quad (8)$$

Conveniently, what has been introduced for the simple linear regression holds for the multiple linear regression case. However, since the number of covariates is usually greater than two, visualization becomes complicated and not possible for more than three dimensions. Also, the algebra becomes more complex and the use of matrix algebra and vector calculus becomes handy. The model is rewritten in matrix notation as follows:

$$y = X\beta + \epsilon \tag{9}$$

Where y is an $n \times 1$ vector of responses, X is a $n \times k + 1$ design matrix (including an intercept), β is a $k \times 1$ vector of parameters and ϵ is an $n \times 1$ vector of errors.

The OLS estimation procedure follows the same steps as in the linear regression case and the estimated vector of parameters is written in matrix notation as:

$$\widehat{\beta}_{\text{OLS}} = (X'X)^{-1}X'y \tag{10}$$

As a result, the regression plane of the multiple linear regression can be written as follows:

$$\widehat{y} = X\widehat{\beta} \tag{11}$$

And the model residuals are therefore defined as $\widehat{\epsilon} = y - X\widehat{\beta}$.

The OLS estimator for the multiple linear regression enjoys the same properties as in the simple linear case, provided the Gauss-Markov assumptions listed in Subheading 3.1 are met. Remember that in the multiple linear regression, assumption 6 about perfect multicollinearity needs to be checked to ensure that the OLS estimator exists.

Multicollinearity takes place when a covariate can be perfectly explained by a linear combination of some other covariates in the model. This indicates that there is redundancy of information in the model. Technically speaking, perfect multicollinearity makes the matrix $X'X$ from Eq. (10) of less than full rank and becomes not invertible. As a consequence, the OLS estimator does not exist in such cases. In reality, however, perfect multicollinearity is rarely seen, but high degree of multicollinearity can still lead to less precise estimates of the individual covariates. It is therefore necessary to identify such an issue and take corrective actions when needed. Subheading 3.3.1 will discuss one method to identify multicollinearity.

Regarding the diagnostic of the model, all the plots we have seen so far can be reproduced and used to check the validity of the model, except the scatter plot of the original data. Indeed, the number of dimensions we are dealing with can no longer be visually represented. This is also a reason why we look at the model residuals, which can always be plotted no matter how many explanatory variables are in the model. Additionally, one should also plot the residuals against each continuous covariates in order to check for uncaptured relationships.

- *Residuals vs. Covariates* $(x_{ij}, \widehat{\epsilon}_i)$
  - With this diagnostic plot, we want to make sure no information was left out unused. We want to observe randomly scattered points around the horizontal line at 0. If a

relationship appears in one of the plot, it may be an indication to rethink the way in which the covariate has been added to the model (e.g., add a quadratic relationship x_{ij}^2 in the model if a parabolic shape appears for one of the plot).

Furthermore, we already discussed R^2 as a common measure of goodness-of-fit and mentioned that in the case of multiple linear regression, an adjusted version of this statistic is to be preferred, namely the Adjusted-R^2. Indeed, R^2 has the disadvantage of increasing with the number of covariates added to the model (independently of whether the covariate is significant or not), while its adjusted version corrects for this caveat. This new statistic is defined as follows:

$$\text{Adjusted} - R^2 = 1 - \left(1 - R^2\right) \frac{n-1}{n-p-1} \tag{12}$$

This statistic is more reliable as it does not necessarily increase with new added covariate unless they are significant in the model. Rather, it penalizes the addition of new variables if not significant. It takes values between $[0; 1]$ and can also be used to compare models.

Lastly, interpretation in the multiple linear regression follows the same logic as in the simple linear regression with one subtle difference, the "ceteris paribus" meaning "all things equal" in latin. In other words, when we interpret the effect of a given covariate of the model, we do it assuming that the other covariates remain the same.

Another point worth pointing out is that covariates do not need to be continuous, but they can be what we call categorical variables or also dummy variables. Dummy variables are discussed in more detail in Subheading 3.2.1.

3.2.1 Dummy Variables

Dummy variables are variables which take on two values, usually 0 or 1. It takes on the value of 1 when the condition we are interested in is present, and 0 otherwise. The use of dummy variables allows us to introduce in our model explanatory variables that are categorical and therefore cannot be quantified. Dummy variables are extremely useful when we want to understand if the effect of a covariate on an outcome variable is the same for two categories of people. For instance, a common example to illustrate the use of dummy variable would be to ask if a covariate has the same effect on the response variable depending on the gender of the subject. To test such consideration, we include a dummy variable in our regression model which will take on the value of 1 if the subject is a male and 0 if the subject is a female (note that you could as well put 1 for female and 0 for male). Such a regression model would be defined as follows:

$$y_i = \beta_0 + \beta_1 x_{i1} + \delta_1 D_{i1} + \epsilon_i, i = 1, \ldots, n \quad (13)$$

In Eq. (13), we can notice three important elements:

1. δ_1 which is the coefficient associated with the dummy variable D_1 which is the gender.
2. There is only 1 dummy variable although we have two categories.
   (a) We create $k - 1$ dummy variables (k = number of categories)
3. Depending on the value of D_1, we obtain two different regression lines.
   (a) $y_i = \beta_0 + \beta_1 x_{i1} + \epsilon_i$ represents the regression line for females (i.e., $D_1 = 0$)
   (b) $D_1 = 1$: $y_i = (\beta_0 + \delta_1) + \beta_1 x_{i1} + \epsilon_i$ represents the regression line for males (i.e., $D_1 = 1$)

Therefore, we can easily see that depending on the sign of δ_1, the regression line for males will be either above the one for females ($\delta_1 > 0$) or below ($\delta_1 < 0$). Figure 5 provides an illustration of the effect of dummy variable in regression analysis.

An important consideration to make regarding dummy variables is the way we interpret the coefficients associated to them. In

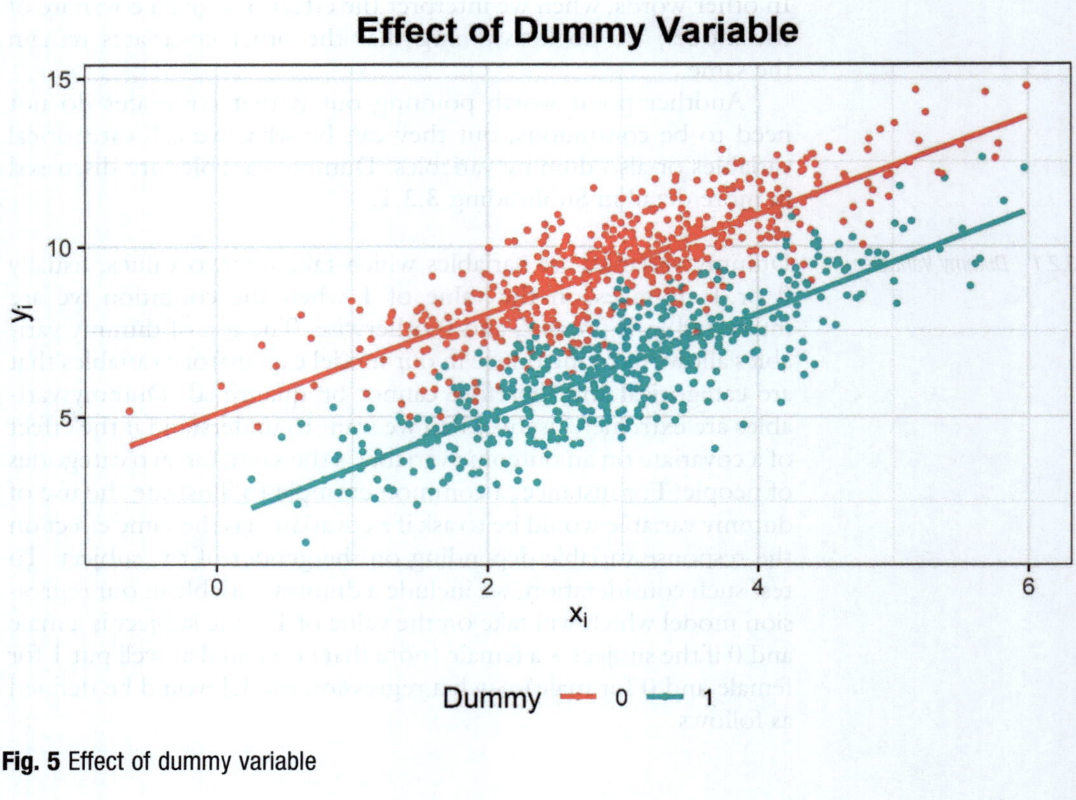

Fig. 5 Effect of dummy variable

this case, the interpretation is to be done relative to a reference category ($D_1 = 0$ in our case). Therefore for a positive value of δ_1, we would say that on average, the effect of x_1 is higher for male compared to female while the opposite holds true for negative values of δ_1. Note that we often refer to $D_1 = 0$ as the control group, while we refer to $D_1 = 1$ as the treatment group.

Essentially, dummy variables have the effect of moving the intercept of the regression line up or down depending on the sign of its coefficient, but they can also be used to test whether the regression slope is statistically different across groups, which we will cover in Subheading 3.2.2.

3.2.2 Interactions: Continuous with Categorical

Interactions are additional terms that we add to a regression model. They are the product of two or more variables. Those variables can be continuous or categorical. In this section, however, we only consider interaction between continuous and categorical variables.

Interaction terms are often used in regression analysis and can enhance one's understanding of our model. We can discover additional information on the existing relationship among the variables by using interactions.

In Subheading 3.2.1, we discussed how dummy variables are used to check if the effect of a covariate on an outcome variable is the same for two categories of people. With this in mind, we now want to understand if the effect that this covariate has on the response changes for different values of the covariate. As an example, we could ask if the effect of education on wage is more or less pronounced depending on whether one is a male or female.

A regression model containing a dummy variable and an interaction can be written as follows:

$$y_i = \beta_0 + \beta_1 x_{i1} + \delta_1 D_{i1} + \gamma_1 D_{i1} x_{i1} + \epsilon_i, i = 1, \ldots, n \qquad (14)$$

Equation (14) contains a new coefficient γ_1 for the first and only interaction term $D_1 x_1$. It becomes now interesting to think about the marginal effect of the covariate x_1 on the response y. To derive this marginal effect, we need to differentiate with respect to the variable of interest. In our case:

$$\frac{\partial y}{\partial x_1} = \beta_1 + \gamma_1 D_1 \qquad (15)$$

We can easily observe that depending on the group the subject belongs to, the marginal effect of x_1 is different. Indeed, if $D_1 = 1$, the effect of x_1 is $\beta_1 + \gamma_1$, but if $D_1 = 0$, the effect of x_1 is β_1 only. Note that to compute the correct marginal effect, x_1 has to be included individually to the model. Now, depending on the value of γ_1 the slope of the regression line will vary. For $D_1 = 1$, if $\gamma_1 < 0$, the regression line will be flatter whereas if $\gamma_1 > 0$, the regression

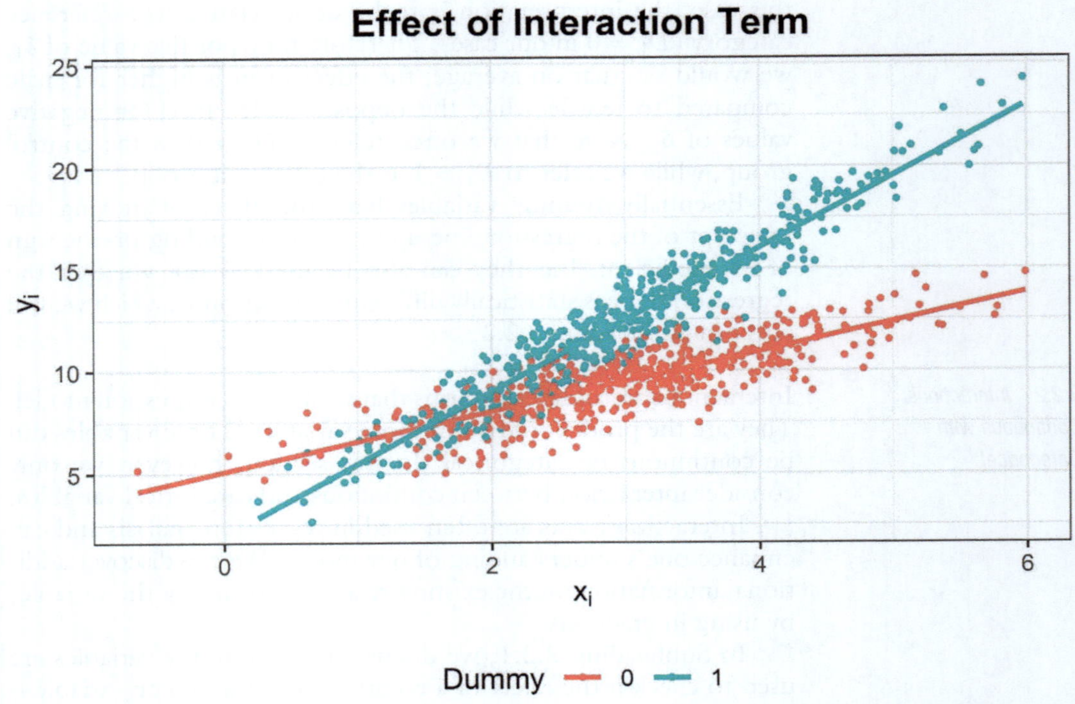

Fig. 6 Effect of interaction term

line will be steeper. In other words, for positive γ_1, we can say that the effect of the covariate is amplified for the treatment group compared to the treatment group. Figure 6 shows the effect of adding an interaction term to a regression model.

Interactions are therefore extremely useful in regression analysis and allows to test more hypotheses the researcher may be interested in.

3.2.3 Inference

So far, we have introduced simple and multiple linear regression models, discussed the necessary assumptions behind it and mentioned how to check its validity through model diagnostic plots. We also discussed the use of dummy variables and interactions in a linear regression model to obtain a more advanced understanding of the relationships we are interested in. However, we have not discussed whether the covariates included in our model are statistically significant or not. Indeed, a covariate which is not significant should, in principle, not be part of the model (unless theory dictates it) and the interpretation of insignificant explanatory variables is not relevant. To check statistical significance, we use statistical inference. Inference is at the heart of statistical modeling. It aims at generalizing results we obtained on a sample to the underlying population. To this purpose, hypothesis testing is extensively used in statistics and econometrics.

For our purpose, we will cover a few classical hypothesis tests that we commonly use in linear regression, however, there exists plenty of them. The necessary ingredients to test an hypothesis are the null hypothesis H_0, the alternative hypothesis H_A (usually what we want to prove), the $\alpha -$ level (the significance level of our test which is usually set at 5%), a test statistic $T$and its distribution (or at least approximate distribution).

Test of Significance

The test of significance on the regression coefficients allows us to statistically test whether a given covariate has a significant effect on the response variable.

The procedure to perform this statistical test can be defined as follows:

1. We start by stating our null and alternative hypotheses
   (a) $H_0 : \beta_j = 0,\ H_A : \beta_j \neq 0$
2. We set the significance level of our test $\alpha -$ level $= 0.05$
3. We define our test statistic and its distribution under H_0: $T = \frac{\widehat{\beta_j}}{se\left(\widehat{\beta_j}\right)} \sim T_{n-2}$ (T-distribution with $n - 2$ degrees of freedom)
4. We compute the sample value of our test, which we denote $t.$ obs

Now that we have all the ingredients to perform our test, we want to check if the observed value of our test statistic is unexpected. To do so, we define a rejection region using the significance level previously defined and the quantiles of the T_{n-2} distribution. Since our significance level is defined as 5% and our alternative hypothesis is two-sided, our rejection region will be the 2.5% and 97.5 % −quantiles of the T_{n-2} distribution. If the observed value of our test statistic $t.\ obs$ falls below the 2.5 % −quantile or beyond the 97.5 % −quantile, we can reject the null hypothesis with 95% confidence and say that $\widehat{\beta_j}$ is different from 0 and that the covariate associated with this coefficient is statistically significant in explaining the response y. Alternatively, if the p-value reported by your statistical software is below your significance level, you can draw the same conclusion ($p -$ value $< \alpha$).

This test of significance can be performed for each covariate in the model. Statistical software commonly report those values when performing linear regression models and therefore the work of the researcher is only to interpret the result of these tests.

Zero Slopes Test

The zero slopes test is also known as the overall significance test, and it allows us to statistically test whether a model containing a set of explanatory variables is better than a model with an intercept only.

The procedure to perform this statistical test can be defined as follows:

1. We start by stating our null and alternative hypotheses
   (a) $H_0 : \beta_1 = \beta_2 = \ldots = \beta_k = 0$, H_A : at least one $\beta_j \neq 0$
2. We set the significance level of our test α − level = 0.05.
3. We define our test statistic and its distribution under H_0: $F = \frac{\text{MSR}}{\text{MSE}} \sim F_{p,n-p-1}$ (F-distribution with p and $n - p - 1$ degrees of freedom)
   (a) *MSR* stands for regression mean square: $\text{MSR} = \frac{\text{Sum of Squares}}{\text{degrees of freedom}} = \frac{\sum_{i=1}^{n} \left(\widehat{y}_i - \overline{y}_i\right)^2}{p}$
   (b) *MSE* stands for mean square error: $\text{MSE} = \frac{\text{Residual Sum of Squares}}{\text{degrees of freedom}} = \frac{\sum_{i=1}^{n} \left(y_i - \widehat{y}_i\right)^2}{n-p-1}$
4. We compute the sample value of our test, which we denote *f*. obs

Following the same logic as for the test of significance on the individual coefficient, if the observed value of our test statistic *f*. obs falls below the 2.5 % −quantile or beyond the 97.5 % −quantile (or the p-value is <0.05), we can reject the null hypothesis with 95% confidence and say that the model containing a given set of covariate performs better than a intercept-only model. An alternative formulation of this *F* statistic is as follows:

$$F = \frac{R^2/(p-1)}{(1-R^2)/(n-p)} \sim F_{p,n-p}$$

Statistical software also commonly report those values when performing linear regression models.

The test statistics covered in this section are basic ones and more sophisticated ones exist depending on the hypotheses of the researcher. Nevertheless, understanding those tests allow you to adapt them to your needs.

3.2.4 Interpretation

We have already discussed how to interpret estimates from linear regression model, including coefficients of dummy variables and interaction terms, however, in econometrics, there are three levels commonly used to model data in regression analysis:

- Linear-linear model:

  This model specification is the one we have been working with so far where none of the variables has been transformed. *See* Eq. (2) for the model formulation.

  Usually in this case, a unit increase in the covariate x_j, leads to a β_j increase in the response variable y

- Log-linear model

  The log-linear model is a model specification where we take the logarithm of the response variable:

$$\log(y_i) = \beta_0 + \beta_1 x_{i1} + \epsilon_i \tag{16}$$

In this case, a unit increase in the covariate x_1 leads to a β_j percent increase in the response variable y

- Log-log model

  The log-log model specification uses the logarithmic transformation on both the response and the covariate:

$$\log(y_i) = \beta_0 + \beta_1 \log(x_{i1}) + \epsilon_i \tag{17}$$

In this case, a 1% increase in the covariate x_1 leads to a β_j percent increase in the response variable y.

Those specifications become quite handy when it makes sense to think in percentage terms instead of unit (linear) terms, for instance. However, make sure that taking the logarithm of your dependent and/or independent variables makes sense from an interpretation point of view as well as from a statistical point of view. Also, taking the logarithm of an observation whose value is close to 0 can lead to large negative values, which can influence the regression estimates.

3.3 Caveats

Throughout this chapter, we have mentioned a couple of caveats to look out for when using linear regression models. Among them, we discussed multicollinearity and omitted variable bias. Another consideration that is important in econometrics is the one of simultaneity. All those conditions imply the violation of the Gauss-Markov assumptions and lead to unreliable estimates. In the following sections, we will discuss those cases in more detail.

3.3.1 Multicollinearity

As introduced in Subheading 3.2, collinearity can lead to less precise estimates. Furthermore, perfect multicollinearity violates assumption 6. Although rare, perfect multicollinearity can happen. One common case which is known as the dummy trap is when k dummy variables are created instead of $k - 1$.

Spotting multicollinearity is therefore important to increase the reliability of our estimates and to potentially reduce the number of covariates we include in our model. Indeed, when variables exhibit high multicollinearity, they are considered redundant as they tend to explain (at least part of) the same effect on the outcome variable. Removing variables that do not add valuable information to the model therefore leads to a more parsimonious model, which is usually easier to interpret.

One way to check for multicollinearity is the Variance Inflation Factor (VIF), which is defined as follows:

$$\mathrm{VIF}_j = \frac{1}{1 - R_j^2} \tag{18}$$

To compute the VIF, we first regress the x_j covariate on the remaining covariates included in our model. We then compute the R^2 of that model and apply the formula from Eq. (18). This needs to be repeated for all the covariates x_j ($j = 1, \ldots, k$).

To identify if we are in the presence of multicollinearity issue, we use cut-off values above which we consider that there is multicollinearity. Common cut-off values are 5 or 10 although there isn't a consensus on which cut-off should be used.

If multicollinearity is observed, one should investigate the reason behind it and not automatically remove the problematic variable. Ultimately, the decision is taken based on domain knowledge and researcher's judgement.

3.3.2 Omitted Variable Bias

Another common trap in regression analysis is to omit a relevant explanatory variable. When building a regression model, one should always consider all the potential variables that could explain the response variable to minimize the chances of leaving one out. Forgetting to include a relevant covariate leads to the omitted variable bias, which in turn leads to the violation of assumption 5.

To better understand why this is the case, let us consider the following linear regression model, which represents the true model:

$$y_i = \beta_0 + \beta_1 x_{i1} + \beta_2 x_{i2} + \epsilon_i \tag{19}$$

Now, imagine that we omitted to include a relevant covariate, namely x_2. Therefore, the estimated model is:

$$y_i = \beta_0 + \beta_1 x_{i1} + \tilde{\epsilon}_i \tag{20}$$

Knowing this fact, the model from Eq. (20) can be rewritten as follows:

$$y_i = \beta_0 + \beta_1 x_{i1} + (\beta_2 x_{i2} + \epsilon_i) \tag{21}$$

Therefore, if x_1 and x_2 are somewhat correlated, then x_1 will be correlated with $\tilde{\epsilon}_i$ which is a violation of the exogeneity assumption. In econometrics, we refer to this issue as the endogeneity issue, which leads to biased and inconsistent OLS estimates. Note that the endogeneity issue is also present when the covariates are measured with errors (measurement errors) and in the case of simultaneity, which is discussed in the next section.

3.3.3 Simultaneity

Another condition under which endogeneity issue is present is when simultaneity takes place. Simultaneity refers to the case where the covariates affect the response and the response affects the covariate as well. In such a case, the covariate will again be

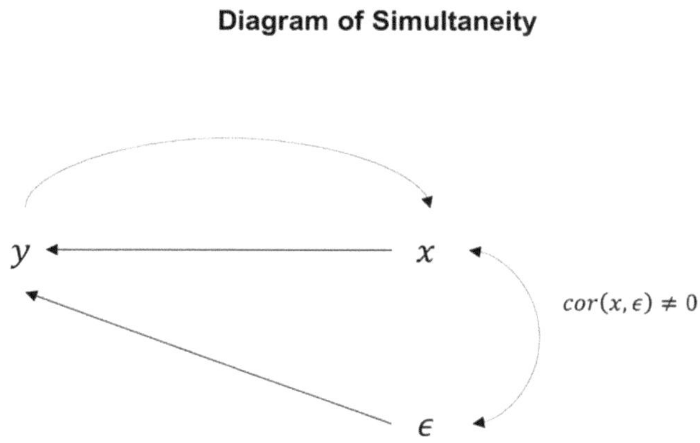

Fig. 7 Diagram of simultaneity

correlated with the error terms which violates the exogeneity assumption as in the omitted variable bias case. The diagram presented in Fig. 7 shows the relationship among the variables of a model.

In economics, simultaneity is a common issue which needs to be addressed to safeguard the Gauss-Markov assumptions. An extensively used method to deal with endogeneity issue in econometrics is the method of Instrumental Variable (IV) which consists in finding a set of exogenous variable that affect the endogenous covariates but that are themselves not affected by the response variable. We will, however, not discuss this method in this chapter.

3.4 Variables that Are Not Quantifiable

Everything we discussed so far relies on data that we either observe or collect, but what happens when we are dealing with a variable that is not quantifiable? There are many variables which may be of interest but are hardly quantifiable. Sometimes, indices are built to quantify them, such as the happiness index proposed by the United Nations Sustainable Development Solutions Network. Another way to measure the effect of not quantifiable variable is to use dummy variables which we already discussed. However, other aspects may be harder to grasp, yet it doesn't mean that they don't have an effect. In such cases, we would fall in the omitted variable bias again.

Let us consider a case where we would like to measure the attitude that individuals have toward a specific food item. Consider also that we collect measurements for several individuals over two time periods (panel data). We therefore use the following regression model:

$$y_{it} = \beta_0 + \beta_1 x_{it1} + \tilde{\epsilon}_{it} \tag{22}$$

In Eq. (22), i refers to the ith observation, while $t = \{1, 2\}$ refers to the time period at which the observation was collected. However, we could argue that a relevant factor influencing the attitude one has toward the food item is culture, which is not observable or quantifiable and thus leads to omitted variable bias. The true model including culture can therefore be written as follows where z_i is the culture variable which does not vary over time:

$$y_{it} = \beta_0 + \beta_1 x_{it,1} + \beta_2 z_i + \epsilon_{it} \tag{23}$$

Variables such as culture rarely change over a short period of time but rather require many decades to evolve. Using this sensible assumption, we could subtract the observations collected at time t by those collected at time $t-1$ and get rid of the unobserved and omitted variable culture:

$$\begin{aligned} \mathrm{y_{it}} - y_{it-1} &= (\beta_0 - \beta_0) + (\beta_1 x_{it,1} - \beta_1 x_{it-1,1}) + (\beta_2 Z_i - \beta_2 Z_i) \\ &+ (\epsilon_{it} - \epsilon_{it-1}) = \beta_1 (x_{it,1} - x_{it-1,1}) + (\epsilon_{it} - \epsilon_{it-1}) \end{aligned} \tag{24}$$

By taking this difference, we eventually get rid of the variable culture and obtain a new regression line, which can also be estimated using OLS. Note that the intercept was also removed when taking the difference. An intercept could be added if we don't want to allow the mean change in response to be 0. We use this method in the blind tasting application below.

Additionally, in the case where data is collected over more than two time periods, fixed-effects models can also be used. We will, however, not discuss them here.

To conclude, variables which are not observable and therefore not quantifiable can also cause omitted variable bias. When those variables can be assumed to be time invariant, the technique presented, which consists in taking the difference, can be used to get rid of them and prevent omitted variable bias.

4 Applications: Practical Examples

In this section, we present three examples of the application of econometrics technique to food science.

4.1 Hedonic Regression (Log-Linear Level): What Determines the Price of a Wine?

In a hedonic regression model, (the logarithm of) a price is usually the dependent variable, and the attributes that are believed to provide utility to the buyer or consumer are the independent variables. Hedonic regression models are commonly used in real estate pricing and quality adjustment for price indexes. However, this method is also extensively used in food science. Satimanon and Weatherspoon [6] study the determination of price premiums for sustainable attributes of fresh eggs by using hedonic analysis. They find that welfare-managed and organic-labeled eggs receive a price premium, but the combination of the two labels does not increase the price premium further, suggesting that egg producer should focus on only one. Smith et al. [8], using a dataset consisting of over 400 beers and including information on calorie content, alcohol content, user rating, and style for each beer, find that a 10-point increase in consumer rating is associated with about a $0.50 increase in the price of beer per unit. A final example is Yang and Dharmasena [11]. This article studies how intrinsic characteristics and differences of such characteristics of dairy alternative beverages and conventional milk products affect the consumers' willingness to pay. Their results show that consumers exert the highest weights and assign the highest evaluation on such qualitative characteristic as nutritional attributes which include calories, protein, fat, vitamin A, and vitamin D in which protein is the most valued attribute and other characteristics such as package size, multi-pack, and brand. This provides essential information for manufacturers to better differentiate their products and develop products catering to consumers' preferred attributes.

We present here, as an example, the results from the paper of Schamel [7]. This paper presents a hedonic pricing model for German-quality wine. The data consist of quality indicators for 4141 wines with the sensory awards received at the annual German wine competition and the legally required quality category, as well as a set of control variables including regional origin, color, style, and their age at the time of judging.

The paper proposes the usual hedonic price equation, where the logarithm of the price of the wine is a linear function of some attributes of the wine. The log-linear function allows to estimate how the presence of a particular characteristic affects the change in the willingness to pay for the wine in percentage terms. Therefore, the proposed regression model is the following:

$$\log(\mathrm{P}_i) = \alpha + \beta_1 \mathrm{D}_{i,\mathrm{Award}} + \beta_2 \mathrm{D}_{i,\mathrm{Quality\ Level}} + \beta_3 \mathrm{D}_{i,\mathrm{Style}} + \beta_4 \mathrm{D}_{i,\mathrm{Color}} + \beta_5 \mathrm{D}_{i,\mathrm{Region}} + +\gamma \mathrm{Age}_i + \delta \mathrm{Bar}_i + \varepsilon_i \quad (25)$$

The dependent variable is the logarithm of the retail price. The independent (explanatory) variables include categorical dummies for award, quality level, style, color, and region. As discussed above, if we have k categories, we need to include $k - 1$ dummies. The category not included is the comparison reference. The author chose to have as a reference for the five dummies the following excluded categories: Gold, dry, white, Spätlese from the Pfalz region. This means that the estimated coefficients β_1 to β_5 will be the price premiums and discounts relative to the contribution of the base category to the price. Finally, the coefficients γ and δ measure the price premiums or discounts paid for older wines and barrique-aged wine respectively.

Equation (25) is estimated using the ordinary least squares (OLS) described above, assuming that the error term ε_i satisfies the Gauss-Markov assumptions mentioned above. The results are presented in Table 1.

Table 1
Estimation results of hedonic price model [dependent variable: log(Price)]

Parameter	Estimate	t-Statistic	P-value	Price Effect (€)[a]
Constant	1.579	68.07	0.00	–
Bronze	−0.078	−5.60	0.00	−0.67
Silver	−0.035	−3.13	0.00	−0.31
Gold Extra	0.106	3.01	0.00	1.01
Qualitätswein	−0.348	−23.89	0.00	−2.63
Kabinett	−0.286	−20.12	0.00	−2.23
Auslese	0.417	27.28	0.00	4.64
Beerenauslese	1.087	43.41	0.00	17.61
TBA	1.512	36.04	0.00	31.68
Eiswein	1.590	69.51	0.00	34.99
lieblich (mild)	−0.123	−9.81	0.00	−1.04
halbtrocken (off-dry)	−0.116	−7.81	0.00	−0.98
Barrique	0.580	24.26	0.00	7.05
Rosé	0.060	1.89	0.06	0.55
Red Wine	0.180	12.15	0.00	1.77
Ahr	0.542	11.36	0.00	6.45
Baden	0.284	16.56	0.00	2.94
Franken	0.471	25.61	0.00	5.39
Hess. Bergstraße	0.350	9.53	0.00	3.75
Mittelrhein	0.123	3.56	0.00	1.17
Mosel-Saar-Ruwer	0.334	17.55	0.00	3.56
Nahe	0.223	7.96	0.00	2.24
Rheingau	0.384	14.63	0.00	4.20
Rheinhessen	0.002	0.15	0.88	0.02
Saale-Unstrut	0.509	8.98	0.00	5.95
Sachsen	0.734	15.19	0.00	9.71
Württemberg	0.185	11.01	0.00	1.82
Age	0.064	3.95	0.00	0.57

Source: Schamel [7]
F-test = 512.66; Rsq = 0.7709; Adj Rsq = 0.7694
[a]Relative to base category (Gold, Spätlese, dry, white, Pfalz) at average price (8.96 €)

The second column in the table represents the estimated coefficients for Eq. (25). A positive coefficient represents a premium with respect to the base category and a negative coefficient represents a discount with respect to the comparison wine. For example, wines with a bronze medal suffer a discount of 7.8% with respect to a wine with a gold medal. On the other hand, a wine with gold extra has a premium of 10.6% with respect to a wine that got a gold medal. The interpretation is similar for the other coefficients. The third column presents the t-statistic for the test of significance previously discussed. The large t-statistics show that most estimated coefficients are significantly from zero both at a 1% and 5% level (fourth column). There are two exceptions, the coefficient associated to Rosé is significantly different from zero only at the 10% confidence level, while the coefficient for the Rheinhessen region is statically equal to zero and therefore cannot be used to predict the price. The last column in the table presents the equivalent in euros of the percentage change in prices coming from the estimated coefficient for each category. The table also presents the F-test of the zero slope test which rejects the null hypothesis that all the estimated coefficients are zero. The R-square and the adjusted R-square show that 77% of the variability in the logarithm of wine prices can be explained by the model.

Overall, the analysis of Schamel [7] confirms that sensory quality awards have a significant and positive price impact. His work also shows that there exist significant wine price differences between quality categories, growing regions, and most of the control variables proposed.

4.2 Blind Tasting: Country of Origin Bias

Imagine we want to study the effects of the country-of-origin on wine experts' sensory perception about wines from established (i.e., France) and relatively new wine-producing countries (i.e., UK and Mexico). Would wine experts have a bias against new wine-producing countries or would they base their assessment in the intrinsic characteristics of the wines? To do this type of analysis, we require two cross-sections. In the first cross-section, we ask a number of experts to do an assessment (including likeability) of different wines from new wine countries and France without providing any information about the wines (blind condition). In the second cross-section, we ask the same experts to do the same assessment, but we provide them with information on the country of origin of the wines (non-blind condition). As we have two repeated cross-sections, we have the simplest panel data (2 periods and N experts). Therefore, we can use two different approaches: a regression in level and a panel data regression. As we will see, both provide similar results.

For the regression approach in levels we run the following empirical model:

$$\text{Likeability}_{ijk2} = \alpha + \beta \times \text{Likeability}_{ijk1} + \gamma' \text{CountryInformation}_k + \delta' X_i + \theta' Z_j + \mu_{ij} \quad (26)$$

Our dependent variable is the likeability score the wine expert i gives to wine j coming from country k in the second phase of the experiment when the information on the country of origin of the wines is revealed. Our first explanatory variable is the likeability score given to the same wines in the first phase of the experiment, which was conducted blind. This variable captures the intrinsic likeability of the wines for each wine expert. The second set of variables, "Country Information," is the main explanatory variable in our analysis and captures the information on the countries of origin of the wines provided in the second phase of the experiment. This is a vector of two dummy variables capturing the "UK" and "Mexico" effects. As discussed above, we use two dummies because we have three attributes (we do not include France as this is the comparison country). A positive (negative) γ would imply a favorable (unfavorable) country bias effect. In our analysis, we will also control for a vector X_i of wine trader characteristics, including years of experience in the wine industry, gender, and whether the person comes from the same country of one of the wines. Finally, we control for wine characteristics (Z_j) that could have not been fully captured by the likeability score in the blind tasting. Here we would typically include information like vintage year, method of production, terroir, medals and experts' scores, price, bottle and label design, producers' name, etc. However, it would be better if the wines are selected considering all these criteria so the wines are similar.

The results of the OLS estimation for the variables of interest are presented in Table 2.

The first regression shows the results for a specification of the model without any control variables, and the second regression presents the results when we consider all our control variables. We present the estimated betas and we specify their level of statistical

Table 2
The country of origin likeability bias

	United Kingdom	Mexico	Blind liking	Own country	Years of experience	Gender	Price
Model 1 in levels (without controls)	−0.73***	−0.87***					
Model 2 in levels with controls	−0.91***	−0.96***	0.42***	−0.17	−0.01	0.11	0.04
Model 3 in differences (panel fixed effects)	−0.71**	−0.73**		0.07			

significance. We use the classical notation to indicate the level of significance. Estimated betas that are significant at 1% are presented with three stars (***), those at 5% with two stars (**) and those statistically different from zero only at the 10% level with one star (*). Finally, those for which the t-tests do not reject the null hypothesis of lack of significance do not receive any star. Then we need to look at the sign and size of the estimated coefficients. Our experiment shows a clear negative bias against the UK and Mexican wines among wine experts. In the second specification, the likeability score of the wines for both countries decreases by almost one point when the information on the country of origin of the wines is revealed. This result is statistically significant at the 1% for both countries. Among the control variables, the only one that seems relevant is the likeability score of the blind test experiment. The estimated coefficient is positive and significantly different from zero at the 1% level. This implies that, despite the bias created by the provision of the information about the country of origin of the wines, the wine experts could still partially recognize the intrinsic likeability of each wine.

A potential problem with a cross-section regression approach is the difficulty to account for all relevant control variables. This could create omitted variable biases. However, the panel structure of our data allows us to control by individual and wine fixed effects (things that do not change over time). Our model 3 specification presents the results of the estimation in differences instead of in levels like in Eq. (26) above. In this case, the dependent variable is the change in the likeability score between the non-blind and the blind experiments. In the second phase of the experiment, the wine experts learn the country of origin of the wines and if the wines come from their own country, so we use those as the explanatory variables to explain the potential change in scoring. Once again, the results show a significant negative bias for wines coming from the UK and Mexico in comparison with French wines.

4.3 Randomized Controlled Trial

Randomized controlled trials have been used over many decades by several scientific disciplines as the most accurate methodology of causal effect estimation in different contexts. This method is still at the vanguard of rigorosity and accuracy to the point that COVID -19 prototype vaccines were tested by this method. Public policy programs, clinical research, nutrition studies, and in many other scientific field do impact analysis that are evaluated under the experimental design that characterizes RCTs. This method allows the researcher to isolate the impact of the intervention from other variables that could impact the outcomes of interest and estimate it with the highest possible level of precision and accuracy.

The essential purpose of the RCT is to determine whether a particular intervention or treatment can reasonably be inferred to cause a change in an outcome variable of interest. To measure the intervention's effect, subjects are randomly assigned to an intervention (treated) or a control (non-treated) group. The randomization process ensures that the differences between the groups if observed at the end of the trial, are only due to the treatment administration.

The estimation of the effects of the intervention is obtained from a simple difference-in-difference analysis. The baseline specification for this type of analysis is the following:

$$\Delta y_{it} = \alpha + \beta T_{it} + \gamma' Z_{it} + \epsilon_{it} \quad (27)$$

where Δy_{it} is the change in the outcome variable of interest for subject i at time t (e.g., the change in the weight of an individual between our baseline period and one year after the exposition to a nutrition course). α is a constant term and β is the coefficient of the variable of interest, T_{it} which is a dummy variable indicating whether a subject i is in the treatment group in a given period. Z_{it} is a vector of control variables and ϵ_{it} is an error term. Therefore, the estimated value of β captures the average treatment effect (ATE).

Many studies in nutrition (*see*, e.g., [1, 3, 4]) and food marketing (*see*, e.g., [2, 5, 10]) use the RCT methodology.

We illustrate the use of RCT with the study by Dubois et al. [2]. The objective of their research is to study if the use of front-of-pack (FOP) nutrition labels improves food purchases in real-life grocery shopping settings. To do this, they use four pre-selected front-of-pack nutrition labels that they put 1.9 million labels on 1266 food products in four categories in 60 supermarkets and analyzed the nutritional quality of 1,668,301 purchases using the FSA nutrient profiling score.

The analysis was conducted using the RCT methodology. In a universe of 174 shops, 60 were randomly selected for the experiment. The research team placed FOP labels on food products in 40 randomly selected supermarkets in France (10 supermarkets per labeling system). This was their treatment group. In addition, 20 supermarkets were randomly chosen as the control group (no additional FOP labeling). These 60 supermarkets belonged to three of the largest supermarket chains in France.

The research identified two outcome variables of interest. The first one captured nutritional quality of food purchased at the purchase incidence level. The second variable captured the purchase quantity level. The empirical strategy chosen was to estimate the average treatment effect (ATE) of nutrition labels using a difference-in-differences (DiD) approach as the one presented above in Eq. (27).

In the purchase incidence regressions, the study included as covariates the difference in the number of store visits and prices between 2016 and 2015 and fixed effects for each tercile and unlabeled product, and food category. In the nutritional basket analysis, the equation included category-specific intercepts and the difference between years of the number of purchases in the category *c* for individual *i* and the difference in the average prices of individual baskets. The research team included in both regression consumers with at least one purchased labeled product and a store visit in any week during the evaluation period. For this study, the coefficients of interest were those of the interactions between the variables coding the intervention and the nutritional quality tier or unlabeled product.

The study found that on average the effect of labeling in real life was 17 times smaller than those found during laboratory studies. The most effective nutrition label of the four tested increased the purchases of foods in the top third of their nutrition-wise category by 14% but failed to have an impact on the purchases of products with medium and low nutrition quality or without any labeling reporting nutrition quality.

The quantitative analyses were supplemented with a survey of shoppers both in the treatment and control stores. The objective of this survey was to provide information on the mechanisms (often a weakness of quantitative analysis) underlying the effectiveness of the nutrition labels. The most effective label was the one that was able to attract more attention and that helped buyers to rank products by nutritional quality.

References

1. Bisimwa G, Owino VO, Bahwere P, Dramaix M, Donnen P, Dibari F, Collins S (2012) Randomized controlled trial of the effectiveness of a soybean-maize-sorghum–based ready-to-use complementary food paste on infant growth in south Kivu, Democratic Republic of Congo. Am J Clin Nutr 95(5): 1157–1164. https://doi.org/10.3945/ajcn.111.028704
2. Dubois P, Albuquerque P, Allais O, Bonnet C, Bertail P, Combris P, Lahlou S, Rigal N, Ruffieux B, Chandon P (2021) Effects of front-of-pack labels on the nutritional quality of supermarket food purchases: evidence from a large-scale randomized controlled trial. J Acad Mark Sci 49:119–138. https://doi.org/10.1007/s11747-020-00723-5
3. Helle C, Hillesund ER, Omholt ML, Øverby NC (2017) Early food for future health: a randomized controlled trial evaluating the effect of an eHealth intervention aiming to promote healthy food habits from early childhood. BMC Public Health 17:729. https://doi.org/10.1186/s12889-017-4731-8
4. Remington A, Añez E, Croker H, Wardle J, Cooke L (2012) Increasing food acceptance in the home setting: a randomized controlled trial of parent-administered taste exposure with incentives. Am J Clin Nutr 95(1):72–77. https://doi.org/10.3945/ajcn.111.024596
5. Roberto CA, Bragg MA, Schwartz MB, Seamans MJ, Musicus A, Novak N, Brownell KD (2012) Facts up front versus traffic light food labels: a randomized controlled trial. Am J Prev Med 43(2):134–141. https://doi.org/10.1016/j.amepre.2012.04.022
6. Satimanon T, Weatherspoon D (2010) Hedonic analysis of sustainable food products. Int Food Agribus Manag Rev 13(4). https://doi.org/10.22004/ag.econ.96336
7. Schamel G (2003) A hedonic pricing model for German wine. Ger J Agric Econ 52(5): 247–253. https://doi.org/10.22004/ag.econ.97975
8. Smith RA, McKinney CN, Caudill SB, Mixon FG Jr (2016) Consumer ratings and the pricing of experience goods: hedonic regression analysis of beer prices. Agric Food Econ 4:24. https://doi.org/10.1186/s40100-016-0068-3
9. Tintner G (1953) The definition of econometrics. Econometrica 21(1) [Wiley, Econometric

Society]:1–40. https://doi.org/10.2307/1906941

10. Velema E, Vyth EL, Hoekstra T, Steenhuis I (2018) Nudging and social marketing techniques encourage employees to make healthier food choices: a randomized controlled trial in 30 worksite cafeterias in The Netherlands. Am J Clin Nutr 107(2):236–246. https://doi.org/10.1093/ajcn/nqx045

11. Yang T, Dharmasena S (2020) Consumers preferences on nutritional attributes of dairy-alternative beverages: hedonic pricing models. Food Sci Nutr 8:5362–5378. https://doi.org/10.1002/fsn3.1757

Chapter 18

Product Portfolio Management: TURF

Anne Hasted

Abstract

TURF stands for Total Unduplicated Reach and Frequency and is a technique developed by the advertising industry in the middle of the last century and first applied to food product lines more than 20 years ago (Miaoulis, Turf: a new planning approach for product line extensions. Marketing Research, 1990; Cohen, TURF analysis. Quirk's Marketing Research Review, 1993; Moskowitz, Consumer testing and evaluation of personal care products. CRC Press, 1996). It aims to identify the best combination of alternative products to reach a maximum number of consumers a maximum number of times.

Food manufacturers are faced with the question: How many products/flavors to offer before capping the assortment?—they have to find the "cut-point" the point at which offering more variants is no longer profitable. This is important for range optimization; when faced with too much choice, consumers default to brands or products that they know and trust or worse still don't make a purchase at all (Iyengar, The art of choosing, 2011).

TURF analysis addresses the problem using simple combinatorial techniques.

Key words TURF, Reach, Frequency, Product portfolio

1 Introduction

TURF stands for Total Unduplicated Reach and Frequency and is a technique developed by the advertising industry in the middle of the last century and first applied to food product lines more than twenty years ago [1–3]. It aims to identify the best combination of alternative products to reach a maximum number of consumers a maximum number of times. This question cannot be addressed by a simple frequency analysis: taking a simple example of three product variants A, B, and C; A appeals to 50% of consumers, B to 40%, and C to 30%. If you could stock only two brands, you would probably choose A and B. If liking of A and B is discrete so people who like A don't like B and vice versa, then this is a good strategy as you would stock something that appeals to 90% of consumers. However, if

Carlos Gómez-Corona and Heber Rodrigues (eds.), *Consumer Research Methods in Food Science*, Methods and Protocols in Food Science, https://doi.org/10.1007/978-1-0716-3000-6_18,

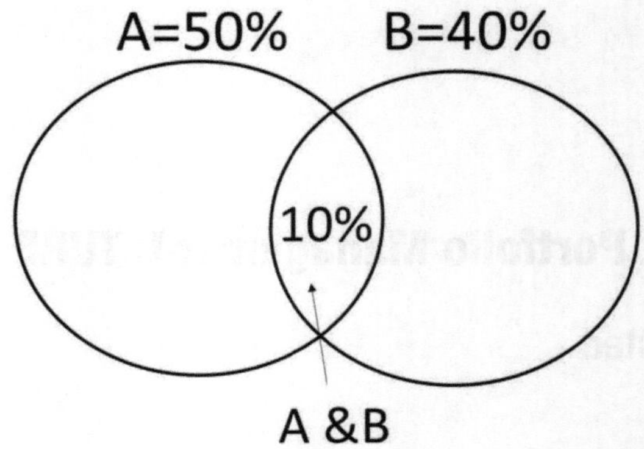

Fig. 1 Venn diagram showing hypothetical purchase frequencies for two products A and B

10% like both, then you are only picking up an extra 30% by stocking B as well as A. For two or three products, this can be illustrated by a Venn diagram (Fig. 1).

The restaurant menu should not be offering just the eight most preferred dishes since some dishes might be equally acceptable to some diners, the manager should be looking for a range that maximizes the chance that a potential diner will see a dish that he is tempted to eat and will visit the restaurant.

Food manufacturers are faced with the following question: How many products/flavors to offer before capping the assortment?—they have to find the "cut-point" the point at which offering more variants is no longer profitable. This is important for range optimization; when faced with too much choice, consumers default to brands or products that they know and trust or worse still don't make a purchase at all [4]. Figure 2 shows a Turf Ladder showing incremental reach (increase in purchasers) as further products are added into a potential range of ten products. Five products show an estimated reach of 63% and there is limited gain in adding more products into the range.

Alternatively, a manufacturer may wish to extend his product line by offering new variants without cannibalizing sales of his existing three flavors.

TURF analysis addresses these issues by collecting data from respondents on their attraction to each of the range of products under consideration and then using a simple combinatorial approach to find the "best" combination of products to meet the research objectives. It has applications in both the development of new product lines and in refreshing existing product ranges and is most appropriate when only one aspect of the product is to be changed.

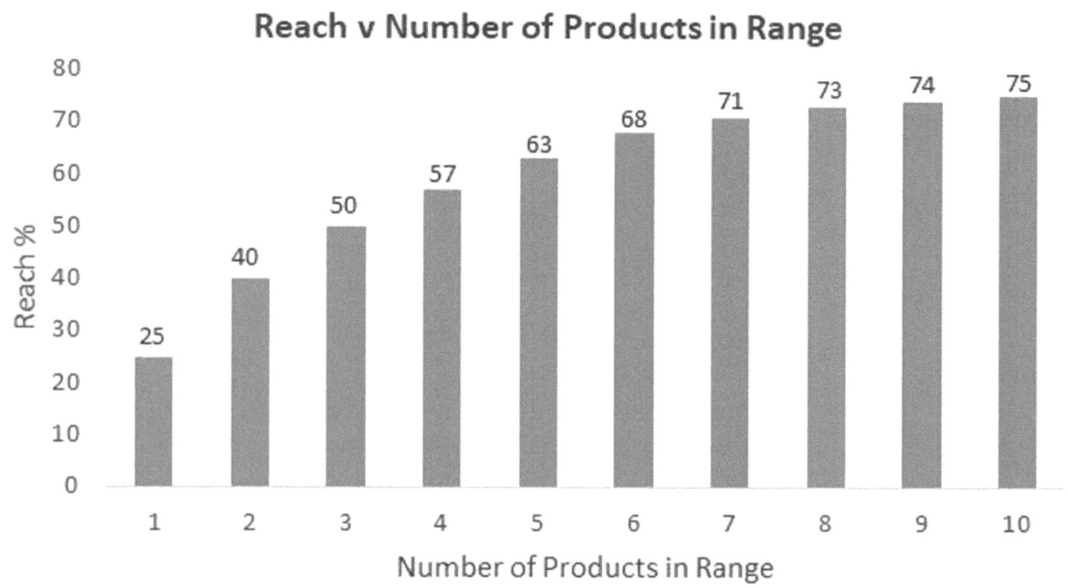

Fig. 2 TURF ladder

2 Method

2.1 TURF Metrics

A line is a subset of products or concepts from the range available.

Reach = Number of respondents for whom at least one of the products/concepts in the line is acceptable

Frequency = Total number of acceptable products in the line counted across all respondents

Reach is more important than frequency for products that are only purchased occasionally, high frequency measures are indicative of variety seeking, consumers wishing to purchase more than one product variant.

2.2 Data for TURF Analysis

A set of product variants covering a range of flavors/fragrances/ product variants are required, typically more than ten but not so many as to make the task fatiguing for the respondent, the author would not recommend more than 30 variants. Usually, these are presented as concepts rather than testing/tasting of actual products. Typically, 200–500 respondents need to be recruited from those who are either existing users or positive to the product concept. The sample size will depend on how many subgroups need to be considered (millennials/females, etc.). Respondents are asked a simple scaled question, which is then used as a simple binary measure of acceptability using top box or top two/three box scores. For example:

Table 1
Percent of top box scores for five products

% Top box scores/product	A (%)	B (%)	C (%)	D (%)	E (%)
Top box (9)	39	24	48	33	15
Top two box (8–9)	55	43	71	55	35
Top three boxes (7–9)	81	65	87	79	52

- How likely would you be to buy? (5-point scale)—score 4 or 5 interpreted as "buy."
- How much do you think you would like the product? (9-point scale)—top 3 boxes (7, 8, 9) interpreted as positive response.
- Would you feed to your child? Yes/No—simplest direct binary scale.

For scaled data, the choice of transformation from the data to a binary measure by top box, top two boxes, top three boxes, etc., will affect the results of the TURF analysis. Aggregating too many scaled scores will tend to flatten the differences between the tested concepts/products and give single favored concepts/products a very dominant reach, or alternatively, using just the top box may give very low reach values. As is so common in statistics, the optimal transformation depends on both the data, the business context of the test and the number of respondents in the study. A simple preliminary frequency analysis should help with the decision. Table 1 shows the percentage of scores in the top boxes for an overall liking scale of 1–9. If top two or top three boxes are used as an acceptability measure, then product C alone has a reach of over 70%; in this case, as the product category generates high levels of perceived liking, a TURF analysis based on top box binary coding is most appropriate.

Respondents answer the question for each variant and the evaluation of products is randomized between respondents to guard against order effects and other confounding variables. Complete data per respondent is required.

The data for analysis is a set of binary codes per respondent indicating their "acceptance" of each variant.

The data is then analyzed to calculate reach and frequency measures according to the study objectives. These could be:

(a) All possible product subsets from 1 to the total number of products in the set.

(b) The optimal subset of 4 products out of 12.

(c) The optimal set of 6 products out of 12, comprising 3 sweet and 3 savory flavors.

(d) The best set of 2 new variants out of 8 to refresh an existing line of 4 products.

3 Examples

3.1 Simple TURF Analysis

A company wishes to launch a new range of snacks for infants. Initial focus group research and internal discussions have produced a list of 12 flavor variants. The company has the budget to launch four flavors initially and wishes to use TURF analysis to help find the best four to launch.

Five hundred respondents were invited to take part in the online research and data is available from 442 who completed the study. They were asked to score likelihood to purchase on a 5-point scale and were defined as finding the product acceptable if they scored 4 or 5. The data was analyzed using XLSTAT software Addinsoft [5].

The input data for analysis is as shown in Table 2 and is converted by the software to binary codes based on the user-specified rule, Table 3.

Table 4 gives the overall count of purchase likelihood by sample; this is the sum of the binary figures in Table 3 by column. Tomato and Honey have the highest proportion of purchase likely scores and broccoli the lowest.

The TURF analysis then looks at the reach for each combination of 4 samples out of 12, this involves calculating the reach of all 495 possible combinations of samples. The top 10 combinations (lines) are listed in Table 5. Line 1 had the highest reach of 375 (85%), 85% of respondents are likely to purchase at least one of the flavors (Cheese, Honey, Raspberry, Tomato). Because many respondents give scores of 4 or 5 on the likelihood to purchase scale for more than one of the products in the line, the frequency of purchase is higher than the reach, 891 for the top line. Lines are ranked by their reach. Notice that line 6 (Cheese, Honey, Tomato, Vanilla) has a higher frequency of purchase than the lines with highest reach. If you are looking for high sales volumes, this may be a better combination of flavors than the first line, it attracts a smaller number of people, but they will buy a larger mix of flavors than other lines.

The breakdown of the reach for Line 1 is shown in Table 6. This helps to understand why so many combinations have a similar reach and a high percentage of respondents are reached by one sample; in this case, Tomato with a reach of 256, an additional 67 respondents who would not buy Tomato flavor are reached by the cheese sample and so on to give the total reach of 375. Sixty-seven respondents would not buy any of the four flavors.

Table 2
Example 3.1 Raw data for TURF analysis (first ten respondents)

id	Apple	Banana	Blueberry	Brocolli	Carrot	Cheese	Honey	Peach	Raspberry	Strawberry	Tomato	Vanilla
101	2	2	2	3	3	2	4	2	3	4	4	4
102	4	4	4	3	4	4	4	4	4	4	4	4
103	4	4	4	4	4	4	3	4	3	3	3	3
104	2	2	4	3	2	2	4	3	2	3	3	3
105	3	2	4	2	3	4	4	4	3	4	4	4
106	4	4	4	4	4	4	4	4	4	4	4	4
107	2	4	3	1	2	1	4	3	3	3	4	4
108	3	3	3	3	3	3	3	3	3	3	3	2
109	4	3	2	2	3	3	4	1	4	2	4	3
110	3	3	3	3	3	3	2	3	3	3	3	3

Table 3
Example 3.1 Data for TURF analysis after binary coding (first 10 respondents)

id	Apple	Banana	Blueberry	Brocolli	Carrot	Cheese	Honey	Peach	Raspberry	Strawberry	Tomato	Vanilla
101	0	0	0	0	0	0	1	0	0	1	1	1
102	1	1	1	0	1	1	1	1	1	1	1	1
103	1	1	1	1	1	1	0	1	0	0	0	0
104	0	0	1	0	0	0	1	0	0	0	0	0
105	0	0	1	0	0	1	1	1	0	1	1	1
106	1	1	1	1	1	1	1	1	1	1	1	1
107	0	1	0	0	0	0	1	0	0	0	1	1
108	0	0	0	0	0	0	0	0	0	0	0	0
109	1	0	0	0	0	0	1	0	1	0	1	0
110	0	0	0	0	0	0	0	0	0	0	0	0

Table 4
Example 3.1 Overall count of likely to purchase scores

Flavour	Count	Count (%)
Apple	174	39.4
Banana	183	41.4
Blueberry	178	40.3
Brocolli	133	30.1
Carrot	195	44.1
Cheese	199	45.0
Honey	229	51.8
Peach	149	33.7
Raspberry	207	46.8
Strawberry	208	47.1
Tomato	256	57.9
Vanilla	220	49.8

Table 5
Example 3.1 Flavor lines after TURF analysis

	Reach	Frequency	Flavour 1	Flavour 2	Flavour 3	Flavour 4
Line 1	375	891	Cheese	Honey	Raspberry	Tomato
Line 2	373	882	Cheese	Raspberry	Tomato	Vanilla
Line 3	373	879	Carrot	Cheese	Honey	Tomato
Line 4	372	855	Cheese	Honey	Raspberry	Vanilla
Line 5	372	830	Carrot	Cheese	Honey	Raspberry
Line 6	371	904	Cheese	Honey	Tomato	Vanilla
Line 7	371	870	Cheese	Raspberry	Strawberry	Tomato
Line 8	371	862	Blueberry	Cheese	Honey	Tomato
Line 9	371	858	Apple	Cheese	Honey	Tomato
Line 10	369	853	Blueberry	Cheese	Tomato	Vanilla

The reach figures for these top 10 lines are very close, varying from 83% to 84% of consumers, this is fairly typical for TURF of product lines which are popular with respondents unless the market is very segmented. An alternative analysis using top box rather than top two box scores might reduce the dominance of the Tomato flavor but will decrease the overall reach.

Table 6
Example 3.1 Detailed build of line 1

Build	Flavour	Count
1	Tomato	256
2	Cheese	67
3	Honey	33
4	Raspberry	19
	Unreached	67

Table 7
Example 3.2 Current and additional flavors

Current flavours	Possible additions
Cold Brew Hibiscus	Kiwi Grape
Pineapple Ginger	Mandarin Mint
Blackberry	Mango Cardamon
Tangerine with Segments	Lemon Alfafa
Lemon with Peel	Mango Cardamon
	Strawberry Grapefruit
	Lemon Tamarind
	Pear Red Fruits
	Strawberry Guava
	Rosemary Orange
	Pear Basil

3.2 Supplementing an Existing Range

A carbonated soft drink manufacturer wishes to supplement an existing range of five flavors. There are funds to add two more flavors to the range from eleven prototypes developed as possible additions to the range. The existing and potential flavors are tabulated in Table 7. A sample of 164 young Mexican consumers familiar with the existing range were shown pictures of the 16 flavors, randomized in order for each respondent, and asked to score their expected liking of the drink on a 1–9 scale. A preliminary frequency analysis of the scores indicated that most of the samples were liked and hence the top box score only was used to create the binary reach measure. The analysis was carried out in two parts: firstly the five existing flavors were analyzed for their reach. $124/164 = 75.6\%$ of respondents were reached (scored liking $= 9$)

Table 8
Example 3.2 detailed build of optimal line (five current flavors + two additions)

Build	Flavour	Reach	% Reach	Cumulative % Reach
1	Blackberry	79	48.2%	48.2%
2	Cold brew hibiscus	26	15.9%	64.0%
3	*Kiwi grape*	17	10.4%	74.4%
4	*Strawberry guava*	7	4.3%	78.7%
5	Tangerine with segments	6	3.7%	82.3%
6	Lemon with Peel	3	1.8%	84.1%
7	Pineapple ginger	2	1.2%	85.4%
	Total reached	140	85.4%	
	Unreached	24	14.6%	
	Additions highlighted in bold			

for at least one of the existing ranges. TURF software that allows grouping of products/concepts into separate groups with different selection rules can then be used to find the best two of the eleven possible prototypes to add to the range to increase the reach, TurfPAQ software (Qi Statistics Ltd.) was used to perform the analysis. Two groups were set up, one group of the five current products with the requirement to select them all and a second group of the new prototypes specifying that two should be selected. Table 8 shows the optimal build. The two additional products that give the greatest uplift in the reach are Kiwi Grape and Strawberry Guava with a total reach of 85.4% an increase of close to 10% from the current flavors. Two of the current products, Blackberry and Cold Brew Hibiscus, make the greatest contribution to the reach, reaching 64% of respondents, adding the two additional products to the range increases the reach by nearly 15%, greater than the other three products in the current range. Adding the Kiwi Grape flavor looks to be justified by the analysis (additional reach = 10%), but from this analysis, the benefit of adding Strawberry Guava to the range does not immediately appear to be justified.

4 Notes

4.1 Drawbacks with TURF

As we have seen in the example, frequently TURF analysis does not provide a clear "winning" combination and there are several combinations with very similar reach and frequency statistics. Alternative metrics have been proposed to overcome this weakness. The most widely used of these are Shapley Values which take into account both the incremental reach and overall appeal of the variant. The Shapley Value for Variant A is a weighted average of the increase in reach if A is added to all the combinations of variants excluding A. For example, (B v AB), (BCD v ABCD), and all other possibilities.

It is tempting to interpret the reach/frequency as a surrogate measure of sales volume, but it must be remembered that TURF is an artificial exercise using concepts and the numbers are highly dependent on the arbitrary choice of how a respondent is reached.

TURF is not appropriate for evaluating multi-featured products or concepts which have been designed using structural experimental designs, for example, 12 pasta packs comprising 3 nutritional claims by 4 flour types. It does not capture information about trade-offs in product features, which can be obtained by using conjoint analysis.

4.2 Data Considerations

Unless the data for TURF is collected as a direct binary measure, the data analyst has to decide (a) which response metric to use and then (b) which rules to apply to transform the data to a binary measure for the TURF analysis. The most widely used responses are perceived liking (overall or specific modality) and purchase intent. Where product concepts are presented without any reference to price or value, perceived liking is usually a better measure to use. The more granular the data scale, the more differentiating the top box scores are likely to be. As purchase intent is usually collected on a 5-point scale, this may be the reason why it does not tend to perform as well as a TURF metric.

There are other indirect measures that can be used as inputs into TURF analysis. MAXDIFF (also known as Best-Worst scaling) is a technique which is increasingly used in product and concept testing. Respondents are presented with subsets of products/concepts (usually four or five at a time) and asked to select the product/concept they like the most and the one they like the least. The analysis generates for each respondent utility scores for each product/concept on a continuous scale, these utilities are used to estimate choice probabilities but can also be used as a surrogate measure of liking and analyzed using TURF analysis.

Another consideration is whether to screen the data before analysis and remove respondents who show little or no variation in the response used for the TURF analysis, as they are reached by all or none of the samples and do not contribute to the differentiation between the samples. If this step is taken, then the number of respondents removed should be reported.

4.3 Challenges with Larger Problems

The example shown in Subheading 3.1 is a relatively small problem requiring enumeration of the reach of 495 lines of 4 out of 12 products. Larger problems require more calculation. For example, choosing 8 from 50 would require enumeration of about

537 million combinations (as well as requiring the respondent to evaluate their attraction to 50 products). However, algorithms have been developed to increase the speed of calculation for these larger problems (they are not all guaranteed to be optimal but are quicker in their calculations). One widely used method is called "the greedy algorithm" this finds the best single product, then adds the product which most increases the reach conditional on the first product selected remaining in the set, building a solution step by step. in a similar way to the forward selection algorithm in multiple regression. Other methods have been proposed including eTURF [6] and Binary Linear Programming [7] and many market research agencies will offer their own special approach.

4.4 Other Applications

TURF can be used to enhance the analysis of the usual consumer product test, where overall liking is measured on a 1–9 or 1–7 hedonic scale. As well as assessing clusters by their differences in average liking, a TURF analysis can be carried out to identify the two or three products that have the highest reach within each cluster.

Kuesten and Bi [8] advocate the use of TURF as a novel addition to CATA data analysis, where the data are already in the form of binary codes, and they propose that a TURF analysis by variant will provide additional insights. TURF can also be used to aid and enhance learnings from discrete choice or MAXDIFF studies, which summarize the hierarchies of choice for each variant for each respondent by utility values. By specifying a threshold value of acceptance on the utility scale, these can be used as inputs into the TURF analysis.

4.5 How to Carry Out a TURF Analysis Yourself

At the time of writing, there are several software packages that offer TURF analysis, including:

- JMP®, SAS Institute Inc., Cary, NC, 1989–2021.
- XLSTAT Addinsoft 2021
- TurfPAQ Qi Statistics Ltd (www.qistatistics.co.uk)
- IBM SPSS Statistics for Windows: IBM Corp
- TurfR, TURF Analysis for R. https://cran.r-project.org/web/packages/turfR

References

1. Miaoulis G, Free V, Parsons H (1990) Turf: a new planning approach for product line extensions. Marketing Research, March
2. Cohen E (1993) TURF analysis. Quirk's Marketing Research Review, July/August, pp 10–13
3. Moskowitz HR (1996) Consumer testing and evaluation of personal care products. CRC Press
4. Iyengar S (2011) The art of choosing. Twelve

5. Addinsoft (2021) XLSTAT statistical and data analysis solution. New York. https://www.xlstat.com
6. Ennis JM, Fayle CM, Ennis DM (2012) eTURF: a competitive TURF algorithm for large datasets. Food Qual Prefer 23(1):44–48
7. Serra D (2013) Implementing TURF analysis through binary linear programming. Food Qual Prefer 28:382–388
8. Kuesten C, Bi J (2021) TURF analysis for CATA data using R package 'turfR'. Food Qual Prefer 91:104201

Chapter 19

Use of Lexicometry in Sensometrics, an Essential Complement to Holistic Methods an Original Methodology

Mónica Bécue-Bertaut, Ramón Álvarez-Esteban, and Joan-Miquel Canals

Abstract

Verbalization tasks contribute to a better understanding of consumers' likes and willingness, which is a requirement in the field of sensory analysis. Lexicometry, also known as textual statistics, text mining, or textual data science, applies a multidimensional statistical approach highly capable of extracting information from free text descriptions of products. These verbalization tasks are particularly effective when used as a complement to holistic methods, such as sorting task or napping. We propose here an original methodology based on multiple-factor analysis for contingency tables (MFACT) that can function even when the judges use different languages. We present this methodology with data from a hall test session where the judges describe eight Catalan wines after performing a sorting task. Part of the panel uses Catalan and another part uses French. No translation is required to apply the method.

Key words Verbalization task, Sensory analysis, Multiple factor analysis for contingency tables, Correspondence analysis, Multiple frequency table, Lexical table, Text mining, Textual data science

1 Introduction

Since the pioneer works by Benzécri [1] and Lebart [2], lexicometry has acquired a new and original dynamic. Also known as textual data science [3], text mining, or textual statistics [4], lexicometry offers a multidimensional statistical approach in a wide range of fields, including sensory analysis, where free verbalization tasks are increasingly used [5]. In particular, they serve as an essential complement to holistic methods such as *sorting task* (or categorization) leading to *labeled sorting* [6–9] or *napping*, producing *ultra flash profile* (UFP) [10, 11]. See Varela [12] and Barbe et al. [13] for a review.

In labeled sorting and UFP, judges are asked to select a few freely chosen words to characterize either the clusters or the products themselves, thus providing information on the characteristics governing the perceived similarities and dissimilarities. This

Carlos Gómez-Corona and Heber Rodrigues (eds.), *Consumer Research Methods in Food Science*, Methods and Protocols in Food Science, https://doi.org/10.1007/978-1-0716-3000-6_19,

produces a collection of free comments—known as the corpus—that is doubly structured by product and by the judge in the case of sensory analysis, or by product and by consumer in the case of consumer research.

We present here an original methodology to analyze this type of corpus on its own, as in [11], where the results of the initial UFP were disregarded and only the comments were taken into account. However, we follow a very different approach, starting with all the different product configurations provided by the judges in order to build the average configuration. Subheading 2 sets out the material we use, derived from a bilingual hall test session, which highlights the method's ability to deal with multilingual comments. Subheading 3 describes the statistical methodology and Subheading 4 contains the results of a bilingual hall test session. Subheading 5 offers some concluding remarks.

2 Material

2.1 Tasting Session

Products Eight Catalan wines are considered, selected according to a full factorial design involving three factors: designation of origin (DO; Priorat or Empordà, both in Catalonia, abbreviated to P and E), dominant grape variety (Grenache or Carignan, abbreviated to G and C), and production year (2005 or 2006, abbreviated to 05 and 06). For each variety, both wine vintages come from the same producer. Hereafter, the wines are identified through the three factors; for example, PG06 indicates that the wine is a Priorat wine made from the Grenache variety produced in 2006.

Judges The wines were tasted first by nine Catalan judges selected by the *Institut Català de la Vinya i el Vi-INCAVI, Vilafranca-del-Penedès (Catalonia, Spain)* and a panel of 15 French judges selected by the *Conseil Interprofessionel des vins du Roussillon-CIVR*, Perpignan (France). First, the Catalan panel and three French judges tasted the wines at the Mercè Rossell School of Enology and Viticulture (*Escola d'Enologia i Viticultura Mercè Rossell)* in Espiells, Catalonia in May 2008. The other 12 French judges then did the same at the CIVR in Perpignan (France) in June 2008. Both panels followed exactly the same protocol described below. The judges, all of whom are associated with the wine sector, did not know the particular tasting method beforehand and tasted the eight wines blind. The judges are identified depending on their country (C or F), the tasting place (E or P), and an order number.

Protocol The wines were presented simultaneously but arranged so that the presentation order and first-order carry-over effects were balanced [14]. The judges were asked to follow the presentation order when tasting each wine for the first time and perform a sorting task (i.e., a categorization) on the eight wines. The division into clusters reflects the similarities/dissimilarities between the wines perceived by the judges but does not inform about the criteria governing them, which can be assumed to differ from judge to judge. To remedy this, in a second step, the judges were asked to describe each cluster with a few freely chosen words (verbalization task).

Specifically, the judges received the following instructions:

Eight red Catalan wines are placed in front of you. You are asked to group them into clusters of wines that you consider alike based on your own criteria. You must form at least two clusters and no more than seven. Then you can associate to each wine or cluster the words that best characterize them.

The Catalan and French judges answered in their own language. The judges then gave the wines a score from 1 to 10. Hereafter, the average score of the French judges is called the "French score," while the average score of the Catalan judges is called the "Catalan score."

2.2 Data Coding

All the occurrences as a whole constitute the corpus. The words used by one judge are called *individual words* and may or may not differ from one judge to another. The coding must maintain the difference between homologous words; that is, corresponding to the same word in the language but used by different judges. It should be noted that one document with the comments of each judge is associated with each wine. The words associated with a cluster are considered to describe all the wines in this cluster. In this case, some judges added certain words to particular wines, which have also been recorded and taken into account. Separately for each judge, an individual documents × words table, called an individual lexical table (LT), with as many rows as document-wines and as many columns as individual words, is built by counting the frequency of the different words in the descriptions corresponding to each wine. Some words (such as *complex*, *round*, etc.) have a hedonic connotation [15]. As observed, even professionals do not reject the use of these terms [10, 11] when giving this type of spontaneous descriptions.

The analyzed data structure consists of a multiple table juxtaposing as many *individual* LTs as judges, here 24. This table has the same number of columns as there are individual words used by all the judges and as many rows as there are wines.

Separately, a sum LT is also built for each language. In this case, each wine document combines all the comments given by all the judges on that wine. The sum tables have as many rows as wines and as many columns as different global words in all the free comments.

In these sum tables, the homologous individual words are merged into a single word. The number of times this wine has been characterized by a word is found at the intersection of a row-wine and a column-word. The sum of a word (column) corresponds to its total count in all the comments; the sum of a wine (row) corresponds to the number of occurrences associated with that wine.

2.3 Initial Study

In a first work [9], the data were analyzed to determine the relative importance of the factors in the perception of the wines for both panels. We also wanted to find out whether textual assessments provide information that is not already contained in the sorting task. We were interested in comparing both the wine configurations produced by either the sorting or the verbalization tasks in each language and the panels. One of the main results was that comments were essential to understand the similarities among the wines. It should be noted that the comments were coded into a single LT in each language. The same approach has been followed with expert and non-expert panels [16] or cross-cultural studies [17–19] and makes it possible to analyze in greater detail whether the sensory, affective, and cognitive dimensions [20] are shared by all groups.

2.4 Objectives

In this secondary use of these data, our main objective is to propose an original methodology within the framework of factorial methods that is able to:

- Work from the free comments alone.
- Provide a global configuration of the wines by averaging the individual configurations issued from only the individual comments.
- Describe this configuration with words which make it possible to understand the criteria used by the judges and to determine the general and particular trends.
- Offer a map of the judges based on the criteria they use in order to study their similarities.

3 Statistical Methodology

3.1 First Option: Correspondence Analysis of the Sum Tables

The first option would have been to apply two successive correspondence analyses (CA) to both sum tables. The CA method is comparable to PCA for use with a frequency table [21–24]. The particular case of a lexical table (LT) is widely used [3, 4, 25], and applications can be found in sensory studies specifically to deal with free-text evaluations given by judges, experts, or consumers. CA compares the documents from their lexical profiles; that is, from the relative frequency of each word in each document. Similarly, the words are compared from their profiles (their distribution among the documents). CA thus displays both the configuration of the wines and that of the words, linked to each other. In the wine configuration, two wines are as close as they are described by the same words and as far as they are described by different words. In the word configuration, two words are as close as they are used to describe the same wines and as far as they are used to describe different wines. Both configurations are jointly interpreted to the extent that they are linked by the transition relationships that establish that the words are, except for one constant, at the center of gravity (CoG) of the wines they describe; and the wines are, except for one constant, at the center of gravity of the words that describe them. In other words, CA explains the variability of the wines in terms of the words and the variability of the words in terms of the wines. So, CA meets the three first objectives listed above. In fact, CA usually gives interesting results when starting from only the free comments associated with either a sorting task or napping, although the comments may be short and few [11]. It offers a configuration of the wines, underlines their similarities and differences, and describes this configuration in words, thereby revealing the criteria used by the judges.

However, CA has several drawbacks. First, this method requires choosing a minimum frequency threshold for the words, as the comparison of the lexical profiles is statistically meaningful only if the words are used with a minimum frequency [4]. It is important to know that the threshold choice does not obey well-defined rules and can lead to a cumbersome pre-study [11]. The occurrences of the same word must be written in exactly the same way for them to be counted together, making it necessary to find the corresponding lemma or dictionary entry for each occurrence (infinitive for a verb form, singular for a noun, masculine singular for an adjective). This operation, called lemmatization, is difficult to automate on a corpus composed of words and not of sentences expressed in correct language. However, it is easy to perform a manual intervention in the case of a short corpus such as the ones in this type of study. It should also be considered that words related to the same sensory perception, such as *fruit* and *fruity*, for example, will continue to be separately counted after the lemmatization step, meaning that equivalencies must be created, although there will always be cases that raise doubts as to the correct intervention. As will be seen, our proposal avoids this cumbersome process and makes it possible to start from the rough corpus, with the requirement for only a spell check and optionally a lemmatization.

In addition, CA neither shows the influence of the judges on the global configuration, which cannot be considered as the average configuration, nor provides a map of the judges for the purposes of comparison. Finally, in the present case where two panels are considered, it would be difficult to compare both panel configurations as they would be located in different spaces.

3.2 Second Option: MFACT for Contingency Tables on the Multiple Table

Multiple factor analysis for contingency tables (MFACT) [26, 27], that is, multiple factor analysis (MFA) tailored for frequency data, applied to the multiple table built from the 24 individual tables, circumvents these problems and allows us to meet the fourth objective. This method also offers very useful auxiliary results, such as, for example, combining liking scores and free-text comments [28]. Below, we set out the principles of MFACT and describe how they are applied to the multiple table built from all the individual tables. After describing the principles of MFA and MFACT, we outline the steps followed to obtain the results.

Multiple Factor Analysis (MFA) is used to analyze a multiple table in which a set of individuals is described by several groups of quantitative or qualitative variables. MFA performs a factorial analysis of the juxtaposed table by overweighting the variables in order to balance the importance of each group. The results provided by this method are:

- Common to all factorial methods, such as principal component analysis (PCA), correspondence analysis (CA), and multiple correspondence analysis (MCA), mainly a representation of the rows-individuals and columns-variables or categories.
- Specific to multiple tables, mainly the superimposed representation of the structures induced by each group of variables on all the individuals (or categories of individuals)—called partial structures—i.e. the representation of the factors obtained during the separate analyses and a synthetic representation of the groups.

3.3 MFACT as a Tool for Building the Average Configuration

MFACT has been proposed by Bécue-Bertaut and Pagès [26] for the simultaneous analysis of a set of contingency tables. This allows the use of a multiple-frequency table while adopting an MFA approach.

When MFACT is applied to juxtaposed lexical tables, the results provided are:

- Similar to those of the CA, primarily a global representation of rows-individuals and columns-words. In this method, the row and column configurations are also linked by transition relationships, so the variability of the wines will be explained by the variability of the words and vice versa.
- Specific to multiple tables. The most interesting result is a synthetic representation of the active groups, in this case, the judges, and the supplementary groups.

In the present application, each judge provides a multidimensional configuration of the wines that could be visualized by applying CA on his or her individual LT. MFACT provides an average configuration in a reduced space. This average configuration is the closest possible to all the individual configurations, as measured by MFA criteria. In the final representation, two wines are close to each other if they are close in most of the individual configurations. The influence of the different LT is balanced out during the computation process.

MFACT offers graphic and numerical results similar to those afforded by MFA, specifically a configuration of the wines and the individual and global words, a map of the judges so they can be compared from the criteria they use, and a representation of the supplementary variables, in this case, the scores. They are presented and discussed below.

4 Results

4.1 Some Lexical Features of both Corpus

First, the spelling of the original comments in either Catalan or French was carefully corrected (Table 1).

A manual lemmatization was then performed and a few compound words such as *non-boisé* (*unwooded*), *peu-consistent* (*inconsistent*), *de-garde* (*for ageing*), and so on were written as a single word. Stopwords, that is, non-informative words such as articles, conjunctions, and prepositions, were deleted.

Catalan panel. The Catalan judges use between 10 and 22 different words each. Overall, they produce 323 occurrences (319 retained after deleting the stopwords) issued from 97 distinct words (95 retained). The most frequent (Table 2) are *confitura/jelly* (13), *fruita/fruit* (12), *tànnic/tannic* (10), and *fusta/wood* (9). Words with a similar meaning, such as *taní/tannin* (4), *roure/oak* (3), and *bota/barrel* (2) reinforce the presence of these perceptions.

French panel. These judges use between 6 and 43 words each. Overall, they produce 655 occurrences (611 retained after deleting the stopwords) issued from 149 distinct words (137 retained). The most frequent are *boisé/woody* and *bois/wood* (30 and 12), *fruit/fruit* (28) and *fruité/fruity* (24), *tanin/tannin* (19) and *tannique/*

Table 1
Excerpt of the free comments

---- *French comments*
Caramel, lactique, souple, gras, fruit en final (judge FE5/ wine PG06)
Curieux, particulier, limite défaut, animal, carton humide, étable (judge FE5/wines EG05 and EG06)
---- *Catalan comments*
Complex, fumat, astringent, dens, cos, toffe (judge CE1/wines EG06 and PG06)
Greix, formatge, mantegós (judge CE2/wines EG5 and EG06)

Table 2
Most frequent words

Most frequent Catalan words			Most frequent French words		
Word	**Count**	**No.docs**	**Word**	**Count**	**No.docs**
confitura (*jelly*)	13	6	boisé (*woody*)	30	8
fruita (*fruit*)	12	6	fruit (*fruit*)	28	7
tànnic (*tannic*)	10	7	fruité (*fruity*)	24	7
vainilla (*vanilla*)	9	7	tanin (*tannin*)	19	8
fusta (*wood*)	9	4	puissant (*powerful*)	17	8
nas (*nose*)	8	8	tannique (*tannic*)	17	7
madur (*mature*)	8	5	mûr(*mature/ripe*)	14	7
aroma (*bouquet*)	7	7	vin (*wine*)	13	8
greix (*unctuous/fat*)	7	6	équilibré (*balanced*)	13	6
astringent (*astringent*)	7	5	très (*very*)	13	6
boca (*mouth*)	7	5	bois (*wood*)	12	5
regalèssia (*liquorice*)	7	5	vanillé (*vanillin*)	11	6
balsàmic (*balsamic*)	6	5	nonboisé (*unwooded*)	9	6
rodó (*round*)	6	5	animal (*animal*)	9	5
secant (*drying*)	6	4	épice (*spicy*)	9	5

tannic (17). It should be noted that *nonboisé/unwooded* (9) is explicitly mentioned by the French judges. Hedonic perceptions are also quoted, such as *puissant/powerful* (17) and *équilibré/balanced* (13).

4.2 Coding Details

From the comments, the same number of *individual* LTs are built as there are judges. This codification preserves the information given by each judge on the wines. Each table has as many columns as different words, called individual words, used in the corresponding *individual* comments. Individual words can be homologous from one judge to another and then correspond to the same *global word*. The individual LTs describe the individual wine configurations as provided separately by each judge, visualized by successive CAs applied to these individual LTs taken one by one.

The 24 *individual* LTs are then juxtaposed row-wise in a multiple-frequency table. All the individual words pronounced by the judges correspond to a column in the multiple-frequency table.

The Catalan sum table has eight row-documents and 95 column-global-words. The French sum table has the same eight row-documents and 137 column-global-words. Both sum tables play a supplementary role in the analysis. The score table, also supplementary, has eight row-documents and two columns of quantitative variables corresponding to the average liking score as given by the Catalan and the French judges, respectively.

The preliminary analysis led us to assign a non-active role to the French judge FP2 because he/she does not share the first global dispersion direction, common to all the other judges, as shown by the very low value of the corresponding canonical correlation computed by MFACT. Thus, only 23 individual tables are kept.

An individual table has eight wine rows and as many columns as the vocabulary of one judge (individual words). The active multiple table that links the 23 individual tables row-wise has eight wine rows (or documents) and 387 columns (the sum of the individual words for all the judges).

4.3 Individual Wine Configurations

CA applied to each individual table provides a multidimensional wine configuration, in this context called the *individual configuration*. Figure 1 shows the individual configuration for judge CE10 mapped on the first factorial plane issued from CA. The wines are clustered by this judge into four sets (preliminary sorting task). In this case, the words qualifying each set differ from one set to another. For this reason, words and wine labels of the same set overlap. They are slightly displaced to obtain a readable representation. The CA first plane keeps 67% of the inertia and therefore provides a good-quality visualization.

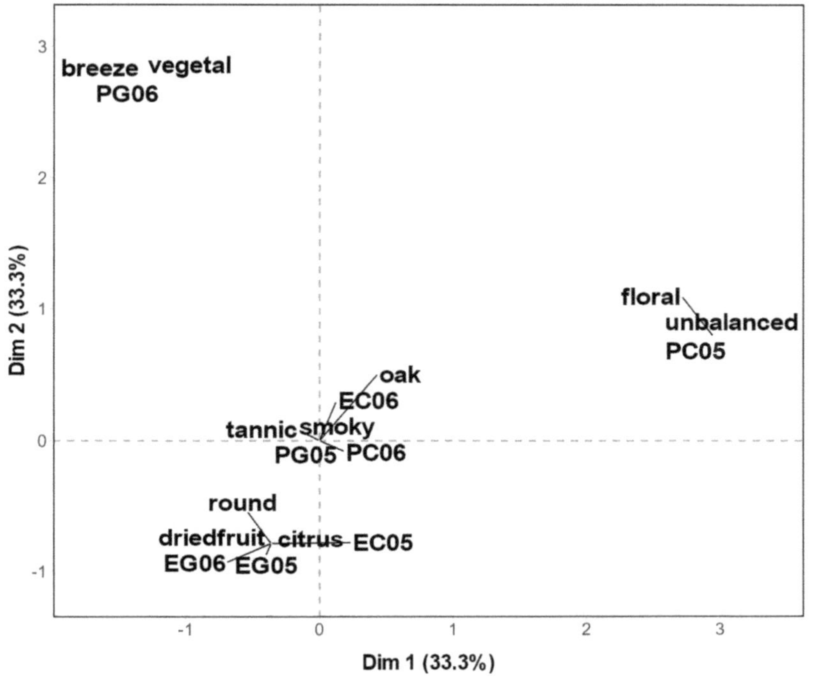

Fig. 1 Individual configuration as provided by judge CE10

Our method starts from these individual configurations and builds the average configuration without considering the equivalences between the words chosen by the different judges, as the sole purpose is to record the individual configurations of the wines in order to build the average configuration.

4.4 MFACT. Global Significance of the Results

MFACT allows us to obtain an average of the individual configurations and to place each one in relation to this average, thereby providing elements for comparison.

The inertia of the first factor is equal to 13.3. It should be noted that the maximum value would be 23, that is, the number of active judges (= active groups). Thus, the first global axis of the MFACT does not correspond to the first direction of inertia in each of the 23 active individual configurations. Nevertheless, the correlation coefficient between the first factor and the projection of the 23 configurations on this axis, called the *canonical correlation coefficient* in MFACT, is over 0.70 for 19 of the judges. This direction of inertia is therefore genuinely present with at least average intensity in the individual configurations that are retained. In the absence of a validation technique, the value of 0.70 is chosen empirically. However, a very small minority of canonical correlations over 0.7 can be seen in the case of the highest rank axes, which can be assumed to be similar to what would be observed in the absence of any structural effect.

The inertia of the second factor is equal to 12.1. The canonical correlation coefficients are over 0.70 for 18 judges, so this second axis also corresponds to the direction of inertia present in the majority of the individual configurations.

Among the following axes, there are several arguments in favor of examining axis 3: its inertia is above the mean (this is also the case for axis 4 but with a very small margin). As usual, it is the interpretability of the results, in particular through words that will be decisive.

One very important consideration is that both French and Catalan wine configurations, as built separately from both sum tables, are practically homothetic to the global wine configuration, issued from the 23 concatenated tables, as measured by the RV coefficient. The RV respective values are 0.96 (French sum table) and 0.98 (Catalan sum table), while 1 indicates a perfect homothety. The interpretation can therefore be supported by both global and individual words, as will be seen below.

4.5 MFACT Plane: Wine and Word Configuration on the First Plane

Figure 2 shows the configuration of the eight wines on the first MFACT plane. If this figure is read according to the axes, the first axis (22.7% of the inertia) can be seen to oppose all the wines from the Carignan grape variety (PC05/EC06 opposed to EC05/ES06), while the second axis (20.6%) opposes all the wines from the Grenache grape variety (EG05/EG06 opposed to PG05/

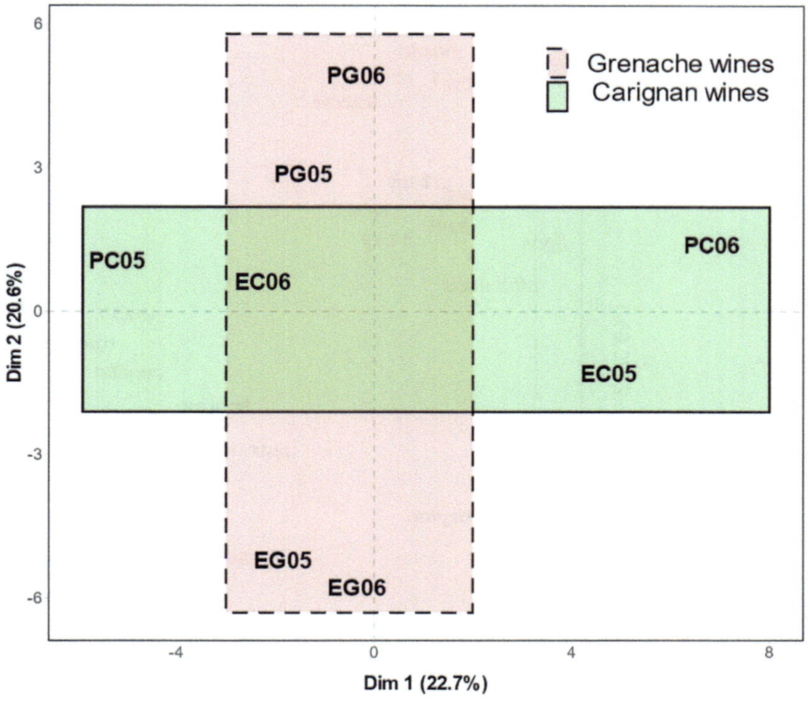

Fig. 2 Global representation of the wines documents

PG06). These oppositions between wines of the same variety are related neither to the DO nor to the year. So, we can say firstly that the vintage does not appear to matter. Moving forward a little on what follows, we can add that Grenache wines are separated according to quality, whereas Carignan wines are separated according to their woody/unwoody character.

However, from now on we prefer to read this first plane according to the three poles in which the wines (and the words, as we will see) are organized: PC06/EC05, PG06/PG05/PC05/EC06, and EG05/EG06. Wines belonging to the same pole share properties that the words will allow us to discover.

The usual transition formulas state that the wine and word representations, either individual words (active columns) or global words (supplementary columns), are linked. This allows the differences or similarities between the wines to be characterized since the words describe their sensory properties. As it is impossible to have a clear map of the words if all of them are represented, we have selected a series of successive maps which complement each other and should be read in parallel. As can be expected, the same tripolar structure is found in the words as in the wines, thus enabling us to easily relate wines and words.

First, Fig. 3 describes the positive characteristic words, namely the words that are significantly overrepresented in one or other of the three poles. See ([3], chapter 5, and [4], chapter 6) for more

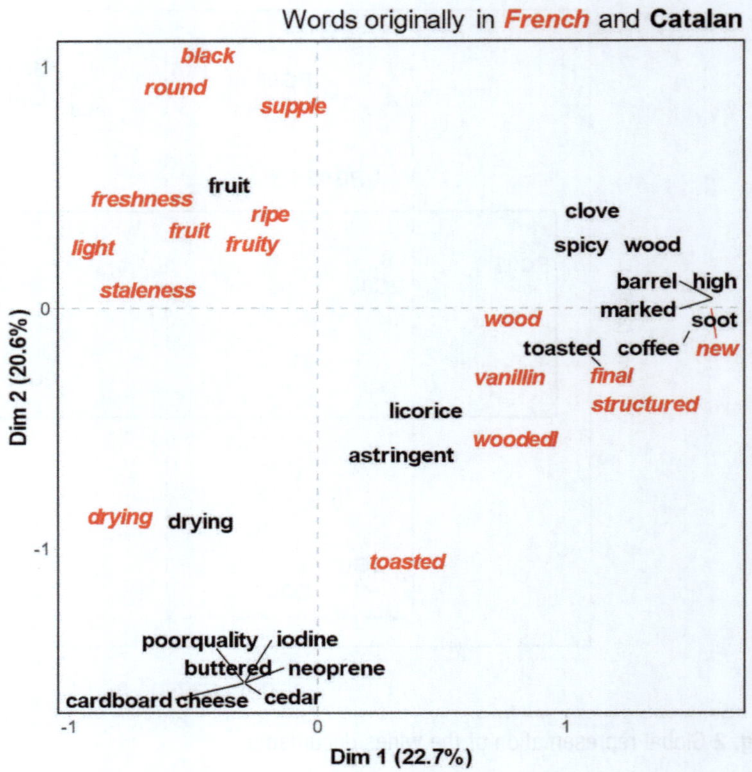

Fig. 3 Characteristic words of the three poles

information about how the positive characteristic words are identified. In this case, each pole is seen as part of the corpus clustering the comments corresponding to its wines. The more a word is overused in a pole, the more it contributes to the chi-2 distance between this pole and the CoG. So, on the CA plane, one part—here a pole—is strongly attracted by its positive characteristic words.

The first pole, to the right of the first axis, brings together Catalan and French words related to the wood perception of woodiness, in the case for both Catalan and French words. In addition to the words *boisé/woody* and *bois/wood* in French and *fusta/wood* and *bota/barrel* in Catalan, we find other sensory perceptions frequently associated with woodiness. The words *charpenté/structured* and *vanillé/vanillin* are cited in French, while *cafè/coffee*, *especiat/spicy*, *torrat/toasted*, *clau d'olor/clove*, *sutja/soot*, and *regalèssia/liquorice* are used in Catalan. Compared to the French judges, the Catalan judges frequently characterize the woody perception through several similar sensory perceptions to the above and less through *fusta/wood* directly.

The words in the second pole, at the positive part of axis 2 clearly allude to the fruity sensation, either directly through *fruit/fruit*, *fruité/fruity* (in French), or *fruita/fruit* (in Catalan), or through perceptions associated with fruit (*souple/supple*, *léger/light*, *mûr/ripe*, *noir/black*, and *fraîcheur/freshness* in French). In this case, the French judges mention more fruit-related perceptions than the Catalan judges.

Finally, the third pole (third quadrant) contains words related to negative assessments and defects. In both French and Catalan, we find *drying* (*asséchant* in French, *secant* in Catalan) along with other negative assessments, in particular directly through *malaqualitat/bad quality*, *iode/iodine*, *cartró/cardboard*, *claudolor/clove*, and *formatge/cheese*. It is slightly surprising to find the word *mantegòs/buttered* among the negative appreciations. To better understand this point, we return to the original comments. Thus, we observe that *mantegòs/buttered* is used only twice, combined with *greix/fat* (positive appreciation) and *formatge/cheese* (negative appreciation) in the same comment by judge CE2 to qualify both wines EG05 and EG06 (*see* Table 1). These two wines are sometimes associated with positive words (positive in the present context), but more often with negative ones. Thus, judge CE1 qualifies wine EG06 with *complex, fumat/smoky, astringent, dens/dense, cos/full-bodied, and toffe/toffee*, while FE5 describes EG05 as *curieux/curious, particulier/particular, limite défaut/borderline defect, carton humide/wet cardboard, and étable/cowshed*. The positive words associated with these two wines, such as *greix/ fat,* are also associated with wines belonging to other more appreciated poles and so lie in another area (*see* the position of *greix/fat* in Fig. 4). In contrast, most of the negative words are only associated with these two wines, although they are little used as each judge chooses his or her own words. However, the redundancy between their negative meanings clearly denotes that these wines are not appreciated. This summary must be completed by reading Figs. 4 and 5, which both map global words.

Figure 4 shows the best-represented words (cos2 over 0.5 on either of the first two axes). These words add aspects or nuances.

In the first pole, we find *gerani/geranium*, *cos/full-bodied, torrat/toasted, and glicerol/glycerol* cited by the Catalan judges, who expand their comments to include perceptions commonly associated with wood. The second pole contains *non-boisé/unwooded*, explicitly mentioned by the French judges. The very significant French word *beau/beautiful* appears in the upper part of this pole, whereas new negative words such as *réduction/reduction, farineux/floury, mou/soft* appear in the lower part. This pole clusters the most highly valued wines and is the most heterogeneous, although all are fruity wines, either of very good quality or a little more limited. Here, the French judges provide more words. Finally, new negative words appear in pole 3.

To complete the information, Fig. 5 maps the words relative to assessments. The positive assessments lie on the upper part of axis 2, while the negative assessments lie on the lower part.

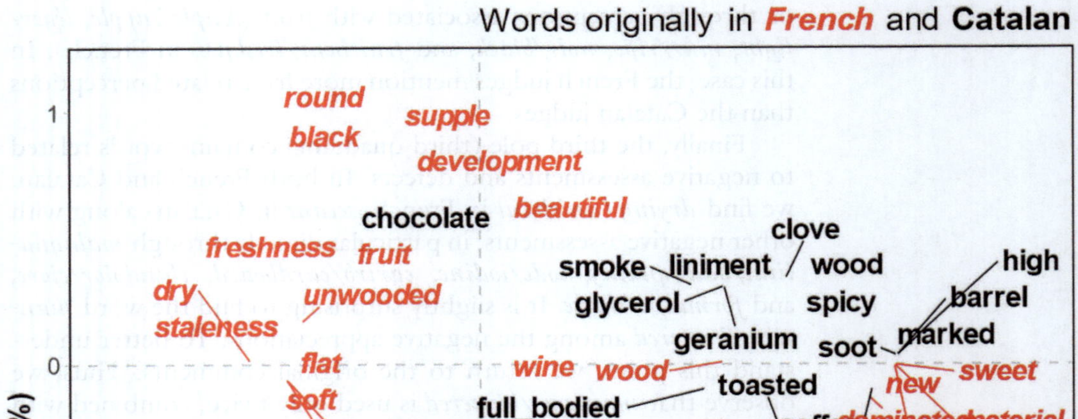

Fig. 4 The best-represented global words on either of the first two axes

Contradictions can appear, such as *desequilibri/imbalance* occurring close to *équilibré/balanced* in the upper part. A frequent finding in sensory analysis is that judges do not always agree; this is the reason for using a panel rather than a single judge.

It should also be noted that:

- There are few words that could a priori be considered positive. For example, from its coordinate on the second axis, *agréable/pleasant* appears to correspond to a rather moderate opinion, something like "not bad but nothing special."
- Some equivalent words in French and Catalan according to the dictionary (*rond/round* in French and *rodó/round* in Catalan, for example) with a somewhat subjective meaning can lie quite far from each other. This means that they qualify different wines and therefore have different meanings in both French and Catalan wine language.

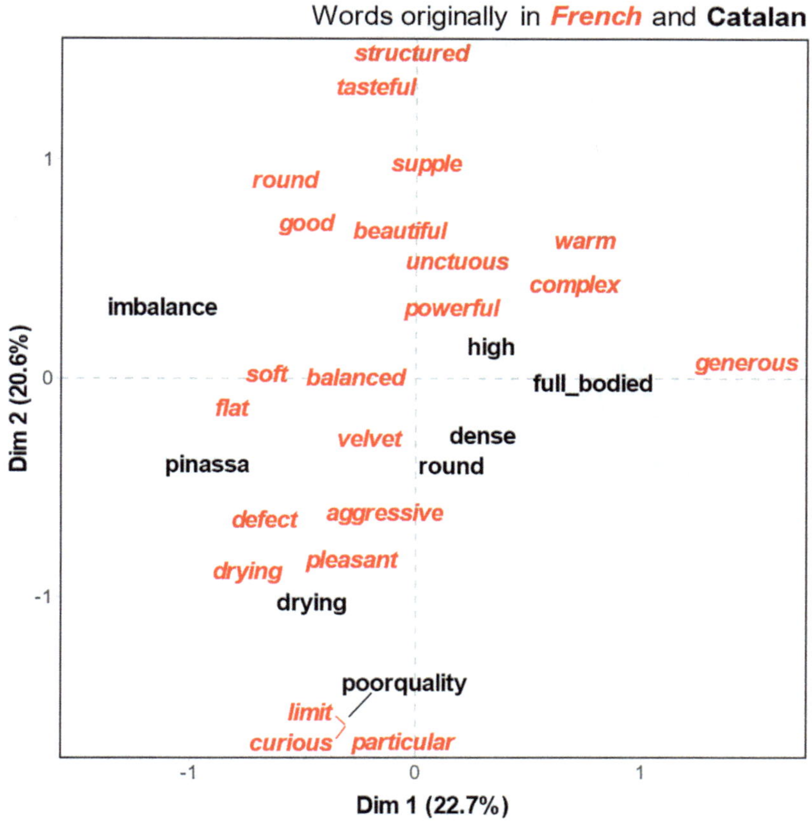

Fig. 5 Global words related to positive and negative assessments

Finally, Fig. 6 maps the individual words related to either wood or fruit, chosen as they are the main perceptions cited. This conveys the dispersal of the same word among the wines and points to the high level of consensus between the judges since the homologous individual words (i.e., corresponding to the same global word) often lie in the same part of the map. The individual words corresponding to fruity (*fruity*, *fruity.1*, *fruity.2*, and so on) all lie to the left of the first axis, in the opposite part to the individual words related to the perception of woodiness. This corroborates that axis 1 opposes fruity to woody wines.

In short, the wines of the first pole (PC06/EC05) (Fig. 1) are therefore clearly perceived as *woody*, *spicy*, *balsamic*, and so on (typical flavors of Carignan varietal wines), while the second pole (PG06/PG05/PC05/EC06) clusters both *non-woody* and *fruity* wines. The most highly valued wines in this group are those at the positive end of the second axis, qualified as *souple/supple*, *rond/round*, and even *beau/beautiful*, while those closer to the CoG are seen as *léger/light* or even *neutre/neutral*. The wines of the third pole (EG05/EG06) are associated with somewhat negative

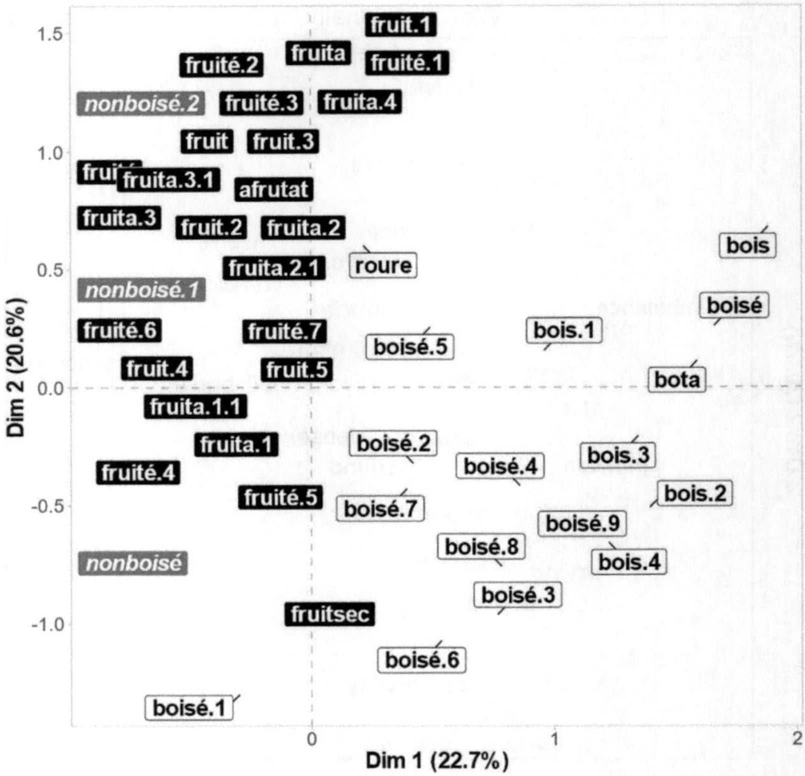

Fig. 6 Individual words related to woody (white rectangle) and fruity perceptions (black rectangle)

features expressed in many different ways and are even qualified as being of *mala qualitat/poor quality.* Axis 2 is in fact a quality axis, due both to the very favorable words found in the positive part of the axis and the strong correlations between the French and Catalan scores and the second axis (*see* next section).

Axes 3 and 4 differentiate wines PG06 and PG05 respectively. PG06 is the wine most appreciated by both French and Catalan judges according to the liking scores, while PG05 ranks second for the French and third for the Catalans. The perceptions of these wines can be seen from the MFACT (3, 4) map (not shown here), when consulting the characteristic words as computed separately for each wine. PG06 is described as *delightful*, *structured, complex,* and *excellent*, highlighting its *young* and *fruity* character and a number of qualities such as *balanced, supple, and round,* while PG05 is seen as *ripe*, *supple,* and *balsamic*; it collects fewer characteristic words than PG06 and is therefore less differentiated from the other wines.

4.6 Representation of the Liking-Scores

Figure 7 represents the liking scores as supplementary quantitative variables. Both the Catalan and French scores are highly and positively correlated with the second axis, thereby confirming that axis

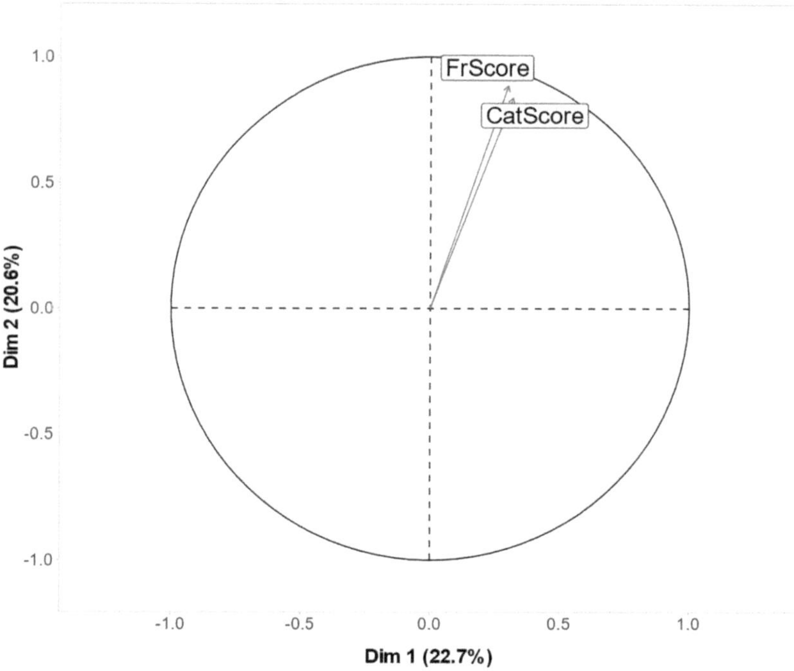

Fig. 7 Characterization of liking scores

2 opposes low-scoring wines (negative part) to high-scoring wines (positive part). In fact, PG06 is the wine that scored highest with both the French and Catalan judges.

Note that the proposed methodology makes it possible to represent supplementary quantitative, qualitative, or frequency-type variables, as in any factor analysis. For example, it would be interesting to use the results of a chemical analysis of the aromatic components in order to compare them with the judges' perceptions.

4.7 Group Representation

In this analysis, the group representation corresponds to that of the judges, as for the active groups, the sum tables (French and Catalan) and liking scores tables for the supplementary groups. This representation (Fig. 8) reveals the following:

- Both the "*woody*" and the "*fruity*" directions may be present with more or less intensity in the individual configurations, thus translating the importance given by the judge to the corresponding sensory perception. For example, the *woody* dimension exists with a high intensity in the individual configuration provided by judge CE4, while the *fruity* dimension is totally absent. The configurations of judges CE2 and CE3 show a high intensity of both dimensions. In the case of judge CE10, the rather low coordinate on axis 1 and high coordinate

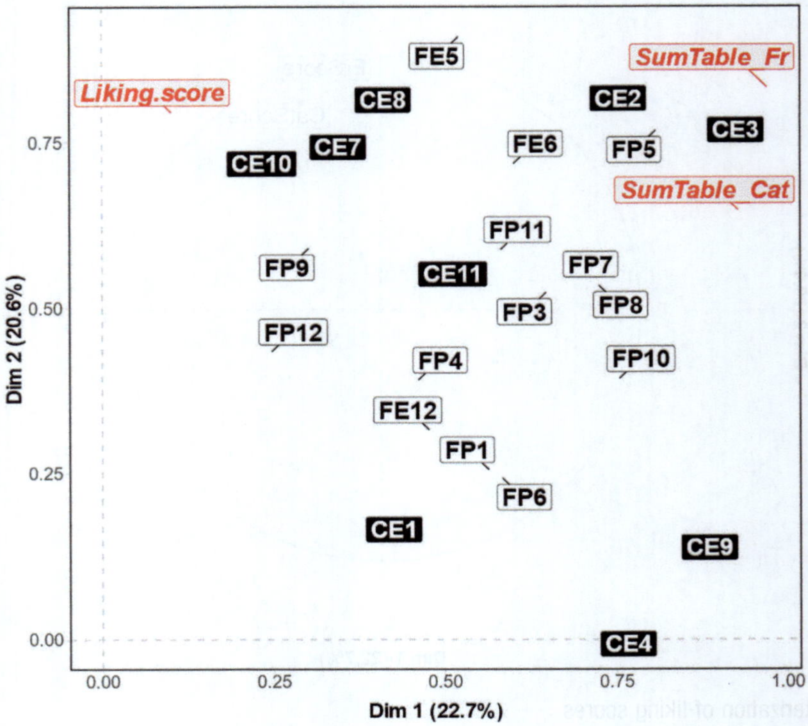

Fig. 8 Map of the groups including the judges

on axis 2 show that this judge is sensitive to the acceptability of wines but does not consider the woodiness features to be important (as can be seen in Fig. 1).

- The sum tables also present these directions of dispersion with high intensity, especially in the case of the first axis (coordinate value of both sum tables is almost equal to 1 on the first axis, the maximum possible value). A high-quality representation of the dispersion of the words is obtained without having to eliminate the infrequent words. This point validates the method. If the tables were analyzed using CA, it would be necessary to select the words due to the excessive influence of infrequent words on the determination of the first axes. As mentioned before, this requirement usually leads to a cumbersome threshold selection process [11]. In MFACT, this influence is limited by the fact that the configurations are averaged, which locks in an excessive influence of infrequent words.
- The second axis is strongly related to both French and Catalan liking scores, as already seen in Fig. 8.

5 Conclusion

Starting from the individual configurations, MFACT builds the average configuration of the wines by solely taking into consideration the individual configurations. The words intervene only at the individual level. This removes the need to look for equivalences between acid and acidity—to give an example—as a judge will rarely use one word and then the other to qualify different wines. Thus, this method functions effectively with free text comments, even with words specific to each judge, which is of great interest. Although a lemmatization is not compulsory, in this case, it is done to limit the number of different words on the maps.

Our methodology provides a broad range of results, which allows us not only to relate the similarities/dissimilarities among wines with the sensory perceptions as indicated by the words but also to map the judges based on the criteria they use in their individual configurations.

We should emphasize that it can be applied even in cases where:

- The judges express themselves in different languages. This eliminates the need for translation and enables international studies.
- The words are directly associated with the products, which is very useful when seeking to generate a lexicon for new products through various methods. This methodology would help uncover descriptors that both discriminate among the different products and are shared by the judges.
- The judges may be consumers, provided they have an adequate knowledge of the language of wine.

6 Statistical Software

The implementation was performed using Xplortext and FactoMineR packages, both available on R Cran. The corresponding data base and script are available on Xplortext web.

References

1. Benzécri J-P (1981) Pratique de l'analyse des données. Tome 3. Linguistique et lexicologie. Bordas, Paris
2. Lebart L (1982) L'analyse statistique des réponses libres dans les enquêtes socio-économiques. Consomm Rev Socio-Economie Paris 29:39–62
3. Bécue-Bertaut M (2018) Textual data science with R. Chapman and Hall/CRC
4. Lebart L, Salem A, Berry L (1998) Exploring textual data. Springer Netherlands, Dordrecht
5. Bécue-Bertaut M (2014) Tracking verbal-based methods beyond conventional descriptive analysis in food science bibliography. A statistical approach. Food Qual Prefer 32:2–15. https://doi.org/10.1016/j.foodqual.2013.08.010

6. Faye P, Br D, Durand M, Courcoux P (2004) Perceptive free sorting and verbalization tasks with naive subjects : an alternative to descriptive mappings. Food Qual Prefer 15:781–791. https://doi.org/10.1016/j.foodqual.2004.04.009
7. Blancher G, Chollet S, Kesteloot R et al (2007) French and Vietnamese: how do they describe texture characteristics of the same food? A case study with jellies. Food Qual Prefer 18:560–575. https://doi.org/10.1016/j.foodqual.2006.07.006
8. Cadoret M, Lê S, Pagès J (2009) A factorial approach for sorting task data (FAST). Food Qual Prefer 20:410–417. https://doi.org/10.1016/j.foodqual.2009.02.010
9. Bécue-Bertaut M, Lê S (2011) Analysis of multilingual labeled sorting tasks: application to a cross-cultural study in wine industry. J Sens Stud 26:299–310. https://doi.org/10.1111/j.1745-459X.2011.00345.x
10. Perrin L, Symoneaux R, Maître I et al (2008) Comparison of three sensory methods for use with the Napping® procedure: case of ten wines from Loire valley. Food Qual Prefer 19:1–11. https://doi.org/10.1016/j.foodqual.2007.06.005
11. Perrin L, Pagès J (2009) Construction of a product space from the ultra-flash profiling method: application to 10 red wines from the loire valley. J Sens Stud 24:372–395. https://doi.org/10.1111/j.1745-459X.2009.00216.x
12. Varela PAG (2014) Novel techniques in sensory characterization and consumer profiling. CRC Press, Boca Raton
13. Barbe JC, Garbay J, Tempère S (2021) The sensory space of wines: From concept to evaluation and description a review. Foods 10. https://doi.org/10.3390/foods10061424
14. MacFie HJ, Bratchell N (1989) Designs to balance the effect of order of presentation. J Sens Stud 4:129–148. https://doi.org/10.1111/j.1745-459X.1989.tb00463.x
15. Lehrer A (1975) Talking about Wine. Language (Baltim) 51:901–923. https://doi.org/10.2307/412700
16. Barton A, Hayward L, Richardson CD, McSweeney MB (2020) Use of different panellists (experienced, trained, consumers and experts) and the projective mapping task to evaluate white wine. Food Qual Prefer 83:103900. https://doi.org/10.1016/j.foodqual.2020.103900
17. Corsi AM, Cohen J, Lockshin L, Williamson P (2017) Testing lexical equivalences for wine flavours in emerging markets: do hawthorns taste like blackberries? Food Qual Prefer 62:296–306. https://doi.org/10.1016/j.foodqual.2017.03.011
18. Rodrigues H, Parr WV (2019) Contribution of cross-cultural studies to understanding wine appreciation: a review. Food Res Int 115:251–258. https://doi.org/10.1016/j.foodres.2018.09.008
19. Jeong S, Lee J (2021) Effects of cultural background on consumer perception and acceptability of foods and drinks: a review of latest cross-cultural studies. Curr Opin Food Sci 42:248–256. https://doi.org/10.1016/j.cofs.2021.07.004
20. Gómez-Corona C, Chollet S, Escalona-Buendía HB, Valentin D (2017) Measuring the drinking experience of beer in real context situations. The impact of affects, senses, and cognition. Food Qual Prefer 60:113–122. https://doi.org/10.1016/j.foodqual.2017.04.002
21. Benzécri J-P (1973) L'analyse des données. Volume II. L'Analyse des Correspondances. Paris
22. Escofier B (2003) Analyse des correspondances: Recherches au cœur de l'analyse des données. Presses Universitaires de Rennes, Rennes
23. Lebart L, Piron M, Morineau A (2006) Statistique exploratoire multidimensionnelle: visualisation et inférences en fouilles de données. Dunod, Paris
24. Escofier B, Pagès J (2008) Analyses factorielles simples et multiples. Objectifs, méthodes et interprétation. Dunod
25. Benzécri J-P et collaborateurs (1981) Pratique de l'analyse des données, Linguistique et lexicologie. Dunod, Paris
26. Bécue-Bertaut M, Pagès J (2004) A principal axes method for comparing contingency tables: MFACT. Comput Stat Data Anal 45:481–503. https://doi.org/10.1016/S0167-9473(03)00003-3
27. Bécue-Bertaut M, Pagès J (2008) Multiple factor analysis and clustering of a mixture of quantitative, categorical and frequency data. Comput Stat Data Anal 52:3255–3268. https://doi.org/10.1016/j.csda.2007.09.023
28. Bécue-Bertaut M, Álvarez-Esteban R, Pagès J (2008) Rating of products through scores and free-text assertions: comparing and combining both. Food Qual Prefer 19:122–134. https://doi.org/10.1016/j.foodqual.2007.07.006

Part V

Methods from Culinary Arts, Design, and User Experience

Chapter 20

Concept and Packaging Testing in the Food Industry

Ignacio Gil-Pérez, Iván Lidón, and Rubén Rebollar

Abstract

The key role of packaging in modulating consumer perception and response has led to a growing scientific interest over the years. However, the multidisciplinary nature of packaging research means that different approaches are being used to study it, which can be intimidating for those seeking to get started in this field. This chapter aims to facilitate this task by summarizing the basic information necessary to prepare an experiment on this topic. To that end, a brief review of the literature devoted to packaging within the field of consumer research is presented, in order to provide readers unfamiliar with the subject with a theoretical framework that encompasses the main insights and lines of research related to new packaging and concept testing. Next, the key points that characterize a packaging research experiment are discussed, summarizing the characteristics, opportunities, and limitations of the most common methods and experimental designs. Overall, this chapter is aimed at both design practitioners and academics interested in packaging research, seeking to provide them with the resources and basic knowledge necessary to start their own research.

Key words Packaging, Industrial design, Consumer research, Perception, Expectations

1 Introduction

Almost all products are marketed in some type of packaging. Packages are especially ubiquitous in the food industry, where the nature of the products (liquid, fragile, perishable, and so on) makes it essential to use packaging to keep the product in good condition until the moment of consumption. This has motivated great scientific interest in the topic, and the scientific literature has approached packaging research from different angles. Entire bodies of knowledge have been devoted to topics such as the development and analysis of new production systems, machinery, and materials; the study of the packaging life cycle, its carbon footprint and the environmental impact of packaging production and waste; or the impact of packaging and its characteristics on food safety and food preservation.

Carlos Gómez-Corona and Heber Rodrigues (eds.), *Consumer Research Methods in Food Science*,
Methods and Protocols in Food Science, https://doi.org/10.1007/978-1-0716-3000-6_20,

In addition, scientific interest in packaging has also increased in the field of consumer research. Packaging is always present at the point of sale, and therefore it acts as a mediator between the product and the consumer at the time of purchase. Research shows that the functions of packaging go far beyond the classic functions of transporting, preserving, and protecting the product, as packaging must also attract the attention of the consumer at the point of sale and effectively communicate the characteristics of the product contained within [1]. Moreover, literature shows that food packaging has an effect on consumer expectations and willingness to buy: specifically, packaging cues such as color, shape, imagery, texture, or even the typeface used in the label may influence how consumers expect the contained product to be, and therefore modulate their response toward it [2].

Thus, packaging designers and practitioners have an important role to play during the design process. Besides looking for a result that meets all functional requirements, they must ensure that their design communicates the intended information in an efficient way and has the desired effect on how consumers perceive the product. Hence, knowing the effect that design decisions will have on consumers' expectations and response is key to ensuring the product's success in the marketplace. Scientific research can shed light on this regard at different stages of the design process, since being familiar with the literature can facilitate decision making in the early stages of conceptualization and ideation. Furthermore, conducting studies and applying the techniques and approaches of scientific research can help during the evaluation stage to validate design proposals.

Thus, the objective of this chapter is twofold. Firstly, it aims to briefly review the literature on packaging, in order to shed light to those unfamiliar with the subject regarding the research questions that have been addressed so far and what the most relevant findings have been. In addition, it focuses on the use of research methods and techniques commonly used in packaging research so that any researcher, designer, or practitioner can integrate them into their workflow.

2 Background Studies

Research dedicated to packaging within the field of consumer research relies on techniques and insights from the fields of sensory science, experimental psychology, neuroscience, and marketing. Although someone unfamiliar with the subject may find this variety of approaches overwhelming, some authors have proposed frameworks that can serve as good starting points for those who want to get started in this field. In this regard, Azzi et al. proposed a general framework useful for conceptualizing the different functions of a package, as well as for suggesting future lines of research to be investigated further [1]. This framework divides the functions of a package into five major categories: safety, sustainability, marketing, logistics, and ergonomics. Although it was developed at the beginning of the great explosion of literature dedicated to the subject of packaging from the field of consumer research, it provides an interesting general framework with which to get started on the subject. Over the years, other authors have developed different frameworks aimed at addressing specific needs. Grönman et al. proposed a framework for designing sustainable packaging [3], while Velasco and Spence developed a framework to analyze and design packaging from a multisensory approach [4]. However, a useful approach from which to begin to delve into this topic when designing a new packaging concept within the field of consumer research is by following the different stages of the packaging life cycle [5, 6]. Thus, the objective of this section is to provide a brief overview of the key insights from packaging research following the main stages of the life cycle of a package from the consumer's point of view: attention capture, categorization, evaluation/purchase, use/consumption, and storage/disposal.

2.1 Attention Capture

The first interaction that occurs between a package and the consumer is usually at the point of sale. In this context, the packaging must capture the consumer's attention in order to succeed in the marketplace: some findings suggest that the main reason why new product launches fail is because they are not able to attract the consumer's attention and go, therefore, unnoticed [7]. This notion, commonly summarized with the truism *unseen is unsold* [8], is highlighted by figures such as two-thirds of new products being ignored, and consumers only examining an average of 1.2 brands when shopping for new products [9]. Food shopping is a low-involvement process in which little time and cognitive resources are devoted to examining products and deciding which to choose, so grabbing the consumers' attention is a prerequisite for the consumer to perceive and process the information the packaging has to convey [7].

Visual attention at the point of sale mainly depends on the consumers' motivation and the salience of the packaging [7]. Consumer motivation refers to the involvement of the consumer in the shopping process itself: for example, a person with a high level of health consciousness will be more likely to be motivated (i.e., will be more prone to pay attention to the characteristics of the products) than one with a low level of health consciousness (who will spend less time examining and deciding which product to buy). In turn, salience is defined as the ability of a stimuli (i.e., the package) to be prominent in a given context. While motivation is a characteristic of the consumer, salience is a characteristic of the package that can be enhanced: increasing the salience of the package through design will increase visual attention regardless of consumer motivation.

Research in this field often relies on eye-tracking studies, where the attention captured by a stimulus is measured by analyzing the eye movements of an observer [7] (*see* also Chapter 19). Regarding the effect of specific packaging cues on attention, findings show that packages depicting visual elements (such as images or texts) grab more attention that plain packages [10], and that images raise higher levels of attention than texts [11]. In addition, results demonstrate that the arrangement of those cues may influence the packaging salience. For example, it has been suggested that in order to catch consumers' attention the image and the logo should be placed in the bottom-left position of the packaging [12]. Moreover, research shows that consumers tend to follow two patterns when visually exploring a package. The first is conditioned by the size of the visual cues, where larger cues tend to attract more attention than smaller ones. The second reflects the consumer's most familiar reading pattern, where there seems to be a tendency to navigate the package following the same path as the one used for reading [13]. However, it is worth noting that packaging salience seems to be ultimately a matter of contrast [7]. In order to make your own package prominent, it is key to know the appearance of the other packaging that will surround it at the point of sale. A package that is visually different from those around it in the supermarket shelf will attract attention more easily than a package that blends in with its context. This contrast can be achieved by the size of the package, its predominant color, the logo, or even packaging imagery: recent research shows that subtle changes such as the orientation of the image displayed on the label can increase its salience and more easily attract the consumer's visual attention [14].

2.2 Categorization

Once the packaging has captured the consumer's attention, the next step is to facilitate categorization (*see* also Chapter 9). Take for instance a yogurt package that is successful in capturing the consumer's attention but fails to easily communicate the category to which it belongs (e.g., sweetened yogurt, skimmed yogurt, flavored yogurt). Since consumers do not devote much cognitive effort to process the information conveyed by the packaging of low-involvement products such as food and beverages, the product will probably be discarded or, even worse, will mislead the consumer and set incorrect expectations [15] (Fig. 1). Thus, correct categorization of new products by the consumer is key to their acceptance and success in the market [16].

Literature shows that consumers categorize a product by relying mainly on three factors: its location at the point of sale, the shape/silhouette of its package, and the information conveyed on

Fig. 1 A bag of nuts depicting an icon of fire on its front may be difficult to categorize correctly, since the ambiguity inherent to the fire icon in this context may lead the consumer to interpret either that the nuts are spicy or that the nuts have been roasted (cf. Gil-Pérez et al. [17]; image source: www.freepik.es)

its label. The context in which a product is located at the point of sale provides the consumer with hints about the product category to which it belongs, especially in cases where it is surrounded by other products (e.g., a supermarket). Thus, the consumer assumes that products are grouped according to common characteristics [18]: if an unfamiliar product is placed in a refrigerated area of the supermarket next to yogurts and other dairy products, the consumer will most likely assume that this new product is a dairy product similar to those around it. A similar process of analogy takes place with regard to the physical appearance of the package. At first glance, consumers rely on the silhouette/shape of the package to identify the category to which it belongs and, if they do not recognize it, they categorize by analogy with other products of similar appearance: for example, in the case of beverages, consumers categorize products based on the width of the cap and the silhouette of the bottle [19]. From a design point of view, this is interesting since it implies that unusual packaging can hinder the categorization process in the case of new products [20].

In addition, the consumer relies on the label to categorize the product. In cases where the other extrinsic cues have led the consumer to think that the product belongs to the desired category (i.e., location, silhouette), the consumer infers the category/subcategory to which the product belongs based on the information on the label. Literature shows that this is not only based on the textual elements that appear on the label, but also on other more subtle visual elements, such as color [21, 22] or imagery [17]. However, it should be emphasized that the label is only examined in cases where the consumer is motivated, there has been a prior match, or the label has caught their attention.

2.3 Evaluation/ Purchase

In addition to capturing the consumer's attention and communicating the product category to which it belongs, the packaging must also persuade the consumer that it is the best choice among all those also available at the point of sale. The literature shows that the packaging influences the evaluation of the characteristics of the product it contains, since the packaging has an influence on the generation of expectations (sensory, non-sensory, and hedonic) and, therefore, on willingness to buy [2, 23]. This effect can be explained in terms of semiotics [24, 25], heuristics [26, 27], or crossmodal correspondences [22, 28], to name just a few.

Although some authors have studied the way in which this process takes place from a holistic point of view (i.e., analyzing the message inferred by the consumer from a global point of view, considering the package as a whole [29]), the most common approach is to separately analyze the role played by the various packaging elements in the communication process. Empirical evidence shows that the consumer infers meaning from each of the packaging elements, which implies that any seemingly trivial design decision can affect the consumer's evaluation of the product. A common approach within this framework (known as the analytical approach) is to group the various elements of packaging into three groups: informational, graphical, and structural [30].

Informative elements are those which, of their own nature, are intended to convey information to the consumer in an explicit way (i.e., texts, seals, or nutrition labels) [31–33]. In a given food package, it is common to find one or more of the following linguistic signs: a description of the product/trade name, which may also include information about the subcategory to which it belongs; claims about its sensory or hedonic characteristics, its functional benefits, or how it has been processed; as well as nutritional information or indications about its origin. Evidence shows that messages conveyed through packaging have higher credibility than those conveyed through advertising [34], and that they might influence consumer expectations and response [35–37].

Graphic and visual elements are those graphical elements that usually make up the label of a product, such as colors, typefaces, or images (for a recent review, *see* [38]). Of these, color is the packaging visual cue that has the greatest influence on consumer expectations and response. Color occupies most of the surface of a package, making it very prominent and serving as an easy diagnostic cue on which the consumer relies to infer product characteristics. The effects of color have a lasting effect on consumers, and also require very little cognitive effort to process [39]. Color has been shown to influence taste expectations [40, 41], healthfulness expectations [42], or food choice [38, 43], to name just a few. However, it is worth emphasizing that while some effects of color on consumer evaluation appear to remain unchanged over time (such as those produced by crossmodal correspondences between basic flavors and some color hues [44]), some of its effects have been demonstrated to be cross-cultural [45].

Images are another graphic element frequently used in food packaging design. Although it is common to use packaging imagery for relatively explicit communicative purposes (think, e.g., of a yogurt package showing the image of a strawberry on its label), their interpretation by the consumer is far from unequivocal: in this example, that image can be interpreted with different meanings (that the yogurt tastes like strawberry, that it contains pieces of strawberry, and so on). Therefore, several studies have been carried out in recent years to try to understand the mechanisms by which consumers interpret the images depicted on a package, finding that aspects such as the congruence between the possible meanings of the image and the possible attributes of the product play a fundamental role in this process (e.g., showing a fire icon on a hot sauce favors the interpretation of that icon as spicy) [17, 46]. In addition, some researchers have investigated the effects of manipulating aspects of an image such as its pictorial style (i.e., whether it is a drawing or a photograph), its rhetorical style (i.e., whether its meaning is supposed to be literal or metaphorical), its location within the package or its orientation/viewing angle, showing that all of them might influence consumer expectations and response [47].

Although not as much as color and images, the effects of the typeface used on packaging have also attracted a great deal of scientific interest. Typefaces are the third of the most important graphic elements of packaging, since all packaging has some text written on its surface. Thus, any packaging designer has to face the dilemma of choosing one of the thousands of existing typefaces to represent those texts. As in the cases of color and images, the findings in this field show that such a decision is not trivial: the graphic characteristics of the typeface, such as its shape (e.g., curved or straight [48]), its width (i.e., light or bold [49]), or its exotism (e.g., its connotations to other cultures [50]) may influence consumer expectations and response.

Other graphical elements that have also been studied include the location of the elements [51], its size [52], orientation [53], symmetry [54], verticality [55], complexity [56], or the camera angle used in imagery [57]. Another relevant question that has recently begun to be explored is how the effects of the various graphic elements of a package interact with each other when applied together [58].

Finally, structural cues are those that have a mainly functional role, such as the size/shape of the package, or its texture/material. The literature shows that even these elements, whose communicative role is often overlooked, are still used by consumers to infer implicit meaning. Thus, the shape of the package can elicit specific brand attributes such as "powerful" or "gentle" [59], and can be used to convey healthfulness [60]. Moreover, the texture and material of the packaging can convey naturalness and sustainability. Products packaged in materials perceived as natural and sustainable are also perceived in this way [61] (although it is worth noting that assessing the sustainability of a product based just on packaging material may not be very reliable [62]).

2.4 Usage/Consumption

Once the product has been purchased, the packaging must facilitate its transport, handling, and opening. While disciplines such as ergonomics have studied how to ensure that the consumer can grasp, control, and hold the package without difficulty and without fear of injury [5], consumer research has mainly focused on investigating how packaging features and design influence the moments of unboxing and consumption. Until now, research devoted to the study of unboxing in the field of consumer research is surprisingly scarce and has mainly focused on the study of how unboxing features can affect consumer perception from a sensory or emotional point of view. Specifically, from a sensory point of view, research has investigated how the sounds produced by the package or its closure system at the time of opening can influence hedonic and sensory expectations of the product [63, 64]. In addition, other studies have analyzed the effect that the design of the unboxing process itself can have on the emotional experience of the consumer, showing that a well-designed unboxing process can enhance the perception of the quality of the product [65] or convey the emotions intended by the design team [66] (Fig. 2).

On the other hand, some authors have pondered on whether the effects produced by packaging on expectations also occur during the moment of product consumption. Given that up to one-third of products are consumed directly from their packaging (such as refrigerated beverages, snack bags, or dipping sauces), it is a relevant research question. The findings show that the effects of packaging at the time of consumption on consumer perceptions are modest compared to those on expectations, but in some cases, not negligible. Thus, it has been shown that the image depicted on the package [67], packaging shape [68], or even how the packaging sounds [69] can alter the way in which the consumer perceives the sensory and non-sensory attributes of the product.

2.5 Storage/Disposal

Once the entire contents of the packaging have been consumed, the consumer must make a decision on what to do with it. The large percentage of single-use packaging available on the market makes disposal one of the most common options (in Europe, each citizen

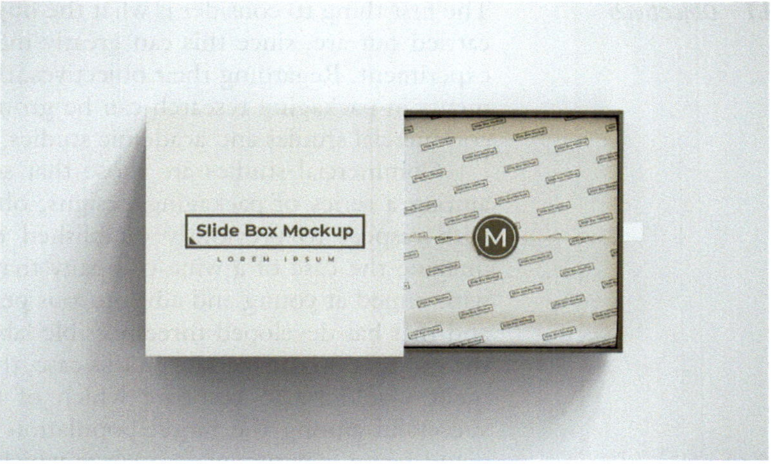

Fig. 2 Recent research has begun to investigate the influence of packaging characteristics during the unboxing process. (Image source: www.freepik.es)

discards on average 174 kg of packaging items per year [70]). The environmental impact of such a decision is leading to a growing number of initiatives that seek to improve the sustainability of packaging by increasing the number of consumption cycles of the same package (e.g., through returnable and refillable packaging services), and by trying to encourage the proper recycling of packaging once the consumer decides to dispose of it. Although this is still an incipient field of research, some researchers have given thought to factors in the acceptance of new materials [71] or reusable food packaging, showing that although consumer attitudes toward returnable packaging systems are promising, aspects such as perceived contamination (due to handling of the packaging by others) or wear and tear produced by continued use (such as dents or scratches) are limitations that could hinder their acceptance as a mass alternative to disposable packaging [72]. The literature also shows that aspects such as the perceived value of the packaging [73], the perceived difficulty of separating materials [73], or even the location of the label within the package [74] can influence the consumer's willingness to recycle packaging correctly or extend its life cycles.

3 Conducting Packaging Research

After having reviewed the existing literature and identified the gaps and insights in the specific topic to be addressed, researchers must prepare an experimental design that allows them to answer the questions they have posed. The following is a brief guide that addresses the most important issues to be defined in order to carry out a packaging experiment.

3.1 Objectives

The first thing to consider is what the objectives of the study to be carried out are, since this can greatly influence the design of the experiment. Regarding their objectives, the most common experiments in packaging research can be grouped into two categories: commercial studies and academic studies.[1]

Commercial studies are those that seek to determine which, among a series of packaging designs, obtains a better evaluation with respect to previously established objectives. For example, imagine the case of a wine company that wants to market a new wine aimed at young and adventurous people with a high income, and that has developed three possible labels with which to launch the product on the market. In this case, the objective of the experiment would be to compare which of the three labels is most successful among the target population cohort. Other examples could be an experiment to analyze which of a series of packaging products attracts more attention at the point of sale, one that investigates which generates higher expectations of sustainability, or one that studies which one best conveys a certain message. These types of experiments allow evidence-based business decisions to be made, and are therefore the most frequently used in the industry. While this type of experiment allows for a realistic setting adjusted to the real context, the specificity of each case study makes it difficult to extrapolate the conclusions obtained to other cases.

Academic studies seek to provide generalizable findings by isolating and manipulating one (or more) packaging variable(s) to see what effects they have on consumer expectations, perception, or response. In contrast to commercial studies, it is not usual to use real packaging, but rather to work with fake packaging that allows the study variables to be easily isolated and thus test the initial hypotheses. This is the most frequent case in academia, and the one in which most of the scientific papers published in this field are circumscribed. Depending on the interests of the research team and the starting point of the experiment, this category can be divided into two subgroups: design-oriented research and response-oriented research. Design-oriented research is that which focuses on the effects of manipulating a particular packaging cue on a given variable (e.g., how packaging color affects taste expectations). Conversely, response-oriented research is that which focuses on a particular outcome, and analyzes how it can be enhanced or reduced through the manipulation of packaging design (e.g., could healthfulness expectations be enhanced through packaging imagery?). Although the findings may be identical in both cases, the starting point may influence the experimental design and the findings of the study.

[1] Please note that this and subsequent classifications are an over-simplification of the various options that exist, and are intended only as a basic guide for those seeking to take their first steps in this field.

3.2 Participants

The type of participants to recruit depends primarily on the objectives of the study. In commercial studies, participants recruited should ideally represent the consumers targeted by the product, so that the response of the study participants can be used to anticipate consumer response in the market. Academic studies usually have fewer restrictions, although the priority is to ensure that the group of participants is sufficiently heterogeneous to represent the average consumer. However, if the objectives of the study include analyzing the specific response of a particular cohort of consumers, the participants should be selected from within that cohort. In any case, in order to eliminate possible biases in their responses, it is critical to ensure that participants are unaware of the purpose of the study during the course of the experiment.

3.2.1 Experimental Design

Regarding the design of the experiment, two types of experiments can be differentiated: those that follow an experimental design within-subjects, and those that follow an experimental design between-subjects. In an experiment following a within-subjects experimental design, a single group of participants sees all the stimuli to be evaluated: at the end of the experiment, each participant has seen each stimulus. In contrast, in an experiment following a between-subjects experimental design, participants are divided into as many groups as deemed appropriate (usually as many groups as there are stimuli), and each stimulus is seen by only one group of participants (who are unaware that there are other stimuli different from the one they see). In this case, it is key to ensure that all groups of participants are homogeneous in those characteristics that are considered relevant to the research, so that any effects found can be attributed to the effect of the design manipulation and not to differences between groups. For example, if one of the groups is composed of older participants than another, this may pose a problem when interpreting the results, since it will not be possible to establish whether the differences found are due to the manipulation of the package design or to differences in the characteristics of the participants in each group.

Thus, in a within-subjects experiment, participants can, consciously or subconsciously, compare the stimuli with each other, so it is easier to find differences between stimuli than in a between-subjects experiment with a smaller number of participants. Contrastingly, a between-subjects experiment more faithfully represents a real situation in which the consumer observes only one version of the same package but involves greater complexity in its execution (more resources, time, and participants). Both types of experiments are frequently used in this field, and the researcher must decide the most appropriate experimental design depending on the objectives, available resources, and type of data analysis to be performed.

3.2.2 Measures

It is always advisable to collect data that enable a basic categorization of the participants (such as gender or age). Such data allow the researcher to compare group characteristics in a between-subjects design (to ensure that the groups are homogeneous with respect to each other), to analyze the results of the experiment by subgroups of participants, or to identify the limitations of the study (as in cases where the participants are mainly university students). In addition, the literature shows that individual consumer characteristics can modulate the effects of packaging, so the information to be collected from each participant will vary depending on the specific objectives of the study (e.g., it may be worthwhile to ask about individual characteristics such as weight, hunger level, or income level if these are individual characteristics to be addressed in the research). In these cases, it is useful to have a control group with which to make comparisons (e.g., a study group of overweight consumers and a control group of non-overweight consumers).

3.2.3 Sample Size

The number of participants to recruit (i.e., the experiment sample size) depends mainly on the method to be used. A very low number of participants will not be representative of the population as a whole, and will be more likely to over-represent outliers or anomalies. However, too high a number may add unnecessary complexity and consume more resources (time, money) than necessary. Although each case is unique and should be analyzed on a case-by-case basis, for reference it is useful to know that survey experiments usually have more than 100 participants, tasting experiments tend to have between 50 and 100 participants per group, and speeded classification tasks typically have between 20 and 40 participants per group.

3.2.4 Participant Recruitment

There are many ways to recruit participants. Although most of the studies published in the literature use voluntary participants, it is not uncommon to offer some kind of incentive to encourage participation (e.g., discount vouchers, movie tickets, or snacks). The type of incentive will vary depending on the resources available to the investigators and the time or effort required to carry out the test. The researcher should be aware that some online services are available to facilitate the task of recruiting participants, such as Mechanical Turk or Prolific, and that some online survey tools (such as Qualtrics or SurveyMonkey) also offer similar services. For a fee, these platforms obtain participants who match the individual characteristics established in advance by the researcher (such as age, gender, nationality), facilitating access to specific consumer cohorts.

3.3 Stimuli

In the context of packaging research, the stimuli are the packages that will be shown to the participants and that are the object of the study. In general, the stimuli have to faithfully represent the packaging to be tested, and depending on the type of experiment, they can be digital images (such as drawings, e.g., [75]; or photo-realistic images, e.g., [76]; mockups of the real package, e.g., [67]; or 3D printed models, e.g., [77]). Also, the stimulus can be the whole package or just one of its parts (such as the brand, the label, or the opening system).

The characteristics of the stimuli will vary depending on the type of experiment and its objectives. Thus, in commercial studies it is common to compare several packaging alternatives in order to decide on the most appropriate one to be marketed, so it is ideal to use as stimuli the designs as they would be marketed. In academic studies, it is common to find two types of stimuli. On the one hand, the stimuli of exploratory studies in which the aim is to analyze, for the first time, a certain effect. In these cases, it is usual to prepare the stimulus in such a way that the element to be studied has great prominence with respect to the rest of the elements of the package (by visually highlighting it or by eliminating other elements of the package unrelated to the interest of the research). Although this type of stimulus may lose in realism, it allows the effect of the cue in question to be analyzed. On the other hand, some studies go a step further and prepare more realistic stimuli in which the studied cue is more integrated within the package as a whole. These cases seek to test whether the effects previously found are maintained in a more realistic setting.

3.4 Setting Types and Ecological Validity

There are three main types of settings in which to conduct a packaging design experiment: online, in-person, and virtual. The choice of one or the other will depend on the specific objectives of the experiment and the resources available to the research team.

The online setting is one in which participants conduct the study via the Internet, generally without being in the same physical space as the researchers. The literature shows that the findings are as valid as those obtained in an in-person setting [78], and is currently widely used due to the advantages it offers (e.g., ability to perform the experiment simultaneously in several countries or continents, or on hundreds of people in a very short time). Its main limitations are its low ecological validity (since the environment and conditions in which the test is performed differ greatly from the real consumer environment) and that it is not suitable for certain types of experiments (such as product tastings or neuromarketing experiments).

An in-person setting is one in which the participants perform the test in the same physical space as the researchers. In this case, two settings can be distinguished: the laboratory and the real environment. While the laboratory offers more controlled conditions as well as easier preparation and access for the researcher, experiments in a real environment have greater ecological validity (e.g., [68]).

Finally, in recent years, some studies have started to be carried out in virtual environments in order to combine the advantages of laboratory settings and real environment settings (e.g., [79]). The creation of a realistic virtual environment aims to provide the participant with an immersive experience with greater ecological validity than a laboratory experiment, but the fact that it is carried out in a laboratory makes it easier for the research team to control the conditions and preparation of the experiment. However, the use of a virtual setting requires hardware and software resources that need to be acquired and which are not often available (such as specific software packages, or virtual reality glasses).

3.5 Most Common Methods: How and When to Use Each One

The interdisciplinary nature of the field of consumer research has favored the multiplication of methods applied to packaging research in recent years. Since the literature offers more exhaustive reviews in this respect [80–82], here follows a brief summary that aims to help the researcher to understand the three main method categories (self-report measures, implicit measures, and physiological and neuroscientific methods) and the opportunities and limitations offered by each one of them. Data analysis varies greatly between methods, so it is recommended to study the specifics of each method in detail before putting it into practice. In addition, it is worth emphasizing that each method offers only partial answers, so if time and resources permit, it is recommended to use combinations of them in order to gain deeper insights.

3.5.1 Self-Report Measures

Self-report measures usually include all research techniques in which respondents are asked to directly report their own thoughts, beliefs, attitudes, or intentions toward the shown stimuli. Within this broad category are classic and widely used methods such as self-report questionnaires, interviews, or focus groups, but also more specific methods from specific domains of consumer research such as PrEmo or EsSense Profile [83].

These techniques are mostly quantitative and record the conscious opinion that the participants report having about the stimulus shown. Thus, the answers given by the participants do not necessarily faithfully represent their true thoughts, since, consciously or subconsciously, they may incur in biases that lower the reliability of their answers (e.g., experimenter bias or social desirability) or may not be able to access their own thoughts (introspective limits). This is a limitation of these type of methods that implicit measures or neuroscience methods try to solve. However, the literature shows that despite these limitations, the insights offered by these methods are for the most part sufficiently reliable to justify their use.

These techniques generally require little preparation and few resources to execute, and their analysis is relatively simple for any researcher with statistical analysis skills. It may be interesting to use any one of these techniques when you want to analyze consumer perception and response in an agile and simple way, and their own limitations do not pose an obstacle to the research objectives.

3.5.2 Implicit Measures

Implicit measure methods include all those techniques that seek to uncover the beliefs, attitudes, or thoughts of respondents without asking them directly, in order to access deeper answers than those achieved through self-report measure methods (*see* also Chapter 10). Within the implicit measure methods, the two sets of techniques most used in the field of packaging research are projective techniques and the speeded classification tasks.

Projective techniques are a group of qualitative techniques, usually not very structured, that allow respondents to project the opinion they have about a certain stimulus on third persons, objects, or situations (cf. [84]). Thus, by not feeling that they are expressing their own opinion, the aim is for participants to give more open and sincere answers. One of the most commonly used techniques in this category is word association, where a participant is asked to say the first thing that comes to mind when presented with a certain stimulus (e.g., [11, 76]). Such methods require little preparation and resources, but their interpretation can be challenging. They can be used as an interesting complement to more traditional quantitative measurement techniques.

Conversely, speeded classification tasks are a group of quantitative techniques in which the participant is asked to associate, as quickly as possible, and according to instructions given beforehand, a series of stimuli that appear on the screen. The common theoretical basis of this type of techniques assumes that the participant will more quickly group the stimuli that in their mind are closely related to each other, while they will be slower in grouping stimuli that they consider unrelated to each other. For example, as per the theory, it will be easier for any participant to associate the word "spider" with concepts such as "fear" or "disgust" than with concepts such as "good" or "attractive." For this reason, this type of technique commonly requires the use of a computer and the response times given by the participant are recorded and analyzed. In this category, the most commonly used technique is the implicit association test (IAT) or any of its variants (cf. [85]; application examples, [17, 22]). This type of method requires more preparation and resources than self-report measures, as well as more time for execution (each test takes longer and it is difficult to perform massive tests simultaneously). In addition, it should be noted that these methods only provide existent associations in the mind of the respondents, but do not provide the reasons for which these associations exist. This is an important limitation that should be addressed when interpreting their results, and the reason for which they are commonly used together with more traditional (and explicit) methods.

3.5.3 Physiological and Neuroscientific Methods

The physiological and neuroscientific methods include those techniques that seek to understand the beliefs, attitudes, or intentions by directly measuring the participants' autonomic nervous system (ANS) responses or the participants' brain activity when exposed to a given stimulus (cf. [80, 86]; application examples, [87, 88]). This set of methods, popularly encompassed under the term neuromarketing, includes heterogeneous techniques of which some are well established while others are subject to ongoing discussion. The most widely used and fruitful technique in the field of packaging research so far has been eye tracking [7] (*see* also Chapter 19), mainly applied to attention capture and gaze measuring studies.

To date, the application of other techniques such as facial electromyography (fEMG), electrodermal activity (EDA), electroencephalography (EEG), or functional magnetic resonance imaging (fMRI) is far from being widespread, as they have three important limitations that have prevented their consolidation in packaging research. Firstly, they are generally invasive techniques that involve equipping each participant with hardware and sensors, thus reducing the ecological validity of the setting. Secondly, they require a large investment of time in the preparation, execution, and analysis stages, as well as financial and material resources that are a barrier for many researchers. The necessary hardware equipment is usually expensive and bulky, and requires extensive technical expertise to operate. In addition, the analysis of the data obtained is very complex and requires specific skills. This means that the number of participants in this type of experiment is usually very small, which makes it difficult to obtain solid conclusions. Finally, the interpretation of the data obtained is not straightforward and there is an ongoing debate about what is really being measured with these techniques, so their insights should be applied with caution.

4 Conclusion

Due to a lack of time and familiarity with the scientific realm, most packaging designers are usually guided by their instinct and experience rather than by insights from the consumer research literature. However, any designer would benefit from having basic scientific skills that would allow them to keep abreast of the latest research findings related to the field and the ability to conduct their own experiments. Research conducted in recent years underlines the importance of packaging during all stages of consumption, and designing with an evidence-based approach can lead to the creation and marketing of more fit-for-purpose packaging, as well as to greater consumer satisfaction.

References

1. Azzi A, Battini D, Persona A et al (2012) Packaging design: general framework and research agenda. Packag Technol Sci 25:435–456
2. Piqueras-Fiszman B, Spence C (2015) Sensory expectations based on product-extrinsic food cues: an interdisciplinary review of the empirical evidence and theoretical accounts. Food Qual Prefer 40:165–179
3. Grönman K, Soukka R, Järvi-Kääriäinen T et al (2013) Framework for sustainable food packaging design. Packag Technol Sci 26:187–200
4. Velasco C, Spence C (2019) The multisensory analysis of product packaging framework. In: Multisensory packaging. Springer, Cham, pp 191–223
5. Mumani A, Stone R (2018) State of the art of user packaging interaction (UPI). Packag Technol Sci 31:401–419
6. Salgado-Montejo A, Velasco C, Ariza L et al (2017) The four moments of experience streamlining the process of packaging development. ESOMAR 70:173–186
7. Duerrschmid K, Danner L (2018) Eye tracking in consumer research. In: Ares G, Varela P (eds) Methods in consumer research, Alternative approaches and special applications, vol 2. Elsevier, Cambridge, MA, pp 279–318
8. Wästlund E, Shams P, Otterbring T (2018) Unsold is unseen ... or is it? Examining the role of peripheral vision in the consumer choice process using eye-tracking methodology. Appetite 120:49–56
9. Burke RR, Leykin A (2014) Identifying the drivers of shopper attention, engagement, and purchase. Rev Mark Res 11:147–187
10. García-Madariaga J, Blasco López M-F, Burgos IM et al (2019) Do isolated packaging variables influence consumers' attention and preferences? Physiol Behav 200:96–103
11. Piqueras-Fiszman B, Velasco C, Salgado-Montejo A et al (2013) Using combined eye tracking and word association in order to assess novel packaging solutions: a case study involving jam jars. Food Qual Prefer 28:328–338
12. Leal Barbosa AA, de Moura JA, de Medeiros DD (2020) Positioning of design elements on the packaging of frozen convenience food and consumers' levels of attention: an experiment using pizza boxes. Food Qual Prefer 87: 104044
13. Rebollar R, Lidón I, Martín J et al (2015) The identification of viewing patterns of chocolate snack packages using eye-tracking techniques. Food Qual Prefer 39:251–258
14. Lacoste-Badie S, Yu J, Droulers O (2020) Small change, big change –increasing attention with product package variations. Food Qual Prefer 86:104007
15. Basso F, Robert-Demontrond P, Hayek M et al (2014) Why people drink shampoo? Food imitating products are fooling brains and endangering consumers for marketing purposes. PLoS One 9:e100368
16. Ackermann C-LL, Teichert T, Truong Y (2018) 'So, what is it? And do I like it?' New product categorisation and the formation of consumer implicit attitude. J Mark Manag 34: 796–818
17. Gil-Pérez I, Rebollar R, Lidón I et al (2019) Hot or not? Conveying sensory information on food packaging through the spiciness-shape correspondence. Food Qual Prefer 71:197–208
18. Gregan-Paxton J, John DR (1997) Consumer learning by analogy: a model of internal knowledge transfer. J Consum Res 24:266–284
19. Arboleda AM, Arce-Lopera C (2015) Quantitative analysis of product categorization in soft drinks using bottle silhouettes. Food Qual Prefer 45:1–10
20. Warren C, Reimann M (2019) Crazy-funny-cool theory: divergent reactions to unusual product designs. J Assoc Consum Res 4:409–421
21. Spence C, Velasco C (2019) Packaging colour and its multiple roles. In: Multisensory packaging. Springer, Cham, pp 21–48
22. Piqueras-Fiszman B, Spence C (2011) Crossmodal correspondences in product packaging. Assessing color-flavor correspondences for potato chips (crisps). Appetite 57:753–757
23. Schifferstein HNJ, Fenko A, Desmet PMA et al (2013) Influence of package design on the dynamics of multisensory and emotional food experience. Food Qual Prefer 27:18–25
24. Celhay F, Remaud H (2018) What does your wine label mean to consumers? A semiotic investigation of Bordeaux wine visual codes. Food Qual Prefer 65:129–145
25. Smith V, Barratt D, Selsøe Sørensen H (2015) Do natural pictures mean natural tastes? Assessing visual semantics experimentally. Cogn Semiot 8:53–86
26. Szocs C, Lefebvre S (2017) Spread or stacked? Vertical versus horizontal food presentation, portion size perceptions, and consumption. J Bus Res 75:249–257

27. Madzharov AV, Block LG (2010) Effects of product unit image on consumption of snack foods. J Consum Psychol 20:398–409
28. Velasco C, Woods AT, Petit O et al (2016) Crossmodal correspondences between taste and shape, and their implications for product packaging: a review. Food Qual Prefer 52:17–26
29. Orth UR, Malkewitz K (2008) Holistic package design and consumer brand impressions. J Mark 72:64–81
30. Magnier L, Crié D (2015) Communicating packaging eco-friendliness. An exploration of consumers' perceptions of eco-designed packaging. Int J Retail Distrib Manag 43:350–366
31. Temple NJ (2019) Front-of-package food labels: a narrative review. Appetite 144:104485
32. Skaczkowski G, Durkin S, Kashima Y et al (2016) The effect of packaging, branding and labeling on the experience of unhealthy food and drink: a review. Appetite 99:219–234
33. Spence C, Piqueras-Fiszman B (2014) Tastes great, but what do we call it? The art and science of food description. In: The perfect meal. Wiley, Chichester, pp 71–107
34. Fajardo TM, Townsend C (2016) Where you say it matters: why packages are a more believable source of product claims than advertisements. J Consum Psychol 26:426–434
35. Yeomans MR, Chambers L, Blumenthal H et al (2008) The role of expectancy in sensory and hedonic evaluation: the case of smoked salmon ice-cream. Food Qual Prefer 19:565–573
36. Sütterlin B, Siegrist M (2015) Simply adding the word "fruit" makes sugar healthier: the misleading effect of symbolic information on the perceived healthiness of food. Appetite 95:252–261
37. Lähteenmäki L (2013) Claiming health in food products. Food Qual Prefer 27:196–201
38. Vermeir I, Roose G (2020) Visual design cues impacting food choice: a review and future research agenda. Foods 9:1495
39. Kauppinen-Räisänen H (2014) Strategic use of colour in brand packaging. Packag Technol Sci 27:663–676
40. Marques da Rosa V, Spence C, Miletto Tonetto L (2019) Influences of visual attributes of food packaging on consumer preference and associations with taste and healthiness. Int J Consum Stud 43:210–217
41. Rebollar R, Lidón I, Serrano A et al (2012) Influence of chewing gum packaging design on consumer expectation and willingness to buy. An analysis of functional, sensory and experience attributes. Food Qual Prefer 24:162–170
42. Plasek B, Lakner Z, Temesi Á (2020) Factors that influence the perceived healthiness of food—review. Nutrients 12:1881
43. Lemos TC, Almo A, Campagnoli RR et al (2019) A red code triggers an unintended approach motivation toward sweet ultra-processed foods: possible implications for front-of-pack labels. Food Qual Prefer 79:103784
44. Spence C, Wan X, Woods A et al (2015) On tasty colours and colourful tastes? Assessing, explaining, and utilizing crossmodal correspondences between colours and basic tastes. Flavour 4:23
45. Velasco C, Wan X, Salgado-Montejo A et al (2014) The context of colour-flavour associations in crisps packaging: a cross-cultural study comparing Chinese, Colombian, and British consumers. Food Qual Prefer 38:49–57
46. Gil-Pérez I, Rebollar R, Lidón I et al (2019) What do you mean by hot? Assessing the associations raised by the visual depiction of an image of fire on food packaging. Food Qual Prefer 71:384–394
47. Gil-Pérez I, Rebollar R, Lidón I (2020) Without words: the effects of packaging imagery on consumer perception and response. Curr Opin Food Sci 33:69–77
48. Velasco C, Woods AT, Hyndman S (2015) The taste of typeface iperception. i-Perception 6:1–11
49. Karnal N, Machiels CJA, Orth UR et al (2016) Healthy by design, but only when in focus: communicating non-verbal health cues through symbolic meaning in packaging. Food Qual Prefer 52:106–119
50. Celhay F, Boysselle J, Cohen J (2015) Food packages and communication through typeface design: the exoticism of exotypes. Food Qual Prefer 39:167–175
51. Togawa T, Park J, Ishii H et al (2019) A packaging visual-gustatory correspondence effect: using visual packaging design to influence flavor perception and healthy eating decisions. J Retail 95:204–218
52. Neyens E, Aerts G, Smits T (2015) The impact of image-size manipulation and sugar content on children's cereal consumption. Appetite 95:152–157
53. Velasco C, Woods AT, Spence C (2015) Evaluating the orientation of design elements in product packaging using an online orientation task. Food Qual Prefer 46:151–159
54. Velasco C, Pathak A, Woods AT et al (2019) The relation between symmetry in food packaging and approach and avoidance words. Q J

Exp Psychol. https://doi.org/10.1177/1747021819887172

55. Bettels J, Haase J, Wiedmann K (2020) Shaping consumer perception: effects of vertical and horizontal packaging alignment. J Consum Mark 37:423–431
56. Favier M, Celhay F, Pantin-Sohier G (2019) Is less more or a bore? Package design simplicity and brand perception: an application to champagne. J Retail Consum Serv 46:11–20
57. Wang F, Basso F (2021) The peak of health: the vertical representation of healthy food. Appetite 167:105587
58. Matthews P, Simmonds G, Spence C (2019) Establishing boundary conditions for multiple design elements congruent with taste expectations. Food Qual Prefer 78:103742
59. Parise CV, Spence C (2012) Assessing the associations between brand packaging and brand attributes using an indirect performance measure. Food Qual Prefer 24:17–23
60. van Ooijen I, Fransen ML, Verlegh PWJ et al (2017) Signalling product healthiness through symbolic package cues: effects of package shape and goal congruence on consumer behaviour. Appetite 109:73–82
61. Magnier L, Schoormans J, Mugge R (2016) Judging a product by its cover: packaging sustainability and perceptions of quality in food products. Food Qual Prefer 53:132–142
62. Steenis ND, van Herpen E, van der Lans IA et al (2017) Consumer response to packaging design: the role of packaging materials and graphics in sustainability perceptions and product evaluations. J Clean Prod 162:286–298
63. Spence C, Wang Q (2015) Sensory expectations elicited by the sounds of opening the packaging and pouring a beverage. Flavour 4:35
64. Spence C, Wang Q (2017) Assessing the impact of closure type on wine ratings and mood. Beverages 3:52
65. Bae J, Self JA, Kim C (2019) Rich unboxing experiences: complexity in product packaging and its influence upon product expectations. J Des Res 17:26
66. Kim C, Self JA, Bae J (2018) Exploring the first momentary unboxing experience with aesthetic interaction. Des J 21:417–438
67. Lidón I, Rebollar R, Gil-Pérez I et al (2018) The influence the image of the product shown on food packaging labels has on product perception during tasting: effects and gender differences. Packag Technol Sci 31:689–697
68. Becker L, van TJL R, Schifferstein HNJ et al (2011) Tough package, strong taste: the influence of packaging design on taste impressions and product evaluations. Food Qual Prefer 22:17–23
69. Wang QJ, Spence C (2019) Sonic packaging: how packaging sounds influence multisensory product evaluation. In: Multisensory packaging. Springer, Cham, pp 103–125
70. Eurostat (2018) Packaging waste statistics. https://ec.europa.eu/eurostat/statistics-explained/index.php?title=Packaging_waste_statistics
71. Magnier L, Mugge R, Schoormans J (2019) Turning ocean garbage into products – consumers' evaluations of products made of recycled ocean plastic. J Clean Prod 215:84–98
72. Magnier L, Gil-Pérez I (2021) Reviving the milk man: consumers' evaluations of circular reusable packaging offers. In: 4th PLATE 2021 virtual conference
73. Langley J, Turner N, Yoxall A (2011) Attributes of packaging and influences on waste. Packag Technol Sci 24:161–175
74. Huang R, Dahl DW, Li S et al (2019) The effect of packaging perceptual cues on consumer disposal behavior of partially consumed products. J Assoc Consum Res 4:352–362
75. Velasco C, Salgado-Montejo A, Marmolejo-Ramos F et al (2014) Predictive packaging design: tasting shapes, typefaces, names, and sounds. Food Qual Prefer 34:88–95
76. Rebollar R, Lidón I, Gil-Pérez I et al (2019) How should I tell you this? The effects of the image used to convey that a natural yogurt is sweetened on consumer expectations and willingness to buy. Food Res Int 126:108721
77. Van Rompay TJL, Kramer LM, Saakes D (2018) The sweetest punch: effects of 3D-printed surface textures and graphic design on ice-cream evaluation. Food Qual Prefer 68:198–204
78. Woods AT, Velasco C, Levitan CA et al (2015) Conducting perception research over the internet: a tutorial review. PeerJ 3:e1058
79. van Herpen E, Immink V, van den Puttelaar J (2016) Organics unpacked: the influence of packaging on the choice for organic fruits and vegetables. Food Qual Prefer 53:90–96
80. Spinelli S, Niedziela M (2016) Emotion measurements and application to product and packaging development. In: Burguess P (ed) Integrating the packaging and product experience in food and beverages: a road-map to consumer satisfaction. Woodhead Publishing, Cambridge, UK, pp 77–119
81. Ares G, Varela P (eds) (2018) Methods in consumer research, New approaches to classic methods, vol 1. Elsevier, Cambridge, MA

82. Ares G, Varela P (eds) (2018) Methods in consumer research, Alternative approaches and special applications, vol 2. Elsevier, Cambridge, MA
83. Spinelli S, Monteleone E (2018) Emotional responses to products. In: Ares G, Varela P (eds) Methods in consumer research, New approaches to classic methods, vol 1. Elsevier, Cambridge, MA, pp 261–296
84. Gambaro A (2018) Projective techniques to study consumer perception of food. Curr Opin Food Sci 21:46–50
85. Kraus AA, Piqueras-Fiszman B (2018) Measuring implicit associations in food-related consumer research. In: Ares G, Varela P (eds) Methods in consumer research, Alternative approaches and special applications, vol 2. Elsevier, Cambridge, MA, pp 203–230
86. Bell L, Vogt J, Willemse C et al (2018) Beyond self-report: a review of physiological and neuroscientific methods to investigate consumer behavior. Front Psychol 9:1–16
87. Moya I, García-Madariaga J, Blasco M-F (2020) What can neuromarketing tell us about food packaging? Foods 9:1856
88. Liao LX, Corsi AM, Chrysochou P et al (2015) Emotional responses towards food packaging: a joint application of self-report and physiological measures of emotion. Food Qual Prefer 42: 48–55

Chapter 21

Measuring Desirability Using The Reaction Cards Method

Charles James Negus and Heber Rodrigues

Abstract

The process of understanding consumers' reaction to a given product, brand, and experience is a good starting point for achieving success in a competitive and challenging market. Different methods can be used to collect individual's reactions to products and/or situations of consumption. This chapter illustrates a relatively *easy, and quick-to-apply* method, through the use of reaction cards. We shall begin with a brief historical retrospective of the method, then we shall explain the general procedural principles of application and data analysis. Finally, we shall illuminate this by illustrating the methodology through the use of an example in wine research.

Key words Consumer behavior, Desirability, Association tests, Wine

1 Introduction

The research on product perception, engagement, and satisfaction have flourished in the last century, being operationalized by the qualitative experience of the meaning of consuming a product. Whereby, this meaning comes from the larger context of the brand in the consumer's life [1]. When engaged with a product, consumers make specific cognitive, emotional, and behavioral investments in these products. For the purpose of gaining a higher valued experience from interacting with the product and/or the consumption of said product. If we take the comparison of the different international iterations of Coca Cola to illustrate this, Coca Cola produced in Mexico has gained a cult following in the USA, for its perceived value over Coca Cola produced in the USA. Thus, understanding consumers reaction to a given product, brand, or experience of consumption seems to be an effective thermometer to point out how engaged and satisfied consumers are with their experiences of consumption. In this way, researchers from the marketing sector historically have borrowed methods from psy-

Carlos Gómez-Corona and Heber Rodrigues (eds.), *Consumer Research Methods in Food Science*,
Methods and Protocols in Food Science, https://doi.org/10.1007/978-1-0716-3000-6_21,

chology to collect consumer reactions. A good example is the use of projective techniques in market research [2] such as the classical free word association methodology, applied to disclose consumers' brand image assessment [3], among others. Another reliable example is the use of Focus Group techniques to understand consumer interaction with consumer goods [4], or even the use of the theory of social representation in order to understand market brand positioning [5]. These methods have been recognized for a long period of time, as those that will produce good-quality data, thus presenting several advantages for their use. However, there are also disadvantages for their implementation. In the forthcoming body we shall highlight a number of difficulties on applying these methods and the complexity of interpreting the data produced when applying these methods. Thus, in this chapter we will delineate the step-by-step instructions of an *easy and quick-to-apply* visual technique—the reaction card methodology—which has attempted to address the shortcomings seen in the methods used previously.

2 The Reaction Card Method

2.1 General Principle

The reaction card method is based on a simultaneous presentation of a set of pre-established cards with words (adjectives, emotions, and descriptors) or images, where participants are asked to select those that best corresponds to their reaction to a certain task and/or question. The method was designed so that experimenters can investigate the intangible aspects of their test subjects' experiences with a product. The cards then provide a way for users to tell the story of their experience and/or subconscious thoughts, choosing the words that have meaning to them as triggers to express their feelings or attitudes (positive or negative). See Mercun and Zumer [6] for a complete review of the method. In order to provide further precision in the interpretation of the results, the method requires for participants who select more than five cards to reduce their selection to the five cards that have the most meaning to them. Then after making their selection subjects are required to provide justification for their choices. This is a useful final task, in order to provide further detail and extra insights into their experience with that particular product/experience/object.

2.2 Application of the Method

Having been developed by Microsoft experts, as part of their "desirability toolkit," the first study that approached the use of this method was in the field of UX, an aspect of IT and Software research [6–8]. UX is used to ascertain the best way in which user and computer can interact with each other [9], defined by its four dimensions: user's perceived hedonic quality (pleasure-producing product qualities), pragmatic quality (user-perceived usability), beauty (aesthetics), and goodness (overall product quality) [10].

Subsequently, the reaction card method has been used to classify engagement in video games [11], evaluating visual designs [12], and biomedical informatics [13].

This method consists of an active task. This has a number of advantages over methods that constituted a passive activity, such as questionnaires or surveys, as it prevents subjects from losing their "in the moment" spontaneity required to investigate their experience with that particular product/experience/object. On top of this, post-task questions will often ask participants to rate their level of a certain emotion or satisfaction with a product [7]. These ratings are often higher than they should be as this method risks being affected by the acquiescence bias of participants. Another aspect that risks this bias is when too many positive emotions and options like in EsSense25 [14] are used. As this method attempts to maintain a near even balance between positive and negative reactions this is avoided. Finally, questionnaires, where individuals are given a scale on certain emotions, are known to confuse participants, thus again preventing the kind of "in the moment" spontaneity required for these studies.

2.2.1 Procedure

In order to be successful in the application of this method, a number of steps needs to be cautiously and correctly performed:

Step 1: Construction of Reaction Card Sets

The original set words have been primed with language specific to software use, as a result some of the reactions cards may not be applicable to food and beverage research, or a particular segment of the food and beverage industry. In order to ensure that the words used have utility, one should perform triangulation tasks [15] on both the original list of words, and words to be added to this list. Typically, words can be taken from support literature on either representations or specific reactions to the product in question. Three experimenters each individually providing their own selection of words, while also removing words that they deem to have the same lemma and theme. Only words that all three experimenters agree upon should be used in a final experiment. A ratio of 40% positive words/reactions to 60% negative words/reactions should be maintained as much as possible.

Step 2: Validation of the Cards

If the practitioner is using a newly constructed set of cards, validation is required. It is recommended to be done if sets have been used before, however it is not required.

Validation of your constructed set should be performed using the same methodology as the rest of the experiment, with a pilot group of individuals. This will provide insights that can be implemented going into collecting the reactions proper. For example, how best to respond to common questions, without biasing the responses of the participants. Or how best to explain certain tasks or aspects of the method to participants.

Step 3: Pre-test

These tasks can create a lot of excitement within a handful of participants at the start of the participation in the research. This may result in their responses and reactions being more positive than they would normally be. To reduce this all participants should perform the same reaction card task, with the same methodology (*see* Step 4), using a similar product to allow the novelty to wear off for them.

Depending on what aspect of consumer experience, the practitioner is looking to research, instructions may be difficult for participantes to initially understand. For example: a participant rarely thinks specifically about how a certain product makes them feel, therefore asking them "how does this product make you feel," may cause some consumers to struggle, or experience cognitive overload until they are in the mindset of thinking about their experience with a product. Allowing the participants the opportunity to get this initial overload out of the way on a product that the practitioner is not interested will help to avoid this.

Step 4: Collecting Participants' Reactions

The procedure for collecting participants' reactions follows this general framework:

1. The practitioner explains the tasks and procedure, and verbally confirms that participants are comfortable with the task.
2. Participants perform the pre-test task with a similar product to ensure that they are prepared fully for the task and the novelty of it does not bias the results.
3. The cards are then shuffled between the pre-test task and the next task (this should occur whenever a new product/experience/stimulus is provided to the participants).
4. The test products as well as any other experimental materials should be placed next to the flashcards on the same side throughout the experiment.
5. Participants were asked then to select the reaction cards that best describes their experiences/reaction/feeling about the product/experience/stimulus. The instructions here should be clear; for example: "please select the cards that best describe how eating this product makes you feel."

   The practitioners should either leave the room or provide the participants with as much privacy as possible, while they choose an initial selection of cards.

6. If participants chose more than five cards, the experimenters should ask them to select the five that have the most meaning to them in their selection.
7. Participants are then asked to justify their selection of each cards, with the experimenters probing this using questions: "*why .....?*" or "*You chose .... What was your reasoning for this?*" This too should be done in isolation from any other participants to avoid bias.
8. Once points 5 to 7 have been completed by the participant with the first product, are the participants able to move on to the next product/experience/stimulus? All attempts should be made up until this point to keep any other products/experiences/stimuli hidden or secret from the participants.
9. Repeat points 5 to 7 for the next product/experience/stimulus until the task has been completed for all products/experiences/stimuli in question.
10. Should the experiment call for participants to first be ignorant (blind conditions) to certain information and then be informed of that information (informed conditions); Then on the completion of the tasks in the blind conditions, participants should be asked to leave and break for a period of time. Upon returning they should be convinced by a second experimenter playing the actor that the products/experiences/stimuli they are being shown next are part of a different experiment and thus not related to the products/experiences/stimuli before the break.
11. Finally at the end of all the reaction card tasks, if it is required for the research, participants should be asked to answer short questionnaires providing further information about their backgrounds.

Step 5: Data Analysis

Word Clouds

Word clouds allow for an initial visualization, and superficial analysis of the data by practitioners, which may help with both reporting the results and informing the use of further analyses in the research. They are images of all the words selected via the reaction cards for a certain product, with the size of each word indicating how many times that reaction card has been selected for that product. To create them, frequency tables of reaction card selections should be created for each product in question, and then imputed into a word cloud generator. The scale of each word cloud in a single study needs to be set at the same level, to allow for comparison of the results.

Correspondence Analysis—Cluster Analysis

Correspondence analysis (CA) allows for an investigation into the frequencies and associations of multiple different variables, in this case, the language selected using the reaction cards, and different food or beverages. The cards selected for the products or experimental conditions need to be tallied into a contingency table, and a CA performed on the table. A graphical representation of the CA is produced, whereby data points that are closer together are more similar than those further apart. Cluster analyses can be performed on top of this to confirm these relative similarities between the different products and the reaction cards selected for them.

Illustration of Experiment

English Sparkling Wine is recognized for being the "King" of the wines produced in the United Kingdom, a brand-new wine frontier. Nevertheless, English Sparking Wine faces a competitive internal market along its counterparts, Champagne, Prosecco, and Sekt. In this example-study, we aim to use the reaction cards methodology to verify how consumers emotionally react to English Sparkling Wine, Champagne, Prosecco, and Sekt during an experience of consumption. Fifty British wine consumers were invited to take part on this experiment in central London, UK. To ensure that the investigation targeted the legal regular wine consumers, participants were asked two pre-test questions. These being: Are you older than 18 years? Have you drunk a glass of wine in the last month? An answer of no to either of these questions meant that these individuals were not invited to participate in the tasting.

Step 1: Construction of Reaction Card Sets

Some terms were included from Urdapilleta et al. [16], a study on the social representation of wines. Finally, additional emotional words were added by including the antonyms (the words with the opposite meanings) of the emotional words used in these papers. This created a list of 259 words.

Not all of these 259 words were relevant to this study. In order to both reduce the number of words chosen and increase the relevance of the set to studying emotional responses in wine, two experimenters removed words if they deemed that the word did not fit the phrase: "*This wine makes me feel.*" For example, the phrase "*This wine makes me feel cutting-edge*" was deemed non-sensical and thus *cutting-edge* was removed. This created a list of 142 words.

These were then also sorted into negative and positive terms. These experimenters then made a joint selection to achieve the ratio of 40% positive to 60% negative. This created a list of 85 words.

Table 1
Final selection of words used

Positive words	Negative words
Agreeable	Annoying
Attractive	Bad
Calm	Boring
Clean	Confusing
Comfortable	Crude
Confident	Destructive
Desirable	Dirty
Energetic	Distracting
Essential	Frustrating
Exceptional	Inconsistent
Fun	Intimidating
Healthy	Irrelevant
Helpful	Ordinary
Innovative	Patronizing
Powerful	Pessimistic
Relaxed	Repulsive
Satisfied	Unattractive
Sophisticated	Unimpressive
Trustworthy	Unnatural
	Unsatisfactory
	Untrustworthy

Finally, to increase the precision of the set, and to reduce the number of words to a workable amount within a wine tasting, they were subsequently subjected to a further selection to reduce their number to fewer than 60 (the amount that could work practically in a wine tasting). This was achieved by triangulation [15] using two additional experimenters and one of the previous experimenters. This resulted in a final set of 40 words (Table 1).

Step 2: Validation of the Cards

Validation of the set of cards was performed using a small group of staff and students at Plumpton College. This provided the following insights:

- A limit on the number of participants to experimenters as 2:1 ensured that experimenters were not overwhelmed with the amount of administration that had to be performed alongside running the tastings.
- How best to respond to common questions, without biasing the responses of the participants. For example, if the participants were to ask what sort of wine they were consuming the response agreed upon was "Whatever sort of wine you think it is."
- The potential location for storing wine (a wine fridge) during the experiment was too out in the open, which had the potential to bias the participants. Thus, modifications and alterations were made to the fridge to make it more private.
- The type of classes were not suitable for the sparkling wine tasting and research, these were swapped for the final experiment to the ISO tasting glass.
- Cards were valid for the research goals and aims of the endeavor.

Step 3: Pre-test

A pre-test consisting of the same methodology but a different sparkling wine was performed before the operations to collect data on the English Sparkling Wine, Champagne, Prosecco, and Sekt began. Participants were asked to taste an Australian Sparkling Wine and asked to select the cards that best describe how drinking that wine made them feel.

Step 4: Collecting Participants' Reactions

The final experiment was performed twice by each participant, firstly in blind condition and then in an informed one. Each participant was given four glasses of wine, one at a time and a set of user experience reaction flashcards. Glasses were placed next to the flashcards, with spittoons, water, and water crackers set to the side of the glasses. Participants were asked then to select the reaction cards that best described how drinking the wine made them feel (Fig. 1). The experimenters left the room while participants chose an initial selection of cards. Once participants had done this, the experimenters took the participants individually to a different room in order to record their answers to the task. If participants chose more than five cards, now was the point in which the experimenters asked them to reduce their selection to five cards. Participants were also asked to justify their selection of each card, with the experimenters probing this using the aforementioned questions: "*why .....?*" or "*You chose .... What was your reasoning for this?*" Participants completed the entire task before being given the next wine.

Once they have completed the task in a blind condition, they were asked to leave the room and take a break for an hour. After they returned and at this moment, a second experimenter playing

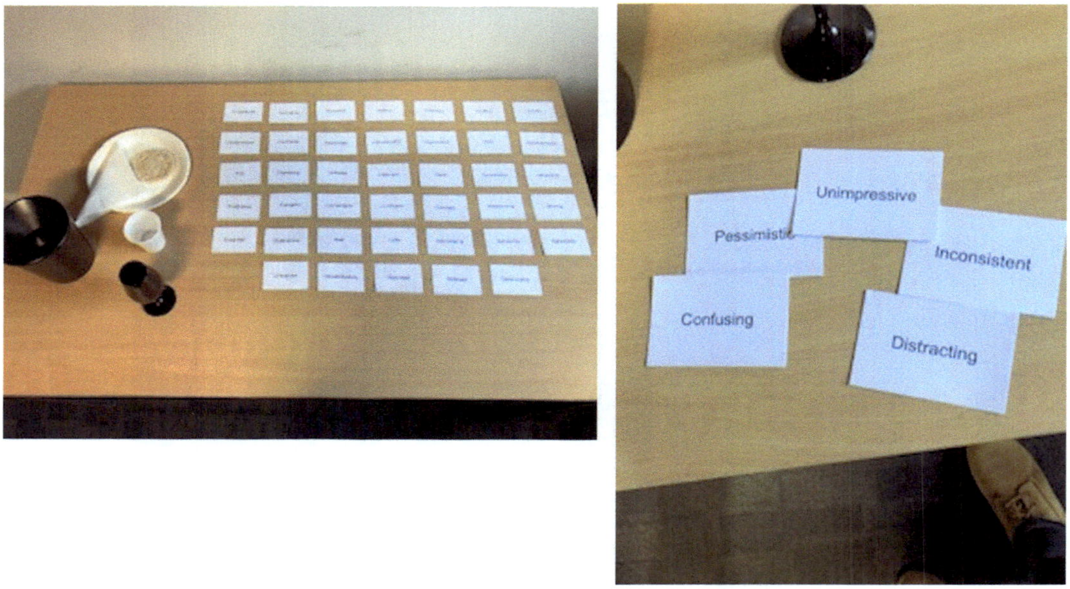

Fig. 1 Materials given to participants

the actor came in and asked the participant if they could participate in a same task for her wines, for work toward another experiment. On these occasions, this second experimenter (actor) presented the same wines to the participants using the same methodology, but now, in an informed condition: the participants knew which wines they were tasting (English Sparkling Wine, Champagne, Prosecco, and Sekt). Again, participants were asked then to select the reaction cards that best described how drinking the wine made them feel.

Having completed all the tasks, the participants were then asked to complete a questionnaire to collect data on their demographic profile.

Step 5: Data Analysis

The following word clouds were created (Figs. 2, 3, 4, 5, 6, 7, 8, and 9) as well as the following Correspondence Analysis (Fig. 10):

The following correspondence analysis was created:

2.3 General Recommendations

- Products should be presented without branding (unless this is what is being investigated) on as neutral a background as possible.
- Words should be printed on the reaction cards, in the center, in the same font, and in black.
- Reaction cards should be white and all the same size.
- Should the experiment call for a blind and an informed state, the longer the break in between the two states the better, as this

Fig. 2 Word cloud showing selections for English Sparkling Wine in the blind tasting

Fig. 3 Word cloud showing selections for English Sparkling Wine in the informed tasting

reduces the possibility that the participant will remember the product/experience/stimulus.

- Participants should be told that they should not discuss the products and/or tasks they are asked to do.
- Any additional tasks with separate methodologies (questionnaires etc.) should come after the reaction card tasks have been completed to ensure that these tasks have maximum spontaneity in their answers.

Fig. 4 Word cloud showing selections for Champagne in the blind tasting

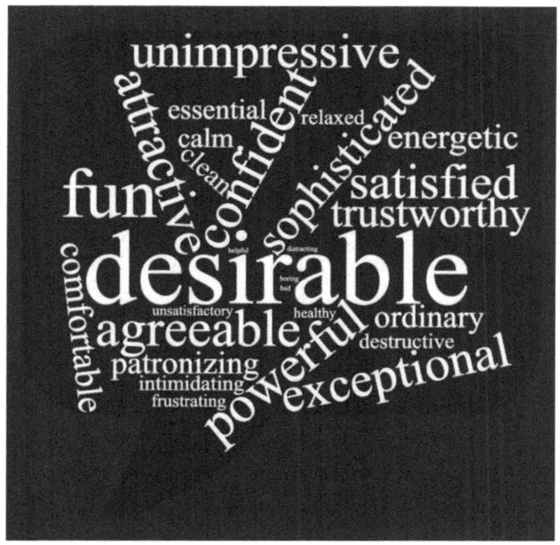

Fig. 5 Word cloud showing selections for Champagne in the informed tasting

Fig. 6 Word cloud showing selections for Prosecco in the blind tasting

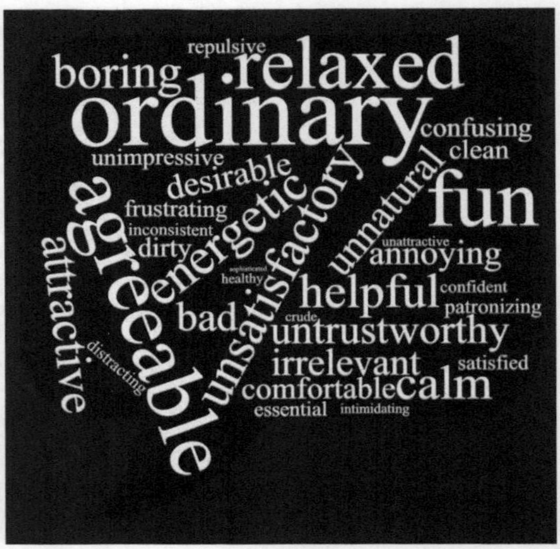

Fig. 7 Word cloud showing selections for Prosecco in the informed tasting

Fig. 8 Word cloud showing selections for Sekt in the blind tasting

Fig. 9 Word cloud showing selections for Sekt in the informed tasting

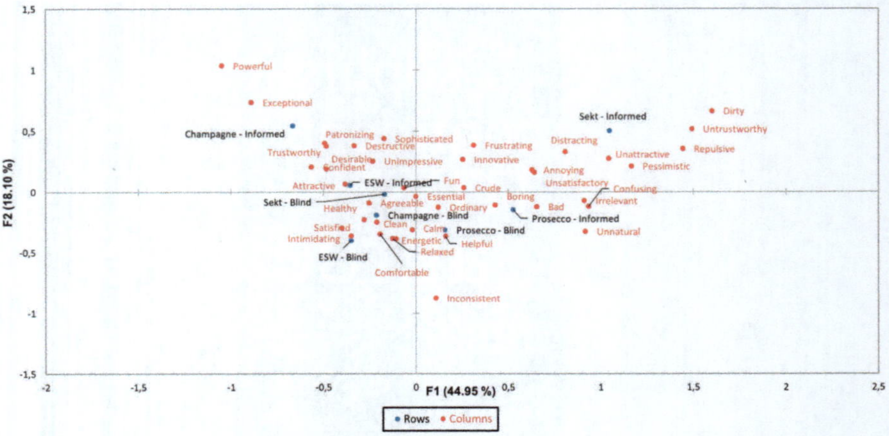

Fig. 10 Representation of emotional reaction for each wine in both the blind and informed tastings derived from the first two dimensions of the correspondence analysis performed on the frequency of participants reaction card selection

References

1. Calder BJ, Hollebeek LD, Malthouse EC (2018) Creating stronger brands through consumer experience and engagement. In: Customer engagement marketing. Palgrave Macmillan, Cham, pp 221–242
2. Zober M (1956) Some projective techniques applied to marketing research. J Mark 20(3): 262–268
3. Prayag G (2010) Brand image assessment: international visitors' perceptions of Cape Town. Mark Intell Plan 28:462
4. Cox KK, Higginbotham JB, Burton J (1976) Applications of focus group interviews in marketing. J Mark 40(1):77–80
5. Lebrun AM, Souchet L, Bouchet P (2013) Social representations and brand positioning in the sporting goods market. Eur Sport Manag Q 13(3):358–379
6. Merčun T, Žumer M (2014) Dimensions of user experience and reaction cards. In: Tuamsuk K, Jatowt A, Rasmussen E (eds) The emergence of digital libraries – research and practices, Lecture notes in computer science. Springer, Cham, pp 365–370. https://doi.org/10.1007/978-3-319-12823-8_38
7. Barnum CM, Palmer LA (2010) More than a feeling: understanding the desirability factor in user experience. In: CHI'10 extended abstracts on human factors in computing systems, CHI EA'10. Association for Computing Machinery, New York, pp 4703–4716. https://doi.org/10.1145/1753846.1754217
8. Benedek J, Miner T (2003) Measuring desirability: new methods for evaluating desirability in a usability lab setting. Proc Usability Professionals Assoc 8:57
9. Hassenzahl M (2008) User experience (UX): towards an experiential perspective on product quality. In: Proceedings of the 20th conference on l'interaction homme-machine, IHM'08. Association for Computing Machinery, New York, pp 11–15. https://doi.org/10.1145/1512714.1512717
10. Law EL-C, van Schaik P (2010) Modelling user experience – an agenda for research and practice. Interact Comput 22:313–322. https://doi.org/10.1016/j.intcom.2010.04.006
11. Schønau-Fog H, Bjørner T (2012) "Sure, I would like to continue": a method for mapping the experience of engagement in video games. Bull Sci Technol Soc 32:405–412. https://doi.org/10.1177/0270467612469068
12. De Guzman ES, Schiller J (2011) How does this look? Desirability methods for evaluating visual design. In: Stephanidis C (ed) HCI international 2011 – posters' extended abstracts, communications in computer and information science. Springer, Berlin, pp 123–127. https://doi.org/10.1007/978-3-642-22098-2_25

13. Fu B, Noy NF, Storey M-A (2013) Indented tree or graph? A usability study of ontology visualization techniques in the context of class mapping evaluation. In: Alani H, Kagal L, Fokoue A, Groth P, Biemann C, Parreira JX, Aroyo L, Noy N, Welty C, Janowicz K (eds) The semantic web – ISWC 2013, Lecture notes in computer science. Springer, Berlin, pp 117–134. https://doi.org/10.1007/978-3-642-41335-3_8
14. Nestrud MA, Meiselman HL, King SC, Lesher LL, Cardello AV (2016) Development of EsSense25, a shorter version of the EsSense profile®. Food Qual Prefer 48:107–117. https://doi.org/10.1016/j.foodqual.2015.08.005
15. Apostolidis T (2003) Représentations sociales et triangulation: enjeux théorico-méthodologiques. In: Abric J-C (ed) Méthodes d'étude des représentations sociales. Erès, Saint-Agne, pp 13–35
16. Urdapilleta I, Demarchi S, Parr WV (2021) Influence of culture on social representation of wines produced by various methods: natural, organic and conventional. Food Qual Prefer 87:104034. https://doi.org/10.1016/j.foodqual.2020.104034

Chapter 22

Meal Evaluations: Studying Foods and Beverages in the Meal Context

Adriana Galiñanes-Plaza and Agnès Giboreau

Abstract

Studying foods and beverages in the meal context is needed to better understand consumers' behaviors toward products. Since Meiselman in 1992, who advocated for the study of real foods in real contexts with real people, researchers try to move beyond the lab. Nevertheless, moving towards more ecological studies is challenging and still needs further investigation due to the hidden variety behind the concept of meal. The present chapter describes the concept and components of a meal, the variables (related to the serving size, food and beverage combinations, and context) that may affect consumers' evaluations, the current methods to be applied (contextual approaches and measurements), and the main aspects to consider when conducting consumer research on this field.

Key words Meal evaluations, Context, Contextual variables, Foodstuffs

1 Introduction

1.1 What Is a Meal?

A meal is "an occasion when people eat food, especially breakfast, lunch or dinner" (Oxford Dictionary). In other words, a meal is not just a sum of food. A meal is defined not only by the content of what is consumed (What?) but also socially (Who eats? Who Serves?), spatially (Where?), temporally (When?) as well as by the way it unfolds (How?). There are many possibilities to consume a given product. In the dining room, various factors influence the choices, the quantities consumed or wasted, the pleasure of eating, etc. In addition to the brand, or name, and the price, as in the case of consumer products, meal factors come into play, such as the composition of the full menu, the information as well as the type of service, the decoration, the atmosphere, etc.

Developing methods to study eating behaviors or to study food and beverage in a meal context needs to control some of the many variables existing in real-life situations and varied responses are possible depending on the expected realism as well as time and

Carlos Gómez-Corona and Heber Rodrigues (eds.), *Consumer Research Methods in Food Science*,
Methods and Protocols in Food Science, https://doi.org/10.1007/978-1-0716-3000-6_22,

budget constraints (*see* [1] for a comparison of laboratory, free-living, and Living Lab approaches). Although contextual factors are very important to qualify an eating situation [2], the design of a product study in a real context needs to start from the product under study itself.

We suggest proceeding in two operational steps. First, the project team needs to define the basic components of the eating situations related to the product of interest and the objective of the study: what is mandatory to control, keeping the core criteria of a classic consumer test and adding contextual factors to be integrated as controlled variables. Second, the team needs to identify the secondary components of the meal, those which are neither controlled nor fixed and will vary as part of the natural differences occurring in the real life.

1.2 The Selection of Under-Control Components of a Meal

There is no universal truth on what the essential elements of a meal situation are, and selecting these core elements depends on the product and the objectives of the study being designed. However, it is essential to select the key elements under the control of the experimenter, from meal preparation to its service and consumption. On the one hand, it is a matter of considering what is consumed. A food considered alone is not a meal. Even a ready-made meal does not constitute a meal by itself. A meal is made up of one or more foods, either one after another or in combination. A meal also includes one or more beverages, and their association throughout the consumption experience contributes to the specificity of the occasion. On the other hand, the restaurant room and the service characterize the moment of consumption. The same menu can be consumed after having been chosen from a buffet, requested from a cook behind a cafeteria-type cooking station, or after having been served at the table by a third party (a waiter, a nurse). The place of consumption also has an important role in the definition of a meal through its atmospheric (temperature, smell, music, light ...), aesthetic, and marketing characteristics.

By observation or through an exploratory Usage and Attitude study, it is possible to describe most of the tangible elements that define the meal of interest (*see* Table 1). This can be done by following the basic working organization of a restaurant: food and beverage on one hand and room and service on the other hand. Thus, it allows to lead to a strategic decision whether the component should be fixed or let flexible.

1.3 The Characterization of Meal Components Related to Individuals

Besides the components under the control of the team, others are linked to the personal experience of the participants. Yet, the occasion of eating is a major variable which determines many other components. This is a complex issue, as there are as many types of eaters as there are individuals. In addition, each eater experiences a variety of eating occasions, as illustrated in Table 2.

Table 1
Tangible components to screen to design a study in an eating context

Category	Criteria	Examples
Foods & beverages	Sequence of food and beverage *(consumed at different steps in the meal)*	Starter Main course Dessert
	Combination of food and beverage for a given sequence *(food-food or food-drink pairing)*	Starter with bread With or without sauce White wine
	Dishes	Size / type of plate Cutlery Glass
Room & Service	Place	Domestic dining room Workplace canteen On the go
	Atmospherics	Light Music Air (Temperature, Odor)
	Delivery	Self-service Waiter

Table 2
Individual components (linked to the participants) to screen to design a study in an eating context

Category of variables	Criteria	Examples
Occasion	Event	Ordinary intake (habitual) Extra-ordinary intake Festive Professional Budget
	Time	Day Hour Duration
Participants, eaters, customers, and guests at the table	Social interactions	Number of guests at meal Type of guests
	Motivation	Emotional status Symbolic value Personal expectations (not related to the food and drink)
	Homeostasis	Hunger Thirst Physical activity

Fig. 1 Different eating contexts: (**a**) School cafeteria at the Living Lab® of the Institut Paul Bocuse Research Center; (**b**) High-end restaurant Saisons

To encounter this difficulty in designing a relevant eating situation to study a product in context, the knowledge of the eating habits of the population of interest is thus essential, at least concerning the product under study. Imagine the product is sometimes consumed as a dessert at the workplace canteen and sometimes for breakfast at home; or sometimes it is consumed at the school canteen and sometimes at the restaurant (where the preparation and presentation would differ due to the consumption situation involved), the conditions are too different to be modeled in a single eating context (Fig. 1). Thus, the eating situation to be selected in the study could be the most frequent one for the target population or, on the contrary, the less frequent one to evaluate the potential of diversification or range extension. Then, the meal occasion and the recruitment criteria are to be selected from the objective of the study. They will condition the other variables, such as the place, the time, and the presence of another guest. The following sections present the variables and approaches that may be considered when studying foods and beverages in a meal context.

2 Variables to Consider When Evaluating Foods and Beverages in a Meal Context

The choice, consumption, and sensory evaluation of food and beverages in standardized conditions are very different from how products are normally consumed [3, 4]. Numerous variables may be considered when evaluating foodstuffs in the context of a meal.

2.1 Bite and Sip Versus Dish and Beverage, and Meal Evaluation

When evaluating foods or beverages, different units of evaluation can be studied: bite or sip, dish or drink, meal or eating patterns during a certain period of time [5]. Several methods (time intensity or temporal dominance of sensations) have already shown changes in sensory perception of products in terms of seconds underlining that products' sensory evaluation evolves over time and that different sensory attributes may be perceived slightly different across the consumption [6, 7]. Depending on the type of product (food or drink) and the way this is generally consumed, we should keep as realistic as possible the evaluation and sometimes one bite or sip may not be enough to lead to a rating representing the full consumption of the product. Hence, it is important to understand what the objective behind each evaluation is and how that evaluation should be conducted. Moreover, a realistic product preparation, including the right pairing: jam with bread, gin and tonic, etc., should be also pondered over to keep as realistic as possible the natural situation of product consumption [8]. Finally, if the product is tested within a meal context, it is important to define in which type of dish the product is going to be included (starter, main course, or dessert) and what time it should be consumed, as the sensory evaluation and choice may differ [9, 10]. Figure 2 shows a food evaluation conducted at the Living Lab® of the Institut Paul Bocuse Research Center, where different target products were included as part of a meal as a starter, main course, dessert, and drink to create a whole menu. As a matter of fact, the sequence and appropriateness of mealtimes when evaluating foodstuffs have shown to have an impact on consumer product evaluation and intake [11, 12]; being this impact more or less prominent depending on the product category (savoury, sweets, drinks) and the type of eating (meal or snack). One key parameter is thus to consider the consumption of a full portion rather than a few bites or sips.

2.2 Food and Beverage Combinations in Meals

Foodstuffs are rarely consumed in an isolated manner, making it necessary to study the sensory experience of food-beverage and food-food combinations. Several studies have been conducted on this matter trying to explain how products should be combined in a win-win manner [13, 14]—underlining that such combinations may occur within individual dishes, at the level of ingredients, and even at the level of chemical reactions. Besides, cultural backgrounds have also proved to influence such combinations, showing differences in terms of product evaluation across countries [4, 15, 16].

Regarding the rules of combination, most of the research on consumer behavior and sensory evaluation has been focused on individual products rather than on food item combinations. Nevertheless, in the past decades, researchers have shown that the study of suitable foodstuff combinations results in more pleasant eating experiences and, therefore, higher overall meal acceptance and

Fig. 2 Food evaluation at the Living Lab® of the Institut Paul Bocuse Research Center

choice [17–19]. However, due to the complexity of meal differences, more consumer research is needed when it comes to evaluating ingredients or foods in more natural consumption situations, such as food-beverage or food-food combinations.

The selection of the relevant food-food and food-beverage combinations is thus a key step that should be considered to evaluate food in a realistic meal situation, not only based on volatile compatibilities but also based on consumers' cultural background and eating habits.

2.3 The Context of Foods and Beverages Evaluation in a Meal Context

Previous sections have focused on the variables related to the product (units of evaluation, combinations). This section focuses on external variables—for example, serving, plating, or environment—in relation to the evaluation of food in the context of a meal. Traditionally, foods and beverages are tested in standard sensory laboratory conditions that significantly differ from the regular way products are consumed. However, cutlery, plating, and even table linen have been shown to impact consumer evaluation and quality perception of food and drinks. In the study of [20] similar yogurts were evaluated by using different spoons materials (stainless steel vs. metallic plastic spoon). Results showed that depending on the perceived quality of the spoon, the quality and the overall liking of the yogurt was different. On this matter, [21], also proved the effect of metallic-coated ceramic cups on the perception of two different types of coffee. Their results showed that higher sensory and hedonic ratings were given to the Brazilian coffee when it was tasted from the bronze cup and to the Kenyan coffee when sampled from the gold cup.

Besides, the colors of serving ware have also proved to impact consumers' sensory evaluation. In another study of [22], several desserts were evaluated by changing the color of the plates (black and white). Cross-modal interactions were observed between color perception and flavor, confirming the effect of color on consumers' expectations and food perception. The color of the plates had an impact on the liking of desserts' presentation and on how appetizing the dessert looked. Besides, the perception of the flavor and sweetness intensity was also affected by the interaction between the type of dessert and the color of the plate. Another example concerning the impact of serving ware on consumers' evaluation is the study of [23]. These researchers compared carbonated, aromatized non-alcoholic beers in receptacles that differed in terms of their color (red or black) and weight. The color of the receptacle influenced the perceived carbonation, with the beverages tasted from the red receptacles and higher weight being rated as tasting more carbonated than when served in black receptacles.

The expression "you eat first with your eyes" clearly sums up how visual factors may contribute to consumers' expectations and influence their evaluation. Chefs and researchers have shown their interest on the effect of subtle changes in the visual presentation of a dish on sensory perception and consumers' overall liking. Zellner et al. [24] studied the effect of plating by comparing a neat versus a messy salad presentation. Their results showed that neatness increases adults' liking for the taste of the food and also their willingness to pay. However, [25, 26] showed that more artistic food presentations were more liked than neat and classical ones. In their studies, a gastronomic salad presented as one of the Kandinsky's paintings was more appreciated, and participants were also willing to pay more when the salad followed an artistic pattern. Nevertheless, the "centered" plating was still preferred to the "offset" one for the main dish [26, 27]. also studied the effect of plating on consumers' preferences and even on portion size perception by varying the orientation of the food on a plate. Their results demonstrated the influence of both plating arrangement (horizontal vs. vertical) and centrality (centered vs. offset plating) on consumers' liking, willingness to pay, and on the perception of portion size. In a hospital context, Navarro et al. [28] showed the aesthetics of the plate dressing increased patients' satisfaction as well as their intake.

The physical environment has also been proven to have an impact on consumers' expectations and, therefore, on the evaluation of foodstuffs. At the beginning of the twenty-first century, several researchers [29–32] underlined that by changing the physical context, the acceptability of a similar product differed. Knowing which environmental variables influence consumers' evaluation and behavior is a major challenge when studying foods in the context of a meal [3]. Within these variables, the color of the room [33], the temperature and light [34, 35], the decoration [36, 37], the presence or absence of music [38], of odors [39, 40], of external distractions, or of the social interaction [41, 42] have been confirmed to modulate consumers' behavior and then the evaluation of foods and beverages.

Hence, when studying foods and beverages in the context of a meal, consumer and sensory researchers should pay careful attention to those variables as consumers' expectations towards the environment, the presentation, or the expected quality of the food may be anchored beforehand.

3 Methods to Study Foods and Beverages in a Meal Context

The previous section illustrated how multiple variables can influence consumers' evaluation of foods and beverages in the meal context. Thus, when looking at available methods to studying meals, it is necessary to define what the objectives are behind them: understanding consumers' acceptance, choice, intake, etc. Several methods have been described in the literature and can be then applied.

3.1 Contextual Approaches

As external factors can influence consumers' evaluation of foods and beverages, different approaches have been developed in order to bring more natural context to the laboratory and provide a baseline in consumer research improving data interpretation: the context can be evoked through a preliminary instruction; prior to testing, it can be reconstructed using immersive technologies or by using a living lab approach and actually reproducing the desired environment.

3.1.1 Evoked Context

"Imagine that you arrive late at home, from work, and you have to prepare the dinner." This is one example of a written scenario that could be used to evoke a specific eating situation or setting. Using the criteria described in Tables 1 and 2 could help define the critical variables that are to be included in the scenario, depending on the objective of the study. This methodology of evocation is easy to implement, and the evoked situations can be as specific or general as the required objectives of the study [43]. Here, the consumer is placed in a particular situation where he/she should keep that situation vividly in mind before answering the questions related to the product. Decorative elements or specific tasks (setting a table) can be included in order to improve the context of consumption. However, experimenters do not have access to the evocation itself and conclusions could be difficult to interpret to distinguish between variables.

3.1.2 Immersive Technologies

Mixed reality and virtual reality are currently used in consumer research to improve the context of product testing with the possibility to control the contextual variables between scenarios. Through the use of multisensory stimuli (odors, music, lights, visuals, etc.), consumers can be immersed in different settings or eating situations that may help to project themselves in a specific consumption situation. For example, Fig. 3 shows a multisensory

Fig. 3 Food and drink evaluation at the Multisensory Immersive Experience Hall (MIXH®) of the Institut Paul Bocuse

immersive restaurant concept where clients were immersed in a library atmosphere. This type of methodology is expensive compared to the previous one but has shown to have a positive impact on consumer engagement [44, 45]. Different sensory stimuli can be diffused and controlled in order to observe how external factors to the product can interact with consumer evaluation. Moreover, in some cases, the use of video recording can help to analyze certain consumer's behaviors besides the declarative data. Immersive technologies have the advantage of offering the control of contextual variables being experimented. However, one limitation is to be somehow artificial, sometimes needing an actual eating situation for confirmation.

3.1.3 Living-Lab

This approach allows for the collection of data in a real usage/consumption situation while controlling different contextual variables (food presentation, service, quantities served, etc.). This approach is based on active user involvement (meal consumption), a real-life setting (consumers pay for their meal), multi-stakeholder participation, an interdisciplinary approach, and co-creation (https://enoll.org/).

In this type of approach, observational measurements can be also applied in order to keep as realistic as possible the consumption experience. As in the case of immersive technologies, participants in a Living Lab study should be informed prior to the recordings (video or sounds).

3.2 Measurements

3.2.1 Observational Measurements

This type of method helps to generate insights that may lead to possible solutions when it comes to the use or combination of certain foods and beverages in a meal context. To do so, the experimenters need to set up an experimental design focused on a given number of possible options. For example, the consumption of different types of bread during a meal, the use of sauces when consuming a salad or French fries, etc., the consumption of water or fruits when they are visible and/or accessible, etc. Observation coding and analysis can be complex and time-consuming.

Recently, a new observational analysis of behavior has been proposed in order to model meal dynamics across time and consumers: Temporal Dominance of Behavior (TDB) [46]. By the use of the Noldus' Observer XT 12.0 Behavioural Analysis Software (Wageningen, the Netherlands) consumers' behaviors in terms of intake, actions, and social interactions can be coded and then analyzed in time, duration, and frequency applying a similar methodology as the Temporal Dominance of Sensations (TDS). Through this method, the main events during the meal consumption are identified and represented as a dynamic graphical synthesis of the subjects' eating behavior.

3.2.2 Questionnaires

Regarding declarative methods, several questionnaires have been developed to provide pre- and post-food and beverage intake evaluation. In order to keep the consumption situation as natural as possible, the use of apps, tablets, and mobile devices helps to collect consumers' responses in a more natural way. In the last decades, research on meal evaluations evolved beyond liking, and nowadays, consumers' experiences and expectations about food provide further insights into the understanding of the motives and emotions that can be triggered by sensory products.

As it occurs with the TDB, multiple moments during a meal are going to set the overall evaluation. Therefore, it is possible to track the course experience over time by looking at aspects such as hedonic evaluation (like or dislike) and intensity (mild to extreme) along the meal or after each course to keep a more natural pace of eating, only suspending the meal between courses. In this way, it has been shown that certain courses can contribute more to the global evaluation of the meal than others, being better predictors for the wish to repeat a meal [47].

Consumers' expectations about foods are another factor to consider in the context of a meal. It has been shown that expectations can set a reference point from where consumers evaluate a product [48]. As context plays a key role in consumers' experience and food evaluation, it is interesting to collect this type of data before the test. For instance, we used an expectation questionnaire before a meal study in order to identify the type of products consumers expected to find in a particular restaurant [49]. Two weeks before the actual meal evaluation, consumers answered an online questionnaire about the type of environment, products, product presentation, etc., they expected to find. We showed that a negative evaluation of the product can be associated with a mismatch between the type of restaurant and the quality of the served product when consumers' expectations are not met. Therefore, in order to better explain the results, consumers' expectations can contribute to this task.

4 Main Aspects to Consider When Conducting Consumer Research in a Meal Context

In this section, the main aspects to consider when conducting consumer research in a meal context are presented:

- When conducting consumer research in meal contexts, it is important to standardize the contextual variables that may influence the consumer's evaluation (portion size, presentation, serving ware, etc.). The amount of served food or beverage should be kept as realistic as possible to ensure that the evaluation of the product corresponds to a regular portion size.

- When considering the context of evaluation, it is important to define the number of tested samples and questions before selecting a contextual approach. Certain types of contextual approaches (immersive technologies and Living Lab) may not be appropriate as the ecological validity of the study could be compromised due to the length of the experiment.
- When designing a particular setting, depending on the research question, congruent or incongruent elements can be injected into the scenario in order to understand product performance and product-context appropriateness. Nevertheless, it is necessary to define what consumers perceive as appropriate for a consumption context.
- When selecting the order of product presentation, the concept of hedonic contrast should be evaluated. Some studies have shown the impact of appetizers on main courses, highlighting the importance of the menu design [10]. Depending on the type of tested product (food or beverage), it is suggested to test it at the beginning of the meal as an appetizer. This aspect could be discussed with chefs to ensure the quality of the menu.
- During the meal, consumers should be asked to not discuss about the products they are testing to avoid bias among consumers' responses. Nevertheless, the intraclass correlation coefficient (ICC) can be used to measure the effects of social interaction in a meal experience; this coefficient quantifies the homogeneity of meal experience measures within groups (e.g., diners sharing the same meal), in contrast to evaluations across groups [50]. In relation to this point, consumers may be proposed to join the test together with some relatives or friends in order to facilitate the social aspects of the meal.

5 Conclusion

Studying foods and beverages in a meal context is a challenge for consumer research due to the variety hidden in the term "meal." Meal research should be conducted through actual meals, where people actually eat them, and not isolated foods and/or beverages in standardized conditions such as laboratories. This requires accepting a certain level of "freedom" in the control of variables and carefully moving beyond the lab. The intent of this chapter was to provide readers with an overview of the research conduted in this field in terms of variables to consider and methods to apply and, to outline the mistakes to be avoided. Future research should also involve chefs, cooks, and waiters, as these professionals have valuable knowledge on consumer experience and they may bring an added value to meal research.

References

1. Dougkas A, Saulais L, Giboreau A (2019) Studying natural meals: what are the benefits of the living lab approach? In: Meiselman H (ed) Context: the effects of environment on product design and evaluation. Woodhead Publishing
2. Giboreau A (2018) Situational factors and the design of in situ evaluations: Living Lab and Real-life approaches. In: Ares G, Varela P (eds) Advances in methodological approaches in consumer research. Elsevier, pp 103–116
3. Galiñanes Plaza A, Delarue J, Saulais L (2019) The pursuit of ecological validity through contextual methodologies. Food Qual Prefer 73 (November 2018):226–247. https://doi.org/10.1016/j.foodqual.2018.11.004
4. Lahne J (2018) Evaluation of meals and food pairing. In: Methods in consumer research, volume 2: alternative approaches and special applications, vol 2. Elsevier Ltd. https://doi.org/10.1016/B978-0-08-101743-2.00004-2
5. Rozin P, Tuorila H (1993) Simultaneous and temporal contextual influences on food acceptance. Food Qual Prefer 4(1):11–20. https://doi.org/10.1016/0950-3293(93)90309-T
6. Pierguidi L, Spinelli S, Monteleone E, Dinnella C (2021) The combined use of temporal dominance of sensations (TDS) and discrete time-intensity (DTI) to describe the dynamic sensory profile of alcoholic cocktails. Food Qual Prefer 93(May):104281. https://doi.org/10.1016/j.foodqual.2021.104281
7. Thomas A, van der Stelt AJ, Prokop J, Lawlor JB, Schlich P (2016) Alternating temporal dominance of sensations and liking scales during the intake of a full portion of an oral nutritional supplement. Food Qual Prefer 53:159–167. https://doi.org/10.1016/j.foodqual.2016.06.008
8. Bleibaum RN, Kern MJ, Thomas H (2019) Contextual product testing for small to medium sized enterprises (SMEs). In: Meiselman H (ed) Context: the effects of environment on product design and evaluation. Woodhead Publishing, pp 501–520
9. de Beukelaar MFA, Zeinstra GG, Mes JJ, Fischer ARH (2019) Duckweed as human food. The influence of meal context and information on duckweed acceptability of Dutch consumers. Food Qual Prefer 71(June 2018): 76–86. https://doi.org/10.1016/j.foodqual.2018.06.005
10. Lahne J, Pepino R, Zellner D (2017) You'll spoil your dinner: attenuating hedonic contrast in meals through cuisine mismatch. Food Qual Prefer 56:101–106. https://doi.org/10.1016/j.foodqual.2016.10.002
11. Boutrolle I, Delarue J (2009) Studying meals in the home and in the laboratory. In: Meals in science and practice. Woodhead Publishing Series in Food Science, Technology and Nutrition, pp 128–165
12. McLeod CJ, James LJ, Witcomb GL (2020) Eating rate and food intake are reduced when a food is presented in an 'unusual' meal context. Appetite 154(July):104799. https://doi.org/10.1016/j.appet.2020.104799
13. Eschevins A, Giboreau A, Allard T, Dacremont C (2018) The role of aromatic similarity in food and beverage pairing. Food Qual Prefer 65(July 2017):18–27. https://doi.org/10.1016/j.foodqual.2017.12.005
14. Scander H, Monteagudo C, Nilsen B, Tellström R, Yngve A (2018) Food and beverage dinner combinations, patterns among Swedish adults. Int J Gastron Food Sci 14 (April):20–26. https://doi.org/10.1016/j.ijgfs.2018.08.003
15. Arellano-Covarrubias A, Varela P, Escalona-Buendía HB, Gómez-Corona C (2022) A food and beverage map: exploring food-beverage pairing through projective mapping. Food Qual Prefer 96(October 2021):104431. https://doi.org/10.1016/j.foodqual.2021.104431
16. Galmarini MV (2020) The role of sensory science in the evaluation of food pairing. Curr Opin Food Sci 33:149–155. https://doi.org/10.1016/j.cofs.2020.05.003
17. Donadini G, Spigno G, Porretta S (2021) Preschooler liking of meal components: the impact of familiarity, neophobia, and sensory characteristics. J Sens Stud 36(3). https://doi.org/10.1111/joss.12649
18. Parizel O, Labouré H, Marsset-Baglieri A, Fromentin G, Sulmont-Rossé C (2017) Providing choice and/or variety during a meal: impact on vegetable liking and intake. Appetite 108:391–398. https://doi.org/10.1016/j.appet.2016.10.027
19. van Stokkom VL, de Graaf C, Wang S, van Kooten O, Stieger M (2019) Combinations of vegetables can be more accepted than individual vegetables. Food Qual Prefer 72 (October 2018):147–158. https://doi.org/10.1016/j.foodqual.2018.10.009
20. Piqueras-Fiszman B, Spence C (2011) Do the material properties of cutlery affect the perception of the food you eat? An exploratory study.

J Sens Stud 26(5):358–362. https://doi.org/10.1111/j.1745-459X.2011.00351.x

21. Carvalho FM, Spence C (2021) Do metallic-coated cups affect the perception of specialty coffees? An exploratory study. Int J Gastron Food Sci 23(September 2020):100285. https://doi.org/10.1016/j.ijgfs.2020.100285
22. Piqueras-Fiszman B, Giboreau A, Spence C (2013) Assessing the influence of the color of the plate on the perception of a complex food in a restaurant setting. Flavour 2(1):24. https://doi.org/10.1186/2044-7248-2-24
23. Mielby LA, Wang QJ, Jensen S, Bertelsen AS, Kidmose U, Spence C, Byrne DV (2018) See, feel, taste: the influence of receptacle colour and weight on the evaluation of flavoured carbonated beverages. Foods 7(8). https://doi.org/10.3390/foods7080119
24. Zellner DA, Siemers E, Teran V, Conroy R, Lankford M, Agrafiotis A, Ambrose L, Locher P (2011) Neatness counts. How plating affects liking for the taste of food. Appetite 57(3): 642–648. https://doi.org/10.1016/j.appet.2011.08.004
25. Michel C, Velasco C, Gatti E, Spence C (2014) A taste of Kandinsky: assessing the influence of the artistic visual presentation of food on the dining experience. Flavour 3(1):1–11. https://doi.org/10.1186/2044-7248-3-7
26. Michel C, Velasco C, Fraemohs P, Spence C (2015) Studying the impact of plating on ratings of the food served in a naturalistic dining context ☆. Appetite 90:45–50. https://doi.org/10.1016/j.appet.2015.02.030
27. Rowley J, Spence C (2018) Does the visual composition of a dish in fl uence the perception of portion size and hedonic preference ? Appetite 128(April):79–86. https://doi.org/10.1016/j.appet.2018.06.005
28. Navarro DA, Boaz M, Krause I, Eli A, Chernov K, Giabra M, Frishman S, Levy M, Giboreau A, Kosak S, Mouhieddine M, Singer P (2016) Improved meal presentation increases food intake and decreases readmission rate in hospitalized patients. Clin Nutr 35:1153–1158
29. Edwards JSA, Meiselman HL, Edwards A, Lesher L (2003) The influence of eating location on the acceptability of identically prepared foods. Food Qual Prefer 14(8):647–652. https://doi.org/10.1016/S0950-3293(02)00189-1
30. King SC, Weber AJ, Meiselman HL, Lv N (2004) The effect of meal situation, social interaction, physical environment and choice on food acceptability. Food Qual Prefer 15(7–8):645–653. https://doi.org/10.1016/j.foodqual.2004.04.010
31. King SC, Meiselman HL, Hottenstein AW, Work TM, Cronk V (2007) The effects of contextual variables on food acceptability: a confirmatory study. Food Qual Prefer 18(1):58–65. https://doi.org/10.1016/j.foodqual.2005.07.014
32. Meiselman HL, Johnson JL, Reeve W, Crouch JE (2000) Demonstrations of the influence of the eating environment on food acceptance. Appetite 35(3):231–237. https://doi.org/10.1006/appe.2000.0360
33. Stroebele N, De Castro JM (2004) Effect of ambience on food intake and food choice. Nutrition 20(9):821–838. https://doi.org/10.1016/j.nut.2004.05.012
34. Bschaden A, Dörsam AF, Cvetko K, Kalamala T, Stroebele-Benschop N (2020) The impact of lighting and table linen as ambient factors on meal intake and taste perception. Food Qual Prefer 79(July 2019):103797. https://doi.org/10.1016/j.foodqual.2019.103797
35. Sester C, Deroy O, Sutan A, Galia F, Desmarchelier J, Valentin D, Dacremont C (2013) "Having a drink in a bar": an immersive approach to explore the effects of context on drink choice. Food Qual Prefer 28:23–31. https://doi.org/10.1016/j.foodqual.2012.07.006
36. Sinesio F, Saba A, Peparaio M, Saggia Civitelli E, Paoletti F, Moneta E (2018) Capturing consumer perception of vegetable freshness in a simulated real-life taste situation. Food Res Int 105:764–771. https://doi.org/10.1016/J.FOODRES.2017.11.073
37. Liu J, Petit E, Brit AC, Giboreau A (2019) The impact of tablecloth on consumers' food perception in real-life eating situation. Food Qual Prefer 71:168–171
38. Kontukoski M, Paakki M, Thureson J, Uimonen H, Hopia A (2016) Imagined salad and steak restaurants: consumers' colour, music and emotion associations with different dishes. Int J Gastron Food Sci 4:1–11. https://doi.org/10.1016/j.ijgfs.2016.04.001
39. Biswas D, Szocs C (2019) The smell of healthy choices: cross-modal sensory compensation effects of ambient scent on food purchases. J Mark Res 56(1):123–141. https://doi.org/10.1177/0022243718820585
40. Hathaway D, Simons CT (2017) The impact of multiple immersion levels on data quality and panelist engagement for the evaluation of cookies under a preparation-based scenario.

Food Qual Prefer 57:114–125. https://doi.org/10.1016/J.FOODQUAL.2016.12.009

41. Cruwys T, Bevelander KE, Hermans RCJ (2015) Social modeling of eating: a review of when and why social influence affects food intake and choice. Appetite 86:3–18. https://doi.org/10.1016/j.appet.2014.08.035
42. Di Monaco R, Giacalone D, Pepe O, Masi P, Cavella S (2014) Effect of social interaction and meal accompaniments on acceptability of sourdough prepared croissants: an exploratory study. Food Res Int 66:325–331. https://doi.org/10.1016/j.foodres.2014.10.001
43. Piqueras-Fiszman B, Jaeger SR (2019) Evoked consumption context matters in food-related consumer affective research. In: Meiselman H (ed) Context: the effects of environment on product design and evaluation. Woodhead Publishing, pp 545–564
44. Andersen INSK, Kraus AA, Ritz C, Bredie WLP (2019) Desires for beverages and liking of skin care product odors in imaginative and immersive virtual reality beach contexts. Food Res Int 117:10–18. https://doi.org/10.1016/J.FOODRES.2018.01.027
45. Zandstra EH, Kaneko D, Dijksterhuis GB, Vennik E, De Wijk RA (2020) Implementing immersive technologies in consumer testing: liking and just-about-right ratings in a laboratory, immersive simulated café and real café. Food Qual Prefer 84(September 2019): 103934. https://doi.org/10.1016/j.foodqual.2020.103934
46. Miele NA, Giboreau A, Almli VL (2021) The temporality of eating behavior as a discriminant tool to characterize consumers: temporal dominance of behavior applied to bread consumption during a restaurant meal in France. Food Qual Prefer 92(January):104225. https://doi.org/10.1016/j.foodqual.2021.104225
47. Garbinsky EN, Morewedge CK, Shiv B (2014) Interference of the end: why recency bias in memory determines when a food is consumed again. Psycho- logical Science 25:1466e1474. https://doi.org/10.1177/0956797614534268
48. Galiñanes Plaza A, Saulais L, Blumenthal D, Delarue J (2019) Eating location as a reference point: differences in hedonic evaluation of dishes according to consumption situation. Food Qual Prefer:103738. https://doi.org/10.1016/j.foodqual.2019.103738
49. Galiñanes Plaza A, Saulais L, Delarue J (2022) Associations between prior expectations towards meal experience and hedonic responses in the restaurant: the role of information (in preparation)
50. Muñoz F, Hildebrandt A, Schacht A, Stürmer B, Bröcker F, Martín-Loeches M, Sommer W (2018) What makes the hedonic experience of a meal in a top restaurant special and retrievable in the long term? Meal-related, social and personality factors. Appetite 125: 454–465. https://doi.org/10.1016/J.APPET.2018.02.024

Chapter 23

Studying Future Food Scenarios with Design Fiction

Olivier Wathelet and Nicolas Minvielle

Abstract

Design fiction is an approach that cuts across design and *la prospective* (a French school of thought that relates to future studies and strategic insight) (Michel Godet, Creating Futures, Scenario Planning as s Strategic Management Tool, 2006; Michel Godet, To Predict or to build the future? Reflections on the field and differences between Foresight and La Prospective, The Futurist, May–June 2012). It aims to explore possible futures related to a given sector, cultural practice, ingredient, or market in order to define preferable scenarios for relevant stakeholders involved (e.g., companies, advocacy groups, states, consumers, NGOs, broader society, etc.). Design fiction is an emerging professional practice, and its boundaries and tools have yet to be standardized. However, it always involves creating tangible fiction to generate debate among selected stakeholders. Several types of design fiction interventions ranging from creative to analytical approaches exist and seek to tackle three main challenges: building a strategic vision; being thought-provoking to facilitate change; and innovating. In this chapter, we outline a proposed protocol focused on innovation, which puts future imaginaries (e.g., science fiction, expert visions) at the forefront of our approach to reach these goals. We present different examples from the food industry and discuss the methodological issues at stake at each step of the process.

Key words Design fiction, Innovation, Design, Prospective, Insight, New markets, Nouveaux marches

1 Introduction

1.1 The Example of Take-Away Meat

On stage, three prototypes of innovative products are being presented to 300 business leaders, representatives of the political community, and researchers. Among these prototypes is a meat fridge which aims to radically transform everyday eating habits. The biggest strength of the project lies in its ground-breaking packaging, which directly adheres to the meat and disintegrates during the cooking process. A protective film is placed on each piece of meat after slaughtering, thereby providing farmers with a low-waste product which can fit local distribution channels. A few first buyers are already showing interest, and we are expecting some of these meat fridges to be set up in petrol stations in the next 3 months.

Carlos Gómez-Corona and Heber Rodrigues (eds.), *Consumer Research Methods in Food Science*,
Methods and Protocols in Food Science, https://doi.org/10.1007/978-1-0716-3000-6_23,

In reality, this project is a fiction which was developed and presented as part of a program gathering 17 research portfolios to advance cooperation between Belgium and France. Two collaborative design workshops brought together about thirty program participants and led to three deliberately provocative proposals, which were then presented to a large audience at a public event. The goal was to start a reflection process among the project stakeholders about innovation strategies in the food industry and the sustainable development sector. The project was first presented as a real solution, and the audience was invited to provide feedback through a live voting system on their smartphone. The audience was only told that the project was fictional at the end of the evening, in order to spark off a debate and identify future scenarios that were considered preferable compared to those initially envisioned through the project. This type of approach is design fiction.

1.2 The Future of Food and the Role of Imaginaries

Three common approaches can be used to envision future food systems from a consumer's perspective:

(A) Studying consumers to identify any unmet needs and issues to solve [1, 2].

(B) Analyzing trends to identify weak signals / emerging consumer practices [3, 4].

(C) Creating new products and services using professional expertise (by chefs, food designers, etc.) [5, 6].

While these three approaches are all relevant, they provide limited to no space for reflection on whether future supply offers can be defined as preferable. A study of the history of scientific and non-scientific imaginaries in the food industry [7] has shown that these approaches lack neutrality and limit opportunities to develop new solutions to tackle societal challenges (ecological crisis, changes in consumptions patterns, etc.). These scenarios and visual representations are spread across traditional and media industries [8, 9].

Design fiction is a method that aims to foster innovation and transformation and consists of analyzing existing visions for the future in a given sector and creating new ones that are considered preferable.

Given the proven important role of the general public in innovation, it is essential to include this analysis in the imaginaries of the future general public. This is the goal of design fiction.

1.3 Design Fiction: Definition, History, and Variety in Purpose and Forms

The expression "design fiction" was crafted by science fiction author Bruce Sterling [10] and by designer Julian Bleecker. It was first defined as *the deliberate use of diegetic prototypes to suspend disbelief about change, by conflating design, science facts and science fiction* [11]. It is a projective approach that aims to spark off a change of perspective by defining possible yet provoking tales. It currently encompasses a large range of activities and cannot be defined as a discipline per se. It is a creative activity that builds upon learnings from design and from "la prospective," as well as world-building in art and literature. When it comes to learning from design, design fiction is very closely linked to speculative design and critical design [12–14]. As for "la prospective," design fiction is in some aspects similar to scenario planning, a practice developed by Rand Corporation in the United States [15]. In worldbuilding, design fiction uses techniques of both activist artists and media professionals to facilitate dialogue about various scenarios [16].

Companies have started using design fiction since 2015 and have explored at least three ways to implement it:

- Creating a strategic vision for a company (or any other type of organization). Design fiction makes it possible to test scenarios and explore their relevance and effects to guide decision-making for a mid-term (5 years) or long-term (10 years) strategy. For example, we have been using this approach to support the Red Team project by the French military and explore possible future conflicts to reflect on how to prepare for these.[1]
- Sparking off a debate, creating a Shock or thought-provoking event, to raise awareness about the risks of a given possible situation and anticipate its consequences. For example, Franco-Italian company Thales Alenia Space reached out to us to encourage its staff to engage in ecological and social issues related to the space industry.[2]
- Creating innovations that build upon possible transformations in the context in which a company (or any other type of organization) operates. For example, we supported Bouygues Immobilier to design property listings considering possible transformations in the housing legal framework in France.

In this chapter, we present the third case, taking the food industry as an example for implementation.

1.4 A Diversity of Techniques with No Canonical Methodology

While design fiction is becoming increasingly popular, there is no clear method that is widely shared and available in the literature. This chapter is a first attempt at filling this gap [17–19] and should not yet be used as a standard or reference.

The design fiction method involves the following activities:

- Analyzing the imaginaries related to a certain area, group of practitioners, and to the "general public"

[1] https://redteamdefense.org

[2] https://www.thalesgroup.com/en/worldwide/space/news/hellospace-story-continues

- Collecting imaginaries that can be found in science fiction to enhance these tales
- Creating new tales to stage challenges for the client and its partners
- Creating artifacts to illustrate these challenges and provoke reflection in a credible way (feeling of veracity)
- Creating the conditions for a targeted audience (e.g., clients, partners, colleagues) to seize those challenges and identify preferable future scenarios
- Developing responses (e.g., supply, process) to make these preferable scenarios possible and developing a value proposition that can be used right away

The proposed methodology aims to foster innovation.

2 Specific Materials Required in Design Fiction

2.1 Imaginaries: Definition and Sources

Imaginaries are representations of the future, which can take different shapes (Fig. 1):

- Narratives that can be either complex (e.g., scenario, tale) or simple (advertisement imagery).
- Visual materials (e.g., sequence of a comic book, speculative design work), audio-visual materials (e.g., science fiction film clip), or written materials (e.g., excerpt from a social science fiction short story).

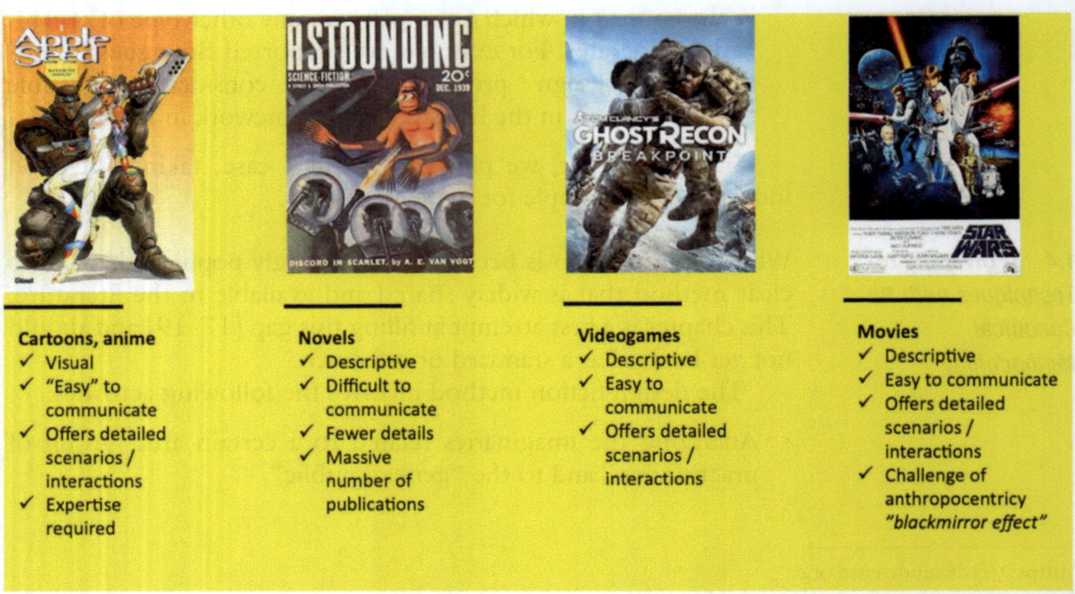

Fig. 1 The benefits and drawbacks of various media

We select visual and audio-visual materials in priority as printed boards and sequences to watch are easier to manage and present to an audience compared to some other forms. As for food, it is much easier for an audience to grasp esthetical decisions made by a product creator if they are visually accessible. Therefore, formal visual representation is key.

- Visual materials consist of one to two European and North American comics or manga boards.
- Audio-visual materials last between 30 s and 1 min and 30 s. Each material is a specific sequence, which shows an imaginary that highlights societal challenges and one or several unusual interactions.
- Written materials are more difficult to show to an audience. For this reason, in addition to a novel excerpt, we usually add drawings or oral descriptions to some sequences to further describe relevant elements.

A project usually involves identifying 30 to 60 imaginaries (the highest number in one of our projects was 100). We always make sure to use various time periods and cultural backgrounds, as all imaginaries are strongly influenced by the context in which they were created [20–22]. We then carry out a first analysis of each imaginary to present its context and identify any thought-provoking elements.

2.1.1 Visual Versus Written Sources

Visual materials (e.g., comic books) and audio-visual materials (e.g., films, series, video games, advertisements) differ from written materials (e.g., novels, short stories), as they enable an audience who is used to fiction to easily understand them and their context. Given the great number of written forms of science fiction, written materials might offer greater content diversity. However, they are more difficult to manage during a workshop: reading takes longer and requires silence. It also does not give much space for collective appropriation. This is why we prefer using visual and audiovisual materials when developing our basic corpus of imaginaries.

Yet, when it comes to creating fiction, we are willing to develop materials such as press articles, which use both text and images to create a combination of assets: fast understanding through images and clear messaging through text. It is an opportunity to collaborate with science fiction authors on a story and manipulate its narrative structure to produce unexpected effects.

2.1.2 Sources and Use of Online Communities to Select Fictions

We use various resources to ensure we can develop ad hoc collections of materials:

- Open calls to participation for staff in the client company
- Imaginaries expert networks (e.g., fan communities, authors, etc.)

- Specialized forums and blogs[3]
- Constant monitoring

Data collection includes a review of the academic and gray literature, interviews with client companies, and interviews with external experts to identify potential challenges and transformations in a company setting. These future tales can also be defined as imaginaries, as they outline possible visions. Both explorations (literature review and imaginaries collection) are carried out at the same time in order to be mutually supportive.

2.2 Forecasting Tools

Several steps in the process include classifying materials (e.g., imaginaries, scenarios, design fiction objects). To do so, we use a number of common tools, such as:

- The futures cone: a tool that invites users to place events on a map according to probability levels (probable, plausible, possible), time (now on the left and future on the right), and whether these events are preferable or not [23].
- The futures wheel: a tool to explore the multiple consequences of plausible scenarios [24].

Other useful tools exist, especially to boost creativity in writing credible scenarios [25]. We have developed various facilitation tools, such as board games, to foster active engagement among workshop participants.

2.2.1 Facilitation Guidelines for "The Futures Cone" Tool

The futures cone is an analytical tool which can also foster creativity. When used for an analytical purpose, it helps to classify items (e.g., imaginaries, fictional scenarios) depending on how probable they are, where they are situated in time, and whether they are preferable or not. Each item is represented by a card or a post-it note and placed on the cone. We suggest the following facilitation approach:

- Introducing every item to all participants, so that everyone is aware of all available topics. Assigning a rapporteur to each item (the group can either appoint a group member as rapporteur or the latter can be chosen randomly).
- Defining a cone and choosing an "end" date that is relevant in the context of the project (e.g., time scale of 5, 10, 20, or more years depending on the project). While the exercise might take participants to go past the initial time scale, it should still be regarded as a key framework.

[3] In the food area, here is an example of a compilation of relevant science fiction movies: https://youtu.be/oZ00Dxbwcyo. See also the example of collaborative website Technonovelgy.com, which includes an inventory of science fiction work. In relation to food: http://technovelgy.com/ct/Science_List_Detail.asp?BT=Food

- Inviting the rapporteur to place the item and explain their choice.
- Once all items have been positioned, initiating a debate to influence the placement of different items. Any decision needs to be unanimous.

The cone offers an analytical vision of various items. It can become a creative tool when adding a number of creative requirements to the exercise. These could include:

- Transforming the item to make it suitable 10 years later on the timeline
- Transforming the item so it becomes an issue (if it is desirable) or becomes desirable (if it is an issue)

2.3 Fiction Templates (e.g., Newspaper)

The scenarios that are produced using the approach described in this chapter to become new imaginaries are called "realist fictions." The use of the adjective "realist" refers to the way these scenarios are staged without revealing that they are fictional (e.g., introducing a fictional new company at a professional fair). A myriad of formats has been tested to date. The success of these formats depends on the nature of the target audience and existing norms in a given sector (e.g., a catalog for mail order works well in home furnishings [26]):

- Advertisement [20, 26]
- Press article or newspaper[4]
- User manual [19, 27]
- Application [28]
- Start-up[5]
- Items sold in a shop [29]
- Items at the R&D stage, presented for a consumer test[6]
- Installation in the public space [30].
- Etc.

The combination of several realist fictions can reinforce how realist it seems and how thought-provoking it is [19]. The objective of the presentation below was to test a scenario about urban air quality (Fig. 2). A fake start-up was created, with its own website, video teaser, application, etc. It was introduced at a professional

[4] Example of an agriculture newspaper in 2040 presented at the 2017 agricultural show in Paris: http://making-tomorrow.mkrs.fr/interpeller

[5] Example of a fictional startup presenting its project to investors: http://making-tomorrow.mkrs.fr/mettre-a-lepreuve-la-deeptech

[6] We use this approach to test how desirable the scenarios are for consumers. This is relevant for consumer goods.

Fig. 2 Fictional materials used in design fiction. AirPollution Revealer is a fictional company, which was introduced as a genuine organization at an event for investors in new technologies

conference among real start-ups. Using multiple templates and displaying them at a real conference greatly helped suspend feelings of disbelief. The audience interacted with the speakers, asking questions as if it was a real company.

Templates to fill in can help structure the approach as part of the fiction creation workshops. Two types of templates exist: for example, leaving blank the front page of a famous newspaper, a social media post (with Twitter, Instagram, Facebook, or other social media features being recognizable), or a bus stop advert space.

In addition, some mockup elements can help boost reflection, visualize possible futures, and provide structure to workshop conversations [31].

2.3.1 Newspaper Format Versus Audio Format

There are many benefits to using a "newspaper" or "advert" format type:

- It combines descriptive (text) and immersive (images) elements.
- It is relatively inexpensive to create.
- It makes it easy to develop a mockup using simple numeric tools during workshops.
- It offers a high credibility level as long as the graphic codes of a newspaper or an advert are applied.

The audio-visual format is more comprehensive and allows for more substantial debates. However, it requires more time and implies high production costs. Yet, we have already used this format in our workshops, building on the participants' expertise, who were equipped with smartphones and editing tools for general use.

Any object or other physical materials can be used to complement the approach and can help make it more concrete. However, to do so, these materials need to be of high quality and therefore require more time and effort dedicated to mocking up.

Finally, an audio-only format can be interesting, like in the example of the War of the Worlds radio broadcast by Orson Welles [32]. However, managing to reach such a high level of engagement among the audience requires large resources. In addition, audio content is linear, which can restrain the audience's capacity to fully grasp it. It usually requires listening several times.

3 Methods

3.1 Initiation Phase: Defining the Mission and Identifying the Thematic Exploration Area

The first step is for the client to identify the objective of the approach (mentioned earlier in this chapter) and the "exploration field"—i.e., the type of issue to address. As we are dealing with forecast, the exploration field will always be broader than the current issue. Conversely, an issue such as "future food" is far too broad and requires refocusing the exploration field to make it operational (e.g., new sources of vitamins; sustainable consumption through transformation; the futures of non-animal protein, to only cite issues covered in our projects). One to two work meetings are needed to define the exploration field. Sometimes, additional interviews with key actors in the company (e.g., leadership, key position holders, etc.) can help strengthen learning.

The exploration field should be determined by identifying five to ten relevant sub-themes to explore and making sure to include at minimum the following categories of elements:

- Technologies and processes
- Consumption
- Professional practices and industry
- Geopolitics and regulations
- Economics and logistics

The next step consists of conducting an analysis of the gray literature, as well as professional and scientific literature, and carrying out expert interviews to identify "change inductors." Change inductors correspond to any possible transformation in the exploration. We then offer a comprehensive mapping of all change inductors.

3.2 Collecting and Classifying Relevant Imaginaries by Subthemes

After the initial collection phase, these imaginaries get sorted and analyzed during one or two workshops. This helps to bring out two types of information:

- Themes relating to societal challenges on the one hand and to how goods and services are used on the other. While this segmentation work is carried out using an inductive approach, it also depends on salient elements identified when defining the "exploration field." To help facilitate discussions at this segmentation phase, we use board games to create a visual mapping.
- Imaginaries are classified according to two elements: when their sources are dated from and common features related to how they represent the way in which a good or service is used. Based on these elements, we then create imaginaries lineages resembling cladistics. With this approach, we can rapidly identify any recurrence and rare occurrence. The former is described as strong lineage. Strong lineages encourage one to take a step back, as they might fall under the main existing visions but could be limiting when it comes to envisioning future perspectives [21, 22]. Taking the example of food, these could be imaginaries related to pills, to the use of algae to replace meat, or to robotization in soil treatment [7]. The latter is described as weak lineage. Weak lineages are anecdotal or don't fall into any trends. They foster curiosity.

In the food industry, these imaginaries might relate to new food shapes, new tools, the consumption of new species, unusual ways to eat, etc. Several interpretative frameworks can be used to analyze these imaginaries. Here is a non-exhaustive list:

- *Food shape*: is it a common shape? If not, what is at stake when considering how users will make it their own? Are there any cultural appropriation issues to consider or is the shape commonly acceptable?
- *Ingestion method*: we use imaginaries because they offer radical ideas. In the case of food, if they offer new ways of swallowing, are these acceptable as well? What are the barriers that require lifting up if they are not acceptable? Is an imaginary conceivable if it introduces smoked food or food that can only be consumed through an implant? Why did the authors envision this imaginary way of consuming food as appropriate?
- *Production method*: as imaginaries can move beyond current technological challenges, they allow new "magical" processes to emerge: making food bigger, "printing" some specific food at a given time, teleporting food, etc. Beyond what is impossible, technologically speaking, what do these representations tell us about our wants and needs? For example, a desire for immediacy, individualized consumption, individual differentiation, etc.

- *The sources of food*: this is one of the most recurrent points in the use of imaginaries. They invite us to envision very diverse sources that might even be shocking. Eating human flesh is, for example, a recurrent trope in science fiction, and so is eating aliens.

This questioning encourages us to move away from a probabilistic approach (will the future be like we describe it or not?) to enhance a creative potential (If that becomes possible, what differences would it make? Are these differences preferable?). A technique called « excuse an imaginary » can help with this. It suggests starting from an unlikely or illogical imaginary and proposes a more relevant one.

3.3 Confrontation Exercise: My Project in a Futuristic World Brings Out Challenges

Based on this analytical work, creative work can begin. Depending on what is at stake (innovation vs. strategy vs. change management), the creative work will either be more deliberately exploratory (innovation, change management) or prospective (strategy).

- When focusing on innovation, we offer various techniques, including projecting the project issues in a fictional world. This phase, consisting of "coring an imaginary" aims to facilitate the emergence of new practices at the crossroads between imaginaries and future visions forecast by experts. A 2–3-hour work session can lead to generating around ten scenarios, which will then be organized and prioritized.

For example, by observing ecological challenges highlighted in many science fiction films and the many scenes of food "appearing" in a quasi-magical way (Star Trek regularly uses this trope), we can combine two types of imaginaries and picture a concept for an edible packaging that would be an integral part of the food. To use the example of meat (this too is a recurrent trope given that with no meat comes no protein contained in meat), we could imagine meat wrapped up in an edible film, which could even be conceived as seasoning. This is a design fiction case study, as explained at the beginning of this article.

- When using a prospective approach, imaginaries get classified according to the type of challenge that they raise for the company. We then explore the consequences that these imaginaries imply for the organization by using various tools, such as the Futures wheel. Eventually, a 2–3-hour work session will allow producing dozens of consequences of many different sorts. These will then get recategorized by theme (inductive approach) and prioritized.

The Red Team project with the French army uses the latter approach. However, it involves the direct participation of science fiction authors who will create imaginaries instead of collecting them [33]. This process includes discussions with army representatives to ensure credibility in this field where confidentiality is paramount.

3.4 Selection of Tensions and Alternatives

A first selection is made depending on the scenarios' capacity to illustrate one or several tensions for and/or within the company. At this stage, the criteria used to evaluate the scenarios include the level of impact and the level of preparedness in the company to face these tensions. If less than five scenarios get selected by the end of this stage, it is considered advisable to go back to the previous exercise to increase the diversity of proposed scenarios.

3.5 Assessment and Identification of Challenges Through the Futures Cone

Between five and ten scenarios are available at this stage. The level of realism of these scenarios is not yet a selection criterion and even the craziest ideas must be preserved. The futures cone, as introduced earlier, is a useful tool to support organizations in classifying scenarios and considering various levels of impact on the organization.

3.6 Creation of Three Alternative Fictions

Once all this work is completed, the next stage should aim to create three fictions to illustrate the challenges in the most realistic way possible. This stage corresponds to a rather classic prototyping phase in design, but it implies making choices that are relatively decisive because the key question is to determine to what extent the fictions that are produced as part of this process are considered plausible when they get introduced to consumers. This stage also allows widening the approach to include an audience that is larger than the project team that carried out the various first stages. While taking part in the fiction creation process increases project buy-in [34], how the fiction is received by a larger audience requires further adaptation.

It is therefore useful to use the templates developed at the beginning of the process (Instagram posts, Twitter accounts, etc.) before attempting to physically represent a concept or presenting it. These templates are not prototypes per se, but they will help "anchor" the future prototype in a plausible world. This exercise can be done at home or in a workshop using prototyping equipment. In either case, collaboration between designers (or food designers) and other types of actors is required to ensure both creativity and production capacity.

In a second time, prototypes are developed to the same completion standards as non-fictional artifact (Fig. 3). However, simulation techniques can be used to make the artifacts more realistic without having to produce objects that match the standards required to put them up for sale. For example:

- A smart milk can be represented by a milk that is commercially available by decanting it into a glass bottle with a fictional label.

Fig. 3 Using the design fiction approach with the Firmenich company, we proposed a product that could be consumed both as a drink and as a chewing paste. The product called "Taste it" was mocked up with the aim of resembling real food. It was then introduced to potential consumers. Although it was not possible for consumers to try the product, it was highly credible (out of the twelve first participants, none manifested any doubt about the veracity of the brand)

- A startup can be presented in an online one-pager with fictional comments from stakeholders.
- Fictional advertisements can be simulated in a very realistic way, especially if they are presented as originating from a foreign market. This helps prevent any cognitive dissonance that could be due to the fact that the product is not known by the audience.

For example, using a milk bottle is not enough to introduce a new innovative milk which would be co-created with consumers. In this specific case, many fake posts on Twitter can be crafted to start creating a narrative around what is considered at stake with the product: discussions between consumers who are for or against the product will help start defining a debate and assessing how relevant the proposed product value is. Using a metaphor for the lean startup prototyping techniques, we could consider this phase as the creation of a "minimum plausible context" in which the final product prototype can be established. The reason why context is so important is because the product is not always sufficiently booming to enable consumers to grasp all the potential issues at stake.

3.7 Showcasing the Fictions

This stage aims to invite the audience to react to the challenges presented in the fiction and to other issues identified by participants. The ultimate goal is to allow collective decision-making in relation to the three objectives cited earlier: developing a strategic vision, managing change, and innovating.

As design fiction plays with reality by blurring its fictional identity for a moment, the framework used to present the fiction is essential. It includes several requirements:

- Kindness and comfort. It should allow space to speak out and manage any disagreements so that they don't create conflict.
- Realism. It should not reveal that the presentation is fictional. It should provide a credible setting to present a brand-new product or service (e.g., professional fair,[7] consumer test). It can also aim to banalize the fiction (e.g., making it available in a shop or online).
- Reflexivity. It should allow space for feedback once the fictional nature of the scenario has been revealed. This step allows further evaluation of the fictions [35].
- Creativity. It should promote various perspectives in exploring the consequences of the audience's reactions: identifying priority challenges for strategy development; defining actions to reduce perceived risks; creating preferable alternatives that can be implemented starting tomorrow to foster an innovative approach.

In the case of design fiction, the prototype is the focus of attention. It is presented to and manipulated by the audience. When a fiction involves the creation of several prototypes (e.g., a fictional product with a user manual and an advertisement), more staging options are available, including opportunities for performances to highlight salient and important aspects of the fiction.

3.7.1 Creating a Debate Versus Initiating Action

The design fiction practice is not yet clearly defined. The most common use of design fiction is to spark off debate [12, 13]. Yet it is possible to also use it to prompt action. In this perspective, its value resides in its capacity to spark off a debate using scenarios that show the limits and risks of societal, economic, or political choices. Fiction plays a provocative role which aims to spark off debate. What makes this exercise difficult is its tendency to radicalize opinions that already exist among the audience, without offering tools to transform them [36]. Our experience rather consists in offering fictions that balance the benefits and the drawbacks of the proposed scenarios to better balance opinions and widen the debate. In any case, research is lacking to support the development of protocols in this area.

[7] See for example the work of Stuart Candy et the « Naturepod » experiment: https://futuryst.blogspot.com/2016/05/naturepod.html

When we seek to initiate action, the idea is to represent options of possible choices and their consequences and present possible solutions and interventions right away. The exercise then consists in testing different assumptions and their effects on a final result. This is the approach we use in this chapter.

3.8 Facilitating a Creativity Workshop Based on Fictions

As a result of this process, participants will be able to identify one or more preferable futures (perceived as progress) and potential obstacles to collective buy-in. In addition, as some creative work has already started to identify new potential offers, one of the benefits of these fictions is to further define these new offers.

The approach is creative and belongs to the broader area of "design thinking," [6] which includes a first phase of problematization and ideation, which could potentially stem from the design fiction process. Organizing a workshop with the project stakeholders will therefore facilitate the emergence of concepts that relate to the identified theme and are truly positive. The workshop can gather between five and ten people, although the number of participants can vary depending on how many partners/teams are involved in the project. The goal is to secure buy-in, which helps to determine the number and diversity of participants' profiles to involve in the process.

The workshop will include three distinctive parts:

- Presentation of the fiction, the knowledge produced beforehand, and the analyses developed following the presentation to an audience (§3.7.). This part consists of making sense of previous learnings.
- The fictions are then broadcast one after the other. For each fiction, participations are invited to generate alternatives that are preferable. This can be done either by proposing a competing offer or by developing an offer based on a phenomenon showcased in these fictions. The fiction and the analysis of this fiction can therefore be used as a framework to contemplate new realities. The exercise uses various creativity tools [37] and its relevance is based on its immersion power and its capacity to allow these fictions to be taken seriously.
- Clarification, synthesis, and prioritization of preferable futures identified during the workshop. The futures cone tool [23] is used to situate concepts and categorize them based on how desirable they are for the stakeholders and how likely they are to exist in a given timescale.

The outcome of the workshop is a series of concepts, which will be turned into prototypes and presented to samples of a given market to continue driving improvement as dictated by the iterative approach used in design thinking.

4 Synthesis and Key Learnings

The reactions provoked by the fictions can vary. Depending on the project goals, we will highlight only some of these reactions.

- Definition of one or several preferable futures. These are the items that are considered as preferable by participants, or the alternatives that were formulated in response.
- Obstacles and benefits of potential new offers.
- Identification of alternative options for action plans or offer development.

A document synthesizing the results of the previous steps is developed to detail all key steps in the process, thereby allowing clients to make choices and revisit learnings for future workshops to continue building upon key takeaways from the initial workshop.

4.1 Major Challenges and Potential Mitigation Strategies

Design fiction is a practice that still sorely lacks documentation and a frame of reference to consolidate and validate its effects on a given audience. Our experience has enabled us to identify four main risks.

- A risk to struggle to convince participants of the value of the approach. Using imaginaries and conceiving futures are exercises that might be considered as not serious. Fictions created as part of this process might also be regarded as lacking rigor. To remove these doubts, in addition to fostering direct participation in the process, we also suggest highlighting the points of convergence between science and science fiction [8], which demonstrate how real these connections that we offer to create are.
- A risk to undertake a collection of imaginaries that is too limited. The quality of the collection is important as it allows expanding the speculative reflection. Collecting imaginaries related to North-American culture, film, or series is easier. We suggest paying attention to the diversification of time and location when sourcing imaginaries with an ideal distribution of around 10% from before 1945 and around 10% from the following decade. Also, we include at least 30% with sources from countries other than the North American countries.
- A risk to only focus on desirable topics and avoid exploring the negative aspects of the scenarios. Clients might wish to produce narratives that are inspirational and positive, thereby neglecting negative aspects of a given theme. However, it is important to invest the time needed to define what is considered preferable. Gathering a limited number of participants in a workshop provides suitable environment for this step.

- A risk to produce fictions with limited credibility. This can be due to either their format, which is either unfinished or too similar to science fiction, or their content considered unbelievable. Our experience has taught us that format is key. Sticking to something that is possible is important (e.g., avoiding traveling in time, modifying the size of something, or other magical actions). However, how an event is presented has a tremendous impact on how acceptable it will be considered. Format banalization can be used as a strategy. The less technologically advanced a product is (even old), the more credible it will be. This technique is used to create works of realistic science fiction [38]. On the contrary, using the codes of science fictions can reduce credibility [39].

References

1. Chrzan J, Brett J (2017) Food culture. Anthropology, linguistics and food studies. Berghahn, New York
2. Wathelet O (2019) Can we treat the eater as an abstraction? On design, anthropology and the rise of 'homo usus' in innovation. In: Diasio N, Julien M-P (eds) Anthropology of family food practices. Peter Lang, Bruxelles
3. Horvat A, Granato G, Fogliano V, Luning PA (2019) Understanding consumer data use in new product development and the product life cycle in European food firms – an empirical study. Food Qual Prefer 76:20–32
4. Tarabella A (2019) Food products evolution. Innovation drivers and market trends. Springer Nature, Basel
5. Busse M, Siebert R (2017) The role of consumers in food innovation processes. Eur J Innov Manag 21(1):20–43
6. Veflen Olsen N (2015) Design thinking and food innovation. Trends Food Sci Technol 41(2):182–187
7. Belasco W (2006) Meals to come. A history of the future of food. University of California Press, Los Angeles
8. Kirby D (2011) Lab coats in Hollywood. Science, scientists, and cinema. MIT Press, Cambridge
9. Bassett C, Steinmuller E, Voss, G (2013) Better made up. The mutual influence of science fiction and innovation. Nesta Working Paper No 13/7
10. Sterling B (2005) Shaping things. MIT Press, Cambridge
11. Bleecker J (2009) Design fiction. A short essay. Independent publishing. Available at: http://nearfuturelaboratory.com/2009/03/17/design-fiction-a-short-essay-on-design-science-fact- and-fiction
12. Dunne A, Rabby F (2013) Speculative everything. Design, fiction, and social dreaming. MIT Press, Cambridge
13. Auger J (2013) Speculative design: crafting the speculation. Digit Creat 24(1):11–35
14. Greyson A (2016) Making the futures present. Master thesis in Strategic Foresight and Innovation. OCAD University, Toronto
15. Hines A (2019) When did it start? Origin of the foresight field. World Fut Rev 12(1):4–11
16. Burnam-Fink M (2015) Creating narrative scenarios: science fiction prototyping at *Emerge*. Futures 70:48–55
17. Minvielle N, Wathelet O (2019) Le design fiction comme outil d'aide à l'innovation. Techniques de l'ingénieur. SF2200 V1
18. Minvielle N, Wathelet O, Lauquin M, Audiney P (2021) Design fiction and more for your organization. Independent Publishing. Available at: http://making-tomorrow.mkrs.fr/design-fiction-and-more-for-your-organization
19. Coulton P, Lindley J, Galen J, Cooper R (2018) The little book of design fiction for the Intenet of Things. Lancaster University, Lancaster
20. Lempert W (2014) Decolonizing encounters of the third kind: alternative futuring in native science fiction film. Vis Anthropol Rev 30(2): 164–176

21. Dourish P, Bell G (2011) Divining a digital future. Mess and mythology in ubiquitous computing. MIT Press, Cambridge
22. Watts L (2007) A future archeology of the mobile telecoms industrym, PhD Dissertation. Lancaster University, Lancaster
23. Voros J (2017) Big history and anticipation. Using big history as a framework for global foresight. In: Poli R (ed) Handbook of anticipation. Theoretical and applied aspects of the use of the future in decision making. Springer
24. Bengston D (2016) The future wheel. A method for exploring the implication of social-ecological change. Soc Nat Resour 29(3):374–379
25. Candy S (2018) Gaming futures literacy. The things from the future. Transforming the future. Anticipation in the 21th century. UNESCO, Paris, pp 233–246
26. Brown B, et al (2017) The IKEA catalogue: design fiction in academic and industrial collaboration. Group16. In: Proceedings of the 19th International Conference on Supp Group Work, pp 335–344
27. Ballard S, et al (2019) Taking future seriously. Forecasting as Method in Robotics Law and Policy. WeRobot2019
28. Coulton P, Lindley J, All Akmal H (2016) Design fiction. Does the search for plausibility lead to deception?. DRS 2016. Brighton
29. Montgomerry E, Woebken C (2016) Extrapolation factory. Operator's manual. Independent Publishing
30. Candy S, Dunacan J (2017) Designing an experiential scenario: the people who vanished. Futures 86:136–153
31. Minvielle N, Lauquin M, Wathelet O (2020) Maquetter, ces entreprises qui innovent avec les mains. Diateino, Paris
32. Ball P (2021) The modern myths. Chicago University Press, Chicago
33. Minvielle N, Wathelet O, Denis-Rémis C (2021) Quelles évolutions pour les pratiques de Red teaming. Défense et sécurité internationale 155:70–74
34. Derboven J, Vandeberghe B (2018) Evaluating future home care scheduling technology through design fiction. NordiChi'18 workshop. The role of design fiction in participatory design process
35. Encinas E, Dylan T, Mitchell R (2017) What's the matter with(in) design fiction? In: Proceedings of the 3rd Biennal Research Through Design Conference. Edinburgh, pp 189–195
36. Boyd A, Mitchell O (2013) Beautiful troubles. A toolbox for revolution. OR Books, New York
37. Darbellay F, Moody Z, Lubart T (2017) Creativity, design thinking and interdisciplinarity. Springer Nature, Singapore
38. Brooker C, Jones A (2018) Inside black mirror. Crown Archetype, London
39. Addey D (2018) Typeset in the future. Typography and design in science fiction movies. Abrams, New York

Index

A

B

C

Carlos Gómez-Corona and Heber Rodrigues (eds.), *Consumer Research Methods in Food Science*,
Methods and Protocols in Food Science, https://doi.org/10.1007/978-1-0716-3000-6,

F

G

H

P

Q

R

S

MIX
Papier aus verantwortungsvollen Quellen
Paper from responsible sources
FSC® C105338

If you have any concerns about our products,
you can contact us on
ProductSafety@springernature.com

In case Publisher is established outside the EU,
the EU authorized representative is:
Springer Nature Customer Service Center GmbH
Europaplatz 3, 69115 Heidelberg, Germany

Printed by Libri Plureos GmbH
in Hamburg, Germany